Optical Fiber Communication Systems

Optical Fiber Communication Systems

Edited by **Marko Silver**

$\mathcal{CL}$LANRYE
$\mathcal{I}$NTERNATIONAL

New Jersey

Published by Clanrye International,
55 Van Reypen Street,
Jersey City, NJ 07306, USA
www.clanryeinternational.com

Optical Fiber Communication Systems
Edited by Marko Silver

International Standard Book Number: 978-1-63240-402-2 (Hardback)

Printed in the United States of America.

Contents

Preface

This book is an important resource elaborating recent developments achieved in fiber communications systems. It consists of a compilation of research works on the essential technologies and mathematical concepts underlying optical fiber communications and devices of our age. The book encompasses various topics like the topologies and architecture of these networks, PONs, WANs, LANs, secure optical communication among others. Therefore, it presents an all-inclusive overview on latest research trends and technologies associated with these topics. It integrates contributions by veteran scientists and academicians hailing from renowned universities and research centers associated with the fields of optical communications and photonics. This book will serve as a valuable reference with a wide spectrum of information about this field. It will appeal to practitioners and researchers engaged in the field of photonics and optical communications.

Significant researches are present in this book. Intensive efforts have been employed by authors to make this book an outstanding discourse. This book contains the enlightening chapters which have been written on the basis of significant researches done by the experts.

Finally, I would also like to thank all the members involved in this book for being a team and meeting all the deadlines for the submission of their respective works. I would also like to thank my friends and family for being supportive in my efforts.

Editor

All-Optical Amplitude Multiplexing Through Fiber Parametric Interaction Between Binary Signals

Marcelo L. F. Abbade[1], Jorge D. Marconi[2],
Eric A. M. Fagotto[1], Felipe R. Barbosa[3],
André L. A. Costa[3], Iguatemi E. Fonseca[4] and Edson Moschim[3]
[1]*Pontifícia Universidade Católica de Campinas*
[2]*Universidade Federal do ABC*
[3]*Universidade Estadual de Campinas*
[4]*Universidade Federal da Paraíba*
Brazil

1. Introduction

The high bandwidth and low attenuation provided by optical fibers has turned them into the most extensively deployed transmission medium in communication systems world-wide. This is especially the case for systems that utilize bit rates ranging from hundreds of Mbits/s to several Tb/s and whose span extends from a few tens of kilometers to intercontinental scales. In fact, global networking with the present speed and quality could hardly exist without fibers, which transport more information than all the other transmission media commercially used today combined (Ramaswamy, 2010).

The interest in optical fibers goes far beyond their valuable characteristics for signal propagation. In particular, fiber nonlinearities have been widely considered for the implementation of several all-optical devices. For example, wavelength converters based on cross-phase modulation (XPM) (Olsson et al., 2000) and fiber four-wave mixing (FWM) (Inoue & Toba, 1992) have been investigated. Dispersion compensators and all-optical regenerators FWM have also been implemented (Chavez Boggio et al., 2004a), as well as wide-band tunable amplifiers, known as fiber optic parametric amplifiers (FOPA), relying on third-order parametric processes with one (Hansryd & Andrekson, 2001) and two high-power pumps have been demonstrated (Chavez Boggio et al., 2005b). Recent work (Jamshidifar et al., 2010) shows that fiber tunable filters and demultiplexers can be achieved through parametric interaction in specially designed optical fibers.

Fiber-based nonlinear devices may also be used for all-optical signal processing. In this area, there is a special interest on techniques that provide conversion between different modulation formats. In fact, recent works deal with this subject and propose ways of performing analog to digital (Brzozowski & Sargent, 2001), digital to analog (Oda & Maruta, 2006), non-return-to-zero (NRZ) to return-to-zero (RZ) (Mishina et al., 2007), multilevel to binary (Fagotto & Abbade, 2010) and binary to multilevel conversions (Zhou et al., 2006; Lu & Miyazaki, 1997), among others.

A strong motivation for pursuing such research is that different kinds of optical networks (long-haul, optical packet switching, access and so on) may coexist and need to exchange information with one another (Mishina et al., 2007). Since each of them may present different physical and logical characteristics, signals within the boundaries of each optical network are subject to different types of impairments. Therefore, a modulation format used to minimize the bit error rate (BER) of signals within a given network domain may not be appropriate for other domains. Consequently, all-optical devices that provide modulation format conversion capabilities may be highly attractive to play a major role at the interface between different optical networks.

In this chapter, we focus on four techniques that use fiber nonlinear effects to perform optical amplitude multiplexing (OAM) of binary signals into multi-amplitude ones. We begin by reviewing the fundamentals of FWM and parametric amplification (PA) in Section 2. Then, in Section 3, we discuss two techniques that use FWM (Abbade et al., 2005; Abbade et al., 2006a; Abbade et al., 2006b) and PA (Abbade et al., 2010a; Abbade et al., 2010b; Marconi et al., 2011) to convert two 2-ASK signals into a quaternary amplitude-shift keying (4-ASK) one; named, respectively, OAM-4F and OAM-4P. In Section 4, utilization of FWM and PA (Abbade et al., 2011) to convert two binary signals into a ternary amplitude-shift keying (3-ASK) is approached. Such techniques are, respectively, termed OAM-3F and OAM-3P. It should be noted that OAM-3F is an innovation, presented for the first time in this work. Advantages of multi-amplitude modulation formats encompass higher tolerance to degradations caused by chromatic dispersion and the possibility of transmitting simultaneously two signals within the same optical bandwidth. Applications and a detailed comparison among the four techniques are presented in Section 5. Finally, conclusions are drawn in Section 6.

2. Theory of parametric interactions in optical fibers

An external electric field $\overline{E}$ applied to an optical fiber will cause an induced polarization $\overline{P}$ in the medium that will depend on its electrical susceptibility χ. However, such a dependence will not only rely on the first-order susceptibility $\chi^{(1)}$, but also on higher-order terms. Processes relying on $\chi^{(2)}$ and $\chi^{(3)}$ are known, respectively, as second- and third-order parametric processes.

For an isotropic medium, the second order susceptibility $\chi^{(2)}$ is equal to zero (in dipole approximation), which means that the term proportional to $\overline{E}^2$ vanishes. Therefore, only the contribution from the third order susceptibility $\chi^{(3)}$, proportional to $\overline{E}^3$ will remain. Thus, the polarization can be expressed as,

$$\overline{P} = \varepsilon_0(\chi^{(1)}\overline{E} + \chi^{(3)}\overline{E}^3), \tag{1}$$

where ε_0 is the vacuum electric permittivity.

At this point it is important to comment on two aspects of Eq. (1). First, the notation $\overline{E}^3$ is used as a simple way of writing the triple external product $\overline{E} \otimes \overline{E} \otimes \overline{E}$, which results in a third rank tensor with 27 elements. Second, the third order susceptibility $\chi^{(3)}$ is a fourth rank tensor which contains 81 independent elements, but it should be noted that the inner product of the $\chi^{(3)}$ tensor and the $\overline{E} \otimes \overline{E} \otimes \overline{E}$ tensor leads a simple vector. Furthermore, considering that silica glass is an isotropic material, the number of independent elements of the $\chi^{(3)}$ tensor is reduced to three independent elements by symmetry (Buck, 2005; Butcher & Cotter, 1990). In addition, if the operating frequency range is far from resonances, then the number of independent elements is finally reduced to one (Buck, 2005). The "cubic" term of Eq. (1) is responsible for several nonlinear effects in optical fibers such as self-phase modulation (SPM), XPM, and FWM. It will be seen that, under proper conditions, FWM can be used to amplify a weak signal that propagates through an optical fiber along with a strong signal. To show that, let us start with the wave equation (Buck, 2005) for silica, a non-magnetic material without free charges and currents,

$$\nabla^2 \overline{E} = \mu_0 \frac{\partial \overline{\nabla} \times \overline{H}}{\partial t} = \mu_0 \left(\varepsilon_0 \frac{\partial^2 \overline{E}}{\partial t^2} + \frac{\partial^2 \overline{P}}{\partial t^2} \right). \tag{2}$$

where μ_0 is the vacuum magnetic permeability, and $\overline{H}$ is the magnetic field (Jackson, 1998) that is related with the magnetic induction field $\overline{B}$ as $\overline{B} = \mu_0 \overline{H}$ (then $\overline{\nabla} \times \overline{B} = \mu_0 \overline{\nabla} \times \overline{H}$). Assuming that $\overline{J} = \partial \overline{P} / \partial t$, then the total polarization can be written as a sum of two terms, the linear and the nonlinear polarizations: $\overline{P} = \overline{P}_L + \overline{P}_{NL}$. Therefore, Eq. (2) can be written as

$$\left(\nabla^2 - \frac{n^2}{c^2} \frac{\partial^2}{\partial t^2} \right) \overline{E} = \mu_0 \frac{\partial^2 \overline{P}_{NL}}{\partial t^2}, \tag{3}$$

where $(n/c)^2 = (\mu_0 \varepsilon_0 (1 + \chi^{(1)}))$, being n the refractive index and c the speed of light in vacuum. In the case of single mode fibers that are used in parametric devices, we have that $\Delta = (n_{core} - n_{cladding}) / n_{cladding} << 1$, where n_{core} and $n_{cladding}$ are the refractive indexes of the core and the cladding, respectively (Agrawal, 2001).

Considering a typical value $\Delta \sim 0.003$, for weakly guiding fibers (Gloge, 1971), the longitudinal components of the electric fields are of the order of $\Delta^{1/2}$, which means that they are ~20 times smaller than the transversal components; therefore, they can be neglected in most practical applications. Then, considering that the fiber propagation is along the z-axis, the fields can be written as:

$$\overline{E}_L = \hat{x} \psi(x,y) \frac{1}{2} [A_L(z) e^{-i(\beta(\omega_L)z - \omega_L t)} + c.c.], \tag{4}$$

where c.c. stands for the complex conjugate of the previous term, $A_L(z)$ is the complex amplitude of the electric field, $\beta(\omega_L)$ is the propagation constant for the angular frequency ω_L, and $\psi(x,y)$ is the transverse distribution of the electric field:

$$\psi(x,y) \propto \begin{cases} A \ J_0(\kappa_T \rho) & \rho \leq a \\ B \ K_0(\gamma_T \rho) & \rho > a \end{cases}, \tag{5}$$

where $A = [J_0 \ (\kappa_T a)]^{-1}$ and $B = [K_0 \ (\gamma_T a)]^{-1}$, $\kappa_T = n_{core}^2 k_0^2 - \beta^2$, $\gamma_T = \beta^2 - n_{cladding}^2 k_0^2$, $k_0 = 2\pi / \lambda$, a is the core radius, and $\rho^2 = x^2 + y^2$. Here J_0 and K_0 are the Bessel functions corresponding to the fundamental mode (called HE_{11} or LP_{01}) which is the only one propagating in singlemode fibers.

As previously mentioned, the nonlinear process responsible for parametric amplification is the FWM. To show this let us assume that the refractive index can be written as the sum of a linear term and a nonlinear contribution, $n = n_0 + I\, n_2$, where n_0 is the linear part of the refractive index and n_2 is the nonlinear refractive index (Boyd, 2008). The nonlinear contribution is proportional to the optical irradiance I (in the SI units system) (Boyd, 2008). Then, when two waves at angular frequencies ω_1 and ω_2 are launched together into an optical fiber, the refractive index will be modulated with a frequency $(\omega_2 - \omega_1)$. Now, if a third wave at frequency ω_3 is coupled along, a new wave at frequency $\omega_4 = \omega_3 \pm (\omega_2 - \omega_1)$ will be generated. This new wave is called idler. The relation $\omega_4 = \omega_3 \pm (\omega_2 - \omega_1)$ means that $\omega_4 + \omega_1 = \omega_3 + \omega_2$ and that $\omega_4 + \omega_2 = \omega_3 + \omega_1$.

It is important to mention that when three waves are launched into the same fiber, a total of nine new frequencies can be in fact generated if all the combinations are taken into account (Hansryd et al., 2002). For instance, frequencies such as $\omega_4 = 2\omega_2 - \omega_3$ or $\omega_4 = \omega_3 - \omega_2 + \omega_1$ are also possible. However, not all these frequencies are generated with the same efficiency. Generally calculations reckon only highly efficient processes and neglect the others. For instance, if we consider the case where $\omega_1 < \omega_3 = \omega_2$, we have that $2\omega_2 = \omega_4 + \omega_1$; this process is a degenerate case of FWM that brings about parametric amplification when the wave at ω_2 is a strong pump-signal and ω_1 is a weak signal to be amplified. As a result, a new wave, an amplified copy of the signal at ω_1, will be generated at ω_4.

In order to standardize the notation, we shall denote the angular frequencies for the pump, the signal and the idler as ω_P, ω_S and ω_{id}, respectively. Following this notation, the total electric field can be written as

$$\overline{E} = \overline{E}_P + \overline{E}_S + \overline{E}_{id} = \hat{x}\,\psi(x,y)\frac{1}{2}[A_P(z)e^{-i(\beta(\omega_P)z - \omega_P t)} +$$

$$+ A_S(z)e^{-i(\beta(\omega_S)z - \omega_S t)} + \qquad\qquad ,\qquad (6)$$

$$+ A_{id}(z)e^{-i(\beta(\omega_{id})z - \omega_{id}t)}] + c.c.$$

where all the waves are supposed to have the same mode profile $\psi(x,y)$ and the same polarization on the x-axis. When this total electric field is included in Eq. (3), the Laplacian leads to the following electric fields (pump, signal, and idler)

$$\frac{\partial^2 \overline{E}_j}{\partial z^2} = \hat{x}\,\psi(x,y)\frac{1}{2}[\frac{\partial^2 A_j(z)}{\partial z^2}e^{-i(\beta(\omega_j)z - \omega_j t)} -$$

$$-2i\beta(\omega_j)\frac{A_j(z)}{\partial z}e^{-i(\beta(\omega_j)z - \omega_j t)} - \qquad\qquad (7)$$

$$-\beta^2(\omega_j)A_j(z)e^{-i(\beta(\omega_j)z - \omega_j t)}] + c.c.$$

where $j = P, S$ or id.

The slowly varying envelope approximation is introduced at this point, and is given by

$$\left|\frac{\partial^2 A_j}{\partial z^2}\right| << \left|\beta_j \frac{\partial A_j}{\partial z}\right|, \qquad\qquad (8)$$

or, considering that $\beta_j = \beta(\omega_j) = 2\pi/\lambda_j$, with $\lambda_j = 2\pi c/\omega_j$,

$$\left|\lambda_j \frac{\partial}{\partial z}\left(\frac{\partial A_j}{\partial z}\right)\right| << \left|\frac{\partial A_j}{\partial z}\right|, \tag{9}$$

which means that the slopes of the envelope fields do not vary significantly along a wavelength distance (λ_j) as compared to the envelope magnitude (Buck, 2005). Using this approximation, valid for most practical cases, the term with the second derivative of z can be neglected.

Considering all these conditions and neglecting fiber attenuation, it is possible to obtain a set of three coupled equations for the three electric field amplitudes,

$$\frac{dA_P}{dz} = i\gamma[(|A_P|^2 + 2(|A_S|^2 + |A_{id}|^2))A_P + 2A_S A_{id} A_P^* \exp(i\Delta\beta\, z)], \tag{10}$$

$$\frac{dA_S}{dz} = i\gamma[(|A_S|^2 + 2(|A_{id}|^2 + |A_P|^2))A_S + A_{id}^* A_P^2 \exp(-i\Delta\beta\, z)], \tag{11}$$

$$\frac{dA_{id}}{dz} = i\gamma[(|A_{id}|^2 + 2(|A_S|^2 + |A_P|^2))A_{id} + A_S^* A_P^2 \exp(-i\Delta\beta\, z)], \tag{12}$$

where the symbol (*) stands for the complex conjugate, $\Delta\beta$ is the linear phase mismatch $\Delta\beta = \beta(\omega_s) + \beta(\omega_{id}) - 2\beta(\omega_p)$, and $\gamma = n_2\omega/cA_{eff}$ is the fiber nonlinear coefficient. Here n_2 is the nonlinear refractive index which is related to $\chi^{(3)}$ as $n_2 = 3\chi^{(3)}/4\varepsilon_0 cn_0^2$, and A_{eff} is the effective area (Agrawal, 2001). Note that the right-hand side of Eqs. (11)-(13) includes the terms of SPM, XPM and FWM.

The exact solutions of Eqs. (10)-(12) involve Jacobian elliptical functions as shown in (Chen, 1989). Here we follow an approximate solution that allows us to obtain a simple expression for the parametric gain. This approximation considers that the intensity of the pump is much higher than that of the signal and the idler. Therefore, the energy transferred from the pump to the signal (and the idler) can be considered negligible. For instance, if the ratio between the pump power and the signal power at the fiber input is ~ 10^4-10^5, and the signal gain is ~20-25 dB at the fiber output, the signal (idler) power is still less than 1% of the pump power, which justifies the approximation.

Under such conditions Eqs.(10)-(12) may be written as,

$$\frac{dA_P}{dz} = i\gamma\left(P_p A_P\right), \tag{13}$$

$$\frac{dA_S}{dz} = i\gamma[2P_p A_S + A_{id}^* A_P^2 \exp(-i\Delta\beta\, z)], \tag{14}$$

$$\frac{dA_{id}}{dz} = i\gamma[2P_p A_{id} + A_S^* A_{id}^2 \exp(-i\Delta\beta\, z)], \tag{15}$$

where $P_P = |A_P|^2$ is the pump power. These equations, which are valid for an ideal fiber with attenuation coefficient $\alpha = 0$, have an analytical solution for the signal gain and for the idler conversion efficiencies as,

$$\frac{P_S}{P_S(0)} = 1 + \left(\frac{x_0 \sinh(x)}{x}\right)^2 , \tag{16}$$

$$\frac{P_{id}}{P_S(0)} = \left(\frac{x_0 \sinh(x)}{x}\right)^2 , \tag{17}$$

where $P_S = |A_S|^2$ is the signal power at the fiber output, $P_S(0)$ is the signal power at the fiber input, $P_{id} = |A_{id}|^2$ is the idler power at the fiber output, L is the fiber length, $x_0 = \gamma P_P L$, $x = x_0 \sqrt{1 - (\Delta\beta_T / 2\gamma P_P)^2}$, and $\Delta\beta_T = \Delta\beta + 2\gamma P_P$.

It should be noted that even when $\alpha \neq 0$, previous equations are good approximations if the pump power P_P is replaced by $\overline{P_P} = \frac{1}{L}\int_0^L P_P(z)dz = P_P(0)\left(1 - e^{-\alpha L}\right)/\alpha L$. The phase mismatch $\Delta\beta$ can be calculated by expanding $\beta(\omega)$ in Taylor series around an arbitrary frequency ω_t as follows:

$$\beta(\omega) = \beta(\omega_t) + \left(\frac{\partial\beta}{\partial\omega}\right)_{\omega=\omega_t}(\omega - \omega_t) + \frac{1}{2}\left(\frac{\partial^2\beta}{\partial\omega^2}\right)_{\omega=\omega_t}(\omega - \omega_t)^2 +$$

$$+ \frac{1}{6}\left(\frac{\partial^3\beta}{\partial\omega^3}\right)_{\omega=\omega_t}(\omega - \omega_t)^3 + \frac{1}{24}\left(\frac{\partial^4\beta}{\partial\omega^4}\right)_{\omega=\omega_t}(\omega - \omega_t)^4 + \ldots\ldots \tag{18}$$

Keeping terms up to the fourth order and taking $\omega_t = \omega_P$ then,

$$\Delta\beta(\omega) = \beta_2(\omega_P)(\omega - \omega_P)^2 + \frac{\beta_4(\omega_P)}{12}(\omega - \omega_P)^4 . \tag{19}$$

Within the spectral region where the parameter x is real, the parametric gain is maximum when $\Delta\beta_T = 0$ ($x = x_0$), and its value is $G_{max} = 1 + \sinh^2(x_0)$. On the other hand, the gain has a local minimum (within the region of interest) when $\Delta\beta = 0$, and its value is $G_{min} = 1 + x_0^2$ (Chavez Boggio et al., 2005a). In other words, the parametric gain will be high if $\Delta\beta_T$ is small. This means that the pump must be tuned at some frequency within the fiber anomalous dispersion region, that is, $\omega_P < \omega_0$, with $\omega_0 = 2\pi c / \lambda_0$, λ_0 the fiber zero dispersion wavelength and c is the vacuum light speed. If the approximation $\beta_4 \sim 0$ is valid, then the gain bandwidth can be roughly written as $\Delta\Omega = \omega_P \pm \sqrt{2\gamma P_P / |\beta_2|}$.

In the extreme case of $\gamma P_p << \Delta\beta$ and $P_P \sim P_S >> P_{id}$, Eq. (17) gives the mixing condition without amplification (Stolen & Bjorkholm, 1982). Here we change our notation and designate, the pump, the signal, and the idler powers as P_1, P_2, and $P_{-,+}$, and the angular frequencies change from ω_P, ω_S, and ω_{id} to ω_1, ω_2, and $\omega_{-,+}$, respectively. The extreme case just considered implies that $\Delta\beta_T \cong \Delta\beta$, $x \cong i\,L\,\Delta\beta$, and then $\sinh(x) \cong 2i\sin\Delta\beta L$. Introducing all these results in Eq.(17) we have that

$$P_{-,+} = 4P_{1,2}^2 P_{2,1}\gamma^2 L^2 \left(\frac{\sin(\Delta\beta L)}{\Delta\beta L} \right)^2 .\qquad(20)$$

The notation $P_{-,+}$ refers to the power of the two principal FWM processes that generate waves at frequencies $\omega_{-,+} = 2\,\omega_{1,2} - \omega_{2,1}$.

3. All-optical generation of quaternary amplitude-shift keying signals

This section presents two all-optical techniques for multiplexing two binary ASK signals (ASK-2), traveling at different carrier wavelengths, in a single 4-ASK signal. In the first case, OAM-4F, the four levels of the quaternary pattern are obtained when the two binary signals, which have similar optical power, interact through FWM. The theoretical calculations that allow estimation of the power of the quaternary levels are developed from Eq. (27). The second approach, OAM-4P, used to generate the single 4-ASK signal from two binary ASK signals is based on PA. In this case, one of the signals is a strong optical signal that acts as a pump, with the unusual characteristic of being modulated by binary information. OAM-4F is presented in Section 3.1 and OAM-4P is approached in Section 3.2.

3.1 Optical amplitude multiplexing through fiber four-wave mixing

3.1.1 Theory

The diagram shown in Fig. 1 illustrates the principle of OAM-4F. Two co-polarized input signals at ω_1 and ω_2 are coupled into a fiber, where they co-propagate through a medium that favors the occurrence of FWM. When the fiber attenuation coefficient $\alpha \neq 0$, Eq. (20) can be rewritten as:

$$P_{-,+} = \eta\gamma^2 P_{1,2}^2 P_{2,1} \exp(-\alpha L)\left[\frac{1-\exp(-\alpha L)}{\alpha}\right]^2 ,\qquad(21)$$

where P_1, P_2, P_-, and P_+ are the respective optical powers of the channels at frequencies ω_1, ω_2, ω_-, and ω_+, L is the fiber length, γ is the fiber nonlinear coefficient, and η is the wavelength and intensity-dependent FWM generation efficiency, well described in the literature (Mussot et al., 2007), which is given by:

$$\eta = 4L^2 \left(\frac{\sin(\Delta\beta L)}{\Delta\beta L} \right)^2 .\qquad(22)$$

If the channels at ω_1 and ω_2 are codified with ideal on-off keying (OOK) modulation, then $P_{-,+}$ is null whenever one of these input channels transmits a 0-bit. Here, however, we assume that these channels are codified by a binary amplitude-shift keying (2-ASK) scheme, where the 0-bit powers of are intentionally offset. In this case, the extinction ratios (ER) corresponding to the channels at ω_1 and ω_2 are:

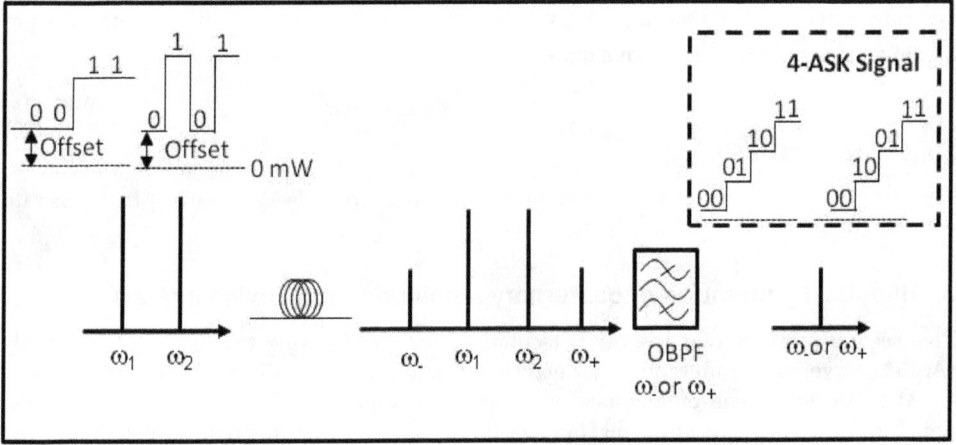

Fig. 1. Scheme illustrating the principle of operation of OAM-4F.

$$r_1 = P_1^1(0) / P_1^0(0) \tag{23a}$$

$$r_2 = P_2^1(0) / P_2^0(0) \tag{23b}$$

where, $P_i^j(0)$ designates the power of bit j (j= 0 or 1) at the channel at ω_i (i= 1 or 2). Eq. (21) indicates that the signals filtered at ω_+ may assume four different power levels given by:

$$P_{out+}^{00} = k_+ P_1^0(0) \left[P_2^0(0) \right]^2 \tag{24a}$$

$$P_{out+}^{01} = k_+ P_1^0(0) \left[P_2^1(0) \right]^2 \tag{24b}$$

$$P_{out+}^{10} = k_+ P_1^1(0) \left[P_2^0(0) \right]^2 \tag{24c}$$

$$P_{out+}^{11} = k_+ P_1^1(0) \left[P_2^1(0) \right]^2 \tag{24d}$$

where P_{out+}^{mn} is the power of the signal envelope at ω_+ when the signal at ω_1 transmits a bit m (m= 0 or 1) and the signal at ω_2 carries a bit n (n= 0 or 1), and

$$k_+ = \eta \gamma^2 \exp(-\alpha L) \left[\frac{1 - \exp(-\alpha L)}{\alpha} \right]^2 \tag{25}$$

Eqs. (24) clearly show that the signal formed at the fiber output, and selected by the optical band-pass filter (OBPF) centered at ω_+, is a quaternary amplitude-shift keying (4-ASK) one. It should be noted that P_{out+}^{00} is always the lowest power whereas P_{out+}^{11} is always the highest one. On the other hand, P_{out+}^{01} may be lower or higher than P_{out+}^{10} depending, respectively, on

whether $P_1^0(0)\left[P_2^1(0)\right]^2 < P_1^1(0)\left[P_2^0(0)\right]^2$ or $P_1^0(0)\left[P_2^1(0)\right]^2 > P_1^1(0)\left[P_2^0(0)\right]^2$. In case $P_1^0(0)\left[P_2^1(0)\right]^2 = P_1^1(0)\left[P_2^0(0)\right]^2$, the quaternary signal degenerates into a ternary one.

The four-levels of quaternary signals give rise to an eye-diagram structure that comprises three eyes. We identify the eye made up of the two lowest power levels with the subscript "*low*"; analogously the subscripts "*int*" and "*up*" are utilized for the eyes that involve the two intermediate and the two higher power levels, respectively. For OAM-F4, it is possible to find the relative extinction ratios (RER) of such eyes, r_{low}, r_{int}, and r_{up}, by substituting (32) in (33). When $P_1^0(0)\left[P_2^1(0)\right]^2 < P_1^1(0)\left[P_2^0(0)\right]^2$:

$$r_{low} = P_{out+}^{01}\Big/P_{out+}^{00} = r_2^2 \tag{26a}$$

$$r_{int} = P_{out+}^{10}\Big/P_{out+}^{01} = r_1\Big/r_2^2 \tag{26b}$$

$$r_{up} = P_{out+}^{11}\Big/P_{out+}^{10} = r_2^2 \tag{26c}$$

Similarly, if $P_1^0(0)\left[P_2^1(0)\right]^2 > P_1^1(0)\left[P_2^0(0)\right]^2$:

$$r_{low} = P_{out+}^{10}\Big/P_{out+}^{00} = r_1^2 \tag{27a}$$

$$r_{int} = P_{out+}^{01}\Big/P_{out+}^{10} = r_2\Big/r_1^2 \tag{27b}$$

$$r_{up} = P_{out+}^{11}\Big/P_{out+}^{01} = r_1^2 \tag{27c}$$

Eqs. (24), (26) and (27) reveal some important properties of the generated 4-ASK signal. First, its powers do not depend on the phase of the input signals. Second, the power level distribution depends solely on the ERs of the two input signals. Finally, such power level distribution cannot be arbitrarily chosen. For instance, in the case where Eq. (26) hold, if one increases r_2, both r_{low} and r_{up} are enhanced; however, r_{int} is simultaneously decreased.

The analysis above can be repeated for the signal at ω. In this case, Eqs. (24)- (27) would be modified, but the general properties of the generated 4-ASK signal would not change. Such analysis is left for the interested reader.

It is important to understand how information of the input binary signals may be recovered from the quaternary-amplitude one. To achieve this goal, it is assumed that the 4-ASK signal is photodetected by a circuit such as the one illustrated in Fig. 2.

Initially the signal is optically amplified and filtered at ω_+; then, it is photo-detected by a PIN photo-diode with responsivity R_S, low-pass filtered and submitted to an electronic

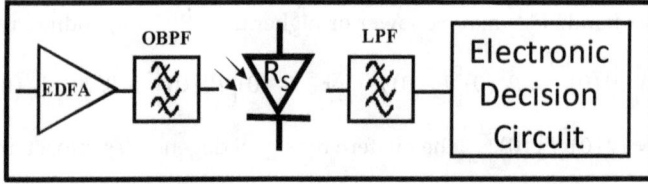

Fig. 2. Multiamplitude signal detector.

decision circuit (EDC), whose purpose is to indicate which bits were transmitted by the signals at ω_1 and ω_2. It is assumed that noise at EDC obeys a Gaussian distribution.

There are then two possibilities. The first one occurs when $P_1^0(0)\left[P_2^1(0)\right]^2 < P_1^1(0)\left[P_2^0(0)\right]^2$ and it is illustrated in the left part of the inset of Fig. 1. A simple inspection of this figure indicates the following two detection rules should be utilized by the EDC: a) ω_1 transmitted a bit 0 (1) whenever the two lower (upper) levels are detected; and b) ω_2 transmitted a bit 0 (1) whenever the lower and third (second and fourth) power levels are detected.

These detection rules may be used to estimate the BERs of the binary signals extracted from the 4-ASK signals. To accomplish this goal, we first consider that the noise fluctuations between consecutive levels are much larger than the ones between non-consecutive levels. This hypothesis must hold for practical situations where even the noise fluctuations between adjacent levels must be low to keep the BERs at acceptable levels. Then, we observe from rule (a) and the left part of the inset of Fig. 1, that the BER for the signal at ω_1, BER_1, is equivalent to the one of a binary signal with threshold level between the second and third levels. In this way:

$$BER_1 = Q\left(\frac{i_{10} - i_{01}}{\sigma_{10} + \sigma_{01}}\right) \tag{28a}$$

where, $Q(x) = \left(1/\sqrt{2\pi}\right)\int_x^\infty e^{-y^2/2}dy$ is the complementary error function, $i_{xy} = R_S\, P_{out+}^{xy}$ is the average electronic current associated with power level P_{out+}^{xy}, and σ_{xy} is the i_{xy} standard deviation. From rule (b) and the left part of the inset of Fig. 2, it is observed that the BER for the signal at ω_2, BER_2, is equivalent to the average BER of three binary signals with decision thresholds between the first (lower) and second; the second and third; and the third and fourth (highest) levels:

$$BER_2 = \frac{1}{3}\left[Q\left(\frac{i_{01} - i_{00}}{\sigma_{01} + \sigma_{00}}\right) + Q\left(\frac{i_{10} - i_{01}}{\sigma_{10} + \sigma_{01}}\right) + Q\left(\frac{i_{11} - i_{10}}{\sigma_{11} + \sigma_{10}}\right)\right] \tag{28b}$$

The second possibility occurs when $P_1^0(0)\left[P_2^1(0)\right]^2 > P_1^1(0)\left[P_2^0(0)\right]^2$ and it is illustrated in the right part of the inset of Fig. 1. Following a procedure similar to the one described above and inspecting this figure, it is easy to verify that the detection rules are: c) ω_1 transmitted a bit 0 (1) whenever the lower and third (second and fourth) power levels are detected and d)

ω_2 transmitted a bit 0 (1) whenever the two lower (upper) levels are detected. In this way, BER_1 is now the average BER of the three eyes of the quaternary signal whereas BER_2 may be estimated from the BER of the intermediate eye:

$$BER_1 = \frac{1}{3}\left[Q\left(\frac{i_{10} - i_{00}}{\sigma_{10} + \sigma_{00}}\right) + Q\left(\frac{i_{01} - i_{10}}{\sigma_{01} + \sigma_{10}}\right) + Q\left(\frac{i_{11} - i_{01}}{\sigma_{11} + \sigma_{01}}\right) \right] \tag{29a}$$

$$BER_2 = Q\left(\frac{i_{01} - i_{10}}{\sigma_{01} + \sigma_{10}}\right) \tag{29b}$$

We note that other reports suggest that the input binary signal may also be optically recovered from the 4-ASK signal with the use of transfer functions generated by self-phase modulation (Oda & Maruta, 2006) or FWM (Fagotto & Abbade, 2010) effects. However, a discussion concerning such all-optical approaches is beyond the scope of this chapter.

3.1.2 Results and discussion

Fig. 3 illustrates the experimental setup used to perform amplitude multiplexing through FWM. Two 1 Gb/s 2^{12}-1 pseudorandom bit sequences (PRBS) directly modulate the optical carriers at f_1= 193.00 and f_2= 193.15 THz, where f_i= $2\pi\omega_i$ (i= 1, 2). Previous to being coupled, these signals are co-polarized and then amplified by an Erbium-doped fiber amplifier (EDFA), EDFA1, up to an average peak power of 12 dBm. In the sequence, they are launched into a dispersion-shifted fiber (DSF) with α= 0.20 dB/km, λ_0= 1550 nm, dispersion slope S_0= 0.074 ps/(nm.km), γ= 2.0 (W.km)$^{-1}$, and L= 25.0 km. Since the powers at the fiber input are relatively low, it is not necessary to use any mechanism to prevent Brillouin backscattering.

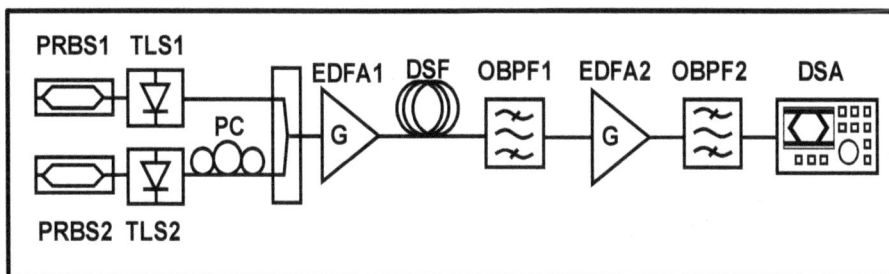

Fig. 3. Experimental setup for all-optical multiplexing.

Actually, since both signals are modulated, the Brillouin backscattering threshold should be a few dB higher than in the case of continuous-wave (cw) operation. Fiber FWM generates two sidebands, one at f_-= 192.85 THz and other at f_+= 193.30 THz. The latter is filtered by OBPF1, amplified by EDFA2, and then filtered again by OBPF2. Next, the signal is received by a digital signal analyzer (DSA). The double filtering is required because the first filter OBPF1 is not enough to eliminate effectively the input signals. Therefore, the OBPF1 output needs to be amplified and then filtered again by OBPF2 before being inputted to the DSA.

The signal power spectra at (a) EDFA1 input, (b) DSF output, and (c) OBPF2 output are plotted in Fig. 4. An optical signal-to-noise ratio (OSNR) of 27 dB is achieved at the output of the second optical filter.

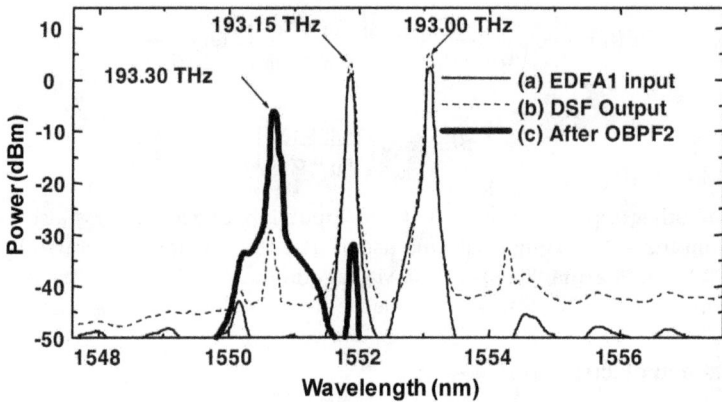

Fig. 4. Power Spectra.

Fig. 5 exhibits two unsynchronized PRBSs used to modulate the signals (a) at f_1 with r_1= 4.0 dB and (b) at f_2 with r_2= 1.7 dB, and the (c) quaternary signal obtained at f_+. In this situation, $r_1 > r_2^2$ (which is equivalent to $P_1^0(0)\left[P_2^1(0)\right]^2 < P_1^1(0)\left[P_2^0(0)\right]^2$) and the quaternary signal is governed by (26).

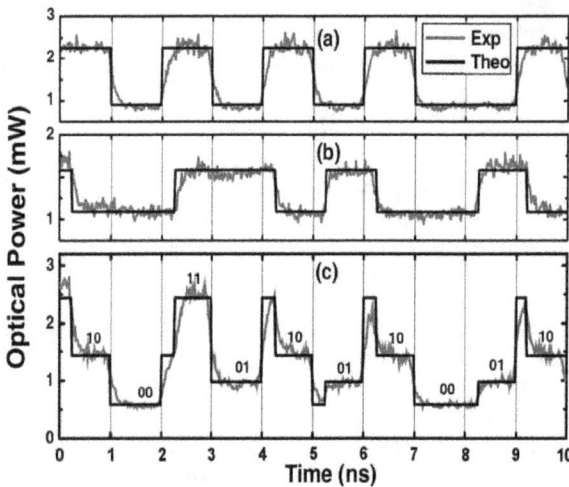

Fig. 5. (a) Binary input sequence at f_1 and (b) at f_2 and (c) quaternary output signal.

Fig. 6a shows the eye diagrams for r_2= 2.6 dB, and r_1= 2.6 dB. In this case, again $r_1^2 > r_2$ ($P_1^0(0)\left[P_2^1(0)\right]^2 < P_1^1(0)\left[P_2^0(0)\right]^2$) and so the two intermediate powers, in increasing magnitude of power, represent levels 01 and 10 (where, as before, level ij stands for the bit i transmitted by the channel at ω_1 and for the bit j transmitted by the channel at ω_2). When r_1 is increased to 4.8 dB, r_{int} in (26b) becomes close to unity and the two intermediate eyes get very close; this is shown in Fig. 6b. If r_1 is further increased to 7.7 dB, then $r_1{}^2 < r_2$

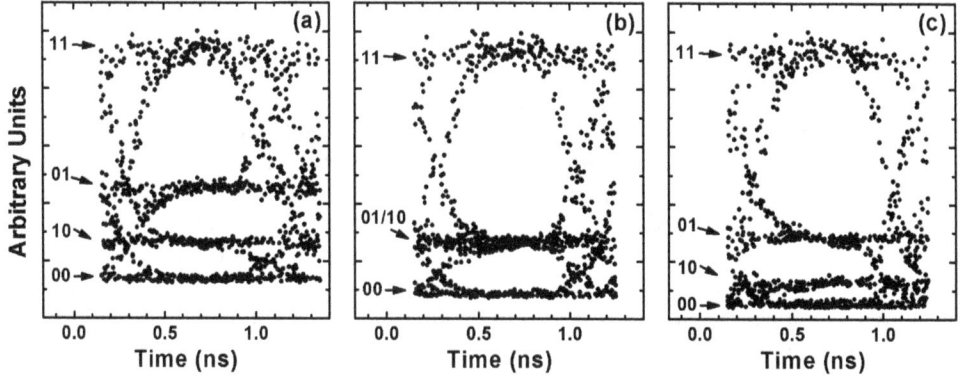

Fig. 6. Eye-diagrams for r_2=2.6 dB and (a) r_1=2.6 dB, (b) r_1=4.8 dB, and (c) r_1=7.7 dB.

$(P_1^0(0)[P_2^1(0)]^2 > P_1^1(0)[P_2^0(0)]^2)$ and the position between levels 01 and 10 is exchanged; Fig 6(c) shows the eye diagrams for such situation.

To complete such analysis, graphs of (a) r_{up}, (b) r_{int}, and (c) r_{low} as a function of r_1 are plotted in Fig. 7 for experimental data and theoretical curves, for r_2= 2.6 dB. The agreement between such results is quite good. As predicted by (27), increasing r_1, initially causes r_{int} to decrease.

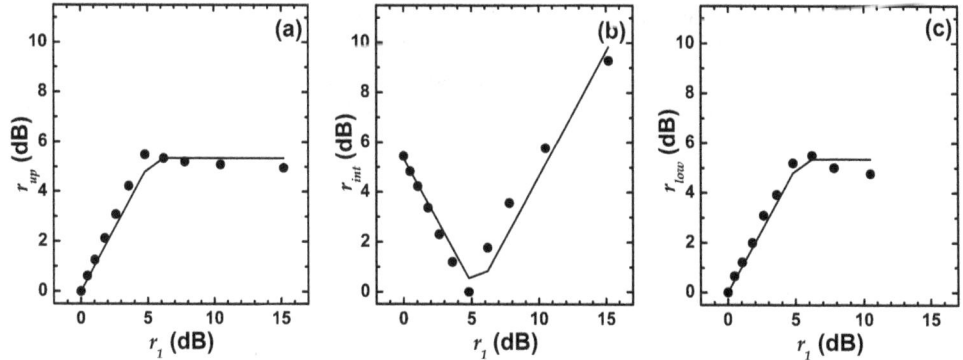

Fig. 7. Theoretical (line) and experimental (dots) ERs for the quaternary signal. When $r_1 \approx r_2{}^2$, r_{int} achieves a minimum value; this is the point where the quaternary signal degenerates into a ternary-amplitude one and where levels 10 and 01 exchange their positions. As r_1 is further increased, (27) is no longer valid and (26) needs to be applied; in this region r_{int} increases again. It is also observed that $r_{up} \approx r_{low}$ for any value of r_1. As r_1 is increased, r_{up} and r_{low} initially also increase, in agreement with (27). Then, for $r_1 \geq r_2{}^2$, r_{up} and r_{low} remain at a constant and maximum value in agreement with (26).

It should be noted that the FWM equations presented in Section 2 assume that the involved signals are in *cw* regime. The accordance between theoretical results and experimental data observed in this subsection suggest that such FWM equations are also valid when the involved signals are modulated.

We reckon that the first experimental results concerning this technique were reported in (Xu et al., 2005). There, two 10 Gb/s signals were multiplexed in a 20 Gb/s 4-ASK one, which was then transmitted through a 160 km single mode fiber (SMF) link; after such propagation distance, chromatic dispersion was compensated. The eye-penalties experienced by the quaternary signal were relatively small. In another experiment (Abbade et al, 2006), a 2 Gb/s 4-ASK was generated and propagated through a 20-km long SMF link in a field-trial network (FAPESP, n.d.). Eqs. (28) and (29) were utilized to estimate the BERs for signals with different values of r_1 and r_2. Although chromatic dispersion was not compensated and no special care concerning the bandwidth of filters was taken into account, $BER_1 < 10^{-9}$ and $BER_2 < 10^{-12}$ were obtained for several combinations of values of r_1 and r_2. This indicates that the 4-ASK signal generated by FWM amplitude multiplexing could be properly utilized in practical applications.

BER performance of multi-amplitude signals depends on the dominant kind of noise. Generally speaking, different types of noise require different power level distributions to minimize signal BER. Theoretical and experimental analyses of such optimum distribution for 4-ASK signals are presented in (Walklin & Conradi, 1999). In particular, it is shown that under the dominance of amplified spontaneous emission (ASE) noise, the power levels of a quaternary amplitude signal should be in proportions of 0: 1: 4: 9.

Concerning the 4-ASK generated by OAM-4F, such optimal distribution would apply to the cases described by Eqs. (28b) and (29a), where consecutive power levels represent complementary bits (*i.e.*, 0101). Unfortunately, the proportions of 0: 1: 4: 9 cannot be achieved by FWM amplitude multiplexing. This may be understood by noting in Eq. (24a) that P^{00}_{out+} is never null; moreover, $r_{up} = r_{low}$ ((26) and (27)). In the cases described by Eqs. (28a) and (29b), the two lowest power levels stand for bit 0, whereas the two highest ones represent a bit 1. Therefore, for a given average power, BER minimization would require the two lowest power levels and also the two highest power levels to coincide. But this degenerates the 4-ASK signal into a 2-ASK one, which does not allow information from the signals at ω_1 and ω_2 to be recovered. This discussion shows that minimizing BER_1 does not lead to a minimization of BER_2.

3.2 Optical amplitude multiplexing through fiber parametric amplification

3.2.1 Theory

As detailed in Section 2, fiber optics parametric amplifiers are realized by coupling both a weak probe signal at angular frequency ω_S, and a high-power pump signal at angular frequency ω_P into an optical fiber that acts as a nonlinear medium. Usually the performance of parametric amplifiers is analyzed for binary signals with two intensity levels representing bits "0" and "1" (Kalogerakis et al., 2006; Mussot et al., 2007; Chavez Boggio et al., 2004b; Peucheret et al., 2006) as sketched in Fig. 8(a). The power of these levels at the fiber input are indicated by $P^0_S(0)$ and $P^1_S(0)$, respectively; the ideal $P^0_S(0)$ being null. Figure 8(b) shows a

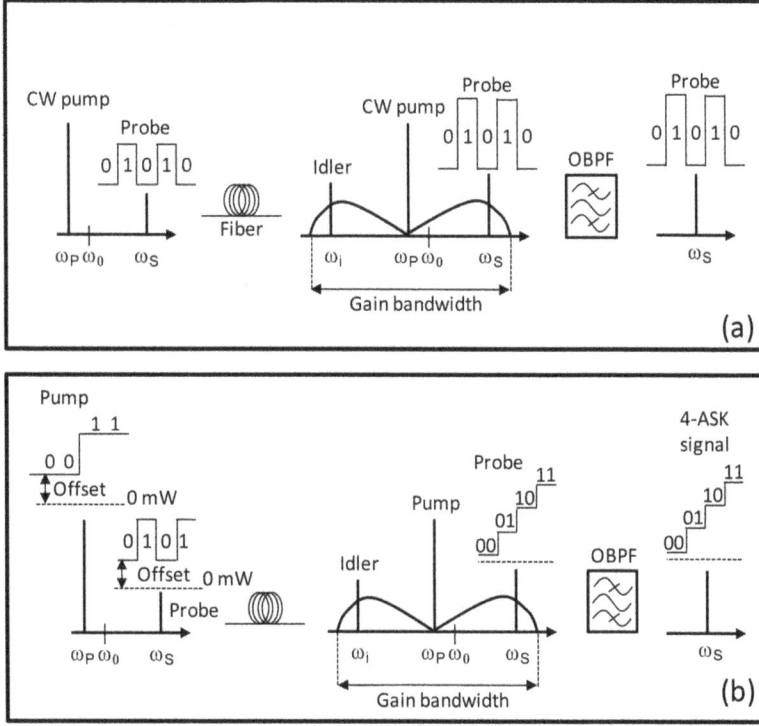

Fig. 8. Scheme for (a) PA and (b) ASK-4 generation. (ω_k with k = P, S, i represents the angular frequencies of the pump, the signal and the idler, and $\omega_0 = 2\pi c / \lambda_0$, where λ_0 is the zero dispersion wavelength of the fiber).

scheme of the theoretical principle of OAM-4P. Essentially, this principle is equal to the PA, but two new features are added to the signals involved.

First, as in OAM-4, the power of the probe signal is intentionally offset from zero. Thus, the ER of the signal is no longer infinite (in contrast to the ideal OOK signals), and is given by:

$$r_S = P_S^1(0) / P_S^0(0) . \tag{30}$$

Second, the pump is also modulated with binary information, in an analogous way to that of the probe signal. The powers of the "0" and "1" bits transmitted by the pump at the fiber input are designated as $P_P^0(0)$ and $P_P^1(0)$, respectively. It is assumed that $P_P^0(0)$ is considerably higher than $P_S^1(0)$. Once the pump is modulated, it will provide two amplification factors, G_P^0 or G_P^1, depending on whether the pump information is a "0" or "1" bit. Consequently, each of the binary levels of the signal at ω_S will have a lower or a higher gain, splitting into two new power levels at the fiber output.

For example, consider that the signal at ω_S has the "1" bit power level, then the pump modulation will split it in two power levels "01" and "11", depending on whether the pump information is a bit "0" or "1", respectively. This argument is also valid when the probe signal transports a bit "0" power level. In such case, the probe signal will either assume a "00" or "10" power level at the fiber output, depending again on whether the pump information is a bit "0" or "1", respectively. Therefore, the power levels of the signal envelope at ω_S and at $z = L$ will be:

$$P_{out}^{00} = G_P^0 P_S^0(0) \tag{31a}$$

$$P_{out}^{01} = G_P^0 P_S^1(0) \tag{31b}$$

$$P_{out}^{10} = G_P^1 P_S^0(0) \tag{31c}$$

$$P_{out}^{11} = G_P^1 P_S^1(0) \tag{31d}$$

where P_{out}^{ij} represents the power of the signal at ω_S and at $z = L$, when bits i and j (i, j = "0" or "1") at $z = 0$ are assigned to the pump and the probe signal, respectively. G_P^0 and G_P^1 can be obtained through Eq. (16) with pump powers P_P^0 and P_P^1, respectively.

Note that Eqs. (31) clearly show that the 4-ASK signal contains information from both the pump and the probe signal. As in Section 3.1, we identify the eye made up of the two lowest power levels with the subscript "*low*"; analogously, the subscripts "*int*" and "*up*" are utilized for the eyes that involve the two intermediate and the two higher power levels, respectively. By using this convention and considering $P_{out}^{10} \geq P_{out}^{01}$ ($G_P^0 P_S^1(0) \leq G_P^1 P_S^0(0)$), the ERs for each one of these eyes can be written as:

$$r_{low} = \frac{P_{out}^{01}}{P_{out}^{00}} = r_S \tag{32a}$$

$$r_{int} = \frac{P_{out}^{10}}{P_{out}^{01}} = \frac{r_G}{r_S} \tag{32b}$$

$$r_{up} = \frac{P_{out}^{11}}{P_{out}^{10}} = r_S \tag{32c}$$

where r_G is defined as

$$r_G = \frac{G_P^1}{G_P^0}, \tag{33}$$

and it is the ER associated with the amplification factors G_P^0 and G_P^1. In the case $P_{out}^{10} \leq P_{out}^{01}$ ($G_P^0 P_S^1(0) \geq G_P^1 P_S^0(0)$), the ERs are given by:

$$r_{low} = \frac{P_{out}^{10}}{P_{out}^{00}} = r_G \tag{34a}$$

$$r_{\text{int}} = \frac{P_{out}^{01}}{P_{out}^{10}} = \frac{r_S}{r_G} \tag{34b}$$

$$r_{up} = \frac{P_{out}^{11}}{P_{out}^{01}} = r_G \tag{34c}$$

Eqs. (32) and (34) show that the power level distribution of the generated 4-ASK signal does not depend on the powers of the probe signal at ω_S (only on its ER), but on the powers utilized in the pump. This is quite reasonable because in our analytical model the power of the probe signal was supposed to be much lower than the pump power, and so it was neglected (Song et al., 1999a). Besides that, it is clear that by controlling only the parameters r_S and r_G it is possible to regulate the power of the quaternary levels.

The pump ER $r_P = P_P^1(0) / P_P^0(0)$ is controlled by setting the values of the pump average powers, and then it is possible to control the value of r_G. For the ideal OOK modulation in the pump ($P_P^0(0) = 0$), Eqs. (31), (32) and (34) are still valid, and in this special case $G_P^0 = 1$ and $r_G \cong G_P^1$.

The theoretical model for 4-ASK presented here and applied to modulated signals is actually based on a set of equations generated for cw signals. In the next two subsections the applicability of the technique to practical situations is tested through simulations and experiments, showing that our assumptions work fine.

3.2.2 Results and discussions

Fig. 9 shows the experimental setup used in both simulations and experiments. The lasers (two DFB lasers) that were used as the pump and the signal were tuned at λ_P = 1553.5 nm and λ_S = 1552.2 nm ($\lambda_k = 2\pi c / \omega_k$, k = P or S), respectively.

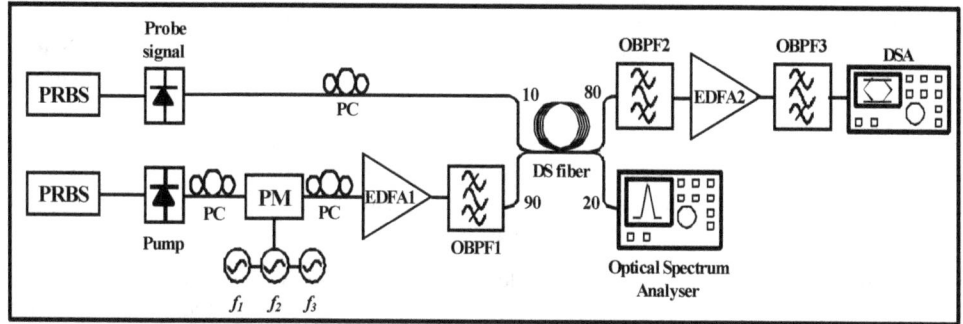

Fig. 9. Experimental setup.

A pseudo-random bit sequence at 1 Gb/s was used to modulate the lasers through direct modulation. In order to suppress stimulated Brillouin scattering (SBS), the pump linewidth was broadened by phase modulation using a phase modulator (PM) driven by three RF signals. After the Erbium-doped fiber amplifier (EDFA 1), the pump was filtered with an optical band-pass filter (OBPF) to suppress most of the amplified spontaneous emission (ASE). A 90/10 coupler was used to couple the pump and the signal to a 7 km long segment of dispersion shifted fiber (DSF).

The other fiber parameters were: the zero-dispersion wavelength λ_0 = 1550.2 nm, the zero-dispersion wavelength variation $\Delta\lambda_0$ = 0.08 nm, the dispersion slope S_0 = 0.074 ps/nm²/km, the nonlinear coefficient γ = 2.1 W⁻¹km⁻¹, and the attenuation coefficient α = 0.2 dB/km. The maximization of the parametric gain was achieved by aligning the pump and the signal states of polarization using polarization controllers (PCs). The optical power was divided with an optical coupler at the fiber output for spectral and systemic characterization. A fraction of ~20% was delivered to an optical spectrum analyzer (OSA) for recording the spectra. The remaining power (~80%) was sent to an OBPF centered at the signal frequency ω_S. The signal was then amplified and filtered again to reduce the ASE power accumulated in the amplification stage. Finally, the signal with an average power of approximately 0 dBm entered a digital oscilloscope to characterize the resulting ASK-4 eye-diagrams. The quaternary signal was also analyzed after being transmitted through a 75 km-long spool of standard fiber. To compensate for the dispersion due to the propagation, the signal was also passed through a -68 ps/nm compensating fiber.

An accurate control of the ERs of both the pump and the signal was a necessary item for the proper evaluation of the proposed technique. Thus, with the equipment available in our labs, we found that the required accuracy was more easily obtained by programming the modulation index of directly modulated lasers than by varying the bias voltage of the external modulator. However, the technique should also hold for signals with external modulation if specific modulators were available. It is anticipated that due to the femto-second response of PA (Grudinin et al., 1987) the technique should also work well for higher bit rates.

The simulated results presented here were obtained using exactly the same setup as used for the experiments. To perform the simulations as close as possible to the experiments, the laser linewidth was set at the maximum value of the equipment at ~30 MHz. Also the same phase modulation scheme used experimentally was added to our simulations. This point is particularly important because the PM induces noise at the level "1" of the binary eye-diagrams. Consequently, additional noise is also expected at quaternary levels. The DSF was divided into 10 segments to perform the simulations in order to consider the influence of the zero dispersion wavelength variations along the fiber. All segments had the same length and the $\Delta\lambda_0$ was distributed within the range $\lambda_0 - \Delta\lambda_0/2$ to $\lambda_0 + \Delta\lambda_0/2$. The segments were randomly ordered. The propagation along each segment was handled by solving the non-linear Schrödinger equation using the split-step Fourier method (Agrawal, 2001). A commercially available software program was used to perform the simulations.

Fig. 10 shows spectral and temporal results. Fig. 10(a) shows a comparison between the simulated and the experimental spectra at the fiber output. Note that the main difference between them is in the noise region of the spectra. The principal reason for such a difference is that the ASE originated in the pump Erbium booster amplifier, that is not completely suppressed by the band-pass filter, is then amplified by the parametric amplifier. It is important to note that ω_S is placed within the frequency region where the signal performance is not affected by the phase modulation used to suppress the SBS (Boggio et al., 2005a). This region for conventional (non-modulated pump) PA is the region of maximum

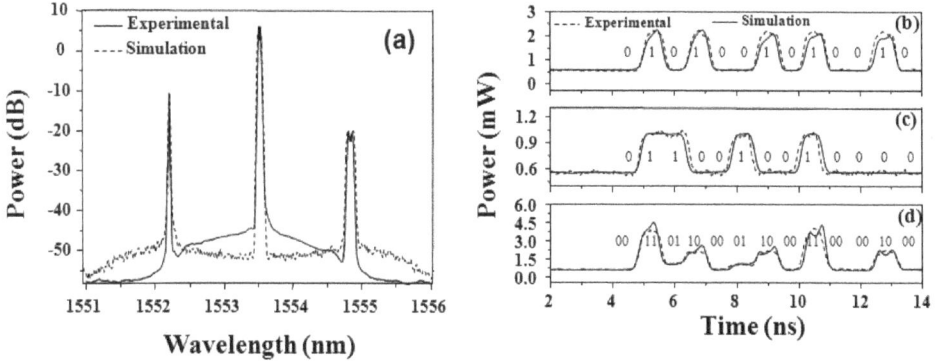

Fig. 10. (a) Experimental and simulated optical spectrum at the fiber output. (b) Binary pump. (c) Binary signal before amplification. (d) ASK-4 signal obtained after PA

gain; however, the gain bandwidth does not remain the same when the pump transmits either a bit "1" or "0", thus the change in bandwidth also modifies the spectral regions where the PM affects the signal performance. Consequently, some influence on the OSNR performance of the quaternary signal is expected.

Two bit sequences carried by the pump and the probe signal at the fiber input are illustrated in Figs. 10(b) and (c). The resulting quaternary signal is illustrated in Fig. 10(d). The pump and the probe signal ERs are $r_P = 6.0$ dB and $r_S = 2.6$ dB, respectively. As expected the signal of Fig. 10(d) presents four well-defined power levels. Note that the same power levels are obtained when the bits of the pump and the probe signals are repeated, and in the experimental case, the probe signal is slightly delayed in relation to the pump, which does not occur in the simulated case. This fact explains the slight delay between the quaternary signals obtained in the experimental results.

Fig. 11 shows the values of r_{up}, r_{int} and r_{low} as a function of r_S, for r_G varying from 3 to 7 dB. These values were calculated using the sets of Eqs. (32) and (34), and the analytical, the simulated, and the experimental values were plotted in all of these cases, showing a rather good agreement for r_{up} and r_{int}.

However, in the case of r_{low}, a difference between the experimental and the analytical/simulated values for $r_S > 6$ dB appeared. Such difference increases when the value of r_G increases. Moreover, the experimental values are always smaller than those of the simulated ones and the difference is always smaller than 1.4 dB. This difference can be explained by the fact that when r_G is significantly high, the quaternary signal ER, given by $r = r_{low} r_{int} r_{up} = r_G r_S$ is also high. Thus, even when the measurements are taken with a fixed average power, the power of level "00" can reach a value around -10 dBm; this value is comparable to the power level generated by the DSA photodiode dark current. Consequently, measurements of r_{low} are not as accurate as those for smaller values of r_G.

To measure the r_S values, the pump was simply turned off. Then, the ER of the binary signal at ω_S was measured from the eye-diagram at the DSA. A continuous-wave (cw) at ω_S and a modulated pump with ER = r_P were used to obtain r_G. The PA gave two gains to the cw

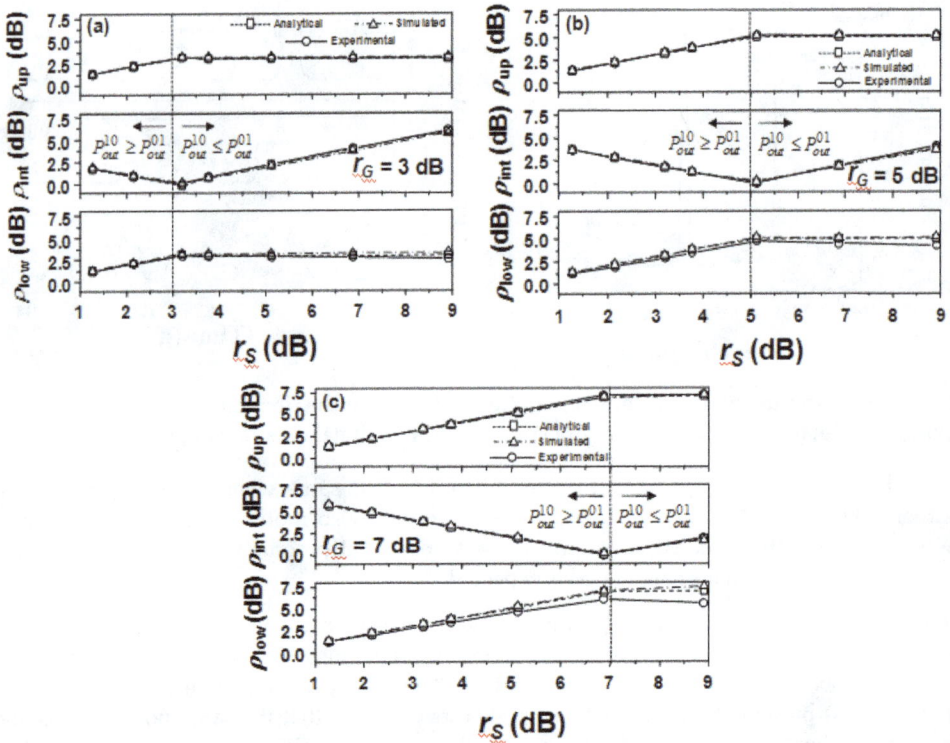

Fig. 11. Values of r_{up}, r_{int} and r_{low} as a function of r_S for: (a) $r_G = 3.0$ dB, (b) $r_G = 5.0$ dB, (c) $r_G = 7.0$ dB.

signal, generating a binary signal whose eye diagram can be seen on the DSA. The ER of such signal is precisely r_G.

Two sets of three ASK-4 eye-diagrams for the signal at frequency ω_S are exhibited in Figures 12 and 13. The labels "00" and "11" used to identify the quaternary pattern levels always correspond to the lowest and uppermost levels, respectively. On the other hand, the labels "01" or "10" are always used to identify the two intermediate levels, but in this case the label of each specific level depends on the value of the signal extinction ratio r_S. For instance, for $r_S = 2.2$ dB (as in Fig. 12(a) the second and third power levels of the quaternary eye diagram correspond to labels "01" and "10", respectively. This can be explained as follows. When $r_S = 2.2$ dB and the pump level is "1", the parametric gain obtained by the signal level "0" is higher than that of the signal level "1" when the pump level is "0".

Fig. 12(b) shows a particular case when the parametric gain given by the pump level "1" to the signal level "0" coincides with that given by the pump level "0" to probe level "1". In such a situation, $r_S \approx r_G$ (3.8 dB ≈ 4 dB), and the quaternary signal degenerates into a ternary one. This situation must be avoided in practical applications. Fig. 12(c) presents the case $r_S = 6.8$ dB $> r_G$, which is the inverse case of the one shown in Fig. 12(a). Now, the power level of the label "10" is lower than that of the label "01". This is because the gain of pump level "1" given to level "0" is higher than that given by the pump level "0" to signal level "1".

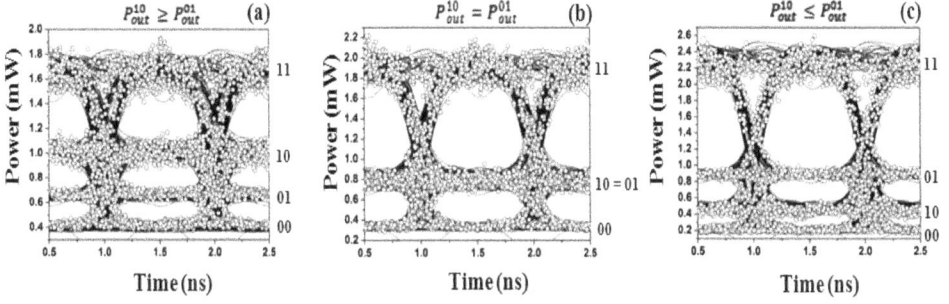

Fig. 12. Simulated (full lines) and experimental (white circles) eye diagrams for (a) $r_S = 2.2$ dB, (b) $r_S = 3.8$ dB, and (c) $r_S = 6.9$ dB. In all the cases $r_G = 4$ dB.

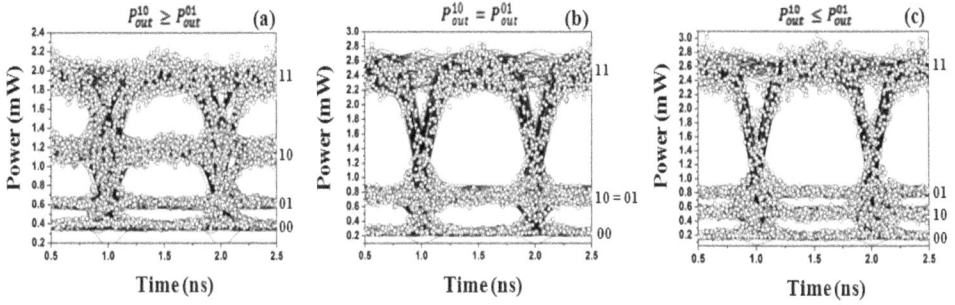

Fig. 13. Simulated (full lines) and experimental (white circles) eye diagrams for (a) $r_S = 2.2$ dB, (b) $r_S = 5.1$ dB, and (c) $r_S = 6.9$ dB. In all the cases $r_G = 5$ dB.

A very good agreement between the simulated and experimental eye diagrams is observed in Figs. 12 and 13 for the two values of r_G used, and similar results were observed for other values of r_G.

The eye diagrams for the quaternary signal that propagated through 75 km of standard fiber are shown in Fig. 14. The agreement between experimental eye diagrams and numerical results is also very good. The experimental BERs are lower than 10^{-13} and 10^{-11} before and after transmission, respectively. In principle, it is possible to improve such values by using a narrower optical band-pass filter or by increasing the quaternary signal average power. Table 1 presents estimations for the BERs before and after propagation by using Eqs. (29).

	Before Propagation		After Propagation	
	Experiment	Simulation	Experiment	Simulation
BER_{ω_p}	1.1×10^{-14}	1.7×10^{-14}	7.4×10^{-12}	3.2×10^{-12}
BER_{ω_s}	8.0×10^{-14}	4.8×10^{-14}	7.6×10^{-12}	2.8×10^{-11}

Table 1. BER estimation before and after fiber propagation.

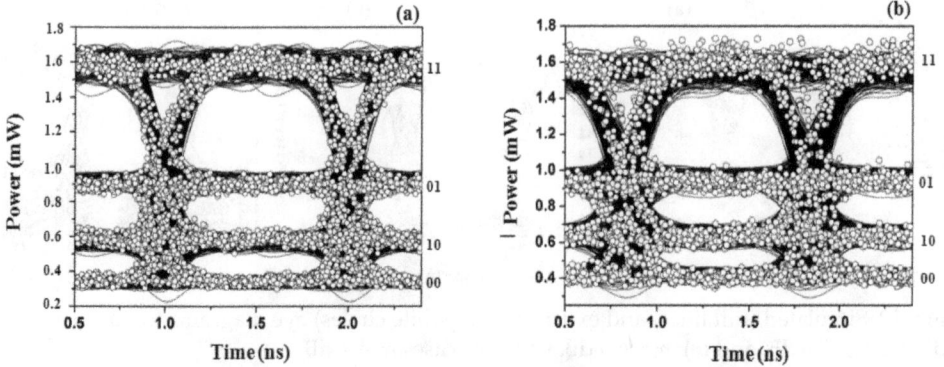

Fig. 14. Simulated (full lines) and experimental (white circles) eye diagrams: (a) before propagation, (b) after a 75 km standard fiber propagation. r_P = 0.86 dB (r_G ≈ 2.5 dB), r_S = 4.4 dB.

The recovery of the original binary information from the quaternary-amplitude signal can also be performed optically by using some known techniques (Oda & Maruta, 2006), or by developing especial optical devices with the S- and U-shaped transfer functions (Fagotto & Abbade, 2010).

4. All-optical generation of ternary amplitude-shift keying signals

In the previous section we showed how to multiplex two 2-ASK signals into a single 4-ASK one. To perform such operation, it was necessary to provide a power offset to both input binary signals. As a result, the lowest level power of the of the 4-ASK signal was not null, which degrades the signal BER performance.

In this section, we discuss two other techniques that multiplex two binary signals and generate a 3-ASK signal rather than a 4-ASK one. The first one, OAM-3F, is based on FWM; it requires a power offset on just one of the input binary signals. The second, OAM-3P, utilizes PA and holds for two OOK input binary signals. For the same average power, ternary-amplitude signals present lower BERs than quaternary ones. Besides, the 3-ASK signal lowest power level is null for both techniques; this also contributes to reduce the BER of the multiplexed signal and it is an important advantage for the techniques presented here. However, recovering information relative to two binary signals from a single ternary one requires the use of some special signal characteristics. All of these aspects are detailed below.

4.1 Optical amplitude multiplexing through fiber four-wave mixing

4.1.1 Theory

The notation employed here is the same utilized in Section 3.1. The principle of OAM-3F is illustrated in Fig. 15a. Two co-polarized OOK signals at ω_1 and ω_2 are coupled and propagated through an optical fiber with nonlinear and dispersion parameters appropriate for favoring FWM; however, it is assumed that the signal at ω_2 is an OOK, so $P_2^0(0) = 0$.

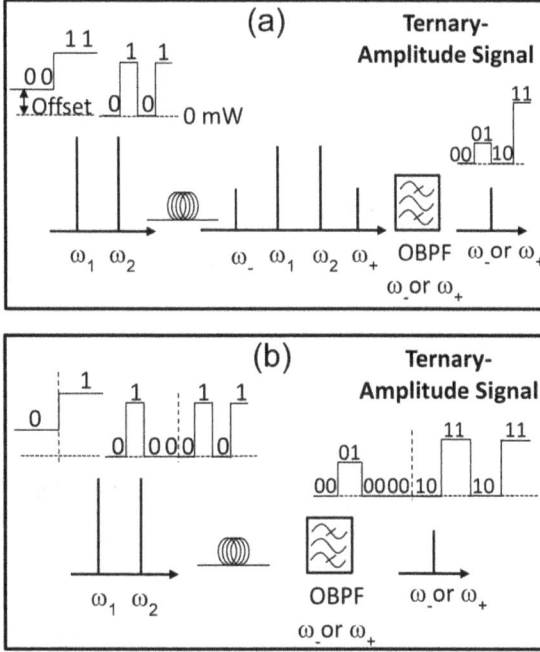

Fig. 15. Scheme illustrating the principle of operation of OAM-3F.

The OBPF is centered at ω_+; therefore, when the signal at ω_1 conveys a bit i (= "0" or "1") and the signal at f_2 carries a bit j (= "0" or "1"), the signal at the fiber output presents the following four power levels, $P^{ij}_{out\,+}$:

$$P^{00}_{out\,+} = k_+ P^0_1(0)\left[P^0_2(0)\right]^2 = 0 \tag{35a}$$

$$P^{10}_{out\,+} = k_+ P^1_1(0)\left[P^0_2(0)\right]^2 = 0 \tag{35b}$$

$$P^{01}_{out\,+} = k_+ P^0_1(0)\left[P^1_2(0)\right]^2 \tag{35c}$$

$$P^{11}_{out\,+} = k_+ P^1_1(0)\left[P^1_2(0)\right]^2 \tag{35d}$$

Since $P^{00}_{out\,+} = P^{10}_{out\,+} = 0$, the output signal now is a 3-ASK one. Recovering information transmitted by the ω_2 signal from the generated ternary amplitude signal is a straightforward task. If one detects power levels corrspeonding to $P^{01}_{out\,+}$ or $P^{11}_{out\,+}$, the signal at f_2 clearly transmits a bit "1"; otherwise it transmits a bit "0". This may be promptly verified by inspecting Fig. 15a or Eqs. (35).

On the other hand, recovering the pump information from the generated ternary amplitude signal is, in principle, not possible because $P^{00}_{out\,+} = P^{10}_{out\,+}$. This power level ambiguity implies that when the lowest (null) power level is detected, the signal at f_1 may either convey a "0" or a "1"-bit. However, it is possible to solve this problem if the following assumptions are made:

a.   The line code used by the signal at ω_2 guarantees a maximum of $(N\text{-}1)$ bits 0 in a row, *i.e.* in a sequence of N bits there is at least one bit "1". In fact, this characteristic is implemented by most practical line codes to minimize the chance of the receiver loosing synchronism to the transmitter. For example, the line code utilized in Gigabit Ethernet standard, called 8B/10B, assures that no bit sequence presents more than four bits "0" in a row of ten bits;

b.   the signal bit rate at ω_2, R_{b2}, is higher than the signal bit rate at ω_1, R_{b1}:

$$R_{b2} = NR_{b1} , \tag{36}$$

where N is the same integer number considered in assumption (a), and

c.   to recover information conveyed by the signal at ω_1, such a signal is oversampled at a rate R_{b2}. Because of (b), this oversampling process establishes that N samples of the signal at ω_1 are obtained for each of its bits. However, following (a), at least in one of these samples, the signal at ω_2 shall transmit a "1"-bit. Therefore, the detection rules for the signal at ω_1 are the following. In a sequence of N samples:

i.   if only power levels P^{11}_{out+} and P^{00}_{out+} are detected, then the signal at ω_1 sent a "1-bit";

ii.   if only power levels P^{01}_{out+} and P^{00}_{out+} are detected, then the signal at ω_1 sent a "0-bit";

iii. if any other combination of power levels is detected, then there is an error and information should be discarded.

The detection rules just described may be verified by simple inspection of Fig. 15b, where rules (a), (b), and (c) were considered for N= 4.

As stated before, power level optimization for optical multi-amplitude signals depends on the kind of the dominant noise and it is analyzed in (Walklin & Conradi, 1999). In case ASE noise is dominant, such optimal distribution must follow a quadratic law; for a ternary-amplitude signal, this means that power levels must be distributed following proportions of 0: 1: 4. Eqs. (35d) show that such distribution may be easily obtained by setting:

$$\frac{P^{11}_{out+}}{P^{01}_{out+}} = \frac{P^1_1(0)}{P^0_1(0)} = r_1 = 4 \tag{37}$$

Therefore, power level optmization may be easily achieved when using this technique. In fact, OAM-3F is able to generate 3-ASK signals with arbitrary power level distributions, by solely setting the ER of the signal at ω_1.

After inspecting Fig. 15b and following a reasoning similar to the one considered in Section 3.1, it is easy to conclude that:

$$BER_2 = Q\left(\frac{i_{01} - i_{00}}{\sigma_{01} + \sigma_{00}}\right) \tag{38a}$$

To estimate BER_1, we assume that an EDC, such as the one illustrated in Fig. 2, is programmed to decide which bit was sent by the signal at ω_1 based only on the first non-null sample of the acquired N samples (assumption (c) above). In this case, the EDC interprets that the signal ω_1 sent a '0'-bit or '1'-bit when, respectively, the intermediate or the highest power level is detected. In this case, one can write:

$$BER_1 = P(\varepsilon \mid 'l') P('l') + P(\varepsilon \mid 'i') P('i') + P(\varepsilon \mid 'h') P('h')$$

$$= \frac{1}{2} Q\left(\frac{i_{01} - i_{10}}{\sigma_{01} + \sigma_{10}}\right) \cdot \frac{1}{2} + \left[Q\left(\frac{i_{01} - i_{10}}{\sigma_{01} + \sigma_{10}}\right) + Q\left(\frac{i_{11} - i_{01}}{\sigma_{11} + \sigma_{01}}\right)\right] \cdot \frac{1}{4} + Q\left(\frac{i_{11} - i_{01}}{\sigma_{11} + \sigma_{01}}\right) \cdot \frac{1}{4} \qquad (38b)$$

$$= \frac{1}{2}\left[Q\left(\frac{i_{01} - i_{10}}{\sigma_{01} + \sigma_{10}}\right) + Q\left(\frac{i_{11} - i_{01}}{\sigma_{11} + \sigma_{01}}\right)\right]$$

where 'l', 'i', and 'h' stand, respectively, for the lowest, intermediate, and highest amplitude levels, $P(x)$ is the probability of occurrence of level x (x= 'l', 'i', and 'h'), and $P(\varepsilon \mid x)$ is the conditional probabylity that an error occurs given that level x (x= 'l', 'i', and 'h') is detected.

Although the theory presented above was derived for a signal with finite ER at ω_1 and an OOK signal at ω_2, the proposed technique could also be applied for an OOK signal at ω_1 and a signal with finite ER at ω_2. Also, the FWM component selected by the OBPF could be at ω_-. Eqs. (35)-(38) should be adapted in all of the possible combinations. Nevertheless, the general aspects treated here, like the power level optimization would still hold.

4.1.2 Results and discussion

Considering the system shown in Fig. 15b, Fig. 16 presents simulation results for bit sequences for the (a) signal at ω_1= 193.1 THz with r_1= 2 and R_{b1}= 2.5 Gb/s, (b) signal at ω_2= 193.2 THz with R_{b2}= 10 Gb/s, and (c) signal at the OBPF output at ω_+= 193.3 THz. Fiber parameters are: λ_0 = 1552.52 nm, S_0 = 0.017 ps/nm²/km, γ = 5.3 W⁻¹km⁻¹, α = 0.2 dB/km, and L= 3.0 km. The signal observed in Fig. 16c clearly represents a ternary-amplitude one. Moreover, the same power levels are obtained when the bits of the pump and the probe signals are repeated.

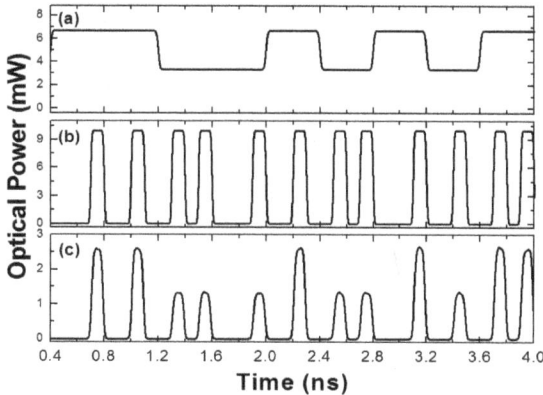

Fig. 16. Input signals at (a) ω_1 and (b) ω_2 and output signal at (c) ω_+.

Fig. 17 plots the eye diagrams for the ternary-amplitude signals with average power of 1 mW, for r_1= (a) 2.0, (b) 4.0, and (c) 6.0. Eye-diagrams present $P^{11}_{out+} \big/ P^{01}_{out+}$, respectively, of 1.9, 3.9, and 5.8, which are in good agreement with the values predicted by the theory previously presented. In these simulations, ASE noise was deliberately added to the signal in order to provide a typical OSNR of 65 dB at the receiver. In the three presented cases,

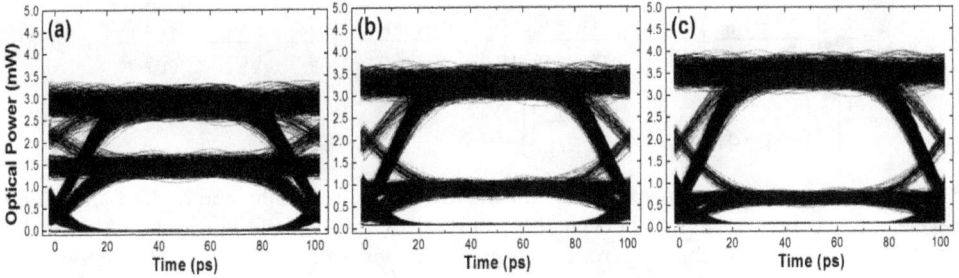

Fig. 17. Eye diagrams for r_1= (a) 2, (b) 4, and (c) 5.

BER_2 is inferior to 10^{-15}, which may be considered as error-free. BER_1 achieves a maximum of 4.6 10^{-12} for r_1= 2.0. Such low BER values suggest OAM-3F viability in practical situations. The uppermost amplitude level is the noisier one; this is a consequence of FWM combining the noise from two "1"-bits of the input binary signals, which are more affected by ASE noise than the "0"-bits.

4.2 Optical amplitude multiplexing through fiber parametric amplification

4.2.1 Theory

Here, we use the same notation as in Section 3.2. Fig. 18 illustrates a diagram for implementing OAM-3P. Two co-polarized OOK signals at optical carriers ω_P and ω_S are coupled and co-propagated through an optical fiber that provides the appropriate dispersion regime and nonlinear parameter to support parametric amplification. However, both input signals are OOK; hence, $P^0_P(0) = P^0_S(0) = 0$. It is assumed that $P^1_P(0) >> P^1_S(0)$. In this case, three power levels are possible for the signal at ω_P at the OBPF output. The first one occurs when both signals transmits a "1"-bit and is given by:

$$P^{11}_{out} = G^1_P P^1_S(0) \tag{39a}$$

Obviously, this corresponds to the situation where the probe signal is amplified by the pump. In the second case, the probe signal transmits a "1"-bit whereas the pump transmits a "0"-bit. In this situation, the probe signal is not amplified by the pump and, assuming low fiber dispersion, it is solely attenuated by the fiber. In this situation, the power at the OBPF output is:

$$P^{01}_{out} = P^1_S(0)e^{-\alpha L} \tag{39b}$$

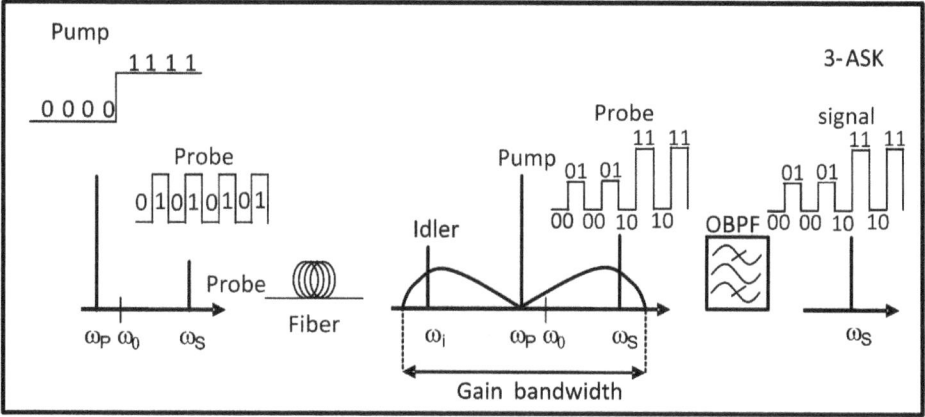

Fig. 18. Scheme for 3-ASK generation (ω_k with $k = P, S, i$ representing the angular frequencies of the pump, the signal and the idler, and $\omega_0 = 2\pi c/\lambda_0$, where λ_0 is the zero dispersion wavelength of the fiber).

Finally, the third case corresponds to the one where the probe signal transmits a "0"-bit. Then, no matter the bit set by the pump, the power at the OBPF output will be null:

$$P_{out}^{00} = P_{out}^{10} = 0 \tag{39c}$$

We have already treated a situation where information from two input binary signals need to be recovered from a ternary-amplitude one in the previous sub-section. In fact, here we may apply the same reasoning used in Section 4.1. Particularly, the detection rule for recovering information of the signal at ω_S from the photo-detected ternary-amplitude signal is the same considered for the signal at ω_2 in Section 4.1. So, it is interpreted that the signal at ω_S transmits a bit "1" if the power levels corresponding to P_{out}^{01} or P_{out}^{11} are detected and that it transmits a bit "0" if a null power is received.

Because of the ambiguity expressed by (39c), recovering the signal information at ω_P from the photodetected ternary-amplitude requires the use of assumptions equivalent to the ones utilized in Section 4.1:

a. the line code used by the signal at ω_S guarantees a sequence of N bits contains at least one bit "1".

b. the bit rate of the signal at ω_S, R_{bS}, is related to the bit rate of the signal at ω_P, R_{bP} by:

$$R_{bS} = NR_{bP}, \tag{40}$$

c. the signal at ω_P is oversampled at a rate R_{bS} and the following detection rules are used:

i.   if only power levels P_{out}^{11} and P_{out}^{00} are detected, then the signal at ω_P sent a "1-bit";

ii. if only power levels P_{out}^{11} and P_{out}^{00} are detected, then the signal at ω_P sent a "0-bit";

iii. if any other combination of power levels is detected, then there is an error and information should be discarded.

Due to the similarities between the ternary-amplitude signals generated by OAM-3F and OAM-3P, it is also possible to write the bit error rates associated to the signals at ω_S and ω_P, respectively, as:

$$BER_S = Q\left(\frac{i_{01} - i_{00}}{\sigma_{01} + \sigma_{00}}\right) \tag{41a}$$

$$BER_P = \frac{1}{2}\left[Q\left(\frac{i_{01} - i_{10}}{\sigma_{01} + \sigma_{10}}\right) + Q\left(\frac{i_{11} - i_{01}}{\sigma_{11} + \sigma_{01}}\right)\right] \tag{41b}$$

It should be observed that, as OAM-3S, OAM-3P also offers arbitrary power level distribution. In fact, since $P_{out}^{00} = 0$ the ratio between the two other power levels may be regulated by simply setting G_P^1. This may be easily verified by dividing (55a) by (55b):

$$\frac{P_{out}^{11}}{P_{out}^{01}} = G_P^1 e^{-\alpha L} \tag{42}$$

Thanks to this characteristic, power level distribution of ternary-amplitude signals generated by OAM-3P may be set to minimize BER under the dominance of any kind of noise.

4.2.2 Results and discussion

Fig. 19 illustrates an experimental setup utilized for generating 3-ASK signals through OAM-3P with N= 4. Pump consists of a modulated signal at f_P= 192.3 THz. It is phase-modulated by radio-frequencies of 601 and 983 MHz to prevent the deleterious effects of Brillouin backscattering. This pump is further amplified and filtered in such a way that its power at the fiber input can be varied from 9.3 to 10.8 dBm. A $2^{15}-1$ pseudo-random bit sequence (PRBS) externally modulates a cw at f_S= 193.2 with R_{bS}= 1.0 Gb/s. Its power at the fiber output is of -19 dBm. The highly nonlinear dispersion-shifted fiber (HNL-DSF) is characterized by λ_0 = 1555.4 nm, $\Delta\lambda_0$ = 10 nm, S_0 = 0.017 ps/nm²/km, γ = 9.1 W⁻¹km⁻¹, α = 0.83 dB/km, and L= 3.0 km. Signals at fiber output are filtered, amplified and filtered again. Both filters are centered at f_S. The 3-ASK signal at f_S is then either analyzed by a DSA or transmitted through a 40-km long SMF link in KyaTera Network (FAPESP, (n.d.)).

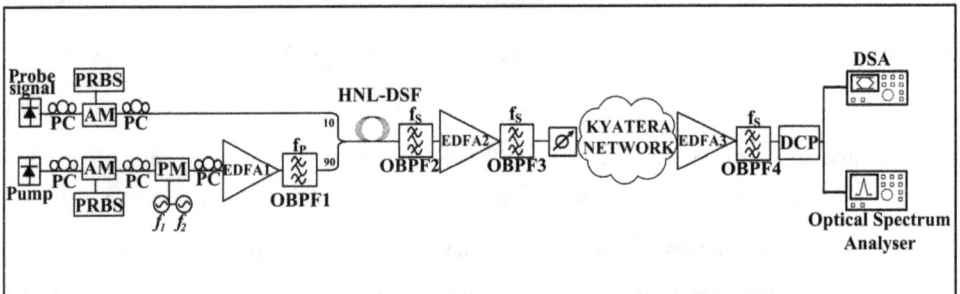

Fig. 19. Experimental setup.

In this second situation, the signal is further amplified, filtered and goes through a dynamic polarization controller (DPC) before being analyzed by the DSA. The DPC ensures the output signal state of polarization (SOP) will be the same, no matter its input SOP. This is important because the utilized link uses aerial cables that frequently introduces SOP changes due to environmental condition variations. Signal spectra at the (a) HNL-DSF output and (b) at the OBPF3 output, for an OSNR of ~31 dB, are shown in Fig. 20.

Fig. 21 plots bit sequences for the signals at (a) f_P and (b) f_S, and (c) for the correspondent generated OAM-3P. It is clearly verified that the signal in Fig. 21c represents a ternary-amplitude one. Moreover, as with the other presented techniques, the same power levels are obtained when the bits of the input binary signals are repeated.

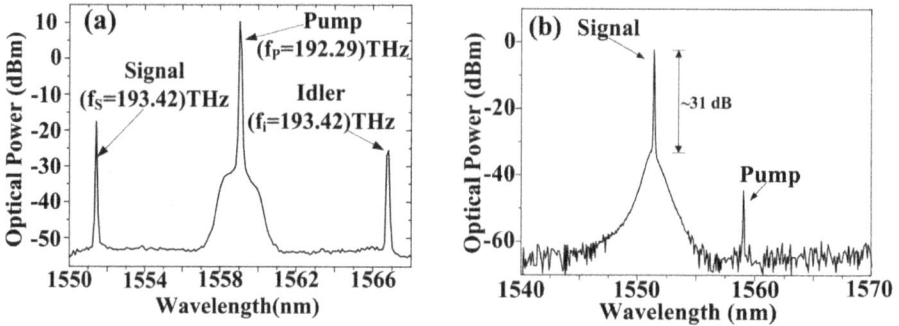

Fig. 20. Power spectra at the (a) HNL-DSF output and (b) OBPF3 output.

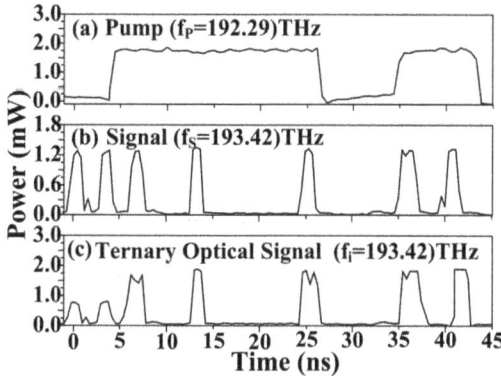

Fig. 21. Binary sequences (a) Pump (b) Signal (c) Ternary Amplitude Optical Signal for r=2.5.

Finally, Fig. 22 presents eye-diagrams for ternary-amplitude signals before and after network propagation with $P_{out}^{11}/P_{out}^{01}$ = 3.0, 3.5, and 4.0. Such values were obtained by varying the pump power, i.e., by controlling the parametric gain provided by the modulated pump. Using (41) we find that the lowest BERs are obtained for $P_{out}^{11}/P_{out}^{01}$ = 2.0 and BER_S= 3.5 10⁻¹⁵ and BER_P= 4.7 10⁻¹⁴. These low values, show that the technique may be properly applied to practical applications.

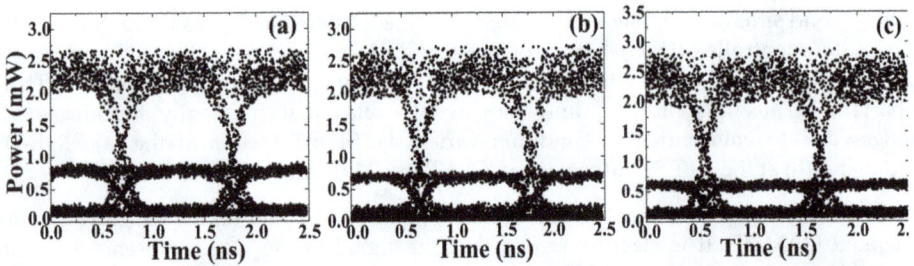

Fig. 22. Eye diagrams of ternary-amplitude signals after 40 km for $P_{out}^{11}/P_{out}^{01}$ = (a) 3.0, (b) 3.5, and (c) 4.0.

5. Applications and comparisons

In Sections 3 and 4 we have presented four techniques that convert two input binary signals into either a 4-ASK or a 3-ASK one. Due to the fast response of parametric interaction in fibers, all of these techniques are bit-rate independet for practical bit rates. Futhermore, all of them are able to transmit information from both input signals simultaneously and in the same bandwidth. Therefore, they could be used, for instance, to multiplex data from two different wavelengths into a single one, which results in bandwidth savings (Abbade et al., 2005).

Another possible application is to use one of the input signals to introduce a label to the second one. This optical labeling operation is illustrated in Fig. 23 and would be useful when binary data is entering the domain of an optical packet switching network (OPSN) and needs to be converted to an optical packet (Abbade et al., 2006a; Abbade et al., 2010a).

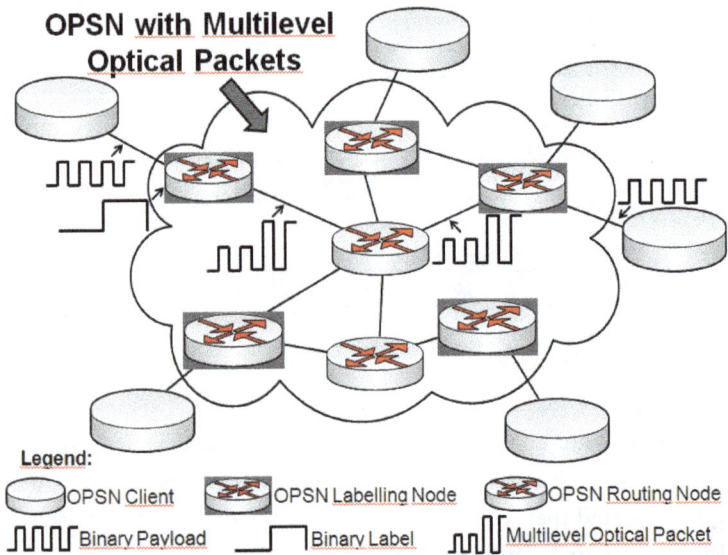

Fig. 23. Optical labeling application

Yet another possible application is to use OAM-4F and OAM-4P in several stages in a way that binary signals are successively converted into an analog signal (Abbade et al., 2010a). For example, OAM-4F could be used to multiplex two binary signals into a quaternary one, which could be combined to another binary signal into an octary one and so on, until the number of levels is high enough to make the signal resemble an analog one.

In spite of being employed to the same applications and utilizing very similar setups, there are marked differences among the analyzed techniques. Some of them have already been mentioned; here, it is interesting to summarize them and present some further distinctions.

One of these is that two of the considered techniques, OAM-4F and OAM-4P, generate 4-ASK signals and the other two, OAM-3F and OAM-3P, produce 3-ASK signals. Both of these modulation formats are advantageous over OOK modulation because they provide higher tolerance to chromatic dispersion degradations. The considered 3-ASK signals present lower BERs than the 4-ASK not only because their number of levels is smaller, but also because their lowest level has null power. In fact, the power level distribution of these 3-ASK signals may be arbitrarily set, which does not happen for the considered 4-ASK signals. OAM-3P still has the advantage of not needing any power offset for its input binary signals, *i.e.*, it can operate with conventional OOK input signals.

On the other hand, OAM-3F and OAM-3P only work under some special conditions, which comprises restrictions to the input signals line codes and bit rates and also demands oversampling the output signal to recover information transmitted by the pump signal. This makes OAM-3F and OAM-3P more suitable for the optical labeling application, where label bit rate is typically inferior to that of payload (otherwise, packet overhead could be very high). By their turn, OAM-4F and OAM-4P are independent of the utilized line codes and bit rates.

Since the techniques based on FWM, OAM-4F and OAM-3F, do not need a pump to transfer power to another signal, they may be accomplished by utilizing lower powers than OAM-4P and OAM-3P require. In fact, the average power of the input signals in experiments of OAM-4F were around ~12 dBm (Abbade et al., 2006b), whereas for the OAM-4P setup an average pump power as high as 20 dBm was necessary (Abbade et al., 2010b). With such high powers, Brillouin backscattering becomes relevant; thus, the experimental setup of OAM-4P and OAM-3P needs some additional equipment to reduce the influence of this effect and becomes more complex than the setup utilized by the other two techniques. In particular, if phase modulators and RF generators are used to prevent Brillouin backscattering, then phase noise may cause further degradations to the generated signal. However, in spite of these drawbacks and in opposition to FWM, PA does not broaden the linewidth of the generated signal. Therefore, if no measures are taken to compensate dispersion, the multi-amplitude signals generated by OAM-4P and OAM-3P can propagate through longer distances than the ones produced by OAM-4F and OAM-3F.

Another important advantage for OAM-4P and OAM-3P techniques is that the generated multi-amplitude signal is at the same wavelength as one of its inputs. This does not occur with OAM-4F and OAM-3F strategies, where it is necessary to know the spacing between the input signals frequencies to set the filter that will select the output signal. Table 2 summarizes the most important differences commented above.

Technique	One of the Input Channels need a higher bit-rate	One of the Input Channels need a higher power	Number of Input Channels with power off-set	Output at a new wavelength	Output with broadened linewidth	Output Signal with Arbitrary Power Level Distribution	Output Signal BER
OAM-4F	No	No	2	*Yes*	Yes	No	High
OAM-4P	No	Yes	1 or 2	*No*	No	No	High
OAM-3F	Yes	No	1	*Yes*	Yes	Yes	Low
OAM-3P	Yes	Yes	0	*No*	No	Yes	Low

Table 2. Comparison among the presented techniques.

6. Conclusion

The utility of optical fibers goes far beyond simple transmission in Optical Communication systems. In this work we have considered fiber applications to the very promising field of all-optical digital signal processing. Particularly, we focused on four techniques that allow the conversion of input binary signals into ternary- or quaternary-amplitude ones. Three of these techniques had already been analyzed in other reports; OAM-3F was, however, proposed and analyzed here for the first time.

Although we considered solely the generation of multi-amplitude signals, some modifications on the presented techniques could allow for optical phase multiplexing, as well. For example, in (Zhou et al., 2006) OAM-4F was adapted to merge two differential phase-shift keying signals into a single differential quadrature phase-shift keying one. Finally, it should be noted that the techniques discussed here could, in principle, be extended to any other nonlinear material that supports FWM and PA, such as semiconductor optical amplifiers and silicon chips. The latter is of special importance because it would allow the aforementioned applications to be performed in chip-to-chip communications, in the emerging field of silicon photonics.

7. Acknowledgment

This work was supported by FAPESP grant 08/57857-2 and CNPq grants 574017/2008-9 and 309031/2008-7. Authors thank Dr. Hugo L. Fragnito for invaluable discussions. Authors also thank VPIPhotonics Inc. for providing academic licenses of VPITransmissionMaker.

8. References

Abbade, M.L.F.; Fagotto, E.A.M.; Braga, R. S.; Fonseca, I. E.; Moschim, E. & Barbosa, F.R. (2005). Optical Amplitude Multiplexing Through Four-Wave Mixing in Optical Fibers. *IEEE Photonics Technology Letters*, vol. 17, no. 1, pp. 151 – 153.

Abbade, M.L.F.; Almeida, F.P.; Branquinho, F. G. G.; Fagotto; Braga, R.S.; Callegari, F.A.; Rocha, M.L.; Rossi, S.M.; Boggio, J.M.C.; Fragnito, H.L. (2006a). Optically Generated Quaternary Packets: Transmission over the KyaTera Network. *Proceedings of the 19th Annual Meeting of the IEEE, Lasers and Electro-Optics Society*, Montreal, Canada, p. 557-558, 2006.

Abbade, M.L.F.; Fagotto, E.A.M.; Braga, R.S.; Barbosa, F.R.; Moschim, E. & Fonseca, I. E. (2006b). Quaternary Optical Packets Generated by Fiber Four-Wave Mixing. *IEEE Photonics Technology Letters*, vol. 18, no. 2, pp. 331 – 333.

Abbade, M.L.F.; Fagotto, E.A.M.; Braga, R.S.; Moschim, E.; Fonseca, I.E. & Callegari, F.A. (2006c). All-optical generation of quaternary amplitude signals. *IEE Electronics Letters*, vol. 42, no. 24.

Abbade, M.L.F.; Costa, A.L.A.; Barbosa, F.R.; Durand F.R.; Marconi, J.D. & Moschim E. (2010a). Optical Amplitude Multiplexing through Parametric Amplification in Optical Fibers. *Optics Communications*, v. 283, p. 454-463.

Abbade, M.L.F.; Marconi, J.D.; Costa, A.L.A.; Barbosa, F.R. ;Moschim, E. & Fragnito, H. L. (2010b). All-optical Generation of Quaternary Amplitude-Shift Keying Signals through Parametric Amplification. *Proceedings of the 12th International Conference on Transparent Optical Networks (ICTON)*,Munich, Germany, June, 2010.

Abbade, M.L.F.; Costa, A.L.A., Marconi, J.D.; Cardoso, V.V., Fragnito, H.L. & Moschim, E. (2011). Optical Labelling through Parametric Amplification. *Proceedings of the 13th International Conference on Transparent Optical Networks (ICTON)*, Stockholm, Sweden, June, 2011.

Agrawal, G.P. (2001). *Nonlinear Fiber Optics*, 3rd ed., New York: Academic Press.

Boyd, R.W. (2008). *Nonlinear Optics*, 3rd ed., Academic Press.

Brzozowski, L. & Sargent, E.H. (2001). All-optical analog-to-digital converters, hardlimiters, and logic gates. *Journal of Lightwave Technology*, vol. 19, no.1, pp. 114-119.

Buck, J.A. (2005). *Fundamentals of Optical Fibers*, John Wiley & Sons, Inc..

Butcher, P.N. & Cotter, D. (1990). *The Elements of Nonlinear Optics*, Cambridge Univ. Press, Cambridge, England UK.

Chavez Boggio, J. M. ; Guimaraes, A. ; Callegari, F. A.; Marconi, J.D.; Rocha, M. L.; de Barros, M. R. X. & Fragnito, H.L. (2004a). Parametric amplifier for mid-span phase conjugation with simultaneous compensation of fiber loss and chromatic dispersion at 10 Gb/s. *Microwave and Optical Technology Letters*, v. 42, p. 503-505.

Chavez Boggio, J.M.; Callegari, F.A.; Marconi, J.D.; Guimarães, A.; Fragnito, H.L. (2004b). Influence of zero-dispersion wavelength variations on cross-talk in single-pumped fiber optic parametric amplifiers. *Optics Communications*, vol. 242, no. 4-6, pp. 471-478.

Chavez Boggio, J.M.; Guimarães, A.; Callegari, F.A.; Marconi, J.D. & Fragnito, H.L. (2005a). Q penalties due to pump phase modulation and pump RIN in fiber optic parametric amplifiers with non-uniform dispersion. *Optics Communications*, vol. 249, no. 4-6, pp. 451-472.

Chavez Boggio, J.M.; Marconi, J.D.; Fragnito, H.L. (2005b). Double-pumped fiber optical parametric amplifier with flat gain over 47-nm bandwidth using a conventional dispersion-shifted fiber. *IEEE Photonics Technology Letters*, vol. 17, no.9, pp.1842-1844.

Chen, Y. (1989). Four-wave mixing in optical fibers: exact solution. *Journal of Optical Society of America B*, vol. 6, no.11, pp. 1986-1993.

Fagotto, E.A.M. & Abbade, M.L.F. (2010). All-optical demultiplexing of 4-ASK optical signals with four-wave mixing optical gates. *Optics Communications*, vol. 283, no. 6, pp. 1102-1109.

FAPESP. (n.d.). KyaTera, July 25 (2011), Available from: http://www.kyatera.fapesp.br/

Gloge, D. (1971). *Weakly Guiding Fibers*. Appl. Opt., 10, vol. 10, 2252-2258.

Grudinin, A.B.; Dianov, E.M.; Korobkin, D.V.; Prokhorov, A.M.; Serkin, V.N. & Khaidarov, D.V. (1987). Decay of femtosecond pulses in single-mode optical fibers. *Journal of Experimental and Theoretical Physics Letters*, vol. 46, no. 11, pp. 221-225.

Hansryd, J. & Andrekson, P.A. (2001). Broad-band continuous-wave-pumped fiber optical parametric amplifier with 49-dB gain and wavelength-conversion efficiency. *Photonics Technology Letters, IEEE* , vol.13, no.3, pp.194-196.

Hansryd, J.; Andrekson, P.A.; Westlund, M.; Li, J. & Hedekvist, P.O. (2002). Fiber-Based Parametric Amplifiers and Their Applications. *IEEE Journal of Selected Topics in Quantum Electronics*, vol. 8, (3), pp. 506-520.

Inoue, K. & Toba, H. (1992). Wavelength conversion experiment using fiber four-wave mixing, *Photonics Technology Letters, IEEE* , vol.4, no.1, pp.69-72.

Jackson, J.D. (1998). *Classical Electrodynamics*, Wiley; 3rd Ed., New York USA.

Kalogerakis, G.; Shimizu K.; Marhic, M.E.; Wong, K.K.Y.; Uesaka, K. & Kazovsky, L.G. (2006). High-repetition-rate pulsed-pumped fiber OPA for amplification of communication signals. *Journal of Lightwave Technology.*, v. 24, no. 8, pp. 3021-3027.

Lu, Guo-Wei & Miyazaki, T. (1997). Experimental Demonstration of RZ-8-APSK Generation Through Optical Amplitude and Phase Multiplexing. *IEEE Photonics Technology Letters*, Vol.20, no. 23, pp.1995-1997.

Marconi, J. D., Abbade M. L. F., Costa, A. L. A., Moschim, E. & Fragnito, H.L. (2011). Experimental Analysis of All-optical 4-ASK Signal Generation through Parametric Amplification, submitted to *Optics Communications*.

Mishina, K., Kitagawa, S. & Maruta, A. (2007). All-optical modulation format conversion from on-off-keying to multiple-level phase-shift-keying based on nonlinearity in optical fiber. *Optics Express*, vol. 15, pp. 8444-8453.

Mussot, A.; Lantz, E.; Durécu-Legrand, A.; Simonneau, C.; Bayart, D.; Maillotte, H. & T. Sylvestre (2007). Simple method for crosstalk reduction in fiber optical parametric amplifiers. *Optics Communications*, vol. 275, no. 2, pp. 448-452.

Oda, S. & Maruta, A. (2006). Two-bit all-optical analog to-digital conversion by filtering broadened and split spectrum induced by soliton effect or self-phase modulation in fiber. *IEEE Journal of Selected Topics of Quantum Electronics*, vol. 12, no. 2, pp. 307-314.

Olsson, B.-E.; Ohlen, P.; Rau, L.; Blumenthal, D.J. (2000). A simple and robust 40-Gb/s wavelength converter using fiber cross-phase modulation and optical filtering. IEEE Photonics Technology Letters, vol.12, no.7, pp.846-848.

Peucheret, C.; Lorenzen, M.; Seoane, J.; Noordegraaf, D.; Nielsen, C.V.; Gruner-Nielsen, L. & Rottwitt, K. (2009). Amplitude regeneration of RZ-DPSK signals in single-pumped fiber-optic parametric amplifier. *IEEE Photonics Technology Letters*, vol. 21, pp. 872-874.

Ramaswamy, R.; Sivarajan, K.; Sasaki, G; (2010). *Optical Networks – a practical perspective*, Morgan-Kaufmann, 3rd.ed., Burlington MA, USA.

Song, S.; Allen, C.T.; Demarest, K.R. & Hui, R. (1999). Intensity-dependent phase-matching effects on four-wave mixing in optical fibers. *Jounal of Lightwave Technology*, vol. 17, no. 11, pp. 2285-2290.

Stolen, R.H. & Bjorkholm, J.E. (1982). Parametric amplification and frequency conversion in optical fibers. *IEEE Journal of Quantum Electronics.*, vol. 18, no. 7, pp. 1062-1072.

Jamshidifar, M.; Vedadi, A. & M. Marhic, M. E. (2010). Continuous-wave parametric amplification in bismuth-oxide fibers, *Optical Fiber Technology, vol. 16, pp. 458-466.*

Walklin, S. & Conradi, J. (1999). Multilevel signaling for increasing the reach of 10 Gb/s lightwave systems. *Journal of Lightwave Technology*, vol. 17, no. 11, pp. 2235–2248.

Xu, Zhaowen ; Zhou, Guangtao & Lu, Chao. (2005). Optical 4-ASK signal generation through four-wave mixing. *Proceedings of the 18th Annual Meeting of the IEEE , Lasers and Electro-Optics Society*, pp.505-506, Sydney, Australia, October, 2005.

Zhou, G. T.; Xu, K.; Wu, J.; Yan, Cishuo; Su, Yikai & Lin, J. T. (2006). Self-Pumping Wavelength Conversion for DPSK Signals and DQPSK Generation Through Four-Wave Mixing in Highly Nonlinear Optical Fiber. *IEEE Photonics Technology Letters*, Vol.18, no.22, pp. 2389-2391.

2

Design and Application of X-Ray Lens in the Form of Glass Capillary Filled by a Set of Concave Epoxy Microlenses

Yury Dudchik

Institute of Applied Physics Problems
of Belarus State University
Belarus

1. Introduction

Glass capillaries are widely used in X- optics. It is a well know fact that a simple glass capillary acts as a waveguide for X-rays because refractive index for X-rays in any medium is less than unity. X-rays transmit glass capillary in the regime of total external reflection instead of visual light that propagates inside glass fiber in the regime of total internal reflection. X-rays also may transmit curved capillaries and, as was proposed by Kumakhov, a bunch of curved capillaries act as a lens that focuses X-rays from a point source into a focal point. This device is known as Kumakhov X-ray lens (Kumakhov & Sharov, 1992). Another well-known X-ray device is a taper or parabolic single capillary that is used to condense or focus synchrotron X-rays into micron-sized spot (Thiel et al., 1992).

Recently a new application of glass capillaries for X-ray optics was proposed: it was demonstrated that capillaries are suitable for designing so named compound refractive X-ray lenses.

Compound refractive X-ray lens at the first time was proposed by A. Snigirev, V. Kohn, I. Snigireva and B. Lengeler (Snigirev et al., 1996) and their idea is based on the following principles. It is a well-known fact that refractive lens for X-rays should be concave instead of convex for visual light. Calculation shown that the focal length F_1 of such biconcave spherical lens is determined by the following ratio:

$$F_1 = R /(2\,\delta),\tag{1}$$

where R-radius of the lens and $(1-\delta)$ is real part of refractive index n. The focal length of the lens is rather large (5-10 m for 5-8 keV X-rays) even when the curvature radius of the lens is equal to hundred of micrometers. The large value of the lens focal length was a reason of the conclusion that there is no any practical interest to focus X-rays by refractive lens. Attempts to reduce focal length of the lens have resulted in creation of a compound refractive lens (CRL) for X-rays with energy 5-30 keV (Snigirev et al., 1996). The lens consists of a large number (10-300) of biconcave lenses, made of material with a low-atomic weight (beryllium, carbon, polymers, aluminium). Focal length F of such lens is defined by the following ratio: $F= F_1 / N$, where N- is the number of lenses. The equation for F shows that the focal length of a compound lens can reach value of 1 m at $R=0.5$ mm. That is quite acceptable for practical applications.

At present compound X-ray lenses are designed by some ways: by using pressing technique for individual lens, by lithographic method, by drilling holes in a plate by a laser (Lengeler et al., 2005). The problem of the lens design is how to produce individual concave lens with a high quality parabolic or spherical shape surface and with curvature radius up to 50 microns or less. Another problem is to stack the lenses coaxially to form compound lens.

The idea of compound X-ray lens was advanced in our work (Dudchik & Kolchevsky, 1999), where it was realised in the form of glass capillary filled by a large number of epoxy drops. The lenses were designed at the Institute of Applied Physics Problems of Belarus State University. The lens was named as microcapillary one and applied at synchrotron SPring-8 for focusing of 18 keV X-rays and as an objective of X- ray microscope (Kohmura et al., 1999). The lens consists of a glass microcapillary, filled by a plenty of biconcave microlenses. The concave microlenses inside the capillary were formed by putting air bubbles into epoxy. The schematic view of the lens is shown in Fig. 1.

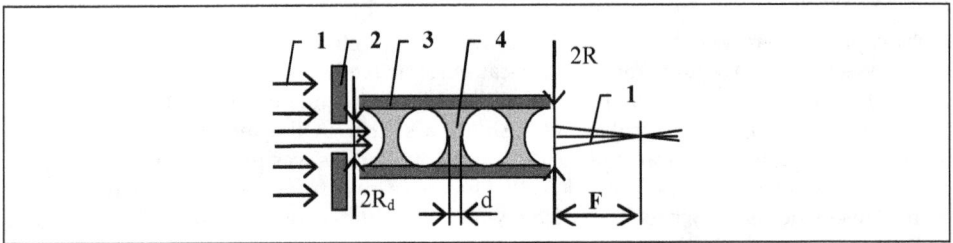

Fig. 1. Schematic view of the microcapillary X-ray lens. 1- X-ray beam; 2- diaphragm; 3- capillary; 4- epoxy lens

It was shown (Dudchik et al., 2000) that the microlenses inside the capillary are spherical ones and its curvature radius is equal to capillary one. This founded dependence of the lens curvature radius on the capillary one leaves a room to decrease the lens focal length. For example the lens in the form of 200 microns in diameter capillary and filled by 103 microlenses has 13-cm focal length for 8 keV X-rays. It was shown experimentally by Adelphi Technology, Inc. using beamline 2-3 at the Stanford Synchrotron Radiation Laboratory (SSRL) (Dudchik et al., 2004). Such short-focal-length lenses are suitable for imaging not only with synchrotron X-rays, but with X-rays from laboratory sources of radiation (Piestrup et al., 2005).

The purpose of the paper is to consider details of microcapillary lens design as the lens application for focusing and imaging of X-rays.

2. Design and application of microcaplillary X-ray lens

2.1 Fabrication technique for the microcapillary refractive X-ray lens

The method of the microcapillary lens preparation consists (Dudchik & Kolchevsky, 1999; Dudchik et al., 2000) in consecutive producing of air bubbles inside of capillary 1, filled by epoxy 2 with using of capillary 4, connected with a cylinder with compressed air 5, as is shown in Fig.2. The growth of the bubble inside of the capillary 1 is supervised by visual light microscope. When the radius of the bubble is becoming equal to the radius of the capillary 1, the capillary 4 is moving to a distance of few microns from the received bubble and the process is repeated. The liquid between two bubbles has a form of biconcave lens. This technique actually has not restrictions in number of lenses. The photo of epoxy lenses, made by the method, is shown in Fig. 3. The diameter of the capillary is equal to 0.2 mm (a) and 0.8 mm (b).The air bubbles between lenses are observed as black ones. Used epoxy consists of carbon, oxygen, hydrogen and nitrogen which are chemically bonded in proportion $C_{200}H_{100}O_{20}N$. The epoxy density is 1.08 g/cc.

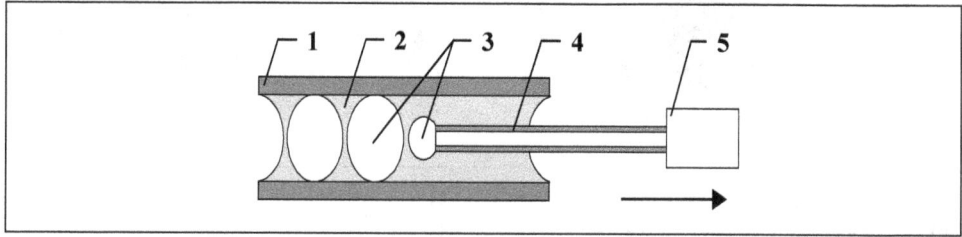

Fig. 2. Schematic view of the setup for microcapillary X-ray lens fabrication. 1- glass capillary tube; 2- glue; 3-air bubbles; 4- injector needle; 5- cylinder with a compressed air

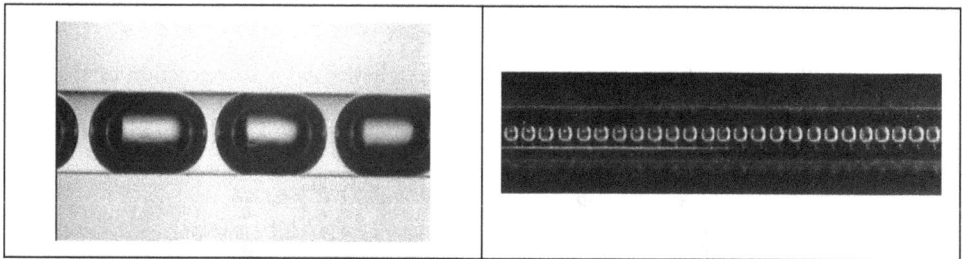

Fig. 3. Visible light microscope image of the microcapillary refractive X-ray lens. The diameter of the capillary is equal to 0.2 mm (a) and 0.8 mm (b).

Important parameter of the lens is thickness d (Fig.1), which for the given material of the lens depends on the diameter of the capillary channel and on the epoxy temperature. We established that for the lens made from epoxy, the lens thickness d might be decreased up to 5-10 microns.

The shape of the lens surface was investigated by an optical method with the help of optical microscope connected with digital camera. The obtained individual lens image was processed by computer. In Fig. 4 (a, b) the images of two lenses are shown. The curve, dividing a light and a dark parts in Fig. 4 was considered as a profile of the lens. At construction of the lens profile we took into account, that the visible lens diameter $2R$ is more than the real diameter of the channel $2R_{real}$ (Fig. 5).

We took into account that the light, that scatters from the inner wall of the lens, does not come directly to the microscope. It is doubly refracted at the (lens-material)-glass and glass-air boundaries. This is a reason why the observed lens profile differs from the real one.

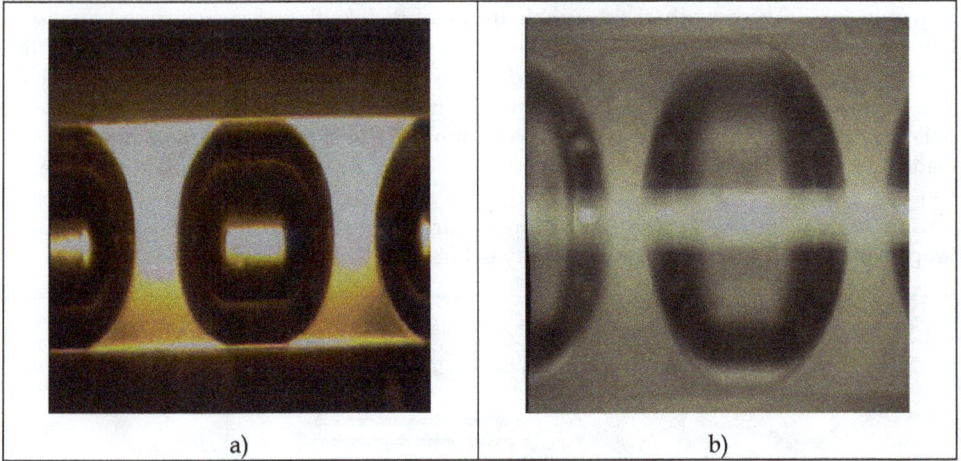

Fig. 4. Visible light microscope images of concave epoxy lenses inside capillary. a) Capillary radius is equal to 0.39 mm; b) capillary radius is equal to 0.21 mm

Formulas for calculating lens profile can be found from a geometrical paths of rays, forming the image of the lens (Fig. 5.). According to the Snell's law:

$$n_{lens} \sin\alpha = n_{glass} \sin\beta; \quad n_{glass} \sin\gamma = \sin\theta, \tag{2}$$

where n_{glass} is the index of refraction of the glass, n_{lens} is the index of refraction of the lens material, $\sin\alpha = R/(n_{lens} R_{chan})$, $\sin\beta = R/(n_{glass} R_{chan})$, $\sin\gamma = R/(n_{glass} R_{cap})$, $\sin\theta = R/R_{cap}$.

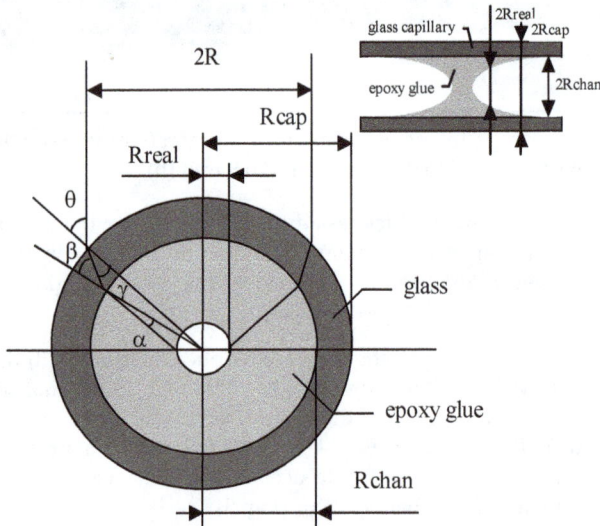

Fig. 5. Schematic view of the transverse section of the microcapillary lens. The rays of visible light forming lens image also are shown. *Rcap*- is the outer radius of the capillary; *Rchan*- is the radius of the capillary channel; *Rreal*- is the measured value of the profile; *R*- visible value of *Rreal*.

From eq.(2) the radius of the channel, shown in fig. 4 as R_{real}, is equal to:

$$R_{real}=R/\left(n_{lens}\cos\left(\alpha+\gamma-\theta-\beta\right)\right).\tag{3}$$

The obtained formula for R_{real} was used for calculation of the lens profile. Fig. 6 shows the profile of lens in comparison to the circle, radius of which is equal to the radius of the capillary. As it can be seen from Fig. 6, the form of the microcapillary lens can be accepted as spherical one, and the radius of lens curvature R is equal to the channel radius R.

This result is in a good agreement with classical molecular theory. The theory states that the form of liquid drop putt into microcapillary can be accepted as biconcave spherical, and the radius of drop curvature is connected by the following ratio to the capillary radius R_{chan} :

$$R=R_{chan}/\cos\varphi,\tag{4}$$

where φ- angle of contact. For epoxy glue located on a glass surface, angle of contact is equal to 0^0.

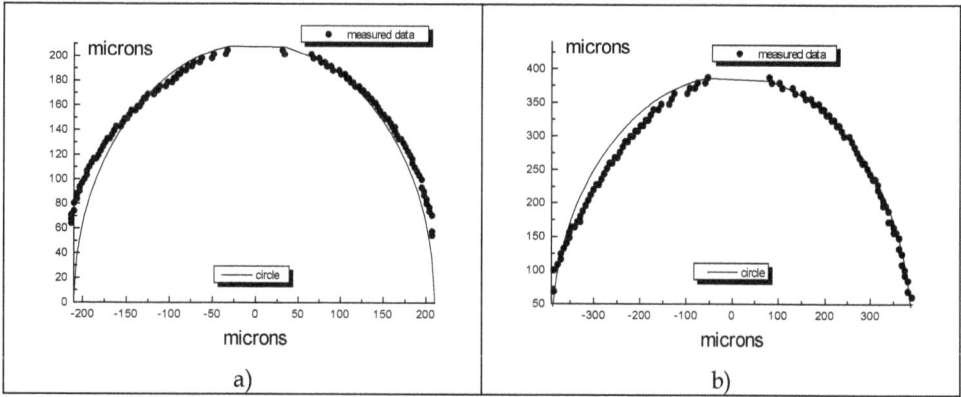

Fig. 6. Measuered profiles of the lenses. a) Capillary radius is 210 microns; b- capillary radius is 390 microns

2.2 Parameters of the microcapillary refractive lens

Lens focal length f of compound refractive X-ray lens is calculated as (Snigirev et al., 1996):

$$f=\frac{R}{2N\delta},\tag{5}$$

where R- curvature radius, N- number of microlenses, $(1-\delta)$- real part of refractive index for X-rays. Parameter δ for used epoxy may be calculated from the epoxy chemical formula as

$$\delta=0.5\left(\frac{22}{E}\right)^2,\tag{6}$$

where E is photon energy measured in eV. Experiments on measuring lens focal length of compound epoxy lenses at Stanford Synchrotron Radiation Laboratory and at Advanced

Photon Source by Adelphi Technology, Inc. shown validity of formula 6 for calculation lens focal length (Dudchik et al., 2004).

Compound X-ray lens consisting of spherical microlenses may be characterized by absorption aperture radius R_a that in a good approximation can be calculated as (Snigirev et al., 1996; Dudchik et al., 2004; Piestrup et al., 2005):

$$R_a = \left(\frac{2R}{\mu N} \right)^{\frac{1}{2}} , \tag{7}$$

where μ is the linear absorption coefficient for the lens material.

The discussed X-ray lens is a linear combination of spherical microlenses and spherical aberrations occur just in the same way as for spherical visual-light lens. To take into account this phenomenon at least two planes around the lens focus may be denoted: they are shown by the lines MS and PP in Fig.7 which shows trajectories of 8-keV X-rays forming focal spot of compound lens consisting of 103-microlenses.

The plane PP represents a focal plane. The plane MS represent the circle of the least confusion (Born @ Wolf, 1975). For spherical lens the size of X-ray beam at MS and PP planes may be decreased by using diaphragm. The beam radius R_{pp} at the PP-plane depends on the radius of used diaphragm R_d and is calculated from the third order aberration theory as (Born @ Wolf, 1975):

$$R_{pp} = fBR_d{}^3 \tag{8}$$

where f is the lens focal length, B is Seidel coefficient, R_d is radius of the diaphragm placed before the lens. Eq. 8 is valid for the case of the point source located at the infinity. As it was shown in (Dudchik et al., 2000), the equation (8) for compound X-ray lens is rewritten as

$$R_{pp} = \frac{1}{2} \frac{R_d^3}{R^2} . \tag{9}$$

The eq. 9 is valid for the case when $R_d < 0.6\ R$ as was estimated in (Dudchik et al., 2000) by numerical calculations. The beam radius R_{ms} at the MS-plane is related to the beam radius R_{pp} at the focal plane as

$$R_{ms} = \frac{1}{4} R_{pp} = \frac{1}{8} \frac{R_d^3}{R^2} . \tag{10}$$

The distance L_{ms} from the lens to MS-plane is calculated as:

$$L_{ms} = f \frac{\left(R_d + R_{ms} \right)}{\left(R_d + R_{pp} \right)} . \tag{11}$$

For example $L_{ms} = 0.917\ f$ when $R_d = 0.5\ R$ and it is illustrated by Fig. 7 where position of MS-plane is shown for the discussed case.

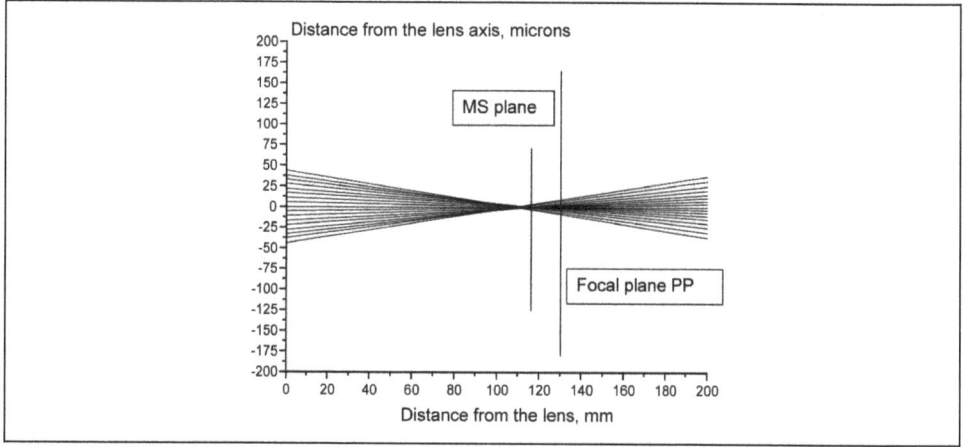

Fig. 7. Paths of 8-keV X-rays forming focal spot of 103- elements lens. Lens radius is 100 microns. Diameter of the diaphragm placed before the lens is equals to 100 microns

Radius R_{ms} of the circle of the least confusion decreases with decreasing of the size of diaphragm and achieves its minimum possible value $R_{ms\text{-}min}$ when a diaphragm with an optimal hole radius $R_{d\text{-}opt}$ is used. In this case the value $R_{ms\text{-}min}$ will be equals to the radius of the first minimum of the Airy diffraction pattern R_{diff} which is $R_{diff} = 0.61\ \lambda\ L_{ms}\ /\ R_{d\text{-}opt}$. The diaphragm radius $R_{d\text{-}opt}$ may be defined by the following equation:

$$\frac{1}{8}\frac{R_{d-opt}^{3}}{R^{2}} = \frac{0.61\lambda L_{ms}}{R_{d-opt}} \tag{12}$$

where λ is the wavelength. The solution of the Eq. (12) for $R_{d\text{-}opt}$ and for $R_{ms\text{-}min}$ under the assumption $L_{ms} = f$ is:

$$R_{d-opt} = \left(\frac{2.44R^{3}\lambda}{\delta N}\right)^{1/4}, \tag{13}$$

and

$$R_{ms-min} = \frac{0.61\lambda f}{R_{d-opt}}. \tag{14}$$

It is interesting to compare above result for $R_{d\text{-}opt}$ (Eq. 13) with the value of so-named parabolic aperture radius R_p. Parabolic aperture radius R_p is the central portion of the spherical lens that focuses X-rays to the same point. From wave approximation it is known (Snigirev et al., 1996; Piestrup et al., 2005) that the value may be calculated as:

$$R_{p} = \left(\frac{2R^{3}\lambda}{\delta N}\right)^{1/4}. \tag{15}$$

Comparing Eq.13 and Eq.15 we can see that there is only a small deference in numerical coefficient for these two values. It may be explained that we used f value instead of L_{ms} when solving Eq.12.

Eq. (13) for $R_{d\text{-}opt}$ and Eq. (12) for $R_{ms\text{-}min}$ are useful to calculated expected X-ray beam size in the MS plane for compound X-ray lenses. In Table 1 and Table 2 the diameter $2R_{ms}$ of the circle of the least confusion is calculated for epoxy spherical lenses with deferent values of number of microlenses N and lens radius R. Table 2 shows result of the same calculations for the case when additional diaphragm is used to decrease size of the beam at MS plane.

From the Table 2 it is seen that spherical compound refractive lens being combined with diaphragm ensures resolution at submicron level.

Lens parameters	Lens focal length f for 8 keV X-rays, mm	Lens absorption aperture $2Ra$, μm	Lms value for 8 keV X-rays beam , mm	Diameter $(2R_{ms})$ of the circle of the least confusion, μm
N=100, R=100 μm	132	117	117.5	5
N=200, R=100 μm	66	82	62	1.72
N=100, R=50 μm	66	82	53.5	6.89
N=200, R=50 μm	33	58	29.4	2.44

Table 1. Parameters of X-ray beam formed by spherical compound epoxy X-ray lens.

Lens parameters	Lens focal length f for 8 keV X-rays, mm	Diameter of the optimal diaphragm aperture $2Ropt$, μm	Lms value for 8 keV X-ray beam , mm	Diameter $2R_{ms}$ of the circle of the least confusion, μm
N=100, R=100 μm	132	62	127.4	0.78
N=200, R=100 μm	66	53.2	64.3	0.46
N=100, R=50 μm	66	37.6	62.7	0.66
N=200, R=50 μm	33	31.6	31.8	0.39

Table 2. Parameters of X-ray beam formed by spherical compound X-ray lens combined with diaphragm.

2.3 Measurement of the microcapillary refractive lens at Stanford Synchrotron Radiation Laboratory

Refractive X-ray lens work as ordinary lens for visual light and lens formula is also valid to describe its operation. The formula is written as:

$$\frac{1}{a} + \frac{1}{b} = \frac{1}{f} , \tag{16}$$

where a is distance from the source to lens, b is distance from the lens to source image, f - lens focal length. The size of the source image S_1, as in the case of visual optics, is related to the source size S by the equation:

$$S_1 = S\frac{f}{a - f}.\tag{17}$$

In the case of synchrotron radiation the distance between the source and the lens is high enough and equals, as a rule, to 10-50 m; the size of the source is also, as a rule, less than 1000 microns. When refractive lens with a focal length equal to approximately 10 cm is used, expected size of source image may be equal to some microns in according to formula 17. This is a way for obtaining micro and nano-sized X-ray beams.

We fabricated and tested some microcapillary refractive X-ray lenses for focusing synchrotron X-rays. The lenses were arbitrarily designated as 1-1, 3-1, 3-4, 3-3, 4-1 and 5-1. The calculated and measured characteristics of these CRLs are given in Table 3 below.

X-ray lens designation	1-1	3-1	3-4	3-3	4-1	4-1
Photon energy, keV	12	12	9	8	8	7
Number of individual lenses	90	196	103	93	102	102
Lens curvature radius, μm	165	250	100	100	100	100
Calculated focal length, cm	52.8	37	15.7	13.8	12.6	9.6
Calculated image distance, cm	54.5	37.8	15.8	13.9	12.6	9.7
Measured image distance, cm	32	36	17.5	13	14	10
Calculated vertical minimum waist diameter, μm	15.1	10.4	4.4	3.9	3.2	2.7
Measured vertical minimum waist diameter, μm	12.8	12	3.9	4.8	2.7	4
Calculated horizontal minimum waist diameter, μm	64.1	44.0	18.8	16.6	13.5	11.5
Measured peak transmission, %	36	30	24	16	27	5
Calculated attenuation aperture diameter, μm	314	262	143	125	119	96.7
Measured attenuation aperture diameter, μm	321	245	147	150	149	149
Calculated 2D-gain	16.6	20.0	25.6	16.9	28.9	6.0
Measured 2D-gain	8.9	3.5	13.4	**	25.5	**

Table 3. Measured and calculated parameters of microcapillary refractive X-ray lens for SSRL BL 2-3 source

Used was the beamline 2-3 on the Stanford Synchrotron Radiation Laboratory's (SSRL's) synchrotron (Dudchik et al., 2004). This beamline possesses a double-crystal monochromator that was capable of giving x-rays from 2400 to 30000 eV with a 5×10^{-4} resolution. The approximate source size (full width half maximum, FWHM) was 0.44×1.7 mm^2, as specified by SSRL. The experimental apparatus is shown in Fig. 8.

The distance from the source to lens was 16.81 meters. The X-ray beam size from this source was approximately 2×20 mm^2 at the entrance to the experimental station; however, this size

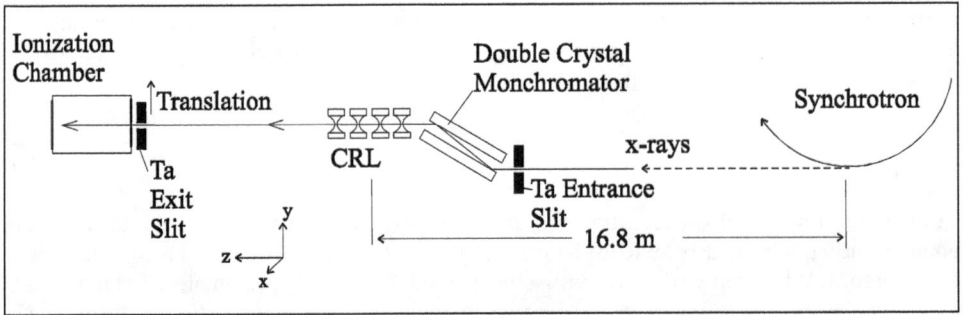

Fig. 8. The experimental apparatus at SSRL for measuring microcapillary X-ray lens

was reduced to approximately 0.4 x 0.4 mm² by Ta slits upstream of the monochromator. The CRL was placed in a goniometer head that could be manually tilted in three axes. The lens could also be remotely translated orthogonally (x and y) to the direction of the x-ray beam to maximize the x-ray transmission though the lens. An x-ray gas ionization detector was placed after a translatable slit for measuring the x-ray beam profile. This Ta slit was adjusted to below 25 μm by using a thin stainless steel shim. It is likely that, as good as the Ta slit was, the jaws are not ideally parallel at these small dimensions, and the slit width was minimally > 3 μm when jaws appeared to be entirely closed.

After the slit width was adjusted, the Ta slit was then translated in the x direction across the focused x-ray beam, and its profile obtained. We then manually moved the slits along the z-axis of the x-ray beam, measuring its vertical widths by scanning the slits in the x direction across the beam at each location.

Using these measured widths, we profiled the beam waist as a function of distance from the lens. Results are shown in Table 3. For example, the source of diameter 0.44 mm was focused by the lens 4-1 to a minimum spot FWHM of 5 microns, at a distances of 13 cm from the CRL. Thus, the demagnification was M = .0114. The spot size in the horizontal direction was measured to be 19 microns, which is larger, because the imaged source diameter was larger (1.7 mm, FWHM) in that dimension.

The CRL has an aperture with a Gaussian absorption profile, which causes stronger absorption of the extreme rays passing through the CRL outer radial regions, as compared to rays that pass through the less absorptive central region. The CRL aperture is much smaller than the source size, especially in the horizontal dimension. We obtained the transmission through the CRLs given in Table 3 by narrowing the x-ray beam to 50 x 50 μm² and translating each CRL through the beam, thereby producing transmission profiles of the lenses. The absorption apertures (e^{-2} points, not FWHMs) were obtained from these figures. The calculated and measured peak transmissions (transmission at the lens axis) for the lenses are given in Table 3.

Given the measured transmissions and profiles, we determined the gains of these lenses. Both measured and calculated gains are given in Table 3 for all these lenses. The gain values varied between 3.5 to 25.5. These gain values are primarily due to the large source size. If the same lens is placed on a beam line using a third generation X-ray source, the gain of the CRL can be substantially larger. These sources can possess spot sizes a factor of 3 smaller (e.g. 0.5 by 0.5 mm²). Also, typical distances from insertion devices to end stations can be approximately 51 m, as compared to 16.8 m in our experiment. For these source parameters, the gain at 11 keV from the same lens is 138, a sizable increase of the intensity over that of the case without a CRL. Although the gains of these CRLs were small, larger gains can be achieved using smaller source sizes, larger CRL apertures, and longer object distances.

2.4 Measurement of the microcapillary refractive lens at ANKA Synchrotron Radiation Source

Parameters of the lens # 5-1 were measured at ANKA Synchrotron Source (Germany) (Dudchik et al., 2007a). The lens was designed in Institute of Applied Physics Problems of Belarus State University. The lens consists of 224 spherical epoxy microlenses formed inside glass capillary with curvature radius equal to 100 microns. Lens length is equal to 69 mm. The individual epoxy lenses inside of the glass capillary are spherical ones with the curvature radius equals to 100 microns. Spherical lenses may be characterized by the following set of parameters: lens focal length f, absorption aperture radius R_a (see formula 7), parabolic aperture radius R_p (see formula 15) , and the diffraction radius $R_{diff.}=0.61\ \lambda f/R_a$, that characterises diffraction blurring of the focused beam, where μ is linear absorption coefficient for the lens material.

The lens # 5-1 was used to focus X-rays with energy 12 keV and 14 keV. Calculations show that for 12 keV X-rays parabolic aperture radius of the lens is R_p=27 microns for the case of the discussed lens (R= 100 microns, N= 224). The absorption lens aperture radius Ra for the lens is equal to 69 microns. The same values for 14 keV X-rays are : R_p=28 microns, R_a=94 microns. Lens focal length f calculated by formula 5 for 12 keV and 14 keV is equal to 133 mm and 180 mm respectively. The lens length is equal to 69 mm and it is "thick enough" comparing to lens focal length. The focal length f_t of a thick lens may be calculated by the next formula (Pantell et al., 2003):

$$\frac{f_t}{f} = \frac{\left(\frac{t}{f}\right)^{1/2}}{\sin\left(\frac{t}{f}\right)^{1/2}} , \tag{18}$$

where t is lens length. Result of calculation of f_t: f_t = 145 mm for 12 keV X-rays and f_t = 192 mm for 14 keV X-rays.

The lens #5-1 has been characterized for 12 keV and 14 keV X-rays at the ANKA-FLUO experimental station situated at a bending magnet of the ANKA Synchrotron Light Source. The energy was monochromatized by a W/BC4 double multilayer monochromator with 2% bandwidth. For the measurement of the beamsize the lens was placed on a five axis positioning device and exactly oriented in the direction of the X-ray beam. The distance a between source and lens was equal to 12.7 m. The size of the source s: 800x250 µm² FWHM. The source size can be reduced by a 0.1mm x 0.1 mm² slit #1 placed at a distance 4.7 m to the source. There was one more slit #2 placed at 1m distance from lens. The slit size was 0.1mm x 0.08 mm². It was also possible to hold slits in opening mode.

Measured were beam size at different distance to the lens and lens transmission. The minimum value of beam size was considered as lens image distance. The beam size was derivated from knife edge scans conducted around the focus position derived with the x-ray camera. A 0.5 µm thin Permalloy structure was chosen and the edges have been scanned with 0.5 µm or 1 µm resolution. Characteristic K_α Fe-atom X-rays emitted by Permalloy structure were registered by X-ray camera. The measured profile of the edge is the convolution of the Fe concentration function (approximated by a step function) and the profile of the x-ray beam. As the step function converts the convolution in to a simple integration, the measured function is the error function if the beam profile is a Gaussian. Thus an error function (Fig. 9) has been fitted to the knife edge data. Fitting a Gauss function to the derivative is equivalent; nevertheless numerical derivating adds a considerable amount of noise to the data.

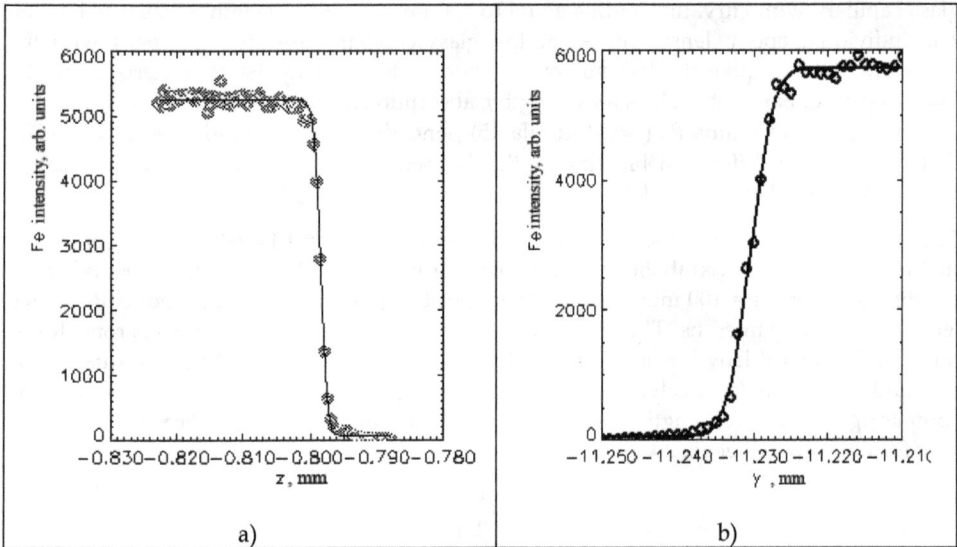

Fig. 9. Fit of error function to vertical (a) and horizontal (b) scan over lithographic structure

To determine gain in intensity of the beam due to focusing by the lens next procedure was applied. The lens was removed and the fluorescence intensity resulting from a Permalloy square of 50µm size was measured. This intensity was compared to the intensity of the focussed beam and the area of the focussed beam was calculated with $A=2\pi\sigma_x\sigma_y$ with being the Gaussian beamsize (FWHM value/2.35). With closed front end slits the gain factor for a smaller source can be obtained. For the ANKA source however closing slits cannot improve the gain. Therefore two values for the gain are given in Table 4 and Table 5.

Investigations shown that tested lens is suitable to focus 12 keV-14 keV X-rays into some microns in size spots (Dudchik et al., 2007a). Calculated lens focal length is in a good agreement with measured one. The lens parameters may be improved my increasing lens transparency. Also lenses with shorter lens focal length may be formed inside capillary with inner diameter equals to 100 microns. In this case the lenses may be used for nano-focusing.

Energy, keV	12	12
Size of slit #1, mm²	1 x 1	0.1 x 0.1
Measured image distance, mm	146	147
Calculated image distance, mm	147	147
Calculated lens focal length f_t , mm	145	145
Measured horizontal focal size, μm	10.4	4.1
Measured vertical focal size, μm	2.2	1.7
Gain	34/31	113/18
Transmission	9.5%	9.5%

Table 4. Parameters of spherical compound X-ray lens for 12 keV X-rays

Energy, keV	14	14
Size of slit # 1, mm²	1 x 1	0.1 x 0.1
Measured Image distance, mm	195	196
Calculated image distance, mm	195	195
Calculated lens focal length f_t , mm	192	192
Measured horizontal focal size, μm	12.2	6.3
Measured vertical focal size, μm	3.0	2.1
Gain	43/40	162 /22
Transmission	21.5%	21.5%

Table 5. Parameters of spherical compound X-ray lens for 14 keV X-rays

2.5 X-ray imaging with compound refractive X-ray lens

X-ray imaging is a power tool to study inner structure of objects and materials. This method is realised with synchrotron and laboratory X-ray sources. A well-known in-line laboratory X-ray projection microscopy and microtomography is based on the using of microspot X-ray tube as a source of radiation. The system for imaging consists of a quasi-point X-ray source and a CCD-camera. The object for investigation is placed at a distance R_1 from the source, and the CCD-camera is placed at a distance R_2 from the sample. The spatial resolution of the method depends on the source size and is in range from 5 to 1 microns. The magnification is determined as $(R_1 + R_2)/ R_1$ and may be 10 or higher. The disadvantage of the method of direct imaging is that the position of the point X-ray source is not stable in time. This disadvantage is remedied by using imaging optics for microscopy. There are some types of imaging X-ray optics: pin-hole, zone plate and compound refractive X-ray lens. We used previously discussed microcapillary refractive X-ray lens as an imaging device. In this case there is no limitation to the source size and ordinary X-ray tubes may be used.

The optical scheme of the system for imaging with refractive X-ray lens is shown in Fig.10. The object for imaging 3 is exposed by X-rays from X-ray tube 1. The lens 2 forms decreasing

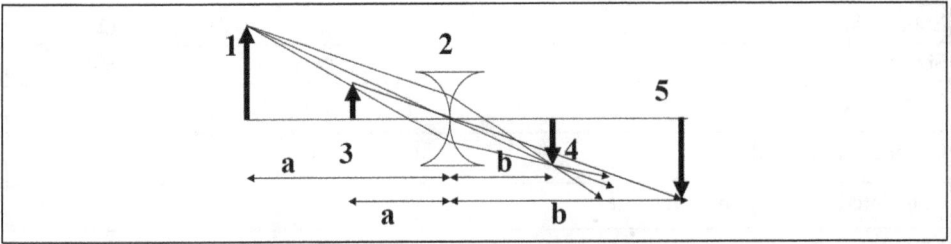

Fig. 10. Schematic of X-ray imaging with refractive lens. 1- X-ray source (tube); 2- compound refractive X-ray lens; 3- object; 4- source image; 5- object image.

image of the X-ray tube focal spot 4 and increased image of the object 5. X-ray CCD-camera is placed at the position of object image. The object, lens and CCD-camera are placed in-line at distances from one another that satisfied the lens formula:

$$\frac{1}{a}+\frac{1}{b}=\frac{1}{f} \ ,$$ (19)

where a is the distance from the object to the lens; b is the distance from the lens to CCD-camera; and f is the lens focal length.

The imaging system (microscope) was designed in the Institute of Applied Physics Problems of Belarus State University (Dudchik et al., 2007b; Dudchik et al. 2007c). The system photo is shown in Fig. 11. The microscope consists of X-ray tube 1, X-ray lens 2 in a holder and goniometer 5, CCD X-ray camera 3. The object for imaging 4 is place between the X-ray tube and the lens.

Fig. 11. X-ray microscope. 1- X-ray tube; 2- X-ray lens in a holder; 3- CCD X-ray camera; 4 – object for imaging; 5- goniometer for X-ray lens.

A water-cooled copper-anode X-ray tube (Russian model # BCV-17) with tube focal spot of 0.6 mm x 8 mm was used as a source of X-rays.

The image of the object was recorded by a Photonic Science camera (model FDI VHR) with 4008 x 2670 pixels, and 4.5 microns pixel size.

The X-ray lens used for imaging consists of 161 individual spherical, biconcave, microlenses, each with 50-microns curvature radius R. The CRL length is equal to 18 mm. The lens photo is shown in fig. 12. The lens focal length, calculated in accordance to the formula 5, is equals to 41 mm for 8 keV X-rays.

Gold meshes #1000 with 5 μm wires separated by 20.4 μm was used as object for imaging.

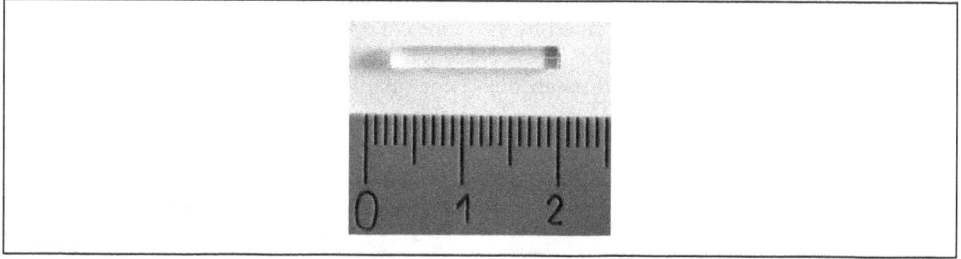

Fig. 12. Photo of microcapillry refractive X-ray lens with 161 concave spherical microlenses inside of glass capillary

The tube voltage was set to 20 kV and the current -14 mA, resulting in a standard bremsstrahlung and 8 keV characteristic-line spectra from the tube without filtering. The mesh was placed at distance a= 45 mm to the lens. The X-ray CCD-camera was placed a distance b= 440 mm to the lens in according to the lens formula (19), magnification M=b/a= 9.8. Fig.13 shows images of mesh #1000 recorded by the CCD-camera at different exposition equals to 5 min and 7 min.

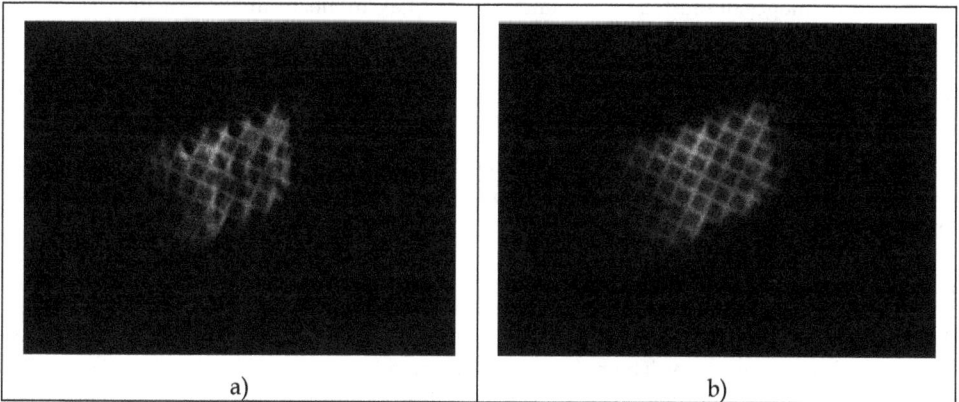

Fig. 13. X-ray image of mesh # 1000 at magnification 9.8. a) 5 min exposition time; b) 7 min exposition time

As it can be seen from Fig. 13, 5 μm gold wires of mesh #1000 are recognized by the CCD-camera, which means the spatial resolution of the simple X-ray imaging system is not worse than 5 μm. In according with calculations of lens parameters, presented in Table 1 and Table 2, better spatial resolution may be achieved by using monochromatic X-rays and diaphragm to decrease spherical aberrations.

To improve spatial resolution of the system imaging experiments were accomplished on the National Synchrotron Radiation Laboratory (China) (Dudchik et al., 2010). The experiments were done on X-ray diffraction and scattering beamline (U7B). Synchrotron radiation (SR) from the Wiggler source was focused by a toroidal mirror. Focused SR was monochromized with a double-crystal monochromator and selected photon energy was 8 keV. The optical scheme of the experiments was the same as is shown in Fig. 10. The only difference was that the torroidal mirror was placed between X-ray source and the lens. Microcapillary X-ray lens in the form of glass capillary filled by 147 concave epoxy microlenses with 50 microns curvature radius each was used. The lens focal length is equal to 45 mm. Gold mesh #1500 with 5.5 microns wires were used as an object for imaging. Fig.14 shows images of gold mesh #1500 obtained with 8-keV monochromatic synchrotron X-rays at magnification M=11.6 (a) and M=18.6 (b).

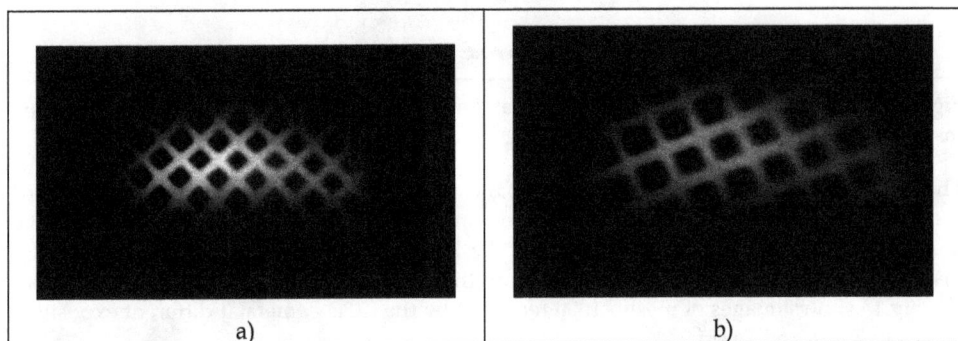

Fig. 14. X-ray images of mesh #1500 obtained with 8-keV monochromatic synchrotron X-rays at magnification M=11.6 (a) and M=18.6 (b).

Comparing images of gold mesh shown in Fig. 13 and Fig. 14 we may conclude that using monochromatic X-rays give significant improvement of spatial resolution of the system.

In conclusion, imaging experiment shows that the spherical compound refractive lens is a promising imaging optical element for hard x-rays, giving better than 2- 5 μm spatial resolution.

3. Conclusion

We have fabricated and tested compound refractive lenses (CRL) composed of micro-bubbles embedded in epoxy. The bubbles were formed in epoxy inside glass capillaries. The interface between the bubbles formed spherical bi-concave microlenses. The lenses were named as microcapillary refractive lenses or "bubble lenses". When compared with CRLs manufactured using other methods, the micro-bubble lenses have shorter focal lengths with higher transmissions for moderate energy X-rays (e.g. 7 – 12 keV). The lenses were tested at the Stanford Synchrotron Radiation Laboratory (SSRL) and ANKA Synchrotron Source. We used beamline 2-3 at the SSRL to measure focal lengths between 100-150 mm and absorption apertures between 90 to 120 μm. Transmission profiles were measured giving, for example, a peak transmission of 27 % for a 130-mm focal length CRL at 8 keV. The focal-spot sizes were also measured yielding, for example, an elliptical spot of 5 x 14-μm^2 resulting from an approximate 80-fold demagnification of the 0.44 x 1.7 mm² source. Experiments at ANKA Synchrotron Source shown that the designed lens with 145 mm focal length focuses 12 keV-rays into 2.2 X 10.4-μm^2 spot.

The lenses are imaging device and may be used as objective for X-ray microscope. A simple microscope consisting of the X-ray tube, microcapillary refractive X-ray lens and X-ray CCD-camera was designed at the Institute of Applied Physics Problems of Belarus State University. The X-ray lens consists of 161 individual spherical, biconcave microlenses, each with 50-microns curvature radius. The lens focal length is equals to 41 mm for 8 keV X-rays. It was shown that the spatial resolution of the microscope is better than 5 microns when unfiltered X-ray beam from cupper anode X-ray tube is used. Better spatial resolution (about 2-3 microns) was obtained in the experiments on the National Synchrotron Radiation Laboratory's (China) were monochromatic 8-keV X-ray beam was used.

The micro-bubble technique opens a new opportunity for designing lenses in the 8-9 keV range with focal lengths less than 30-40 mm.

4. Acknowledgment

I would like to acknowledge my colleague Dr. N.N. Kolchevsky, who spent a lot of time to improve parameters of the microcapillary lenses when he was PhD student in the Institute of Applied Physics Problems of Belarus State University. I would like to acknowledge my colleagues Dr. M.A. Piestrup, Dr. C.K. Gary, Dr. J.T. Cremer from Adelphi Technology, Inc., who did a lot of experiments on testing microcapillary lenses for focusing and imaging with synchrotron and laboratory X-ray sources. Prof. T. Baumbach and Dr. R. Simon were so kind to invite me for taking part in experiments on focusing X-rays at ANKA Synchrotron Radiation Source. Prof. Zhanshan Wang, Dr. Baozhong Mu, Dr. Chengchao Huang, Prof. Guoqiang Pan invited me to take part in imaging experiments with microcapillary lenses at the National Synchrotron Radiation Laboratory (China). I am grateful to all of them for continues interest to this research and useful comments.

5. References

Born, M. & Wolf, E. (1975). Principles of Optics. 5th edition, Pergamon Press, Elmsford, New York.

Dudchik, Yu.I. & Kolchevsky, N.N. (1999). A microcapillary lens for X-rays. Nucl. Instr. and Meth. A 421, pp. 361-364.

Dudchik, Yu.I.; Kolchevsky, N.N.; Komarov, F.F.; Kohmura, Y.; Awaji, M.; Suzuki, Y.& Ishikawa, T. (2000). Glass capillary X-ray lens: fabrication technique and ray tracing calculations. Nucl. Instr. Meth. A, 454, pp.512-519.

Dudchik, Yu.I.; Kolchevsky, N.N.; Komarov, F.F.; Piestrup, M.A.; Cremer, J.T.; Gary, C.K.; Park, H. & Khounsary, A. M. (2004). Microspot x-ray focusing using a short focal-length compound refractive lenses. Rev. Sci. Instr., 75, N.11, pp.4651-4655.

Dudchik, Yu.I.; Simon, R.; Baumbach, T. (2007a). Measurement of spherical compound refractive X-ray lens at ANKA synchrotron radiation source. Proceedings of the 8-

th International conference "Interaction of radiation with solids". 26-28 September 2007, Minsk, Belarus . P. 239-241.

Dudchik, Yu.I.; Komarov, F.F.; Piestrup, M.A.; Gary, C.K.; Park, H.& Cremer, J.T. (2007b) . Using of a microcapillary refractive X-ray lens for focusing and imaging. Spectrochimica Acta, 62B, pp. 598–602.

Dudchik, Yu.I., Gary, C.K.; Park, H.; Pantell, R.H.; Piestrup, M.A. (2007c). Projection-type X-ray microscope based on a spherical compound refractive X-ray lens. Advances in X-Ray/EUV Optics and Components II, edited by Ali M. Khounsary, Christian Morawe, Shunji Goto, Proc. of SPIE Vol. 6705, pp. 670509-1 – 670509-8.

Dudchik, Yu.I.; Huang, C.; Mu, B.; Wang, Z. & Pan, G. (2010). X-ray microscopy with synchrotron source and refractive optics. Vestnik Belorusskogo Universiteta. Physics, Mathematics, Informatics. #2., pp. 24-28.

Kohmura, Y.; Awaji, M.; Suzuki, Y.; Ishikawa,T.; Dudchik, Yu.I.; Kolchevsky, N.N.& Komarov, F.F. (1999). X-ray focusing test and X-ray imaging test by a microcapillary X-ray lens at an undulator beamline. Rev. Sci. Instr., 70, No.11, pp. 4161-4167.

Kumakhov, M. & Sharov, V. (1992). A neutron lens. Nature 357, pp. 390-391.

Pantell, R.H.; Feinstein, J.; Beguiristain, H.R.; Piestrup, M. A.; Gary, C.K. & Cremer, J.T. Characteristic of the thick compound refractive lens. Applied Optics, Vol. 42, pp. 719-724.

Piestrup, M.A.; Gary, C.K.; Park, H. ; Harris, J.L.; Pantell, R.H.; Cremer, J.T.; M. A. Piestrup, C. K. Gary, H. Park, J. L. Harris, J. T. Cremer, R. H.; Dudchik, Yu.I.; Kolchevsky, N.N.& Komarov, F.F. (2005). Microscope using an x-ray tube and a bubble compound refractive lens. Appl. Phys. Lett. 86, pp. 131104-1- 131104-4 .

Lengeler, B.; Schroer, C. G.; Kuhlmann, M.; Benner, B.; Günzler, T. F.; Kurapova, O.; Zontone, F.; Snigirev, A. & Snigireva, I. Refractive x-ray lenses. (2005). J. Phys. D: Appl. Phys. 38, pp. A218-A222.

Snigirev, A.; Kohn, V.; Snigireva, I. & Lengeler, B. (1996). A compound refractive lens for focusing high-energy X-rays. Nature, Vol. 384, N.6604, pp.49-51B.

Thiel, D.J.; Bilderback, D.H.; Lewis, A.; Stern, E.A. & Rich, T. (1992). Guiding and concentrating hard x-rays by using a flexible hollow-core tapered glass fiber. Applied Optics, Vol. 31, Issue 7, pp. 987-992.

3

Multimode Passive Optical Network for LAN Application

Elzbieta Beres-Pawlik, Grzegorz Budzyn and Grzegorz Lis
Institute of Telecommunications, Teleinformatics and Acoustics,
Wroclaw University of Technology, Wroclaw
Poland

1. Introduction

PON's (*Passive Optical Networks*) have gained a lot of interest in recent years. They minimize the number of required optical transceivers, reduce the fiber optic infrastructure and the need to power the intermediate network nodes. Single-mode PON structures are commercially available and presently they are used in LAN's (*Local Area Networks*). Usually these structures are point-multipoint networks which do not require any active components between the signal's source and the destination point. In other words, there's no need to supply electrical power to components such as fibers or optical splitters/combiners etc. Unfortunately, single-mode structures are not cheap, they're not easy to install nor are they customer-friendly.

Point-to-point multimode fiber optic structures have marked their presence in literature. Positive results obtained for different multimode optical fibers have been shown in the past (Gilmore M.C., 2006). The bandwidths and maximum transmission distance have been determined for three types of multimode fibers. Another paper includes a theoretical description connected with multimode transmission line (Pepeljugowski P., 2003). The same work presents an investigation of multimode coupler cascades' transmission bands supported by experimental results (Stepniak G., 2009).

This chapter describes several commercial applications for multimode passive network structures designed and constructed to include N-equivalent nodes. Such networks should be cheaper and more flexible than single-mode structures (part 2). Medium Access Method in PON's is presented in part 3. Some basic communication parameters like bandwidth or bit error have been tested in multimode networks and presented in part 4.

Some new solutions that came through for multimode fiber optic structures are presented here as well.

Measurements taken let us estimate the distance and possibility to obtain desired speed transmission in designed structures (part 5). There are many methods of designing a passive-optical-structure LAN. Each of the methods entail features that need to be considered carefully before designing a network. Two important features are structure of the network and medium access network. They are covered in detail in next two paragraphs.

Finally, possibilities of applying the CWDM method in multimode structures are examined in part 6.

2. Passive LANs network structures

There are two known types of passive optical structures – symmetrical and asymmetrical. The former includes reflective star, directivity star and transmissive star; The latter includes a typical tree structure. Each of them can be built using 3dB 50/50 multimode optical couplers. Examples of different topologies are presented further down in this chapter (Beres-Pawlik E., 2007).

It is assumed that the structures presented below will be built using 1×2(Y) and 2×2(X) couplers based on multimode (graded GI - and step-index SI) fibres. Silica fibre-based couplers or plastic-based couplers may be used. In some of the proposed structures it would be more advantageous to use asymmetric couplers but, to the best of our knowledge, no such elements are commercially available (Pawlik E., 1993).

2.1 Tree structure

In a tree structure, the transmission is of broadcast downward type (from operator to subscriber) and there is a point-to-point link upward (from subscriber to operator). This requires a power thirsty optical transmitter in the tree trunk and sensitive detectors in the receiving devices in the nodes. An exemplary structure with 4 rows and 8 access nodes is shown in the Fig.1.

Y couplers are usually used to construct tree structures but some structures are built using only X couplers. The coupler's extra branch can be used for monitoring a given tree branch.

This type of access service structure is collisionless since the network nodes can transmit only and exclusively in fixed time slots using TDMA (*Time Division Multiple Access*).

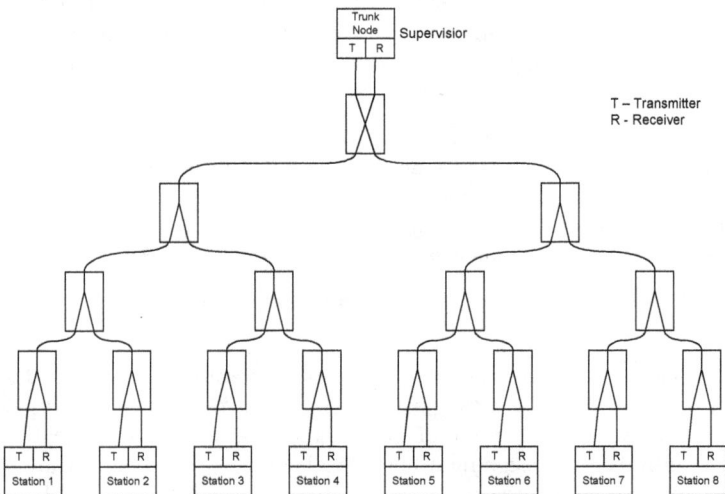

Fig. 1. Tree structure of order 4.

2.2 Transmissive star

A transmissive star is a structure in which a signal fed to one of the inputs on one side will be distributed equally among all the outputs on the other side. Since the structure is symmetric, the signal fed into one of the inputs will obviously be uniformly distributed among all the outputs on the opposite side. The structure requires, however, quite a large number of couplers. The number of nodes in a transmissive star is the power of 2.

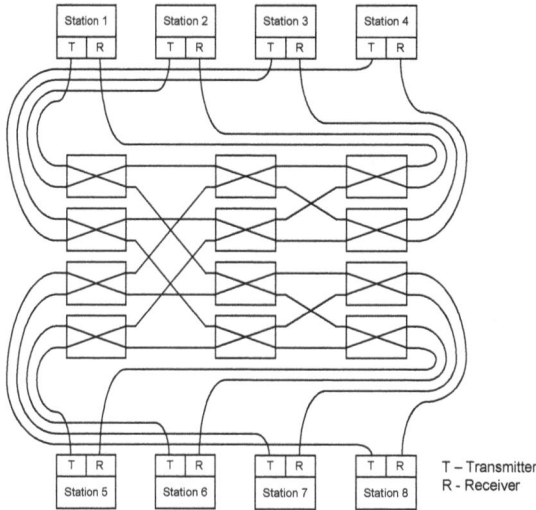

Fig. 2. Transmissive star with 8 nodes.

2.3 Reflective star

A reflective star is designed as a structure in which a signal fed to any of the inputs is uniformly distributed among all inputs/outputs. A drawback of this solution is that the light emitted by a given transmitter returns to it, and this adversely affects the operation of the laser transmitter. Because the transceiver notifies itself, the use of effective collision detection methods is impossible.

Y couplers can be used to build such a reflective star. It is a kind of tree structure where the trunk node is looped. An exemplary reflective star structure is shown in the Fig. 3.

2.4 Directional star

A directional star is a structure in which a signal sent from the transceiver of a selected network node reaches all the other transceivers without notifying itself. This is achieved by appropriately connecting directional couplers. The directionality of the couplers is an important parameter of the star's components. For multimode couplers it is usually at the level of 35 dB. This parameter determines the structure's dynamics and its size if one assumes that the transceiver cannot notify itself. Assuming a coupler loss of 4 dB and a directionality of 35 dB, a directional star structure with 12 nodes can be designed (fig.4). Attenuators were employed to level the attenuation of the optical paths between any two nodes.

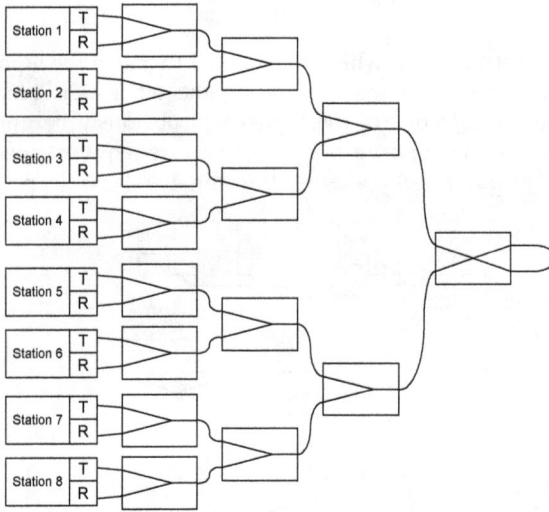

Fig. 3. Reflective star with 8 nodes.

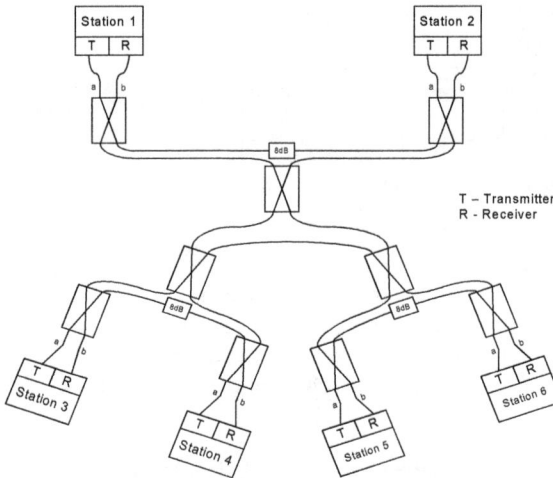

Fig. 4. Directional star with 6 nodes.

The structure can be expanded using asymmetric couplers whereby one can increase the number of network nodes and eliminate the necessity of using attenuators to level the attenuation of all the paths.

2.5 Optical elements in PONs

Each of the structures presented above requires different number of couplers in the network C and different number of couplers in the path between nodes S. The values for the described network types were given in the figures 5 and 6. The largest number of couplers in the network C is required for the Transmissive Star structure. On one hand it increases

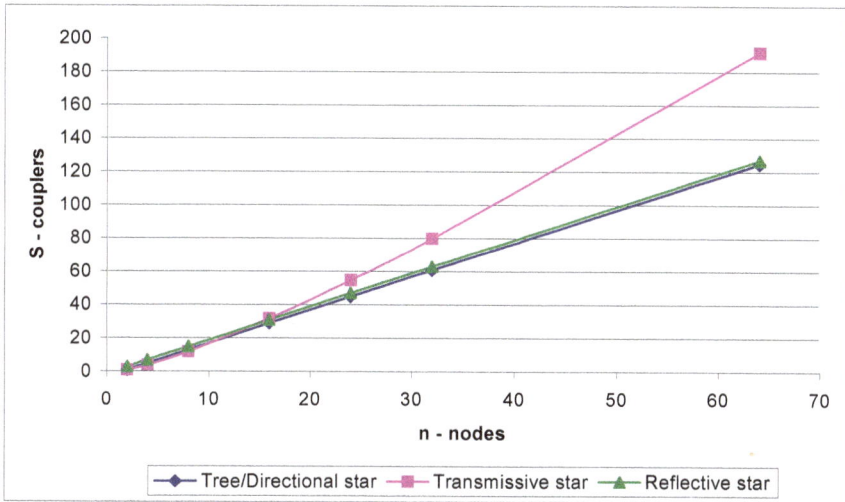

Fig. 5. Number of couplers in the network C

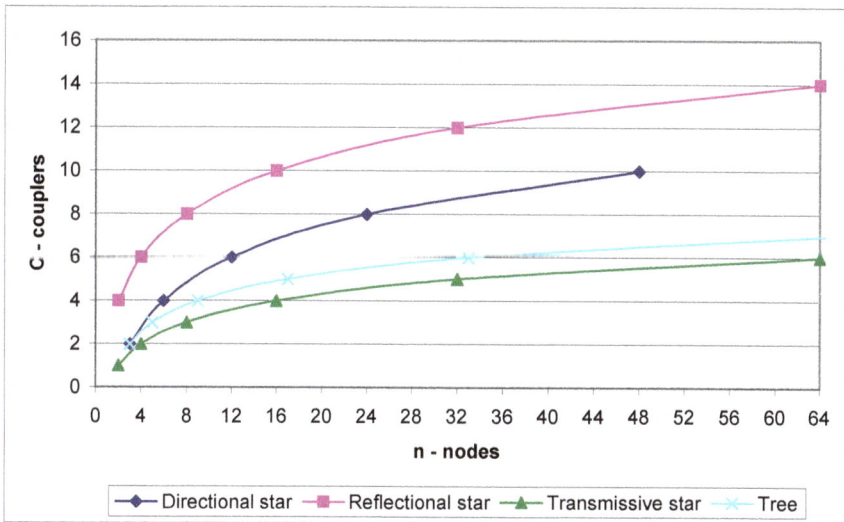

Fig. 6. Number of couplers in the optical path S

the total cost of the network. On the other hand in this structure there is the lowest number of couplers in the optical path S, which makes it possible to improve the signal quality between nodes and to increase the overall network size.

Apart from the number of network nodes, parameters limiting a network size are the dynamic range of the transceivers and the length of the used patchcords. It seems that the number of couplers used has little influence on transmission speed but GI couplers bring losses around 3,5 dB (splitting 50% = 3dB and their intrinsic losses of about 0.5 dB). It can be assumed that losses introduced by inserting less than 7 couplers do not depend on the signal frequency. However, fiber losses depend on the signal frequency. Probably connector losses increase with frequency but that fact is yet to be investigated. It can be proven that the optical power budget changes with the frequency.

3. Medium access methods in PONs

3.1 Introduction

Ethernet networks based on the CSMA/CD (*Carrier Sense Multiple Access with Collision Detection*) protocol have gained widespread popularity thanks to their easy expandability and the simplicity of their node arrangements. Unfortunately, the introduction of the CSMA/CD method into optical networks is complicated because of the peculiar nature of optical signals. Collision detection in conventional copper cabling networks takes place in the electric domain: the voltage level elevated above a certain threshold is measured. Collision detection in the case of optical signals carried in networks is much more difficult. Since fibre optic circuits have different signal attenuation coefficients, bouncing occurs at circuit junctions whereby the power of the carried signals changes. As a result, such simple collision detection methods as the ones used for electric circuits cannot be employed here.

Below, the collision detection methods used in the passive optical networks, their advantages and limitations and the potential for implementing them in proposed specific passive structures are discussed. We show several possible uses for different medium access mechanisms: CSMA/CD, CSMA/CA (*Carrier Sense Multiple Access with Collision Avoidance*), TDMA (*Time Division Multiple Access*) and WDMA (*Wavelength Division Multiple Access*).

3.2 Methods of detecting collisions in CSMA/CD networks

We assume that the network structure is based exclusively on optical fibre circuits and optical fibre couplers (Reedy J.W., 1985). The collision detection methods can be classified as:

- operating exclusively in the optical domain – the solutions presented in sections 3.2.1 and 3.2.2,
- ones in which collision detection is performed after the optical signal has been converted into an electric signal in a network node – the solutions presented in sections 3.2.3 and 3.2.4.

3.2.1 Measurement of average optical power

In the average optical power measurement method a collision occurs when the optical power received in a receiver is higher than the power transmitted by a single network node. The threshold power above which a collision is detectable is higher than the power required to receive data in normal transmission conditions. In order to increase the method's effectiveness the optical transmitter must be switched off when no transmission occurs. This increases the transmitting system's complexity and has an adverse effect on the laser sources. The time needed for switching on the laser reduces the effective speed of transmission in the network. In order to minimize this drawback, one can use power supply systems with two working points. In the idle state the laser current is low but sufficient for lasing to occur. A working point for collision detection requires a much higher laser feed current to emit the higher optical power needed for the proper transmission of data in a network. One should also take into account the laser sources' power level tolerance, the effectiveness of the laser source/optical fibre coupling and the quality of the fibre optic connections. Hence one must determine the allowable optical power level in the network for both transmission and no transmission (Fig. 7).

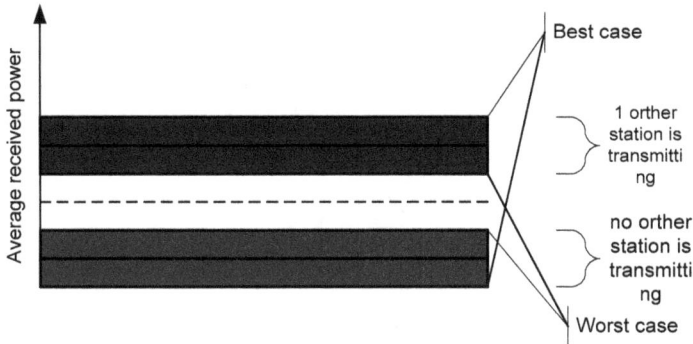

Fig. 7. Power levels at transmission and at no transmission.

3.2.2 Directional coupling

This method makes use of special optical fibre coupling techniques whereby one can create such a system in which a given station can hear all the stations, but not itself. Collisions in this case are detected directly: if a given station transmitting data detects a signal in its receiver, this means that a collision has occurred.

How can such a directional system of connections be built to meet the requirements? An example here is the construction of a duplex bus based on two optical fibres, as shown in Fig.8. In such a system each node transmits data through the lower fibre to the neighbouring nodes on its left and via the upper fibre to its neighbours on the right. As a result, all the stations, except for the transmitting station, receive the data signal.

Since this bus topology is now outdated, efforts are made to build more effective networks using the star topology. A directional star coupler in which optical power from each of the

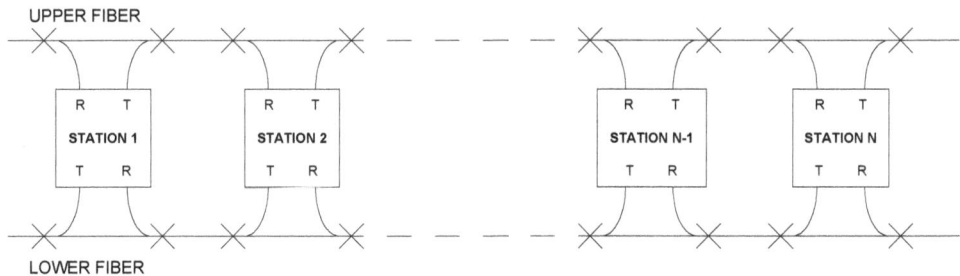

Fig. 8. Duplex fibre optic bus.

inputs is equally distributed among all the outputs except for one is used for this purpose. Such a network can be constructed by connecting 2×2(X) and 2×1(Y) couplers. An M×M star coupler can be obtained by connecting 2*M couplers with M – 1 inputs/outputs. The function of the inputs/outputs is to split or combine optical powers. An exemplary four-port star with sending-receiving systems is shown in Fig. 9.

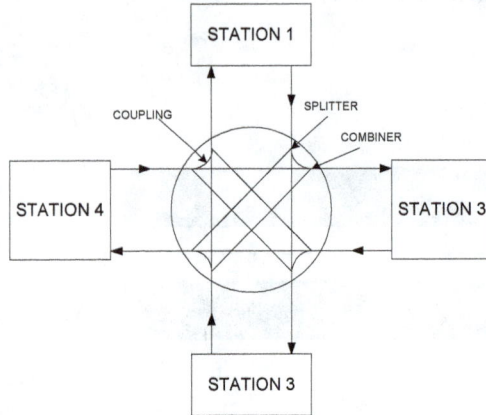

Fig. 9. Directional four-port star.

3.2.3 Pulse width disturbance

The pulse width disturbance measurement method exploits the fact that in the primary Ethernet system a data stream is encoded using the Manchester code so that the information bit is always encoded as transition "01" or "10". Consequently, one can exactly define the data stream pulse width free of any collisions. Modulation at a rate of 20 Mbaud was employed in a system with a throughput of 10 Mbps whereby a single pulse should be nominally below 100 ns. The collision detection system's function is to detect signals exceeding the nominal pulse width (Fig. 10).

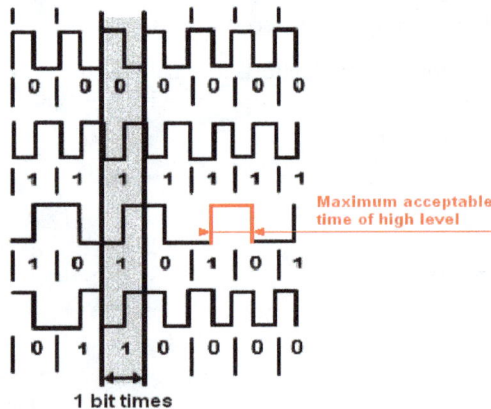

Fig. 10. Manchester coding scheme.

Similarly to other amplitude measurement techniques, this method is limited by situations in which a weaker signal is masked by another stronger signal. Hence, it is essential to build junctions between network nodes having the same attenuation for each optical circuit. Optical attenuators or asymmetric couplers are used for this purpose. Another way of solving this problem is to employ a centralized collision detection unit. If a collision is detected by the central mechanism, the latter sends a strong jam signal to all the sending-receiving devices in the network whereby a change in the pulse width is easily detectable. This represents, however, a departure from the fully passive network concept. The jam signal can be used to amplify collisions in systems with collision detection, in which at least one station informs another transmitting station in the network that a collision has occurred. The method's drawback is the necessity of using Manchester encoding, i.e. modulation twice as fast as the transmission speed. For this reason, the above method is limited to speeds below 10 Mbps, becoming highly ineffective at higher speeds.

3.2.4 Direct comparison of streams

This method consists in comparing the sent binary stream with the received stream in the electrical signal domain within a given sending-receiving device. If the received bit stream does not tally with the sent one (allowing for propagation delay), the system determines that a collision has occurred. This complicates a little the system since it is necessary to install memories buffering the card's outgoing traffic. One should also include a system analysing incoming signal delays relative to the sent signal. The system ought to be able to negotiate the connection parameters by sending test packets when a new card is installed in the system. Obviously, one should also specify the attachment of a sending-receiving device to the network in such a way that the detector of one device could receive the signal from its own transmitter (Fig. 11).

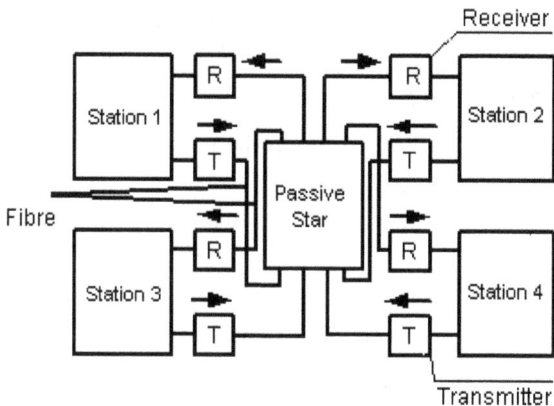

Fig. 11. Device-to-passive-star attachment diagram.

The method's apparent advantage is the system's transparency with regard to coding since the detection of collisions takes place at the level of analysis of individual binary pulses. Moreover, the method does not introduce significant transmission rate limitations and is suitable for transmission rates of 100 Mbps.

3.3 Medium access in CSMA/CA networks

A need of all-optical networks has prompted the design of protocols that could detect the presence or absence of optical signal on a specific channel without regard to the high-bit rate data being transmitted. One of these kinds of protocols is protocol CSMA/CA, whereby nodes using optical carrier-sense capability prevent transmitting a packet at times when it would collide with other packets which are already in transit. Unlike the CSMA/ protocol where collisions are tolerated and the retransmission is required, here the collisions cannot happen.

Below there is a presented exemplary scheme of arbitrary node in a ring network that enables collision detection. Each node receives packets on a single unique wavelength but can transmit packets on any wavelength (Wong E., 2004).

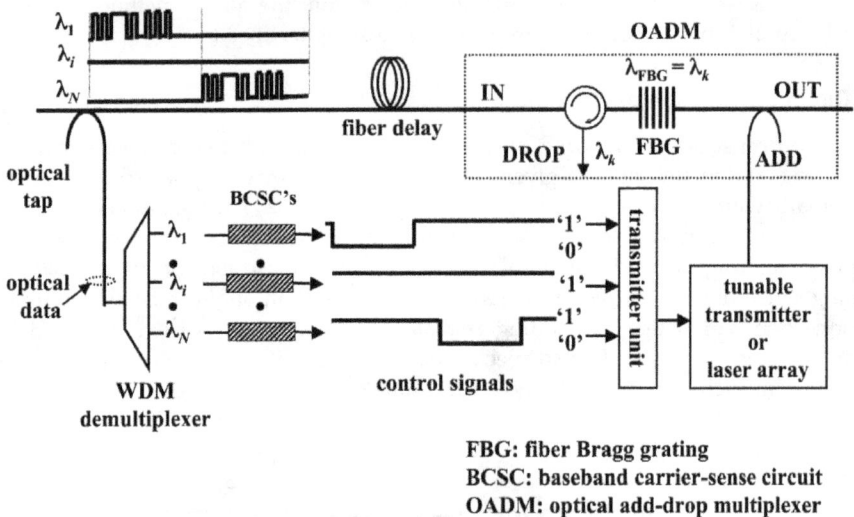

FBG: fiber Bragg grating
BCSC: baseband carrier-sense circuit
OADM: optical add-drop multiplexer

Fig. 12. Architecture of an arbitrary node in a CSMA/CA network.

To prevent collisions at the out ports between the transmitted packets and those that are already in transit, a part of the optical power of all packets arriving at the node is tapped.

The tapped signals are demultiplexed into individual wavelengths, which are then detected by BCSCs (*Baseband Carrier-Sense Circuits*) that perform packet detection. Each BCSC generates a control signal that informs if the channel is occupied or not. Based on this, the transmitter unit evaluates the duration of the transmission gap between adjacent arriving packets and if it is suitably long to send its own packet.

Presented scheme concerns the module based on the single-mode optical fibres. According to our knowledge there are no presented similar solutions for multimode optical fibres so far, but we predict it is potentially possible.

The proposed protocol requires complicated and expensive electronic processing so it can not be used in the planned commercial applications.

The main advantage is a simple management layer (L2) and the main disadvantage is a complicated physical layer (L1).

3.4 Medium access in TDMA networks

TDM *(Time Division Multiplexing)* is a technology that is used mainly in access networks, but it may also be useful in local networks. This technique relies on the assignment of suitable time cells for the input streams. TDMA technique is usually used in tree type structures (Pesavento G., 2003).

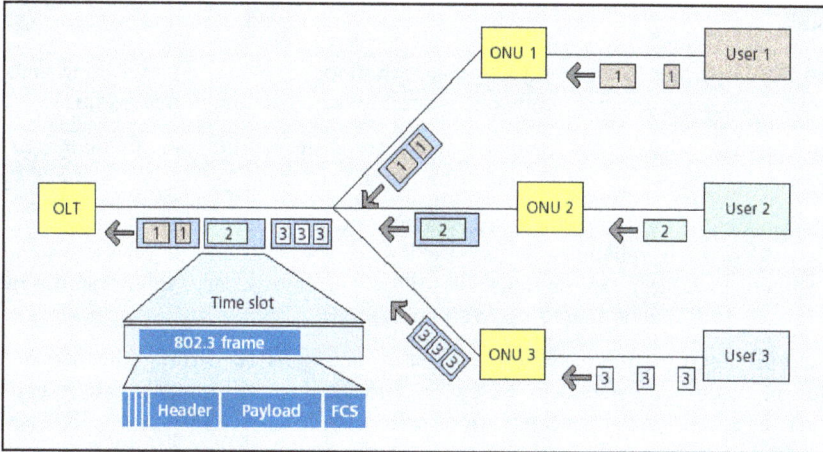

Fig. 13. Architecture of TDMA network.

The main advantages of TDMA protocol are:

a.   possible larger network span at higher efficiency than in CSMA/CD
b.   management algorithms adaptable from EPON *(Ethernet Passive Optical Network)* networks
c.   centralised management
d.   very easy to prioritise traffic
e.   QoS support

The main disadvantages of TDMA protocol are:

a.   required complicated algorithms for traffic management
b.   efficiency dependent on network size and network load
c.   central node much more complicated than other ones

3.5 Medium access in WDMA networks

Although PON's provide higher bandwidth than traditional cooper-based access networks, there exists the need for further increasing the band of the PON's by employing WDM *(Wavelength Division Multiplexing)* so that multiple wavelengths may be supported in either or both upstream and downstream directions. Such a PON is known as a WDM-PON. Fiber optical networks, working on the basis of WDMA technique, are natural evolution of optical fiber links working in point-to-point topology using WDM. WDMA network development can also be considered as abilities to increase the effect of one wavelength-based passive optical networks (Banarjee A., 2005).

The ability of data sharing between users, when a common transmission medium is being used, is an important feature of these kinds of networks.

Data streams are transmitted, using different wavelength multiplied optical transmitters, to all network nodes. When the detector receives information, it selects a desirable signal from all transmitted signals in one fiber using selective optical filter. In order to meet the above-mentioned requirements co-shared medium currently requires the star topology network architecture.

Standard PON operates in the "single wavelength mode" where one wavelength is used for upstream transmission and a separate one is used for downstream transmission.

Different sets of wavelength may be used to support different independent PON subnetworks, all operating over the same fiber infrastructure.

Even though they provide the highest capacity, optical WDMA networks are usually too expensive. Also, their reliability is usually low due to the use of active systems (e.g. multiplexers or switches). Access networks still require inexpensive solutions in which the costs of the network will be shared between all users.

In the world literature there are no interesting solutions concerning the use of WDM technique in multimode networks based on wavelengths 850 or 1300 nm. We propose installation of several sources with various wavelengths (1310, 1330, 1350, 1370 nm) and passive filters in nodes, which would increase the transmission speed but decrease the number of users. We must use supplementary couplers for connecting several sources and detectors with CWDM (Coarse *Wavelength Division Multiplexing*) multimode couplers.

As far as we know, an interesting solution can be achieved for wavelengths 1300 –1550nm in multimode optical fiber (there can be used the fiber elements which are commercially available).

Based on the preliminary measurement of the passive structures, one can assess parameters of the network built presented above and working with 1Gbps transmission speed. Parameters presented in Table 1 were determined for optical path with 100m fiber optical patchcords connecting nodes with the structure. In the table, based on the date from our measurement, we present projects of structures and parameters possible to achieve. There are also proposals of suitable protocols for the chosen structures.

The number of nodes in the networks depends significantly on the dynamics of available electro-optical converters. For the 850nm bandwidth the normal off-the-shelf transceivers usually offer dynamics only slightly better than 15dB, while in 1300nm windows the dynamics can reach beyond 25dB.

The main advantages of WDMA protocol are:

- possibility of building a few "logical networks" on top of only one physical structure
- "logical networks" can be invisible to each other (depends on the central node)
- efficiency depends on the access mechanism used in "logical networks" (usually TDMA)
- more wavelengths = better utilised fibre
- ease of adding a special channel for network management

- most elastic with tunable receivers and transmitters

The main disadvantages of WDMA protocol are:

- complicated and expensive in most configurations
- efficiency depends on the access mechanism used in "logical networks" (usually TDMA)
- most flexible with tunable receivers and transmitters

4. Measurements of base transmission parameters

4.1 BER measurement

Special systems were designed in order to measure BER (*Bit Error Rate*) in multimode passive optical networks based on a FPGA programmable logic combined with electro-optical transceivers for 850nm and 1300nm wavelengths. The transmitters used in the 850nm transceivers were VCSEL lasers whereas in the 1300nm transceivers there were DFB lasers. The spectra of both are presented in Fig.14. The dynamic of the AFBR-53D5Z was 13dB and HCDTR-24 was 22dB. The built system allows for the selection of a number of transmitted bits in the range between $10^6 - 10^{12}$ as well as the transmission speed. Communications with the FPGA setup was carried out using standard LVPECL differential signals. The measurements were performed in two speed ranges: 100Mbps and 1Gbps.

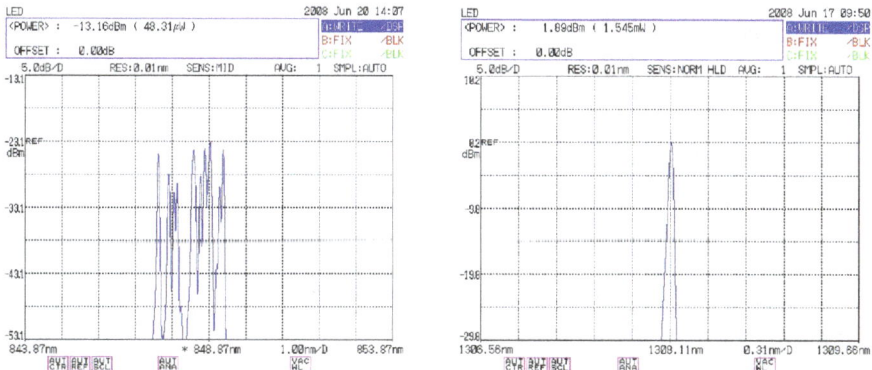

Fig. 14. Spectra of used VCSEL (850 nm) and DFB (1300 nm) lasers.

The network configuration the measurements were carried out in are presented in Fig. 15.

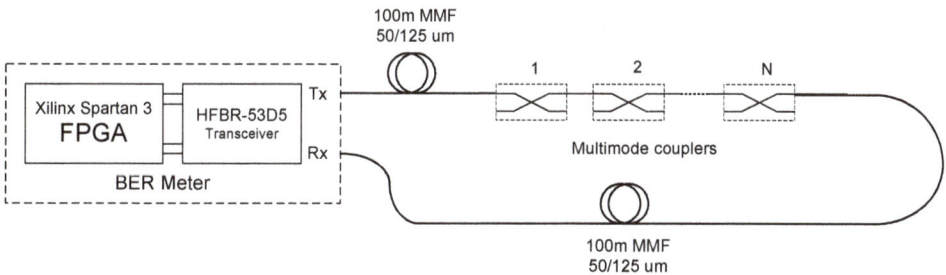

Fig. 15. The tested optical path.

The tested optical path included a cascade of GI optical couplers and two 100m GI patch-cords at the start and the end of the cascade of couplers. The obtained measurement results for two different wavelength BER in 100Mbps range are presented in Fig.16.

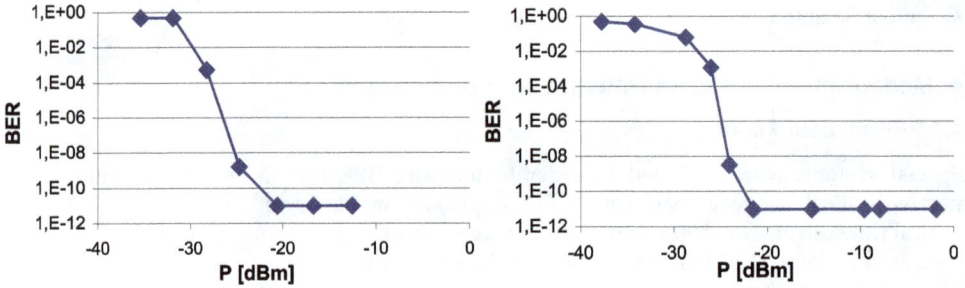

Fig. 16. BER measurements for two different wavelengths 850 nm (a) and 1300 nm (b) for speed transmission 100 Mbps as a function of attenuation obtained by including following couplers in optical path.

In order to construct the electro-optical transceiver working in 1GHz range we chose the byte method. The block diagram of the E/O transceiver working in the byte mode was shown in the Figure 17.

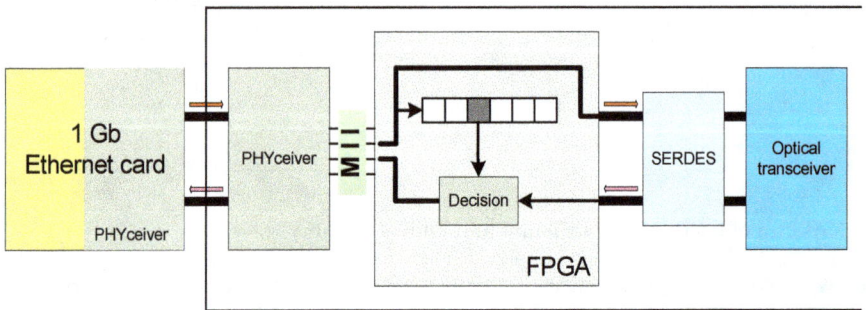

Fig. 17. E/O transceiver working in the byte mode.

In the byte mode, the Ethernet frame is decoded only into small pieces, i.e. nibbles for 100 Mbps and bytes in 1Gbps network speed. The bytes are sent in parallel to the SERDES (*serializer/deserializer circuit*). Although this makes frame end detection more troublesome (5 bit or 10 bit long words have to be analyzed), it offers faster collision detection, higher network throughput and a possibility of using the XC3S200 chip in 1Gbps networks.

Fig.18 shows WER (*Word Error Rate*) as a function number of coupler. Multimode GI coupler cascade measurements show that the tested off-the-shelf transceivers make it possible to build 1Gb optical networks with up to three coupler levels in optical path (Fig.18).

4.2 Bandwidth measurement

We've designed media converters for bandwidth measurement purposes. Transceivers AFBR-53D5Z, HCDTR-24 and PIN diodes have been used to build media converters. The bandwidth measurements were performed for different network configurations, including the GI patchcord cascade, GI coupler cascade and a complete optical path. The measurements were performed in the measurement setup shown in Fig.19 (for patchcord cascade).

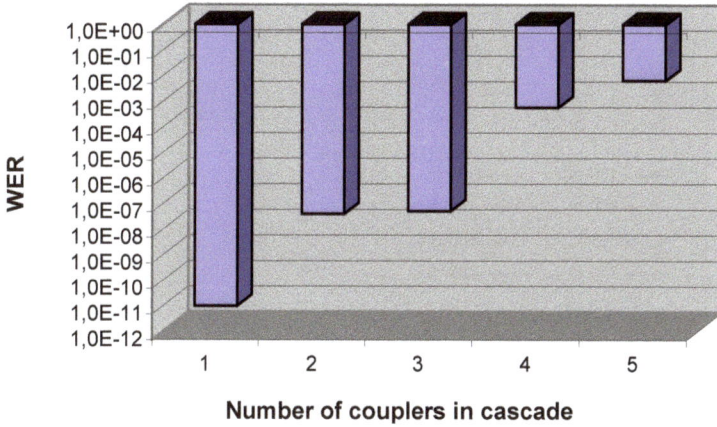

Fig. 18. WER of transmission through cascade of couplers at 1,25GHz.

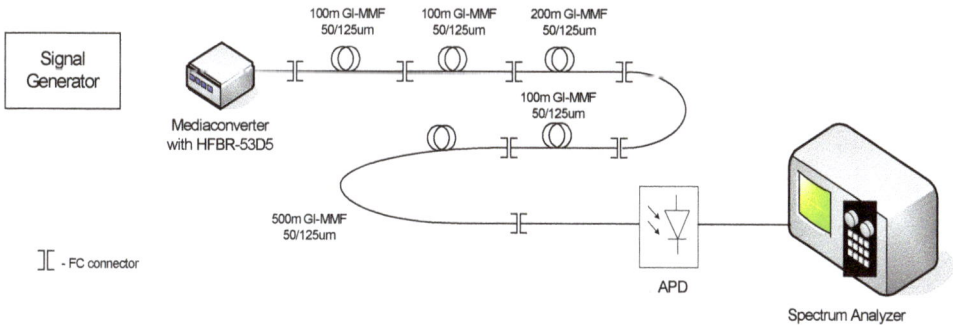

Fig. 19. Bandwidth measurement setup – a patchcord cascade.

Measurements were performed at various frequencies within the 10 MHz -1 GHz range. Similar to the setup where BER measurements were taken the network elements (GI patch-cords and GI couplers) in this setup were also joined using FC connectors. Fig.20 presents the transmission spectrum for a patchcord cascade and Fig.21 - for the complete optical path. One can notice that if the path attenuation does not exceed the dynamics of the transceivers, then the attenuation does not depend on the transmitted signal's frequency. On the other hand, if the optical path attenuation exceeds the system's dynamics, then the transmission bandwidth becomes significantly reduced.

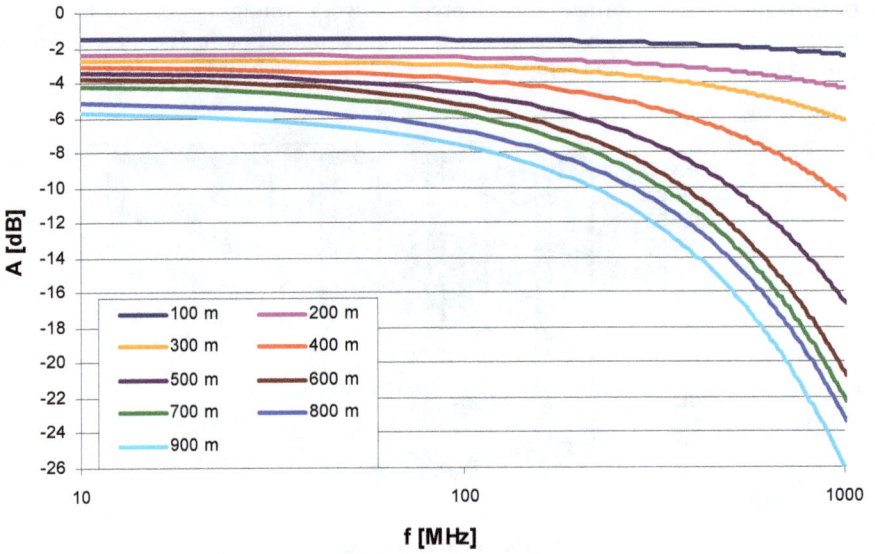

Fig. 20. Frequency response of the fiber optic patchcord cascade (850 nm).

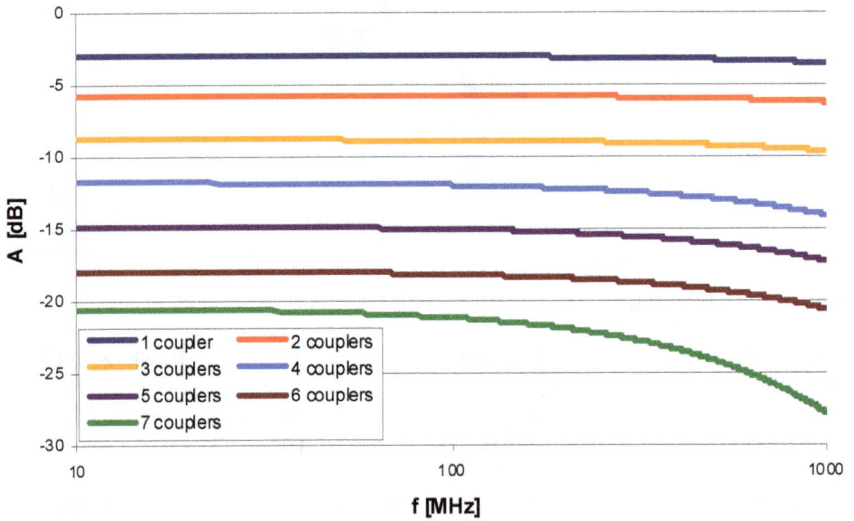

Fig. 21. Frequency response of the optical path (850 nm).

5. Network's size estimation

Parameters for a 1Gbps PON network can be assessed based on the passive structures measurements. The parameters presented in Table 1 were determined for an optical path with 100m fiber optic patchcords connecting nodes with the structure. In the table, we present projects of structures and parameters possible to achieve based on the data from our measurements. There are also proposals of suitable protocols for the chosen structures.

The number of nodes in a network depends significantly on the dynamics of available electro-optical converters. For the 850nm bandwidth the normal off-the-shelf transceivers usually offer dynamics only slightly above 15dB, while the dynamics in 1300nm and 1550nm windows can exceed 25dB.

Dynamic	15 dB		25 dB	
Type of structure	Transmissive star	Directional star	Transmissive star	Directional star
Number of nodes	16	6	64	12
Number of couplers in the network	32	9	192	21
Bandwidth	1 Gbps	1 Gbps	1 Gbps	1 Gbps
Method of collision detection	Comparison of streams	Directional coupling	Comparison of streams	Directional coupling
Necessary Hardware	Ethernet card with MII interface + FPGA programmed structure	Standard Ethernet card + converter NRZI <=> MLT-3	Ethernet card with MII interface + FPGA programmed structure	Standard Ethernet card + converter NRZI <=> MLT-3

Table 1. Parameters of the network topologies

6. The possibilities of applying CWDM method in a multimode structure

Transmission of different wavelengths in the optical passive structure allows obtaining more effective and quicker communication between nodes. We propose installation of a few different wavelength transmitters (T1-1310 nm, T2-1330, T3-1350 and T4-1370 nm) and receivers with CWDM multimode couplers in receiving stations, working as optical filters. Such installation allows the given structure to simultaneously transmit different wavelengths while giving transmission speeds proportional to the number of introduced wavelengths.

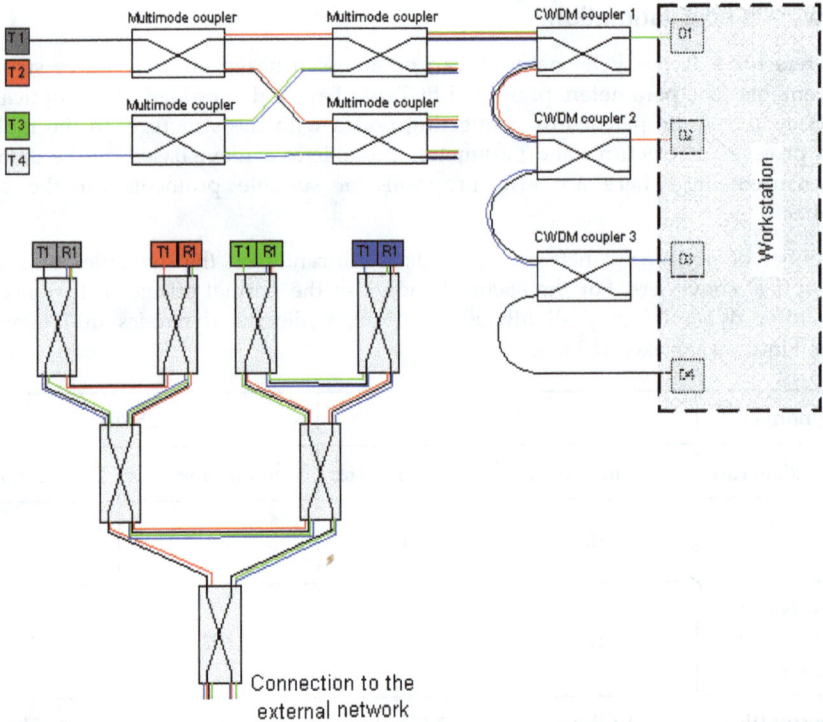

Fig. 22. Proposal of various CWDM PON structures

7. Conclusions

In this chapter we've proven feasibility of constructing Ethernet type local area networks based on multimode fibers working in either 850nm or 1300nm windows. Bandwidth, BER and WER analyses demonstrated possibility of building efficient networks with many GI 50/50 couplers in the optical path. Within the analyzed network types, the most promising for implementation at this point seems to be the transmissive star structure. Although the electrical part is complicated, much higher network capacity at relatively low cost per unit is well worth the tradeoff.

The available network structure size also greatly depends on the structure type. Out of the structures chosen for analysis, the transmissive star seems to offer much larger network size than the directional star structure. The advantage of transmissive star is its simpler collision detection method. Implementing directional coupling can be faster than the other streams because it's easier. Nodes based on directional coupling will be slightly cheaper, also there is no need for additional, complicated logic.

8. References

Aronson L., Lemoff B., Buckman L., Dolfi D. *Low-Cost Multimode WDM for Local Area Networks Up to 10 Gb/s" IEEE Photonics Technology Letters*, Vol. 10, No 10, 1998, pp. 1489-1491

Banarjee A. et al. *Wavelength division – multiplexed passive optical network (WDM-PON) technologies for broadband access: a review* [invited] 2005 Optical Society of America, November 2005, /vol. 4, No.11/ Journal of Optical Networking

Beres-Pawlik E., Budzyn G., Lis G., Napierala M., Nawrot K., *Multimode fiber passive optical network*, Progress Report France-Telecom, February 2007.

Beres-Pawlik E., Budzyn G., Lis G., Duchiewicz T., *Multimode fiber passive optical network-practical realization of designed network*, Finish Report France-Telecom, November 2007.

Beres-Pawlik Elżbieta, Budzyń Grzegorz, Lis Grzegorz, Napierała Marek: *Passive multimode fibre optic structures*. 8th ICTON, Vol. 4. Conference proceedings IEEE, 2006 pp. 72-75.

Beres-Pawlik E., Budzyń G., Lis G., Duchiewicz T., *Estimation of Multimode Optical Fiber Network Size* ICTON 2008 Mediterranean Winter, Marrakech, December FR3A.5

Budzyn G., Lis G., Głąb J., Bereś-Pawlik E., *An analysis of transmission of signals above 150MHz in passive optical multimode networks* Proceedings of Conference ICTON 2005, Barcelona 3-7 July 2005,

Budzyn Grzegorz, Lis Grzegorz, Bereś-Pawlik Elżbieta.: *Methods of Building Ethernet – type Passive Optical Networks*, 8th ICTON, Conference proceedings IEEE, 2007. Vol. 4, pp.35- 38.

Budzyn G., Lis G., Duchiewicz T., Beres-Pawlik E., *Realization of the stream comparison mechanism in a CSMA/CD type multimode passive optical network node-algorithm* ICTON 2008, Conference Proceedings vol.4, pp 332-334.

Chae C., Wong E.,. Tucker R, *Optical CSMA/CD Media Access Scheme for Ethernet Over Passive Optical Network*, IEEE Photonics Technology Letters, Vol. 14, No. 5, May 2002. pp. 711-713.

Chae C., *Multi-wavelength Ethernet Passive Optical Network with Efficient Utilization of Wavelength Channels*, ECOC 2005 Proceedings – Vol. 3 Paper We4.P.067. pp 635-636.

Gilmore M.C., *Multimode fibre bandwidth – its true value for high bit rate networks within plug-and-play data centre infrastructures*, The Fibreoptic Industry Association (FIA), 16 April 2006.

Hakamada Y., Oguchi, K. *32-Mbit/s Star Configured Optical Local Area Network Design and Performance*, Journal of Lightwave Technology, Vol. LT-3(3), ©1985 IEEE, pp. 511-524.

Hu S., Jack Ko J., Hegblom E., Coldren L. *Multimode WDM Optical Data Links with Monolithically Integrated Multiple-Channel VCSEL and Photodetector arrays*, IEEE Journal of Quantum Electronics, Vol.34, No 8 1998, pp.1403-1413.

Huang X., Ma M., *Efficient scheduling with reduced message delay for passive star coupled WDM optical networks*, IEEE Proc.-Commun., Vol. 152, No. 6, December 2005. pp. 765-770.

Kramer G., Pesavento G., *Ethernet Passive Optical Network (EPON): Building a Next-Generation Optical Access Network*, IEEE Communications Magazine, February 2002

Lee S.E., Boulton P., *The Principles and Performance of Hubnet: A 50 Mbit/s Glass Fiber Local Area Network*, Journal on Selected Areas in Communications, Vol. SAC-1(5), ©1983 IEEE, pp. 711-720.

Lemoff B., Aronson L., Buckman L., *Multimode WDM LAN using novel wavelength demultiplexer*, CLEO 98, Monday May 4 pp. 27-28

Musa S. et al. *Multimode fiber matched arrayed waveguides grating-based (de-) multiplexer for short distance communications* ECOC, 2002. PLC 6.2.1.

Nadarajah N., Manik A., Ampalavanapillai N., Wong E., *A Novel Local Area Network Emulation Technique on Passive Optical Networks*, IEEE Photonic Technology Letters, Vol. 17,No. 5, May 2005. pp. 1121-1123.

Olshansky R., Keck D., *Pulse broadening in graded-index optical fibers*, Applied Optics, Vol. 15(2), February 1976, pp. 483-491.

Pawlik E., Juszczak R., *Asymmetrical multimode couplers for local area networks*, Optica Applicata, Vol. XXIX, No. 1-2, 1999,

Pawlik E., Słowiński Z., *Polish Patent no 162049*, 1993

Pepeljugoski P. et al., *Modeling and Simulation of Next-Generation Multimode Fiber Links*, Journal of Lightwave Technology, vol.21, No 5, 2003, pp.1242-1255.

Pesavento G. *Ethernet Passive Optical Network (EPON) architecture for broadband access*, Optical Networks Magazine, January/February 2003,

Piotorowski A., Bereś-Pawlik E., *Passive Optical Multimode Fibre Network*, ICTON 2004, Wroclaw 4-8 July, Proceedings Conference.

Podwika D., Stefański D., Witkowski J., Pawlik E.., *Computer Networks based on optical Passive Couplers* , ICTON 2000, Gdańsk 5-8 June, Proceedings Conference pp.123-126,

Radovanović I., *Ethernet - Based Passive Optical Local-Area Networks for Fiber-to-the-Desk Application*, Journal of Lightwave Technology, vol. 21, no. 11, November 2003,

Rainer M. *Four-Channel coarse WDM 40 Gb/s Transmission of short –Wavelength VCSEL Signals Over High-Bandwidth Silica Multi-Mode Fiber*, Annual Report 2000, Optoelectronics Department, University of Ulm.

Rawson E., Metcalfe R., *Fibernet: Multimode Optical Fibers for Local Computer Networks, IEEE Transactions on Communications*, Vol. COM-26(7), ©1978 IEEE pp. 983-990.

Reedy J., Jones J. R., *Methods of Collision Detection in Fiber Optic CSMA/CD Networks IEEE Journal on Selected Areas in Communications*, vol. 3, no. 6, November 1985,

Schmidt R., Rawson E., Norton R., Jackson S., Bailey M.D., *Fibernet II: A Fiber Optic Ethernet*, Journal on Selected Areas in Communications, ©1983 IEEE, pp. 291-300.

Stallings W., *Local Network Performance*, IEEE Communications Magazine, Vol. 22(2), © 1984 IEEE, pp. 27-35.

Stępniak G., Ł.Maksymiuk, J Siuzdak, *Bandwidth analysis of multimode fiber passive optical network (PONs)*, Optica Applicata, vol XXXIX, No2, 2009, pp.239 – 233.

Tamura T., Masuru Nakamura M., Ohshima S., Ito T., Ozeki T., *Optical Cascade Star Network – A New Configuration for a Passive Distribution System with Optical Collision Detection Capability*, Journal of Lightwave Technology, Vol. LT-2(1), ©1984 IEEE, pp. 61-66.

Shimada T. et al. *WDM Access system Based on Shared Demultiplexer and MMF Links* Journal of Lightwave Technology , vol.23, No.9 2005, pp. 2621-2628.

Wong E., Chang-Joon Chae, *CSMA/CD-based Ethernet passive optical network with shared LAN capability*, 2000 Optical Society of America, Fiber optics communications; Networks.

Wong E., Chang-Joon Chae, *Performance of Differentiated Services in a CSMA/CD-based Ethernet over Passive Optical Network*, 2004 IEEE. pp. 641-642.

Wong E., Summerfield M., *Sensitivity Evaluation of Baseband Carrier-Sense Circuit for Optical CSMA/CA Packet Networks*, Journal of Lightwave Technology, Vol. 22, No. 8, August 2004 pp.1834-1843.

Zheng J., Hussein T. Mouftah, *An Adaptive MAC Polling Protocol for Ethernet Passive Optical Networks, 2005 IEEE.* pp.1874-1878

Zheng J., Hussein T. Mouftah, *Media Access Control for Ethernet Passive Optical Networks: An Overview*, IEEE Communications Magazine, February 2005, IEEE. pp. 145-150.

Zhu Y., Maode Ma, Tee Hiang Cheng, *A Novel Multiple Access Scheme for Ethernet Passive Optical Networks*, GLOBECOM 2003, pp. 2649-2653.

4

Effects of Dispersion Fiber on CWDM Directly Modulated System Performance

Carmina del Río Campos and Paloma R. Horche
Escuela Politécnica Superior, Universidad San Pablo CEU
ETSIT, Universidad Politécnica de Madrid
Spain

1. Introduction

The Coarse Wavelength Division Multiplexing technology (CWDM) enables carriers to transport more services over their existing optical fiber infrastructure by combining multiple wavelengths onto a single optical fiber. CWDM is technologically simpler and easier to implement and is a good fit for access networks and many metro/regional networks. ITU-T G.694.2 defines 18 wavelengths for CWDM, using the wavelengths from 1270 nm through 1610 nm with a channel spacing of 20 nm. This channel spacing allows to use, in CWDM systems, low-cost and uncooled lasers, e.g. direct modulated laser (DML).

The high output power of commercial 1.55- μm DMLs can provide a power budget that allows for amplifier/regenerator spacing of 80–100 km. However, the frequency chirp characteristics of DMLs significantly limit the maximum achievable transmission distance over standard single-mode fibers (SMF) [Cartledge 1989].

A number of approaches have been used to improve transmission performance using directly modulated lasers, including cutting down the chirp externally using a narrow band-pass filter [Yan 2005] and the deployment of a negative dispersion fiber [Tomkos 2001a, Tomkos 2002]. However, typical metro and access networks are made up of a conventional single-mode fiber (SMF) and because of the cost and difficulty (or lack of feasibility) in changing embedded fiber links, a method that enhances system performance requiring only the modification of one or both of the endpoints of a link is a critical requirement.

In this chapter, by means of a commercial Optical Communication System Design Software, we evaluate the interaction of fiber dispersion with the laser chirp in context of positive and negative dispersion coefficient. We evaluate two types of laser, with different characteristics and we determine optimal optical output power in every case. We have demonstrated that enhanced system performance, which uses a positive dispersion fiber, can be achieved if positive chromatic dispersion in the optical fiber is equalized by SPM, whereas laser transient chirp can be compensated using a negative dispersion fiber.

2. Dispersion in optical fibers

Dispersion occurs when a wave interacts with a medium or passes through an inhomogeneous geometry. It causes pulses to broaden in optical fibers, degrading signals over long distances.

If dispersion is too high, a group of pulses representing a bit-stream will spread in time making the bit-stream unintelligible. This limits the length of link or the information capacity of the fiber without regeneration. There are different types of dispersion, which all involve the dependence of the phase velocity or phase delay of light in some medium: intermodal dispersion, polarization mode dispersion and chromatic dispersion.

Intermodal dispersion results from different propagation characteristics of higher-order transverse modes in waveguides and can limit the possible data rate of a system for optical fiber communications based on multimode fibers.

Polarization mode dispersion results from polarization-dependent propagation characteristics. In a waveguide, different polarizations of light travel at different speeds due to imperfections and asymmetries, causing spreading of optical pulses.

Chromatic dispersion is the result of the wavelength dependence of the group velocity v_g.

The most commonly used chromatic dispersion parameter is D, defined as

$$D = \frac{d}{d\lambda}\left(\frac{1}{v_g}\right) = -\frac{2\pi c}{\lambda^2}\beta_2 \tag{1}$$

Where c is the velocity of light in vacuum, λ is the wavelength and β_2 is the Group Velocity Dispersion parameter (GVD).

The reason for defining the dispersion in this way is that $|D|$ is the temporal pulse spreading Δt per unit bandwidth $\Delta\lambda$ per unit distance travelled, commonly reported in ps/nm km form optical fibers. D is generally used to indicate the amount of dispersion in fiber specifications. If D is less than zero, the medium is said to have positive dispersion and if D is greater than zero, the medium has negative dispersion.

There are two sources of chromatic dispersion: material dispersion and waveguide dispersion.

Material dispersion occurs because an optical pulse emitted by a light source has a certain spectral width. The index of refraction of a material is dependent on the wavelength, so each frequency component travels at a slightly different speed. As the distance increases, the pulse becomes broader as a result. *Waveguide dispersion* refers to differences in the signal speed depending on the distribution of the optical power over the core and cladding in an optical fiber.

Material dispersion and waveguide dispersion have opposite effects (see Figure 1) and due to the difference of signs, there is a lambda (λ_{ZD}) for which the chromatic dispersion coefficient is zero.

Fiber manufacturers can manipulate these effects to change the location and slope of the chromatic dispersion curve so it is possible to shift the zero dispersion wavelength to the 1.5-µm region (C band).

All fibers with λ_{ZD} near to 1550 nm are called **dispersion-shifted Fibers** (DSF) and fibers with λ_{ZD} outside C band are called **non-zero dispersion shifted** (NZ-DSF). Figure 2 shows the dispersion curves for different models of optical fiber.

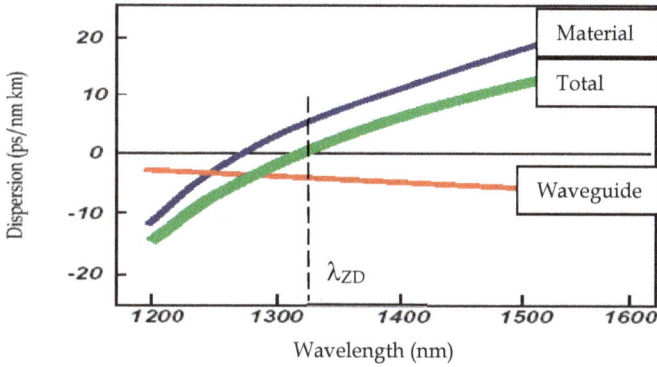

Fig. 1. Chromatic dispersion curve

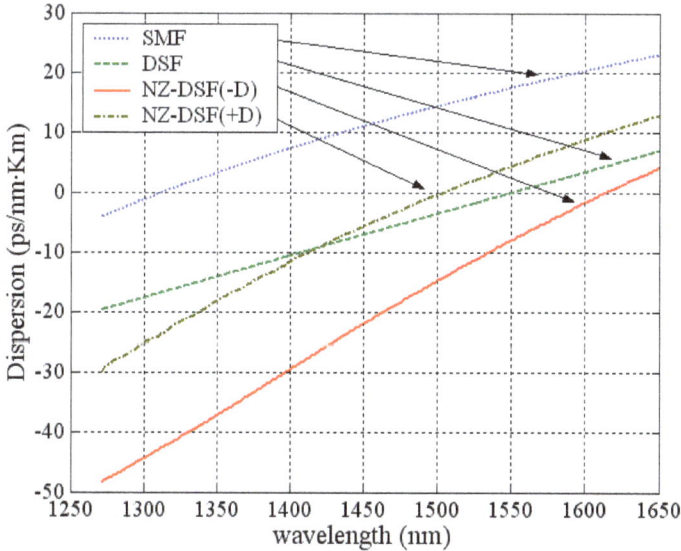

Fig. 2. Dispersion coefficients for different types of fibers.

For wavelengths longer than the zero dispersion wavelength (λ_{ZD}), called **anomalous dispersion region**, the rate of change of the index of refraction with respect to the wavelength changes sign, in which case high frequency components travel faster than the lower ones, and the pulse becomes *negatively chirped* or *down-chirped*, decreasing in frequency with time.

By the other hand, if a light pulse is propagated through a **normally dispersive medium** (for wavelengths no longer than the zero dispersion wavelength) the result is the higher frequency components travel slower than the lower frequency components. The pulse therefore becomes *positively chirped* or *up-chirped*, increasing in frequency with time.

3. The effect of dispersion on directly modulated systems

Currently DML transmitters are used to build low-cost optical communication architectures.

A block diagram of the externally modulated transmitter is illustrated in Fig. 3, in contrast with directly modulated transmitter, in Figure 4. It consists of a bit random generator, which determines the sequence of bits, a_k, that will be sent to an electric pulse generator, with NRZ format, which injects a modulation current, $I(t)$ to a Mach-Zehnder modulator. The CW Laser output port is connected to the Mach-Zehnder Modulator.

In both cases, the injected laser current is given by the expression:

$$I(t) = I_b + \sum_{k=-\infty}^{\infty} a_k I_p(t - kT) \tag{2}$$

where I_b is the bias current, T is the period of the modulation pulse, the sequence of bits transmitted (a_k = 1 (0) if a binary one (zero) is transmitted during the k_{th} time), and $I_p(t)$ is the applied current pulse.

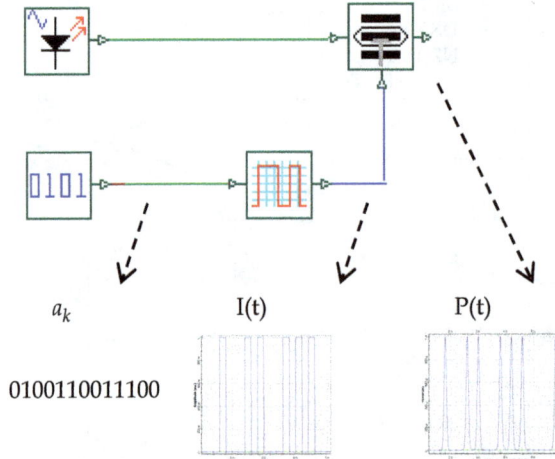

a_k I(t)   P(t)

0100110011100

Fig. 3. Block diagram of the externally modulated transmitter

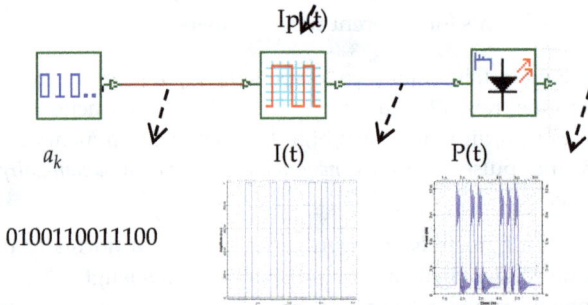

a_k I(t)   P(t)

0100110011100

Fig. 4. Block diagram of the directly modulated transmitter

In an externally modulated laser, the optical power output pulse of the laser, P(t), is given by

$$P(t) = \eta_0 . \frac{hv}{q} . \sum_{k=-\infty}^{+\infty} a_k I_p(t - kT)$$ (3)

where η_0 is differential quantum efficiency of the laser, hv the photon energy at the optical frequency v, q is the charge of an electron and $Ip(t)$ the applied current pulse.

However, expression (3) is not applicable in case of directly modulated sources where the injected current that modulates the laser introduces a shift in the emission frequency (*chirp frequency*). As a consequence, the optical power output pulse is not a linear transformation of the applied current pulse.

The optical power and chirping response of the semiconductor laser to the current waveform I(t) is determined by means of the large-signal rate equations [Tomkos 2001b], which describe the interrelationship of the photon density, carrier density, and optical phase $\phi(t)$, within the laser cavity.

Thus, the emission frequency associated with the output of the laser (*chirp*) can be formulated by:

$$\Delta v(t) = \frac{1}{2\pi} \frac{d\phi}{dt}$$ (4)

where ϕ is the optical phase, which variation with the time depends on

$$\frac{d\phi}{dt} = \frac{1}{2} \alpha \left[\Gamma v_g a_0 (n - n_t) - \frac{1}{\tau_p} \right]$$ (5)

n_t represents the carrier density at transparency, α the line-width factor or Henry's factor, Γ the confinement factor, vg the group velocity, a_0 the gain coefficient, and τ_p is the photon lifetime.

To analyze the interaction of dispersion with the chirp generated in directly modulated systems we initially study systems where pulse have got Gaussian shapes and finally pulses with others forms.

3.1 Gaussian pulses

Even though the shape of optical pulses representing 1 bits in a bit stream is not necessarily Gaussian, one can gain considerable insight into the effects of fiber dispersion by focusing on the case of a chirped Gaussian pulse with the input field.

In Fig. 5 the pulse chirp and optical phase, is plotted together with the pulse intensity. As it can be seen quadratic changes in the optical phase corresponds to linear frequency variations, for this reason, such pulses are said to be linearly chirped [Agrawal 2010].

In a linear chirp, the instantaneous frequency f(t) varies linearly with time:

$$f(t) = f_0 + Ct$$ (6)

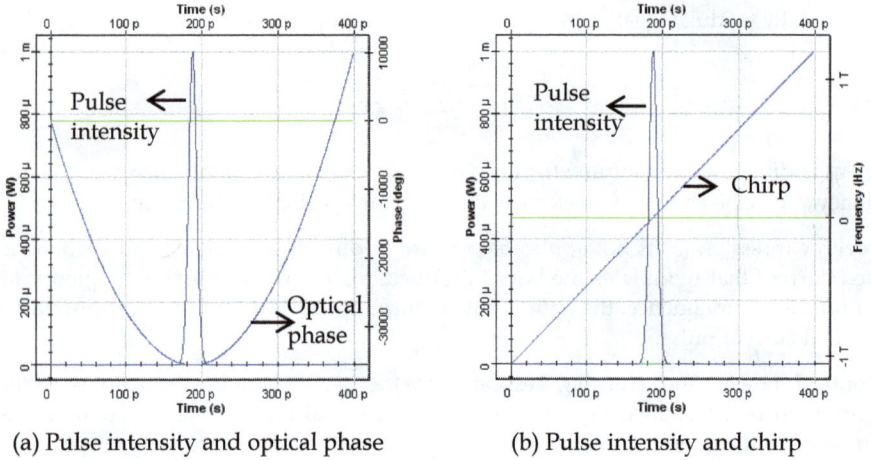

(a) Pulse intensity and optical phase   (b) Pulse intensity and chirp

Fig. 5. Pulse intensity, optical phase and chirp in an optical Gaussian pulse.

where f_0 is the starting frequency (at time $t = 0$) and C is the rate of frequency increase or chirp rate. In a **chirp-free Gaussian** pulse, the optical frequency remains constant throughout the pulse, as shown in Figure 6a, whereas an **up-chirp (down-chirp)** pulse means that the instantaneous frequency rises (decreases) with time (Figures 6b and 6c).

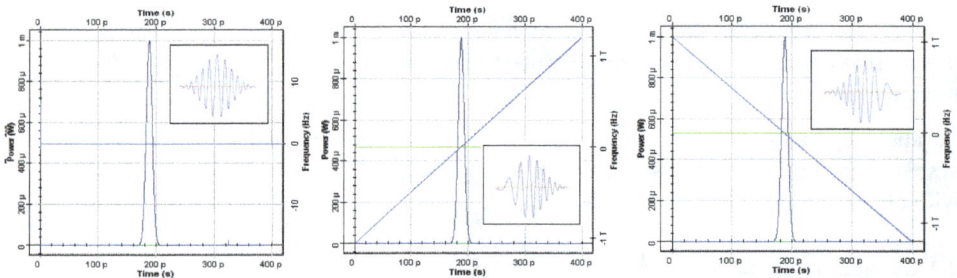

(a) Chirp-free gaussian pulse   (b) Up-chirped pulse   (c) Down-chirped pulse

Fig. 6. Pulse intensity, optical phase and chirp for Gaussian opical pulses.

3.2 Non Gaussians optical pulses

In directly modulated sources, optical pulses are often non-Gaussian and may exhibit considerable chirp caused by carrier-induced changes in the mode index.

In these cases, the chirp is not linear and therefore can not be characterized by the value of the chirp rate constant (C) and it can be calculated from the following expression [Coldren 1995]

$$\Delta v(t) = \frac{\alpha}{4\pi} \left(\frac{1}{P(t)} \frac{dP(t)}{dt} + \left[\frac{P(t)\varepsilon}{\tau_p} - \frac{\beta\Gamma n(t)}{\tau_n P(t)} \right] \right) \quad (7)$$

This equation evaluates the chirp for Fabry-Perot semiconductor laser, but is used as a good approximation for DFB lasers [Morgado 2003]. By the other hand, this expression can be simplified whenever working with modulation frequencies above 100 MHz, then comes to a new equation that depends on two laser intrinsic parameters: the adiabatic chirp coefficient (κ) and the Henry factor (α).

$$\Delta v(t) = \frac{\alpha}{4\pi} \left(\frac{1}{P(t)} \frac{dP(t)}{dt} + \frac{P(t)\varepsilon}{\tau_p} \right) = \frac{\alpha}{4\pi} \left(\frac{1}{P(t)} \frac{dP(t)}{dt} + \kappa P(t) \right) \quad (8)$$

The **adiabatic coefficient**, κ, depends on the laser structure, being the figure that takes into account the output power and the generated chirp. The relationship to the photon energy, hν, optical frequency, υ, laser quantum efficiency, η_0, confinement factor, Γ, cavity volume, V_0 and gain compression factor, ε, is in agreement with the expression:

$$\kappa = \frac{2\Gamma}{\eta_0 h\nu V_a} \varepsilon \quad (9)$$

Its units of measurement are (W S)$^{-1}$ and takes values around 10^{12}.

Henry factor, normally called linewidth enhancement factor, quantify the effect of the change of refractive index with carrier density on the dynamical properties of semiconductor lasers, [Henry 98]

$$\alpha = -\frac{\partial \chi_r / \partial N}{\partial \chi_i / \partial N} = -\frac{4\pi}{\lambda} \frac{dn}{dg} \quad (10)$$

where n is the refractive index, g is the gain per unit length, λ is the wavelength and N is the carrier density.

A lot of experiments have reported a great variety of values for the α-factor ranging from 0 to values of the order of those commonly observed in bulk and quantum well materials (1.5-3). Larger values for the α-factor than those published for quantum well devices have also been reported and a value for α as high as 60 has been published.

The first term of (8), called *transient chirp*, causes variations in the pulse width and is scaled by α. The second term (*adiabatic chirp*) produces a frequency shift proportional to the instantaneous optical power, and scaled, in addition to α, by parameter κ [C.del Rio 2008]. Figure 7 shown two types of chirp in DML. We can observe a transient chirp corresponds to the chirp that occurs during the transition between steady states and an adiabatic chirp corresponds to the frequency offset between steady state output powers of the laser.

(a) Optical pulse

(b) Adiabatic chirp DML-A

(c) Transient chirp DML-T

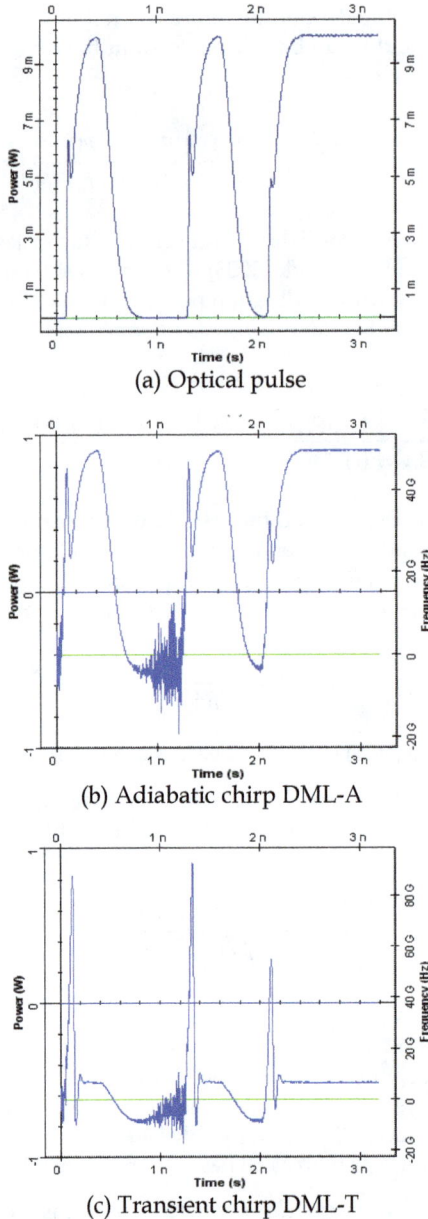

Fig. 7. Optical pulses shapes (a) and chirp in an adiabatic (b) and transient (c) dominated laser.

Depending on the term in equation 8 that predominate over the other DMLs can be classified as transient (DML-T) or adiabatic (DML-A) chirp dominated. This terminology was first used the Hinton by 1993 [Hinton 93].

Each extreme category has significantly different optical spectra. DML-T lasers have higher bandwidths to DML-A, as shown in Figure 8.

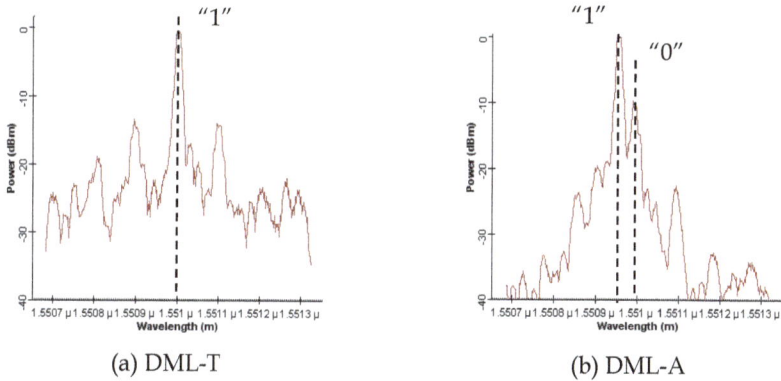

(a) DML-T (b) DML-A

Fig. 8. Spectrum of transient chirp dominated (a) and adiabatic chirp dominate laser (b)

Besides the difference in the spectral width, is observed in lasers DML-T that the emission peak is centered at the nominal frequency of emission, which corresponds to the peak frequency during continuous vawe transmission (Figure 8a). However, the laser DML-A there are two different frequency peaks corresponding to carrier frequencies of the bits "1" and "0" (Figure 8b). The separation between the wavelengths of the carriers of bits 0 and 1 is directly related to the adiabatic chirp of the signal and, therefore, depends on the output power.

Figure 9 represents the chirp of bit "1" when the output power varies from 0.1 to 15 mW. As can be seen, the higher the laser emission power the higher the adiabatic chirp generated. In this particular case, the peak-to-peak chirp generated when the value of output power is 0.1 mW is 8 GHz and 35 GHz if the power change to 15 mW.

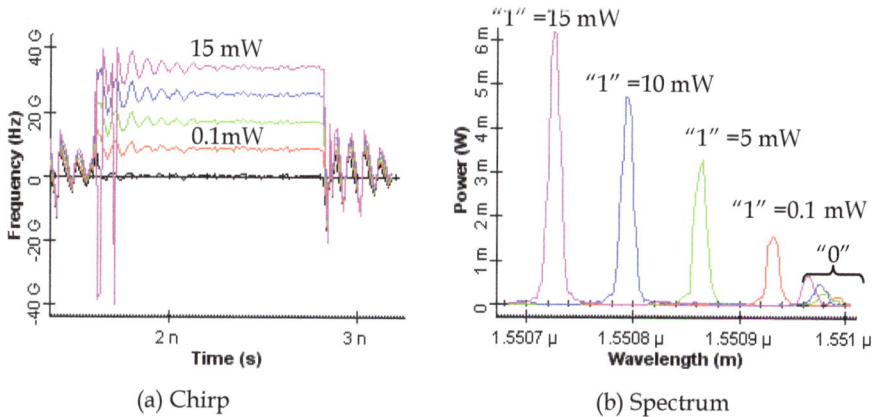

(a) Chirp (b) Spectrum

Fig. 9. Chirp (a) and spectrum (b) for 10 output power DML for 0.1 to 15 mW

The frequency chirp, due to the adiabatic chirp and shown in Figure 9a, implies a shift in the wavelength emission, as it shown in Figure 9b. This optical frequency shift from the actual laser output frequency is the principal cause of system performance limitations.

3.3 Interaction dispersion-chirp-SPM

Gaussian pulse remains Gaussian during propagation in a transparent medium but its chirp and with change due to the effects of chromatic dispersion and nonlinearities (e.g. self-phase modulation arising from the Kerr effect).

Therefore, when a pulse with an initial chirp C, crosses a distance ξ in a dispersive media, the chirp will acquire a new value given by C_1 and the pulse will be broadened by a factor represented by bf

$$bf\ (\xi)= [(1+sC\ \xi)^2+\ \xi]^{1/2} \tag{11}$$

$$C_1(\xi)= C+s(1+C^2)\ \xi \tag{12}$$

Where s is the sign of β_2 (+1 if the pulse propagates in the normal dispersion and -1 in the anomalous dispersion) and ξ is related to β_2 and FWHM (full width at half-maximum).

Namely, the shape of the signal at the output of the optical fiber is affected by the sign of both β_2 (fiber parameter) and C (transmitter parameter).

An **unchirped pulse** (C=0) broadens monotonically by a factor of $(1 + \xi^2)^{1/2}$ and develops a negative chirp across the fiber. Chirped pulses, on the other hand, may expand or compress depending on whether β_2 and C have the same or contrary signs.

When $\beta_2C > 0$, a chirped Gaussian pulse broadens monotonically at a rate faster than that of the unchirped pulse. The reason is related to the fact that the dispersion-induced chirp adds to the input chirp because the two contributions have the same sign. The situation changes for $\beta_2C < 0$. In this case, the contribution of the dispersion-induced chirp is of a kind opposite to that of the input chirp.

In Figure 10, the pulse chirp is plotted together with the pulse intensity. Initial narrowing of the pulse for the case $\beta_2C<0$ can be explained by noticing that the faster frequency components are in the trailing edge, and the slower in the leading edge of the pulse. As the pulse propagates, the faster components will overtake the slower ones, leading to pulse narrowing. At L= L_2 full compensation between dispersion induced chirp and the initial one

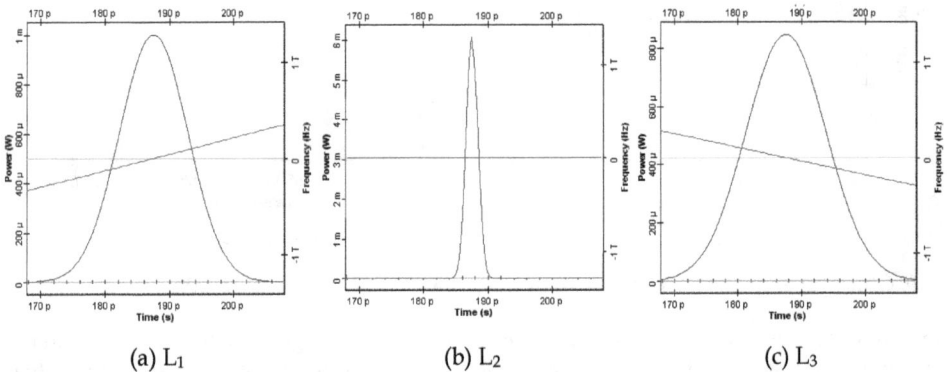

(a) L_1 (b) L_2 (c) L_3

Fig. 10. The output optical pulse signal for the different transmission length (L_1, L_2 and L_3)

will occur. At this length $L_2 = |C|/(1+C^2)$, C_1 becomes zero and the pulse becomes unchirped. We define this situation as optimal.

Finally, with further propagation, the fast and the slow frequency components will tend to separate in time from each other and pulse broadening will be observed.

On the other hand, the SPM alone leads to pulse chirping, with the sign of the SPM-induced chirp being opposite to that induced by anomalous GVD.

In Figure 11, the leading edge of the pulse becomes red-shifted and the trailing edge of the pulse becomes blue-shifted. If the effects of anomalous dispersion were present, with the chirp induced by SPM some pulse narrowing would occur. This means that the effect of SPM counteracts GVD.

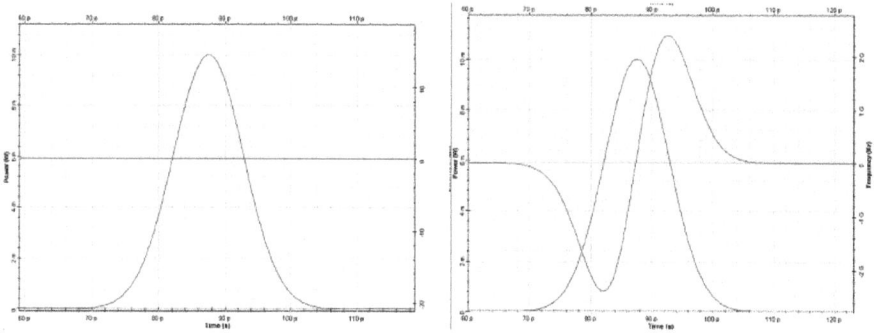

Fig. 11. Input (left) and output (right) pulse shape and chirp.

The effect of GVD on the pulse propagation depends, mainly, on whether or not the pulse is chirped, the laser injection pulse shape, [del Rio 2010], and also on the fiber SPM (Self Phase Modulation [Hamza, M. Y., Tariq, S. & Chen, L. 2006, 2008]. With the correct relation between the initial chirp and the GVD parameters, the pulse broadening (which occurs in the absence of any initial chirp) will be preceded by a narrowing stage (pulse compression). On the other hand, the SPM alone leads to a pulse chirping, with the sign of the SPM-induced chirp, being opposite to that induced by anomalous GVD. This means that in the presence of SPM, the GVD induced pulse-broadening will be reduced (in the case of anomalous), while extra broadening occurs in the case of normal GVD.

4. Enhancing the performance of systems using negative and positive dispersion fibers

In this section, we study that the transmission performance depends strongly on dispersion fiber and DML output power. We demonstrated that systems using SMF fibers can achieve a good performance if the DML output power is properly chosen. Finally, we have found a mathematical expression that make an estimation for a power value to fix the laser power output for each channel in WDM systems.

In order to study the CWDM system performance a simple arrangement is proposed, as can be seen in Figure 12. We have selected 16 output channels with wavelengths , in agreement with Recommendation ITU-T G. 694.2. The pulse pattern was a periodic 128-bit OC-48 (2.5 Gb/s) nonreturn-to-zero (NRZ). After transmission through 100 km of fiber, channels are demultiplexed and detected using a typical pin photodiode.

We have used two kinds of optical fibers; the already laid and widely deployed single-mode ITU-T G.652 fiber (SMF) and the ITUT-T G.655 fiber with a negative dispersion sign around C band (NZ-DSF). It is well known, SMF fiber dispersion coefficient is positive in the whole telecommunication band from O-band to L-band and the dispersion coefficient of the NZ-DSF fiber is negative in the optical frequency range considered. For our purpose, the same spectral attenuation coefficient of both fibers has been considered whose water peak at 1.38 μm is well suppressed. The dispersion slope, effective area and nonlinear index of refraction are compliant with typical conventional G.652 and G.655 fibers.

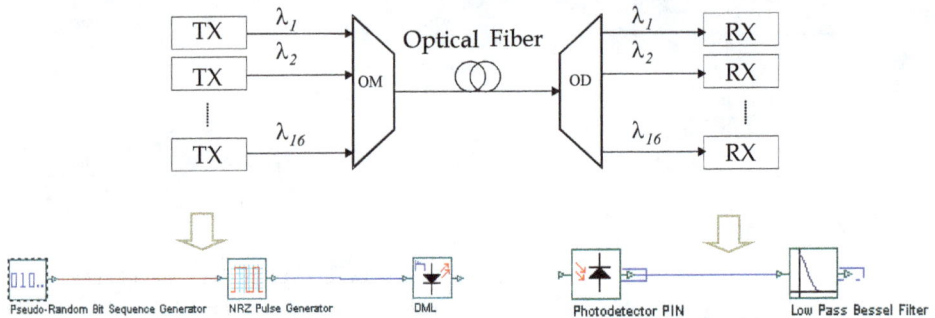

Fig. 12. Arrangement set up of simulated transmission link.

We have to point out that the transmission performance of waveforms produced by directly modulated lasers in fibers with different signs of dispersion, depends strongly on the characteristics of the laser frequency chirp. For this reason, we have modeled two DMLs (made up of DFB-DMLs), by using the Laser Rate Equations in agreement with that reported in [Tomkos 2001b], both DMLs presenting extreme behaviors [Hinton 1993]: DML-A is strongly adiabatic chirp dominated; $\alpha = 2.2$ and $k = 28.7 *10^{12}$ (W.s)$^{-1}$ and DML-T is strongly transient chirp dominated; $\alpha = 5.6$ and $k = 1.5 *10^{12}$ (W.s)$^{-1}$. The α and k values used in our simulation are in agreement with potential commercial devices [Osinki 1987, Peral 1998, Rodríguez 1995].

In this work, we are mainly interested in comparing the system performance based on the type of fiber and DML used; for this reason, the rest of link components have been modeled by considering ideal behavior.

The performance of transmission systems is often characterized by the bit error rate (BER), which is required to be smaller than approximately 10^{-12} for most installed systems. Experimental characterization of such systems is not easy since the direct measurement of BER takes considerable time at these low BER values. Another way of estimating the BER is using the Q of the system, which can be more easily modeled than the BER.

The parameter Q , the signal-to-noise ratio at the decision circuit in voltage or current units, is given by the expression[Alexandre 1997]

$$Q = \frac{I_1 - I_0}{\sigma_1 + \sigma_0} \tag{13}$$

where I_i and σ_i are average values and variances of the "1" and "0" values for each pattern. Q factor can be considered just a qualitative indicator of the actual BER and it can expressed as

$$BER = \frac{1}{2} erfc\left(\frac{Q}{\sqrt{2}}\right) \tag{14}$$

This parameter guarantee an error-free transmission of *Q-Factor* higher than 7, corresponding to a BER lower than 10^{-12}.

In order to study the transmission performance of DMLs presenting extreme behavior on a fiber with positive or negative dispersion, a set of simulations were carried out; called Cases A, B, C and D, as shown in Table 1. The quality of transmission between them has been compared. Thus, Case-A deploys DML-A lasers and SMF fiber, Case-B: DML-A/NZ-DSF, Case-C: DML-T/SMF and Case-D: DML-T/ DSF.

Case	DML	Optical Fiber
A	DML-A	SMF
B	DML-A	NZ-DSF
C	DML-T	SMF
D	DML-T	NZ-DSF

Table 1. Different configurations for the simulated system

The DML output power of all channels was varied from -10 dBm to 10 dBm (0.1-10 mw), and the performance, in terms of *Q-Factor*, is analyzed for each transmitted channel.

Figure 13 shows the *Q-Factor* dependence on channel power for the wavelength channel centered at 1551 nm.

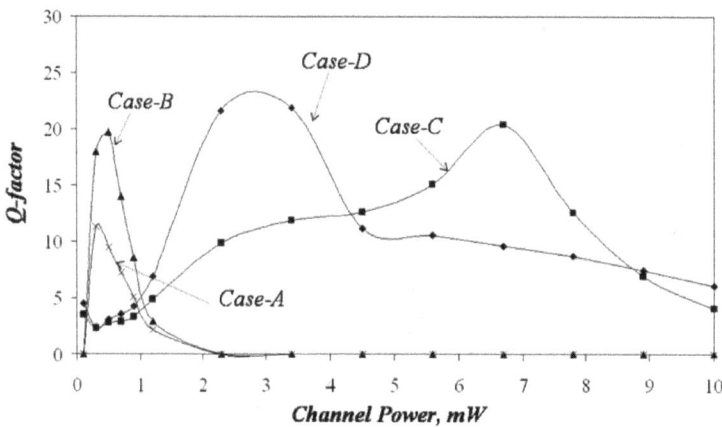

Fig. 13. Simulated results for the transmission performance, *Q-Factor*, at 1551 nm wavelength after transmission over 100 km of positive and negative dispersion fiber

Independently of the Case and wavelength channels, the *Q-Factor* always presents a maximum value for a specific DML output power [Horche 2008]. This behaviour demonstrates the existence of an optimum channel power that will have to be considered during the system design.

This optimum value corresponds with the power value that allows compensating the laser chirp with the fiber dispersion and it depends on the combination of components used in each case.

4.1 A and B cases: Adiabatic dominated laser

A and B Cases use adiabatic chirp dominated DML-A lasers. The Qmax value is reached at 0.3-0.46 mw, independently of the fiber type. Over this value the function drastically gets worse when increasing the output laser power. In both cases the type of the laser used in the simulation is an adiabatic chirp dominated, so for values over 0.4 mW the filter reduces partially the spectrum and this phenomenon closes the eye diagram.

Fig. 14 shows the spectrum of adiabatic chirp dominated laser together with the transfer function of a Gaussian filter. The shift of the spectrum towards blue would cause a bigger reduction of the peak emission of bit "1" than the one produced on the peak of bit "0". This would bring both "1" and "0 peak emission power closer and the eye diagram be closed.

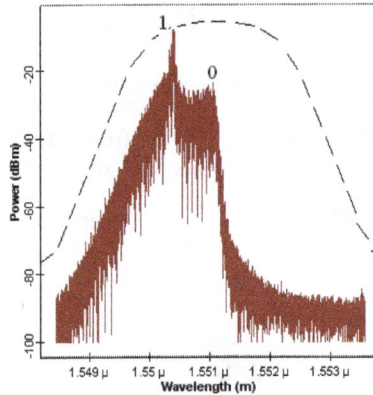

Fig. 14. Spectrum of adiabatic chirp dominated laser together with the transfer function of a Gaussian filter

On the other hand, the power waveform coming from DML suffers a deformation when getting through the dispersive media. In the case of DML-A, the result of the interplay of the dispersion with the specific chirp characteristics will result in a high intensity spike at the front of the pulses for transmission through a fiber with positive dispersion (SMF) and at the end for negative dispersion (NZ-DSF) [Krehlik06], as can be seen at the top of the Figure 15.

The absolute value of the dispersion (and not its sign) will play a major role in the transmission performance. Thus, the performance corresponding to transmission through an SMF fiber will be worse than that corresponding to transmission through an NZ-DSF fiber because of the larger absolute value of the dispersion.

Figure 15 represents the power waveforms for five different optical output powers (from 0.5 to 4 mw) after transmission through 100 km NZ-DSF fiber.

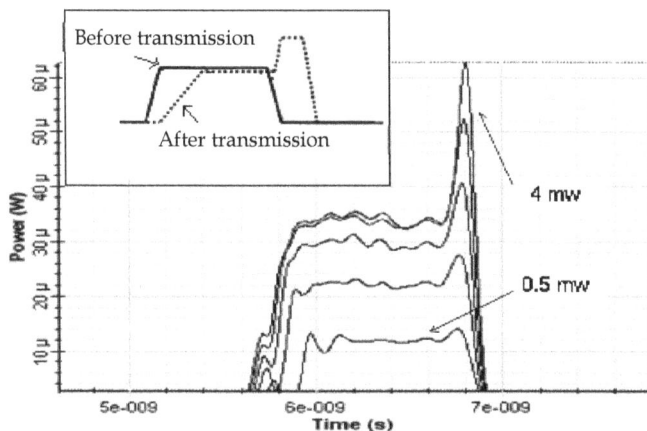

Fig. 15. Shapes of optical pulses for different DML-A output powers, after transmission through 100 km negative dispersion fiber

The increment of P_{ch} will result in a higher intensity spike at the trailing edge of the pulse. As consequence the eye pattern after transmission will be severely closed.

In Fig. 16 the eye diagrams are shown for the case of the adiabatic chirp dominated transmitter after transmission over 100 km of a negative dispersion fiber for (a) P_{ch} = 0.46 mw (optimum power) and (b) P_{ch} = 1 mw. For P_{ch} = 0.46 mw, the eye pattern is clearly open, while for P_{ch} = 1 mw eye pattern experiencing more than 3dB eye closure.

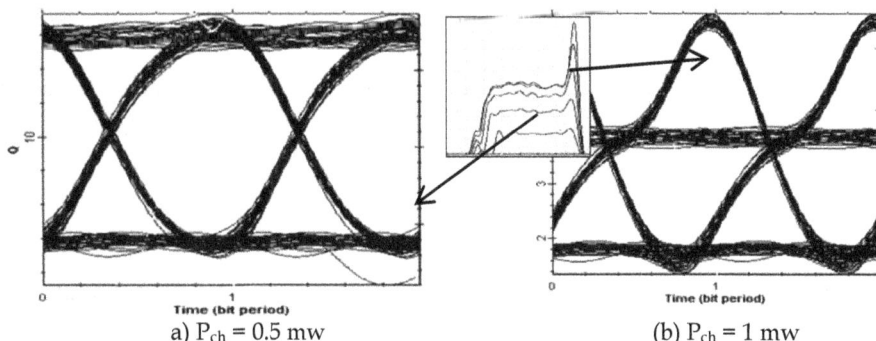

a) P_{ch} = 0.5 mw   (b) P_{ch} = 1 mw

Fig. 16. Eye diagrams for the case of the adiabatic chirp dominated transmitter after transmission over 100 km of a negative dispersion fiber for (a) P_{ch} = 0.46 mw and (b) P_{ch} = 1 mw.

For small powers, the Q-Factor increases with P_{ch} because a large amount of power reaches the detector. For higher P_{ch} the optical pulse deformation arising from chirp induced by DML becomes too large and causes an error in pulse reconstruction.

4.2 C and D cases: Transient dominated laser

C and D Cases use transient chirp dominated DML-T lasers. For Case-C (DML-T/SMF), the Q_{max} value takes place for an output power of 6.7 mw approximately. In Case-D (DML-T/DSF), the necessary output power to reach the Q_{max} is around 2.3-3.4 mw.

In DML-T, the wavelength shift by laser transient chirp is a blue shift during the pulse rise time and a red shift during the pulse fall time; exactly the opposite effects takes place with SPM (Self-phase-modulation) [Suzuki 1993]. Therefore, the optical pulse chirped by direct modulation is compressed in fibers with negative dispersion (NZ-DSF), while that chirped by SPM is compressed in fibers with positive dispersion (SMF).

As it can be seen in Figure 13, for channel power Pch from 0.1 to 4 mw, the performance of system that uses an NZ-DSF fiber (D-Case) is better than that of an SMF fiber (C-Case). In this power range, SPM magnitude is not enough and the wavelength shift by laser transient chirp is the predominant effect. Thus, the optical pulse chirped by direct modulation is compressed in fibers with negative dispersion (NZ-DSF) and uncompressed in fibers with positive dispersion (SMF). Therefore, case D is better than case C, however, for P_{ch}. from 4 mw to 9 mw, Case-C (DML-T/SMF) presents a better performance than Case D (DML-T/NZ-DSF) because of the increment in the magnitude of the SPM in the optical fiber and, therefore, chromatic dispersion of the positive dispersion fiber is equalized by the SPM as long as the pulses are broadened for negative dispersion fiber. As resulting from this, the eye pattern after the transmission through SMF fiber will be more open than using NZ-DSF fiber when higher output power is used.

Figure 17 shows the eye diagram of Case-C (a) and Case-D (b). In both cases a P_{ch}. of 7 mw was used and the eye diagram is measured for the signal transmission after 100 km of dispersion fiber at 1551 nm wavelength. After the transmission through SMF, the eye look perfectly open (Fig 17a) while the eye pattern after transmission through NZ-DSF is severely closed (see Fig 17b) and intersymbol interference will occur. On other hand, the different dispersion sign will only affect the asymmetry of the eye diagram, as is obvious from the results of Fig. 17.

Therefore, we can conclude that systems using an SMF fiber can have a similar or better performance to those systems that use an NZ-DSF fiber if the DML is transient chirp dominated and its output power is properly chosen.

5. Management of the power channel of to enhance CWDM system performance

In order to analyze the influence of the selected wavelength in a CWDM system, simulations varying the number of channels from 1 to 16 have been carried out, using the same schematic arrangement set up shown in Fig. 12. The channel wavelengths were between 1531 and 1591 nm. In this case, this wavelength range was used due to the system does not need optical amplifiers. Some channels were located at compatibles frequencies with CWDM ITU-T grid in order to, in the future, extend this work to whole useful fiber optic spectral range (1271-1611 nm).

In every case, the *Q-Factor* shows a maximum value for a given optical output power. In A and B Cases, due to small powers of channels, Q_{max} is almost independent of number of

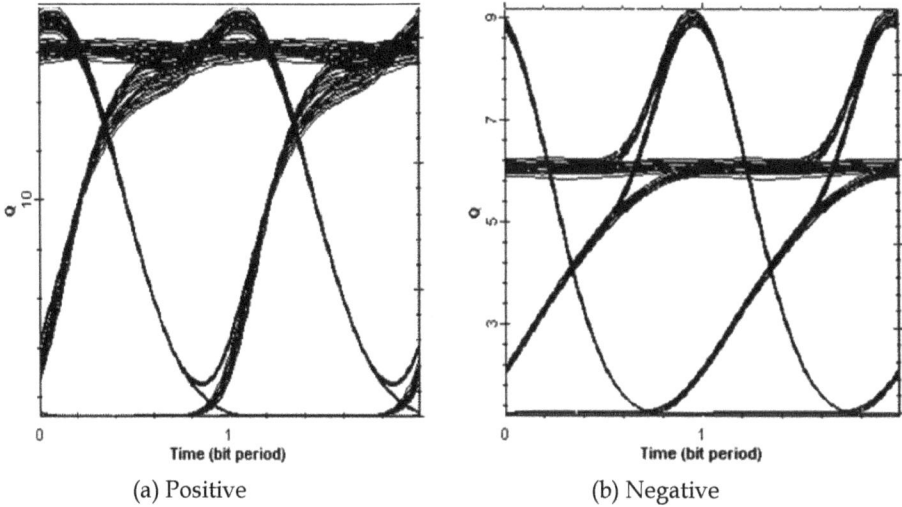

(a) Positive	(b) Negative

Fig. 17. Eye diagram at the receiver side of a 2.5 Gb/s transient chirp dominated transmitter (7 mw of output power at 1551 nm wavelength) over (a) SMF fiber where the dispersion is positive and (b) NZ-DSF fiber where the dispersion is negative

channels. In C and D Cases, this maximum value decreases with the increment of the number of channels used manly due to crosstalk between channels and others no-lineal effects. However, the Q_{max} value, for a given channel, takes place for a very similar output power.

Figure 18 shows the Q-Factor versus channel power for channels centered at 1531, 1551, 1571 and 1591 nm respectively, for 16-Channel WDM system using DML-T/SMF (a) and DML-T/DSF (b).

In both cases, each channel presents a different optimum P_{ch}. Thus, by means of the P_{ch} management of each channel it is possible to reach the Q_{max} and enhances WDM system performance can be achieving.

As an example; if a 16-Channel WDM system is designed using DML-T and SMF with channel powers equal to the optimum channel power Pch. all 16 channels will have a Q higher than 8, corresponding to a BER lower than 10^{-15}. In contrast, if a system design with equal channel power is used some of channels (higher dispersive channels) will fail after propagation through SMF fiber.

In Case D, in order to guarantee a Q-Factor=15, the output power laser of the channels centered at 1531, 1551, 1571 and 1591 nm should be 3.2, 3.5 3.8 4 mW respectively. Such difference is due to the different fiber dispersion coefficients that would be associated to every one of them, as shown in Table 2. Then, the compensation of the dispersion would happen for different chirp values and therefore for different output power values.

From another point of view, if the system were designed with the same value of output power in every laser, there is the risk for the channel with the bigger dispersion value not to exceed the minimum criteria that assure an error-free transmission.

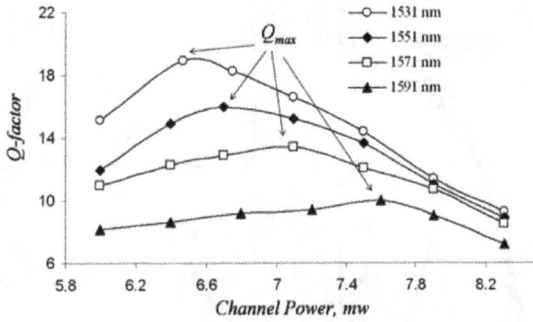

(a) Case C (DML-T/SMF)

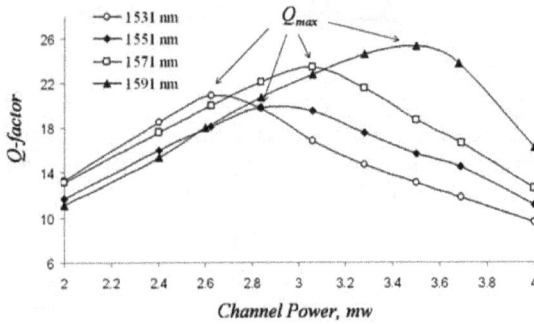

(b) Case D (DML-T/NZ-DSF)

Fig. 18. *Q-Factor* vs channel power for channels centered at 1531, 1551, 1571 and 1591 nm respectively, for 16-Channel WDM system using DML-T/SMF (a) and DML-T/NZ-DSF (b).

Channel	Dispersion
1531 nm	15,21 ps/nm ·km
1551 nm	16,34 ps/nm ·km
1571 nm	17,47 ps/nm ·km
1591 nm	18,56 ps/nm ·km

Table 2. Chromatic dispersion of differents channels (SMF fiber)

Since the optimum power channel depends on the global dispersion of the system, a study including the variation of the accumulated dispersion of the global system will be done. The optimum channel powers (P_{ch} to reach Q_{max}) are plotted as a function of dispersion in Fig. 19 (open circles in the case of transmission through positive dispersion fiber and solid circles for negative dispersion fiber). In Fig. 19, the results for channel centered at 1551 nm after transmission over 100 Km of SMF and NZ-DSF fibers as well as a potential CWDM channel centered at 1391 nm are shown. Attenuation dependence with wavelength was taken account in the calculation of optimum P_{ch} and, in all cases, Q_{max} higher than 7 (BER lower than 10^{-12}) was obtained.

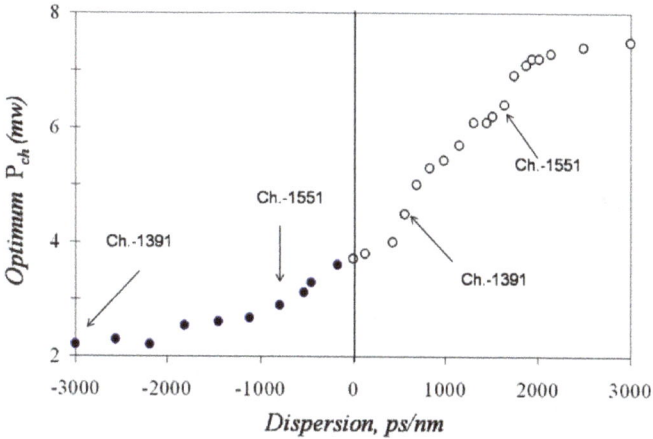

Fig. 19. Comparison of Optimum Channel Powers versus accumulated dispersion for a positive dispersion fiber (open circles) and negative dispersion fiber (solid circles).

In both cases, each channel presents a different optimum $P_{ch.}$ Then, by the $P_{ch.}$ control of each channel it is possible to reach the Q_{max} and an enhancement of the WDM system performance can be achieved. This optimum P_{ch} is the conclusion of the following considerations: for low power levels, below the optimum power, the Q-Factor increases with Pch because a larger amount of power reaches the detector and the performance enhancement will be dependent upon the level power, so that the greater the power in the receiver, higher system performance is obtained; while, for P_{ch} higher than optimum power, the chirp increases with level power and it causes greater frequency shift and linewidth broadening which results in an error in pulse reconstruction.

A mathematical expression that fits this curve would be very useful, since it would make an estimation of the power value to fix the laser output for each channel. For this reason, using the Matlab simulation tool, this function has been estimated from a polynomial expression of degree 4 (Figure 20)

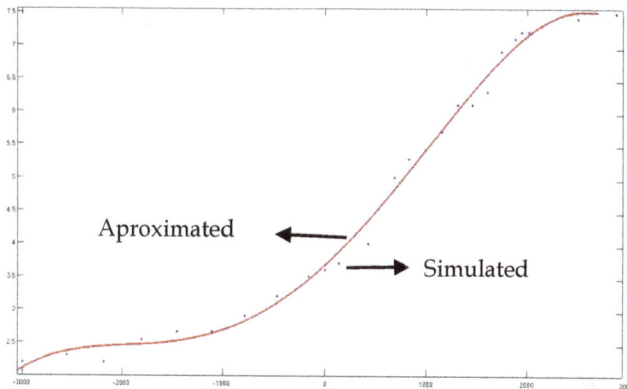

Fig. 20. Estimated and approximated curve

$$f(x) = ax^4 + bx^3 + cx^2 + dx + e \tag{15}$$

a = -3.482 ·10-14; b = -6.588 ·10-11; c = 4.202 ·10-07; d = 0.001435; e = 3.673

where x is the dispersion accumulated across the link.

Thanks to this equation it is possible to optimize the system behaviour reducing the number of simulations needed for the design stage.

6. Conclusions

The performance of fibers relative to positive or negative dispersion characteristics is discussed for the case of directly modulated lasers. The effects of chirp and fiber nonlinearity in a directly modulated 2.5-Gb/s transmission system have been researched by simulation. We have demonstrated that enhanced system performance, which uses a positive dispersion fiber, can be achieved if positive chromatic dispersion in the optical fiber is equalized by SPM, whereas laser transient chirp can be compensated using a negative dispersion fiber. We can conclude that systems using SMF fiber can have a similar or better performance to those systems that use NZ-DSF fiber if the DML is transient chirp dominated and its output power is properly chosen.

Since the magnitude of SPM can be changed by controlling the optical power in the fiber, the balance between SPM and laser transient chirp can be controlled. Therefore, an optimum compensation condition can be achieved by controlling the optical DML output power. This technique is simple, flexible, and applicable to WDM systems.

In order to analyze the effectiveness of this technique for WDM systems, simulations varying the number of channels from 1 to 16 have been carried out and checking. In every case, *Q-Factor* shows a maximum value depending on the optical power of each channel and accumulated dispersion. This maximum value decreases depending on the number of channels used. Also, we have shown that through the management of the Pch. of each channel it will be possible to enhance the performance of each channel as well as the whole WDM system.

7. Acknowledgements

The authors gratefully acknowledge the support of the MICINN (Spain) through project TEC2010-18540 (ROADtoPON).

8. References

G.P. Agrawal (2010) Fiber-Optic Communication System. A John Wiley & Sons Ed.

S. B. Alexander (1997). Optical Communication Receiver Design. *SPIE Press/ IEE.*

J.C. Cartledge, G.S. Burley. (1989). The Effect of Laser Chirping on Ligthwave System Performance. *J. Lightwave Technol.*, vol. 7, no. 3, pp. 568-573.

L.A. Coldren and S.W. Corzine. (1995) Diode Lasers and Photonic Integrated Circuits. *Wiley Series in Microwave and optical Engineering.*

C. del Río and P.R. Horche (2008). Directly modulated laser intrinsic parameters Optimization for WDM Systems, *International Conference on Advances in Electronics and Micro-electronics,* Sep-Oct 2008.

C. del Río, P. R. Horche, A. Martín Minguez (2010). Effects of Modulation Current Shape on Laser Chirp of 2.5 Gb/s Directly Modulated DFB-Laser. *Proceedings of The Third International Conference on Advances in Circuits, Electronics and Micro-electronics.* CENICS 2010. ISBN.: 979-0-7695-4089-4

C. del Río, P.R. Horche, and A. M. Minguez (2010). Effects of Modulation Current Shape on Laser Chirp of 2.5 Gb/s Directly Modulated DFB-Laser. *Proc. Conf. on Advances in Circuits and Micro-electronics,* pp. 51-55, CENICS 2010

GVD effects in fiber optic communications: dispersion- and power-map cooptimization using genetic algorithm, Optical Engg., volume 47, pp. 075003.

Hamza, M. Y., Tariq, S., Awais, M. M. & Yang, S. (2008). Mitigation of SPM and

Henry, C. H. (1982), Theory of the linewidth of semiconductor lasers, *IEEE J.Quantum Electron.,*QE-18(2), 259–264.

K. Hinton and T. Stephens. (August 1993). Specifying Adiabatic Lasers for 2-5gbitls. High Dispersion IM/DD Optical Systems. *Electronics Letters.* Vol. 29 No. 16.

P. Horche and Carmina del Río (2008). Enhanced Performance of WDM Systems using Directly Modulated Lasers on Positive Dispersion Fibers. Optical Fiber Technology. Volume 14, Issue 2, April 2008, Pages 102-108.

Krelik P. (2006)Characterization of semiconductor laser frequency chirp based on signal distorsion in dispersive optical fiber". *Opto-electronics review.* vol. 14, no.2, pp. 123-128.

J.A. P. Morgado, and A.V. T. Cartaxo (October 2003). Directly Modulated Laser parameters Optimization for Metropolitan Area Networks Utilizing Negative Dispersion Fibers. *IEEE Journal of Selected Topics in Quantum Electronics,* Vol. 9, No. 5.

M. Osinski and J. Buss (January 1987). Linewidth broadening factor in semiconductor lasers—An overview, *IEEE J. Quantum Electron.,* vol. 23, no. 1, pp. 9–29,.

E. Peral, W. K. Marshall and A. Yariv (October 1998). Precise Measurement of Semiconductor Laser Chirp Using Effect of Propagation in Dispersive Fiber and Application to Simulation of Transmission Through Fiber Gratings. *Journal of Lightwave Technology,* Vol. 16, No. 10,.

D. Rodríguez, I. Esquivias, S. Deubert, J. P. Reithmaier, A. Forchel, M. Krakowski, M. Calligaro, and O. Parillaud. (February 2005). Gain, Index Variation, and Linewidth-Enhancement Factor in 980-nm Quantum-Well and Quantum-Dot Lasers. *IEEE Journal of Quantum Electronics,* Vol. 41, No. 2.

N. Suzuki, and T. Ozeki. (1993). Simultaneous Compensation of Laser Chirp, Kerr Effect, and Dispersion in 10-Gb/s Long-Haul Transmission Systems. *J. Lightwave Technol,* vol. 11, no. 9, pp. 1486-94.

Tomkos et. al. (2001a). Demonstration of negative dispersion fibers for DWDM metropolitan Area networks, *IEEE J. of Select. Top. in Quan. Elec.* vol. 7, no. 3, pp. 439-60.

I. Tomkos, I. Roudas, R. Hesse, N. Antoniades, A. Boskovic, R. Vodhanel. (2001b). Extraction of laser rate equations parameters for representative simulations of metropolitan-area transmission systems and networks. Optics-Communications. 194(1-3): 109-29

I. Tomkos, R. Hesse, R. Vodhanel, and A. Boskovic. (March 2002). Metro Network Utilizing 10-Gb/s Directly Modulated Lasers and Negative Dispersion Fiber. *IEEE Photon. Technol. Lett.,* VOL. 14, NO. 3.

L.-S. Yan, C. Yu, Y.Wang, T. Luo, L. Paraschis, Y. Shi, and A. E. Willner.(2005). 40-Gb/s Transmission Over 25 km of Negative-Dispersion Fiber Using Asymmetric Narrow-Band Filtering of a Commercial Directly Modulated DFB Laser. *IEEE Photon. Technol. Lett.*, vol. 17, no. 6, pp. 1322-1324.

5

Advanced Modulation Formats and MLSE Based Digital Signal Processing for 100Gbit/sec Communication Through Optical Fibers

Albert Gorshtein and Dan Sadot
Ben Gurion University of the Negev
Israel

1. Introduction

Modern telecommunications infrastructure is based on optical fibers for data transmission, due to their higher bandwidth and low attenuation. Today, optical fiber is starting to reach apartment buildings and even private homes, and delivers wideband services such as Triple Play (a Triple Play service bundles high-speed Internet access, television, and telephone service over a single broadband connection).

There is a real growing need for data rates of 100Gbit/sec and beyond in the access, enterprise, metro, regional and long haul markets. Therefore, next generation transmission systems must be employed, otherwise the optical infrastructure will become overloaded and exceed its current capacity.

Increasing the bit rate, in turn, makes a transmitted data signal be more vulnerable to transmission impairments inherent to an optical fiber, e.g., chromatic dispersion (CD), polarization mode dispersion (PMD), fiber non-linearity (FNL), concatenated optical filtering etc., and noises produced by optical amplifiers and receiver front end electronics. In particular, increasing the bit rate to 100Gbit/sec and above, based on intensity modulation (IM) with direct detection (DD) alone and using a simple hard decision (HD) scheme, leads to dramatic reduction in transmission range and system performance degradation. The main reason for the latter two phenomena is significant shortening of the pulses representing the transmitted bits, which implies broadening of the signal spectrum. It is a well known fact that, the higher the bandwidth of the signal is, the more it suffers from transmission impairments. For example, both CD and PMD cause inter-symbol interference (ISI), making it more difficult to distinguish between the transmitted levels of logical one and logical zero.

Therefore, reliable transmission over sufficient distance at such a high rate (100Gbit/sec and above), can be achieved by both using digital signal post-processing (DSP) and employing advanced modulation formats. Moreover, the quality of the received (and processed) signal may be significantly improved by using (digital) forward error correction (FEC).

Each of these three techniques strives to improve the system performance in its own way. Using advanced modulation formats typically results in a narrower transmitted signal spectrum, effectively increasing the immunity to ISI. This can be achieved, for example, by transmitting more than one bit in each symbol. The main drawback of this approach is, the more symbols are in the alphabet (the more bits are in each symbol), the harder for the receiver to distinguish between them. Furthermore, the more complicated chosen modulation format is, the more complex and expensive the electronic and optical component at both transmitter (Tx) and receiver (Rx) side are.

The DSP reduction of ISI can be divided into two main categories. The first one, termed equalization, uses linear (feed forward - FFE) filters and already made decisions (decision feedback – DFE), in order to undo the combination of all the distortions that caused ISI, according to a predetermined criterion, such as zero forcing (ZF), minimum mean square error (MMSE) etc. Typically such equalizers operate in an adaptive manner, updating the filter(s) coefficients from time to time, such that the adaptation rate is greater than the temporal changes in the channel. As an example, the CD phenomenon is constant for a given link, whereas the PMD effect is stochastic in nature, when the amount of PMD may vary over time constants of 1μsec to 1msec. The second category is called maximum likelihood sequence estimation (MLSE), which uses a different approach. MLSE exploits the statistics of the received signal, whereas the decisions are made on the whole transmitted sequence rather than on every single bit. While the complexity of the MLSE is significantly greater than of linear and nonlinear equalizers from the first category, it is proven to be the best optimal tool for combating ISI in a communication system (Proakis, 1995). MLSE can also be made adaptive, if the statistics that is used for decoding is updated fast enough to follow the changes in the channel. In this chapter we will focus on DSP, that implements MLSE decoders, rather than FFE and DFE.

Prior to performing the DSP, the received signal has to be transferred into the digital domain. This task is done by means of analog-to-digital conversion (ADC), that turns out to be a major bottleneck of communications systems, transmitting 100Gb/sec and more. The common practice is taking two samples per every symbol (also called oversampling) with successive equalization. Since the task of the decision device in the receiver is to estimate the transmitted symbols, it is enough to sample each symbol only once, but at the optimal sampling instants. Oversampling eliminates the sensitivity to the sampling phase, though requires more complex and power hungry ADCs. To the best of the authors' knowledge, the highest ADC sampling rate commercially available today is 64Gb/sec. Even ADCs with the rates of 28-32Gb/sec are still state of the art nowadays. Therefore, it is important to put more effort in the digital domain, and develop such post processing algorithms that enable reduction of sampling rate. Moreover, sampling the received signal at the symbol rate without preceding filtering violates the Nyquist sampling criterion, causing aliasing effect that results in performance degradation. On the other hand, using anti aliasing filter (AAF) prior to symbol rate sampling introduces heavy low pass filtering (LPF) which, in turn, causes heavy inter symbol interference. In this chapter we propose symbol rate sampling with the use of AAF, followed by equalization of transmission impairments and MLSE to compensate for ISI.

The immunity of any communication system can be increased at the expense of adding a few percents (typically 7% to 30%) of overhead symbols that are not part of the information that is needed to be transmitted. This effectively increases the transmitted bit rate and hence the bandwidth of the transmitted signal. This overhead, called code bits, is used by the FEC system to detect and correct errors. Common FEC schemes available today are capable of decreasing the bit error rate (BER) from 10^{-2} or 10^{-3} to 10^{-9} or even 10^{-12}, depending on many factors such as amount of overhead, the coding rules (or coding gain), number of iterations and so on. However, FEC needs a certain BER at its input, in order to be capable of carrying out the correction, called pre-FEC BER. Throughout this chapter, it is assumed that the transmission system incorporates FEC, thus most of the results will be presented in a form of optical signal to noise ratio (OSNR) or optical input power that is required to achieve a pre-FEC BER value of 10^{-3}.

Fiber optics communication systems can be categorized by many different factors like transmission distance, bit rate, modulation and detection scheme, post processing and so on. However, the categories presented above, are not entirely independent from each other. As in any other communication system, the appropriate combination of the above characteristics depends on a particular application. We roughly divide the systems by the required transmission distance: "extended short reach" (up to 80km), metro (up 1000 km) and long-haul systems (more than 1000km). Depending on this division, we propose the system configuration for 100Gb/sec transmission, including modulation format and successive MLSE based DSP, and analyze the expected system performance.

This chapter is organized as follows. Section 2 will briefly discuss the MLSE processing principles. Sections 3, 4 and 5 will present the aforementioned systems analysis, associated DSP algorithms and expected performance of the proposed modulation formats for short reach, metro and long-haul systems respectively. In section 6 we will draw out the conclusions of the presented work.

2. Optical fiber as a communication channel

Propagation of electrical field in the optical fiber can be described by the *non-linear Schrodinger equation* (Agrawal, 2002):

$$\frac{\partial E(t,z)}{\partial z} + \frac{j}{2}\beta_2 \frac{\partial^2 E(t,z)}{\partial t^2} + \frac{\alpha}{2}E(t,z) - j\gamma\left|E(t,z)\right|^2 E(t,z) \tag{1.1}$$

where $E(t,z)$ is the incoming electrical field, α is an attenuation coefficient, γ is a non-linearity coefficient and β_2 is a group velocity dispersion (GVD) coefficient, accounting for chromatic dispersion. When the total input optical power to the standard single mode fiber (SSMF) is not too high (less than 5dBm) the non-linear term in (1.1) is negligible, and it may be assumed that the fiber is operating in the linear regime. Under this assumption, (1.1) is reduced to a *linear Schrodinger equation*:

$$\frac{\partial E(t,z)}{\partial z} + \frac{j}{2}\beta_2 \frac{\partial^2 E(t,z)}{\partial t^2} + \frac{\alpha}{2}E(t,z) = 0 \tag{1.2}$$

Thus, the channel can be modeled with help of transfer functions accounting for two effects: CD and loss.

2.1 Chromatic dispersion

The essence of chromatic dispersion is in the fact that different frequency components of the transmitted signal travel through the fiber at different velocities. The effect of chromatic dispersion is deterministic since it is completely defined by physical parameters of the fiber. The baseband transfer function, describing the CD phenomena for a fiber of length L is obtained by solving (1.2) for $\alpha = 0$ (assuming that loss is compensated by proper optical amplification), yielding:

$$H_{CD}(f) = \exp\left\{-j\beta_2 (2\pi f)^2 \frac{L}{2}\right\} \tag{1.3}$$

Any type of modulated data has a nonzero spectral width, due to the presence of information, which occupies a range of frequencies that is roughly of the same order of magnitude as the bit rate itself. These different spectral components of modulated data travel at different speeds down the fiber. In particular, for digital data modulated on an optical carrier, CD leads to pulse broadening, which, for a given bit rate and modulation format, limits the distance that signal can propagate through, and the appropriate levels can still be distinguished. In other words, the effect of chromatic dispersion is cumulative and increases linearly with transmission distance.

Moreover, it can be shown that this maximal transmission distance L_D (without compensation) is proportional to the ratio of symbol duration T_s and the signal bandwidth BW:

$$L_D = \frac{T_s}{k \cdot D \cdot BW} \tag{1.4}$$

where the coefficient k depends on the allowed performance penalty and modulation format, and D is a dispersion parameter that related to β_2 as follows:

$$D = -\frac{2\pi c}{\lambda_0^2} \beta_2 \tag{1.5}$$

λ_0 being the wavelength of the optical carrier. Given the bit rate and modulation format, the amount of chromatic dispersion depends entirely on the physical characteristics of the channel: the dispersion parameter D and the fiber (or uncompensated span) length L. Therefore, based on (1.4) it convenient to combine these two together, and measure the amount of chromatic dispersion in the link by the CD parameter, defined as:

$$CD \triangleq D \cdot L = \frac{T_s}{k \cdot BW} \tag{1.6}$$

In the sequel, the performance of the examined systems will be presented as a function of CD.

The compensation of chromatic dispersion can be done either in the optical or in the electrical domain. Sometimes, the combination of both techniques is used. In the optical domain the compensation is typically performed with help of dispersion compensation fibers (DCF) and/or dispersion compensation modules (DCM), both having a dispersion parameter with the opposite sign to the corresponding SSMF span to be compensated. DCF and DCM provide constant CD compensation. For 10Gb/sec systems they are typically placed every 80km in the channel, and introduce additional loss, which has also to be compensated by optical amplification. The last span length, however, may vary, depending on specific geography and topology of the field. Hence, for this last lag, two options exist. First one is to put a more complex (and expensive) tunable dispersion compensator (TDC). And a second one is to design the system such that no compensation is required or alternatively, the compensation is carried out in the electronic domain.

2.2 Polarization mode dispersion

As a result of the fiber's birefringence different modes of polarization split and travel in different velocities causing pulse spreading. The effect of light propagation in a birefringent fiber is generally termed polarization mode dispersion (PMD). Birefringence occurs due to changes in fiber geometry (symmetry), which may be caused by either manufacturing imperfections, temperature variations or mechanical vibrations. Therefore, on contrary to CD, PMD has a stochastic nature. In case of polarization multiplexed transmission the effect of PMD can be summed up in the following things. In addition to different delays, the transmitted electrical field undergoes polarization rotation, meaning that the electrical field at the fiber output may have a different state of polarization (SOP), (Kaminow & Li, 2001). Hence, when looking at the received electrical fields in some fixed SOP, there is a linear combination of the incoming fields from both polarization modes. In other words, PMD results in intra-channel crosstalk that cause a pulse spreading. In fact, in this case an SSMF can be viewed as multiple input – multiple output (MIMO) system (with two inputs and two outputs). Therefore, it can be described by the transfer function matrix (in the frequency domain), called the Jones matrix, which in the absence of polarization dependant loss and in case of first order PMD has the form (Ip & Kahn, 2007):

$$\mathbf{H_{PMD}}(f) = \begin{pmatrix} \cos\psi & \sin\psi \\ -\sin\psi & \cos\psi \end{pmatrix} \begin{pmatrix} e^{j2\pi f\frac{\tau}{2}} & 0 \\ 0 & e^{-j2\pi f\frac{\tau}{2}} \end{pmatrix} \begin{pmatrix} \cos\psi & -\sin\psi \\ \sin\psi & \cos\psi \end{pmatrix} \tag{2.1}$$

where τ is the differential group delay (DGD) parameter and ψ is the angle between the SOP of the electrical field at the fiber input and the principal state of polarization (PSP) of the fiber (Kaminow and Li, 2001). Thus, the capacity of transmitted information can be doubled, when sending two independent data streams on different polarization modes of the SSMF.

In systems, where the information is modulated only on a single polarization of the optical fiber, equation (2.1) reduces to (Kaminow and Li, 2001):

$$H_{PMD}(f) = \sqrt{\rho}\exp\{-j\pi f\tau\}\cdot\hat{x} + \sqrt{1-\rho}\exp\{+j\pi f\tau\}\cdot\hat{y} \tag{2.2}$$

where ρ represents the power splitting ratio. Moreover, if in addition to single polarization modulation the signal is detected directly with help of photo-detector (PD), the resulting photo-current would take the form:

$$I_{el}(t) = R\cdot\|\mathbf{E_r}(t)\|_2^2 = R\cdot|E_{r,x}(t)|^2 + R\cdot|E_{r,y}(t)|^2 \tag{2.3}$$

where R represents the PD responsivity, $E_r(t)$ is the incident optical field vector having non-zero $\hat{x}$ and $\hat{y}$ components $E_{r,x}(t)$ and $E_{r,y}(t)$ respectively.

2.3 Optical amplification and amplified spontaneous emission noise

To compensate for the attenuation described by α in (1.2), especially in transmission for long distances (long-haul), the propagated optical field is generally amplified by means of an erbium-doped fiber amplifiers (EDFAs) every L_A kilometers. For example, in a SSMF having $\alpha = 0.2\left[\dfrac{db}{km}\right]$, the EDFAs need to be placed every $L_A = 80[km]$ for 10Gbps transmission.

In addition to the beneficial high gain that remains constant for relatively large wavelength range, EDFA produces and amplifies a spontaneous emission, which is undesirable and can be treated as noise, generally referred as ASE noise. The power spectral density (PSD) of the ASE noise is nearly constant and given by (Agrawal, 2002):

$$S_n(f) = \frac{N_0}{2} = \frac{1}{2}n_{sp}\frac{hc}{\lambda_0}\frac{(G-1)}{G} \tag{2.4}$$

where n_{sp} is the spontaneous emission factor (or population inversion factor), G represents the EDFA gain, $\dfrac{hc}{\lambda_0}$ is the photon energy at the wavelength λ_0, corresponding to the central frequency of the optical carrier f_c, h being the Plank constant and c being the speed of light. The ASE noise in optically amplified systems, particularly in those that have more than one EDFA in the link, is dominant over all other types of noises, produced in the receiver (thermal noise, shot noise, dark current noise), due to the repetitive generation and amplification of ASE noise in each span. To reduce the effect of ASE, optical filters, centered at λ_0 with bandwidth B_O, are used.

3. MLSE processing principles

This section presents MLSE processing principles suitable for fiber optical communication systems. Especially, due to the presence of PMD the optical channel is considered to be non-stationary and adaptive equalization is required. When the histograms, which serve as channel estimators, are updated faster than channel variation rate, successful variations tracking can be achieved. Moreover, since channel estimation without resorting to training sequence is beneficial, a novel blind channel estimation technique is proposed.

3.1 General decoding principle

Maximum likelihood sequence estimation (MLSE) is considered to be non-linear equalization technique. To explain the main idea that is behind the MLSE processing trellis diagrams are often used. The example of 4-state trellis diagram is presented on Fig. 1.

In this example two bits of the channel memory are assumed, i.e. current sample is affected by two previous and the current bit. The two previous bits define the channel state, and a conjunction of a state with a current bit defines a branch. The arrows in the trellis represent

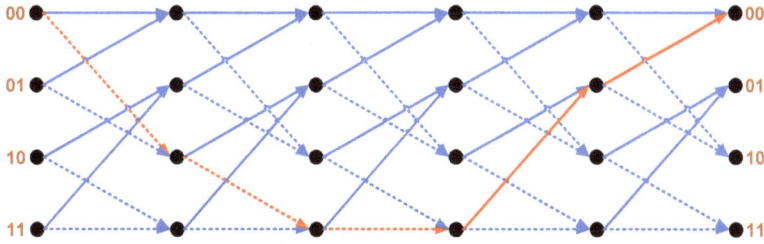

Fig. 1. 4-state trellis diagram example

the transition from one state to another, while a transition corresponding to a '0' bit is drawn with solid line and a transition corresponding to a '1' is represented by a dashed line. To each branch in the trellis we assign the number called "branch metric", which is dependent on channel and noise statistics. Branch metric describes in some sense the probability of a corresponding transition. Furthermore, one can define a "path metric" which is the sum of corresponding branch metrics for a certain path, when a bit sequence with a smallest path metric is the most probable to be transmitted.

The key idea of the MLSE processor is to choose the path with the smallest metric, and produce the most likely sequence by tracing the trellis back (Proakis,1995). For sequence of length N there are 2^N possible paths in the trellis, therefore an exhaustive comparison of the received sequence with all valid paths is inefficient task, becoming non feasible for channels with long memory. However, noting that not all paths have the similar probabilities (or metrics) when proceeding through the trellis, there exists an efficient algorithm, called Viterbi algorithm, that limits the comparison to 2^K "surviving paths", K being the channel memory length, independent of N, making the maximum likelihood principles to be practically feasible. The main idea of the Viterbi algorithm: from two paths entering the trellis node, the path with the smallest metric is the most probable. Such a path is called the "surviving path", and we need to store only surviving paths with their running metrics.

3.2 Metrics generation and update

Maximum likelihood sequence detection is the proven to be the most effective technique for mitigating optical channel impairments such as chromatic dispersion and polarization mode dispersion (Foggi et al., 2006). In order to successfully apply this technique, it is mandatory to estimate some key channel parameters needed by the Viterbi processor.

Channel estimation methods can be classified as parametric and non-parametric. Parametric methods assume that the functional form of the probability density function (PDF) is known, and only its parameters need to be estimated. However, non-parametric methods do not assume any knowledge about the PDF functional form or its parameters. There are two most common methods used in practice for channel estimation: *method of moments (MoM)* and *histogram estimation method*.

Method of moments is considered to be parametric, therefore it assumes that the functional form of the PDF is known and only its moments need to be estimated. When the dominant noise mechanism in the optical system is thermal, like in optically unamplified links, the conditional PDF of the received sample x_n, given that μ_k is transmitted, is assumed to be Gaussian with σ_n^2 being the variance of the noise (Proakis, 1995):

$$f_{channel}^{Gaussian}(x_n \mid \mu_k) = \frac{1}{\sqrt{2\pi\sigma_n^2}}\exp\left\{-\frac{(x_n - \mu_k)^2}{2\sigma_n^2}\right\} \tag{3.1}$$

In this case, only first and second moments need to be estimated. Another case of interest is ASE limited channel. As stated above the noise in such a channel becomes signal dependent and the functional form of the conditional PDF of the received sample x_n, given that μ_k is transmitted, can be approximated by a non-central Chi-square distribution with v degrees of freedom (Agazzi et al., 2005):

$$f_{channel}^{ASE}(x_n \mid \mu_k) = \frac{1}{N_0}\left(\frac{x_n}{\mu_k}\right)^{\left(\frac{v-1}{2}\right)}\exp\left\{-\frac{x_n + \mu_k}{N_0}\right\}I_{v-1}\left\{2\frac{\sqrt{x_n\mu_k}}{N_0}\right\} \tag{3.2}$$

where N_0 is power spectral density of the ASE noise given by (2.4), and $I_{\{\}}$ is the modified Bessel function of the first kind. It is clear that in (3.2) N_0 and v need to be estimated. More information about parametric estimation is given in (Agazzi et al., 2005; Foggi et al., 2006).

Histogram method does not assume anything about the PDF of the received samples. In this method $M^{N_{isi}+1}$ histograms are collected, where M represents the vocabulary size of the transmitted symbols and N_{isi} is the number of the most resent previous symbols that affect the current symbol, i.e. the channel memory length assumed by the algorithm. The received signal is assumed to be quantized to N_{ADC} bits; therefore each histogram consists of $2^{N_{ADC}}$ bins, where N_{ADC} is a design parameter. Notice that each histogram can be uniquely associated with a branch in the trellis diagram of the receiver. Assuming that the number of signal samples collected is large; the histogram (normalized so that the sum of all its bins is unity) is an estimate of $f_{channel}(x_n \mid \mu_k)$. The histogram is updated iteratively, based on the observed samples and the decision bits at the output of the MLSE decoder.

Finally, branch metrics are obtained by taking the natural logarithm of the estimated/assumed PDF. For a transmitted sequence of length N the MLSE decoder chooses between M^N possible sequences that minimize the (path) metric:

$$m_r = \sum_{n=1}^{N} -\ln\left\{f_{channel}(x_n \mid \mu_k)\right\} \tag{3.3}$$

The estimated bit sequence is determined by tracing the trellis (like in Fig. 1) back, based on the minimal path metric (3.3).

In optical fiber systems, the purpose of the MLSE is to combat ISI stemming from CD and PMD. While CD is a deterministic phenomena for a given link, PMD is stochastic in nature, therefore an adaptive equalizer, that performs PMD tracking is required. Moreover, the adaptation properties of the MLSE can be also exploited for CD compensation when the amount of CD is not perfectly known. Basically, as will be explained in the next sections expensive tunable optical dispersion compensation may be replaced by the adaptive MLSE.

This type of operation, without knowing any initial information about the channel is termed blind equalization.

The operation of the proposed blind MLSE equalization scheme can be divided into two main modes: initialization and steady state operating mode. At the initialization mode, there are two main tasks to be fulfilled. First, initial coarse histograms estimation representing the channel is required. This can be achieved by using a set of predetermined histograms representing different channels, as will be explained in the next subsection.

Starting from this initial "guess", an iterative procedure is activated in order to fine tune the channel estimation. At each iteration data decoding is performed by applying the Viterbi algorithm based on a previous estimation of the channel. This iterative histograms estimation is obtained in a decision directed manner; it is assumed that the receiver operates at a sufficiently low BER. The calculation of the channel estimator (histograms) for the next step is then based on the observation-decision pairs available from the current iteration. Consequently, the histograms are obtained from current observation set, while the decoded symbols are used as a 'training sequence' for the attribution process. This process is repeated until a convergence criterion is being met. Even though theoretically the convergence in a decision directed mode is not guaranteed, in practice the proposed method was found to be extremely robust as it converges in a very wide set of channel conditions.

To ensure sufficient tracking, the histograms must be updated fast enough as compared to temporal variations of the channel. Since PMD changes in the scale of $100\mu\sec \div 1\text{msec}$, the histogram adaptation rate must be at least ten times faster, meaning that every $10\mu\sec$ a new set of metrics must be obtained.

3.3 Initial metrics

The key function that enables the *blind* MLSE processing is the proper choice of the initial metrics. In fact, since only *coarse* channel representation is needed, it may be assumed that the branch histograms have nearly Gaussian shape and differ from each other only by the mean and variance. The mean values depend on the channel memory length and the vocabulary size in the Rx side M. To ensure proper operation, the decoder must be designed such that the channel memory length K is at most as the memory length of the decoder N_{isi}. In this case there are $M^{N_{isi}+1}$ branches, whereas the variance of each histogram is associated with the noise power that is present in the corresponding combination describing the branch. In memoryless channel with binary vocabulary there are two histograms, representing the corresponding conditional PDFs, and simple hard decision scheme can be used. When $M = 2$ and channel memory of one symbol $(K = 1)$ there are four distinct histograms, having four different mean values. Generally, when $K < N_{isi}$, the actual number of histograms in the given MLSE decoder is constant, $M^{N_{isi}+1}$, and consists of M^{K+1} different groups, while all the members of such a group are identical. Continuing the example, if $N_{isi} = 4$ there are 32 branches, which can be divided into 4 groups, associated with the four different mean values mentioned above. Thus, the set of initial metrics for the given M and N_{isi} is finite and easy to obtain by distributing M^{K+1} means uniformly, and leaving the fine tuning to the iterative convergence process. Another factor that increases the number of initial metrics in the bank is the variance of the branches. Taking into account that assigning higher variance than actual does not affect the performance, the number of initial metrics can be further reduced by assigning the same (high) variance to all the branches.

Initial metrics determination is summarized on Fig. 2.

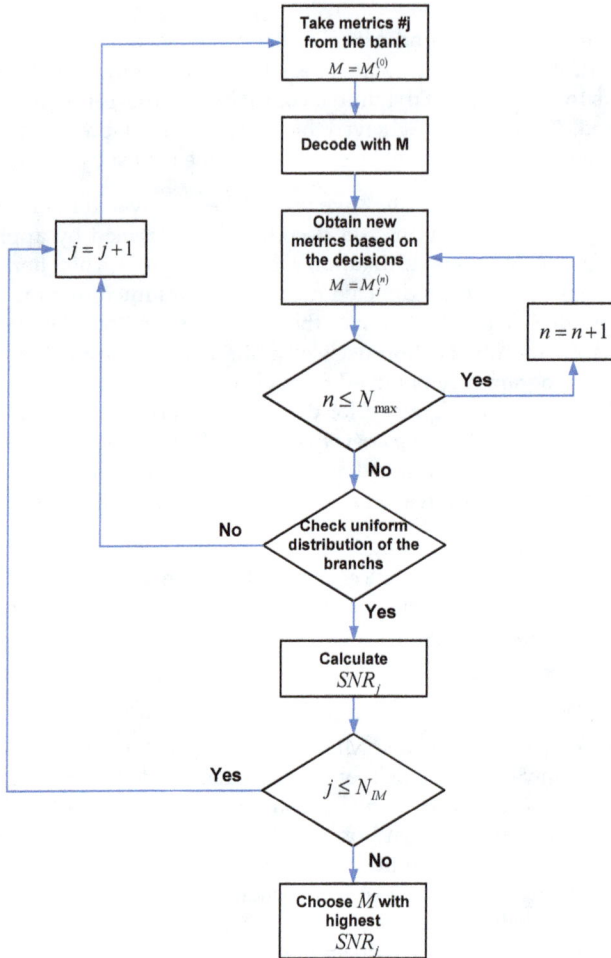

Fig. 2. Initial metrics determination procedure

The initialization phase typically needs to be done "once in a lifetime" during system start up, or after reset, and typically requires only a few iterations, where each iteration uses a couple of thousands of symbols. Once convergence is achieved, the equalizer switches to the steady state operating mode, where the histograms are updated continuously, based on the previously made decisions. In order to verify that the system remains in steady state mode, a uniform distribution of Viterbi trellis branches, represented by the histograms, is required. In addition, a sufficiently low BER value condition must be met during this operating mode. Otherwise, an interrupt signal is generated and sent to the control processing unit (CPU).

To choose the best histogram of all the initial metrics in the bank is the best, the following signal-to-noise ratio (SNR) criterion is used:

$$SNR = \frac{\sum_{i=1}^{N_{br}} \mu_i^2 - \frac{1}{N_{br}}\left(\sum_{i=1}^{N_{br}} \mu_i\right)^2}{\sum_{i=1}^{N_{br}} \sigma_i^2} \qquad (3.4)$$

In (3.4) μ_i and σ_i^2 represent the mean and the variance of branch number i, which are estimated from the histograms, and N_{br} is the total number of branches (histograms). Actually, this is the same criterion as in (Gorshtein et al., 2010), but shown to be working with much wider set of channels describing a signal coming either from coherent or direct detection scheme.

4. "Extended short reach" transmission

4.1 Introduction

In most common 10Gb/sec systems the link length is limited by chromatic dispersion and stands on 80km. When longer reach is desirable, dispersion compensation fibers are used, and the system becomes power limited. This power limitation is been overcome by optical amplifiers, which are typically placed every 80km, and compensate for both attenuation in the fiber and the DCF together with insertion loss (IL) of the DCF. A short reach transmission system is defined here, as a system that does not require any optical amplification. Thus, due to CD limitation typical length of the link (at 10Gb/sec transmission) is about 80km.

Generally, there are two main reasons that lead to a degradation of signal quality: ISI and noise. Due to the lack of optical amplifiers in short reach systems, the dominant noise mechanisms are thermal noise and shot noise (Agrawal, 2002). Both noises are independent of the optical channel length and added to the received signal during the optoelectronic conversion. It is worth to note, that noise, is always present in the frequency band occupied by the signal, and cannot be eliminated or separated from the signal by any means, without hearting the signal itself.

However, the other source of received signal quality degradation, ISI, can be theoretically eliminated completely (given no noise), by employing the MLSE technique, described in the previous section. Moreover, the ISI, being generated mainly by the channel itself (CD, PMD), is an increasing function of fiber length (Agrawal, 2002). Since the received signal quality scales with distance, particularly in short range transmission systems it is desirable to reduce the overall system cost as much as possible. Therefore, it is preferable to use inexpensive optics, leaving in the 'battlefield' only the most simple intensity modulation and direct detection (IMDD) systems. In fact, for a given data rate (of 100Gb/sec in our case), tradeoff can always be done between the overall system cost and complexity versus performance and spectral efficiency.

From equation (1.4) it is clear that the less the bandwidth of a digital signal at a given bit rate, the longer the propagation distance in the fiber. Based on this observation, it is proposed to reduce the bandwidth of the transmitted signals at the transmitter side with use of previous generation 10G optical components. Furthermore, since the receiver bandwidth is lower, less noise will be added to a signal. In addition, 10G avalanche photo diode (APD) can be used for better sensitivity. Note, that to the best of the authors' knowledge APDs for higher transmission rates still do not exist (at least up to June, 2011). However, these potential benefits do not come for free, since such heavy low-pass filtering by itself introduces ISI. To compensate for this effect, together with the ISI stemming from CD and PMD, the MLSE decoder is used in each link. Due to such a short reach and relatively wide transmitted pulses the PMD tolerance is of the second order and only CD tolerance is investigated.

To achieve a target of 112Gb/sec (including 12% FEC overhead), a four-wavelength transmission is proposed, where each lambda (wavelength) carries a data of 28Gb/sec. In light of the aforesaid, the following novel modulation formats (with MLSE detection) will be examined: reduced bandwidth on-off-keying (RBW-OOK) and reduced bandwidth duobinary (RBW-DB) modulation formats.

4.2 System model

Block diagram of the proposed system is presented in Fig. 3.

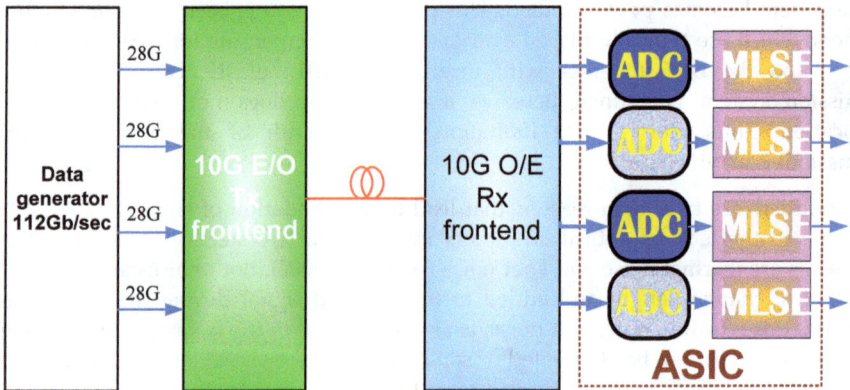

Fig. 3. High- level block diagram of the "extended short reach" transmission system

Four parallel lanes, 28Gb/sec each, are transmitted with help of four different carrier waves, designated by λ_1 to λ_4. The electro-optic (E/O) transmitter (Tx) frontend consists of four identical pairs of *10G electrical drivers and 10G optical modulators*, whereas the modulator type depends on the modulation format.

RBW-OOK can be generated either by modulating the *10G* laser directly or by keeping the laser producing a continuous wave (CW) and using either an external (*10G*) electro-absorption modulator (EAM) or (*10G*) Mach-Zehnder modulator (MZM). DB modulation format is generated with help of *10G* MZM biased at null point. However, there are additional degrees of freedom: the bandwidth and order of the combined filter representing the cascade of modulator and driver filters. The resulting four signals are multiplexed into a short fiber link.

To achieve higher transmission distance with OOK formats chirped MZM or EAM with chirp parameter $c = -0.7$ and $c = -1.2$ are used. In the DB case the chirp has a destructive effect on the encoded signal, hence a zero chirp MZM is used.

The optical receiver (Rx) front end is the same for both OOK and DB. First, the four sub-channels are demultiplexed from the link, and then opto-electronic (O/E) conversion is carried out by four (identical) photodiodes. The resulting photo-currents are sampled at *28Gsamples/sec* rate and quantized with four 5-bit resolution analog-to-digital converters (ADCs). Then MLSE detection, with channel memory of 4, is performed on each lane (in parallel) to recover the transmitted bits. These 4 pairs of ADC and MLSE can be implemented on a single application specific integrated circuit (ASIC).

4.3 Simulation results and discussion

Fig. 4 presents the required optical power (at the Rx input) as a function of SSMF length (with $D = 20\left[\dfrac{ps}{nm \cdot km}\right]$).

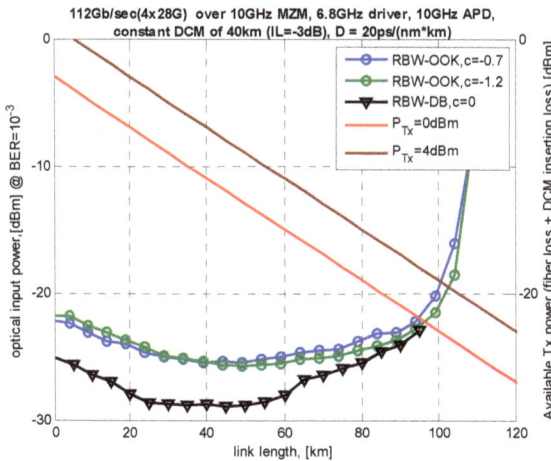

Fig. 4. Required input power and available power vs. link length for RBW-OOK and RBW-DB transmission

The red and brown straight lines show the available optical input power at the APD input for 0dBm and 4dBm transmission respectively. A constant DCM, compensating for 40km fiber with insertion loss (IL) of 3dB is used. It is clear that chirped RBW-OOK (with 10G components at the both sides) allows 80-90km transmission (without amplification), provided that 40km DCM and FEC are used. Moreover, the available power is between 5-9dB higher than the required one. The main reason for this improved sensitivity is using the APD instead of PIN photodiode. The MLSE is capable to compensate for ISI introduced by the reduced bandwidth components and the residual portion of CD that is not compensated optically by DCM.

On contrary to the OOK formats, chirping the duobinary signal will not give any benefit for combating CD. Furthermore, this chirping, in fact, serving as a phase distortion, may even harm the DB signal (Røyset & Hjelme, 1998). In fact, since the transmitted information is also encoded in the phase of the transmitted optical field, chirping crucially destructs the signal structure. Therefore, in the case of DB transmission $c = 0$ is being used. The black curve on

Fig. 4 shows that non-chirped RBW-DB modulation format has a superior performance over both chirped RBW-OOK cases.

As a matter of the fact, the ordinate values presented on Fig. 4, cannot be considered as the absolute values, but rather the representative ones. It is a well known fact, that the required optical power that ensures some nominal BER, in optically non-amplified system is directly related to the sensitivity of the photo-detector being used. Hence, to achieve more accurate results in a specific system, concrete APD parameters should be considered. The above scenario was obtained with APD, having a sensitivity of -30dBm (for $BER = 10^{-9}$), which is typical for 10G APDs.

4.4 Section summary

In this section, a simple and low cost solution for "extended short reach" 112Gb/sec transmission was proposed. To achieve 112Gb/sec 4 different wavelength slots are occupied, whereas the data rate in each slot is 28Gb/sec (including 12% FEC overhead). Two novel modulation formats: RBW-OOK and RBW-DB are examined, where *low cost 10G* components (instead of 28G) are used, and MLSE detection technique is applied to recover the data from each sub-channel (slot). The MLSE compensation is carried out in the digital domain, when the 4 signals are digitized using four 5-bit ADCs, operating at 28Gsamples/sec (sampling at the baud rate). Due to its higher sensitivity, 10G APD O/E front end is used, and a constant DCM compensating for 40km of SSMF was applied. The ADC and MLSE functionality can be effectively implemented on a single ASIC.

Due to such a low transmission distances, major goal of a system design is cost and complexity reduction, at the expense of sub-optimal, but still applicable, performance. Although RBW-DB shows slightly better (2-4dB) performance than chirped RBW-OOK, it requires a more complex and highly accurate transmitter in order to construct the correct DB signal, and a more expensive Mach-Zehnder modulator to allow modulation of the optical phase and zero chirp. MLSE can compensate for the lack of accuracy of the optoelectronic components, e.g., mismatches of cutoff/order of the DB driver and filter, in addition to the CD compensation.

Since this solution is based on 10G components it can be directly implied on the already existing systems, by replacing the simple HD slicer, by the aforementioned ASIC. Due to the use of the 10G APDs, this is the only kind of 4-wavelength 112Gb/sec transmission system that allows extended reach of up to 80km of an SSMF, based on low cost 10G components.

5. Metro transmission

In metro and regional fiber links, the length of the optical channel typically reaches no more than 1000km. Even with fiber's relatively small loss ($\alpha = 0.2 \left[\dfrac{dB}{km} \right]$ for a SSMF), the signal cannot propagate through it without optical amplification. Moreover, at such high rates the uncompensated transmission can be done only for a few kilometres, hence CD compensation is required. CD compensation can be done in the optical domain or in the electronic domain or a combination of both. In the sequel, two metro- transmission systems are proposed. The first can be viewed as an upgrade to the existing direct detection networks, while the latter is based on coherent detection and is suitable for future deployment.

5.1 Multi-wavelength 112Gb/sec transmission with direct detection and MLSE

5.1.1 System description

In most currently deployed systems (brown field), the former approach is used: the channel is divided into spans (or segments) of 80 km, whereas every span consists of a portion of a SSMF, followed by a DCF and an EDFA as shown on Fig. 5.

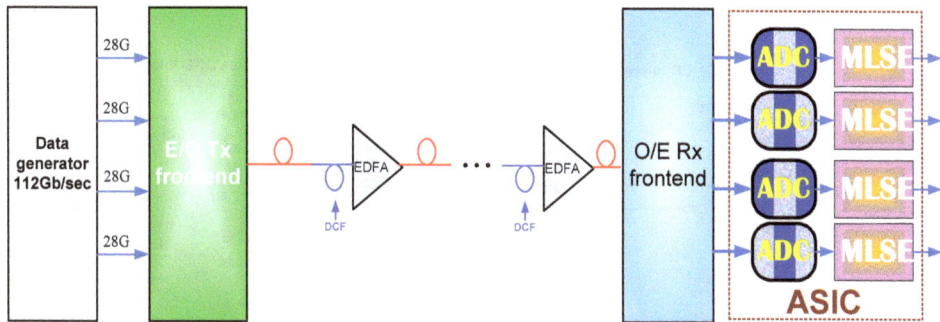

Fig. 5. Multi-wavelength 112Gb/sec direct detection metro transmission system

Pure optical compensation is not financially justified, since different channels have different lengths. DCFs compensate for a fixed amount of dispersion, whereas the residual CD is typically compensated with TDC. It is proposed to employ an electronic MLSE compensation, which adaptively tunes to the uncompensated residual portion of CD, rather than using an expensive TDC. The high level block diagram of the proposed 4-wavelength 112Gbps direct detection metro transmission system is presented on Fig. 5. The O/E and E/O frontends are identical to those from previous section, and simple RBW-OOK and RBW-DB modulation formats are examined.

5.1.2 Simulation results and discussion

To investigate the tolerance of the presented system to the *residual* chromatic dispersion and first order PMD, extensive Monte-Carlo simulations were done. 400,000 bits were randomized at each run to ensure sufficient statistics for a pre-FEC BER value of 10^{-3}.

CD tolerance versus required optical signal to noise ratio (OSNR) of the two examined modulation formats for different Tx and Rx filter configurations is presented in Fig. 6 (a). Conventional hard decision (HD) results (dashed lines) were also added for the sake of comparison.

It is clearly observed from Fig. 6 (a) that the combination of reduced bandwidth components at the Tx together with MLSE extends the tolerance to residual dispersion for all examined modulation formats, as compared to a corresponding conventional HD receiver. The main reason for this is the lower analog bandwidth of the transmitted signals. In addition, it is a well known fact, that DB modulation has inherent improved CD tolerance due destructive interference between the adjacent symbols, smeared by the CD.

However, at zero CD the required OSNR for the nominal BER in RBW-OOK is 4.5dB higher than in DB and RBW-DB. It can be explained by the fact that in DB the combined effect of differential precoding and heavy low-pass filtering at the Tx side is "undone" at the Rx side by the photo detector, effectively eliminating the major portion of (*controlled*) ISI introduced at the Tx. Basically, the square-law operation at the Rx side doubles the bandwidth of the signal, practically removing the *controlled* ISI. Thus, on one hand, in RBW-DB, the analog bandwidth of the received signal is reduced by the 10G frontend. On the other hand, the MLSE recovers the bits back, from this filtered signal.

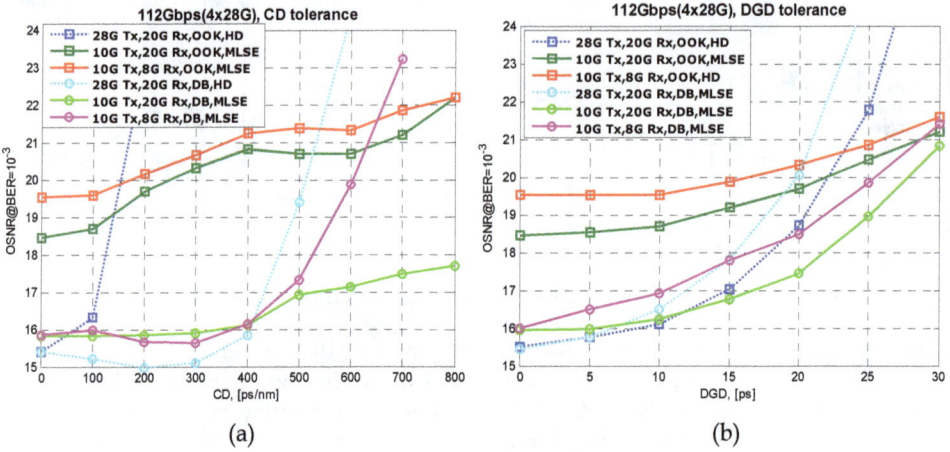

Fig. 6. (a) CD and (b) DGD tolerance of 112Gb/sec 4-wavelength direct detection system

For *CD* of up to $650 \left[\frac{ps}{nm} \right]$, when the overall signal memory length is less than 4 unit intervals (UI), which is also the MLSE memory depth, RBW-DB (magenta circle curve) is more tolerant to chromatic dispersion than RBW-OOK (red square curve). Moreover, CD tolerance can be further improved by extending the bandwidth in the Rx side, as can be seen in the square blue curve. This improvement stems from the fact, that FBW receiver preserves almost the full bandwidth, resulting from Pointing vector calculation during the O/E conversion, thus significantly reducing the overall signal memory. However, extending the Rx bandwidth in the RBW-OOK case provides minor improvement, since ISI introduced at the Tx side is *not controlled* in any manner at the Rx side, on contrary to its duobinary counterpart. In turn, the overall ISI is weaker in the FBW receiver case than in the RBW receiver case, thus the FBW receiver outperforms the RBW one.

DGD tolerance of the above modulation formats is presented on Fig. 6 (b). Similarly to CD tolerance behavior, it is observed that MLSE detection of all proposed modulation formats, provides better immunity to first order PMD, as compared to hard decision.

Fig. 6 (b) reveals that RBW-DB modulation formats are more tolerant to first order PMD, than RBW-OOK format. It is worth to emphasize, that since the ISI that results from PMD 'appears' in the signal only during the photo-detection process, both OOK and DB have to 'deal' with the same effect. However, since in duobinary case the controlled ISI is almost completely removed through square-law detection, the resulting waveform is 'cleaner' (from ISI), than its OOK counterpart. Increasing the analog bandwidth at the receiver, earns another 0.5-1dB of DGD tolerance in both cases.

5.1.3 Summary

In this sub-section, it was shown how to achieve efficient and low cost 112Gb/sec transmission in existing *10G* brown field systems. The tradeoffs between occupied bandwidth (4-wavelength slots), system performance and cost are introduced. It is proposed to use an existing, previous generation, low cost 10G opto-electronic components. Performance implications of this intentional bandwidth reduction are examined for the novel RBW-OOK and RBW-DB modulation formats. Parallel MLSE detection of the four lanes, sampled at 28Gsamples/sec is used, to compensate for both channel impairments (CD and PMD) and the reduced bandwidth optical components. It should be mentioned that in light of the presented performance, *additional constant DCM* module (corresponding roughly to 40km compensation) may still be needed, depending on the amount of residual CD in the link. In fact, a combination of MLSE and constant DCM may replace expensive TDC in current brown field networks.

5.2 "Coherent Metro": Single carrier 112Gb/sec transmission with coherent detection and MLSE

Sub-section 5.1 described a cost effective solution for upgrading existing (brown field) metro systems for 100G transmission, based on MLSE and low cost optical components. However, if a new fiber systems, (green field), an alternative detection scheme should be considered. The latter is called *coherent detection*, and in fact it is a more complex and expensive scheme. On the other hand, it introduces significant advantages. For example, since, on contrary to direct detection, in coherent detection the full information about the phase and the amplitude of the optical field is recovered, entire digital compensation of transmission impairments (e.g., CD and PMD) is possible. This, in turn, means that no optical compensation is required, the number of optical amplifiers is reduced, and the available OSNR in the link is increased. Moreover, higher order modulation formats can be used (without significant increase in cost of E/O and O/E components), resulting in higher spectral efficiency and better performance as compared to direct detection system.

Working at ultra high bit rates of 100Gb/sec and beyond, gives rise to new *bottlenecks* in the electronic domain: *high speed ADCs* and massive parallel *digital signal processing* (DSP) engines are required. To the best of the authors knowledge, the fastest ADC available today (2011) are of 65Gsamples/sec. According to Nyquist criterion, it is able to sample and fully recover signals with 32.5GHz analog bandwidth. Therefore, aiming to a single carrier transmission, the required bandwidth of the received signal should also be 32.5GHz. To fulfill these requirements for 100Gb/sec transmission, advanced modulation formats with richer vocabulary must be adopted.

It is a well known relationship between the vocabulary size and the analog bandwidth of the transmitted signal. The higher the vocabulary size M, for the given bit rate R_b, the lower is the symbol rate R_s (Proakis, 1995):

$$R_s = \frac{R_b}{\log_2 M}$$

(5.1)

Hence the bandwidth of the transmitted signal, which is proportional to the symbol rate, also (inversely) scales with the logarithm of M. On the other hand, when comparing between two modulation formats, one with small $M = M_1$ and the other with larger vocabulary size $M_2 > M_1$, given the same average transmitted power, the former will definitely perform better. The reason for this is that during the decision process, it is harder to distinguish between M_2 constellation points, spread over the constant region, than doing the same task for M_1 points. Thus, for a given bit rate, the optimal vocabulary size is determined by the lowest M, that provides the required analog bandwidth, as dictated by the maximal ADC sampling rate and the shaping pulse.

Another advantage of coherent detection is doubling the data rate per every channel slot. This is achieved by using the two orthogonal polarization axes to transmit two independent complex signals occupying the same frequency band simultaneously. Therefore, the optimal modulation format, in the sense described in the previous paragraphs, for 100Gb/sec transmission is dual-polarization quadrature phase shift keying (DP-QPSK) with $M = 16$. The conclusion of this theoretical discussion was experimentally proven for coherent fiber optical system by (Roberts et al., 2009) for a lower bit rate of 46Gb/sec.

Bearing in mind that the *raw* bit rate is FEC dependent, using the above 65Gsamples/sec ADC, one can transmit 4 independent lanes of net 25Gb/sec yielding 100Gbps with 30% FEC overhead assuming the oversampling ratio of 2. However, focusing on metro transmission ranges (up to 1000km), where the available OSNR in the link is higher than in long-haul links, significant relaxation of ADC sampling rate and DSP operating speed requirements can be achieved, compromising slightly in system performance. Firstly, the powerful, complex and power hungry 30% overhead soft-decision FEC (SD-FEC) can be replaced by the slightly weaker (Optical Interworking Forum [OIF], 2010) but simple 12% overhead hard-decision FEC (HD-FEC) with significantly smaller power dissipation. Secondly, since sampling the received signal at symbol rate (i.e. taking *only one sample per each symbol period*) forms sufficient information (or statistic) for successful recovery of transmitted data (Proakis, 1995), significant relaxation on the ADC operation speed can be achieved. In order not to violate Nyquist criterion, an anti-aliasing filter (AAF) must be used prior the ADC. This AAF introduces additional ISI, which cannot be optimally equalized by means of finite impulse response (FIR) filters. It is shown by (Gorshtein, et al., 2010) that using an MLSE compensates for the AAF effect, enabling the reduction by a factor of two of the ADC sampling rate and DSP processing speed, all together leading to significantly lower complexity and power saving.

5.2.1 System model

High level block diagram of the 112Gb/sec coherent DP-QPSK optical fiber communication system is presented on Fig. 7.

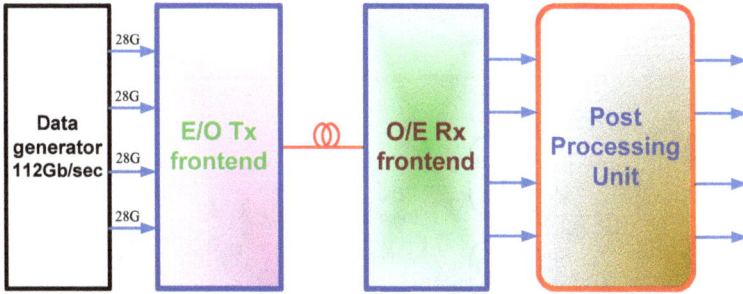

Fig. 7. 112Gb/sec DP-QPSK coherent detection metro transmission system

E/O Tx and O/E Rx frontends are depicted on Fig. 8 (a) and (b) respectively.

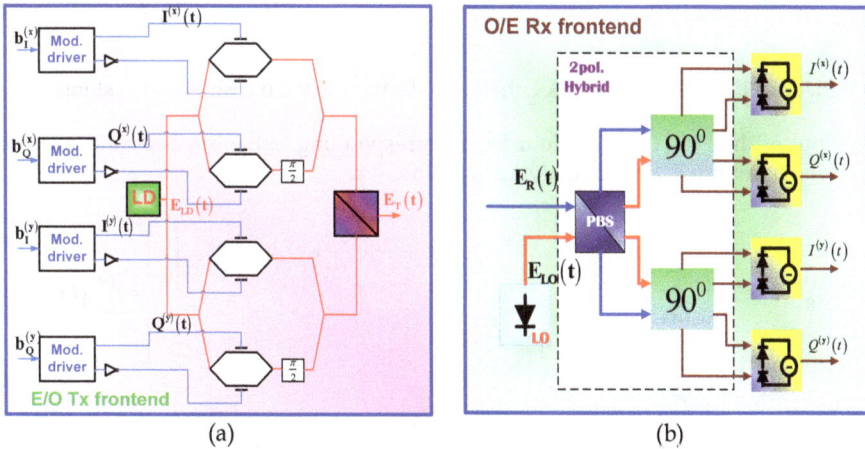

(a)   (b)

Fig. 8. 112Gb/sec coherent DP-QPSK (a) E/O Tx frontend and (b) O/E Rx frontend

Four independent 28G data streams are driving four MZM, operating in a push pull configuration. In two out of four MZM outputs additional $\frac{\pi}{2}$ phase shift is generated, resulting in two QPSK signals, which are combined by a polarization beam splitter (PBS) to construct a transmitted 112Gb/sec DP-QPSK signal, as can be seen from Fig. 8 (a). At the Rx side Fig. 8 (b), the received signal is mixed with the signal coming from local oscillator (LO). The mixing is carried out with help of 2-polarisation optical hybrid (in turn consisting of a pair of 90 degrees hybrids and a PBS), followed by a four balanced detectors. The output photocurrents correspond to the in-phase and quadrature components at both polarizations, conveying the full information about the amplitude and the phase of the received optical field.

5.2.2 Brief description of post-processing unit algorithms

The post processing unit (PPU), for compensation of linear transmission impairments in coherent 112Gb/sec DP-QPSK system, is shown on Fig. 9.

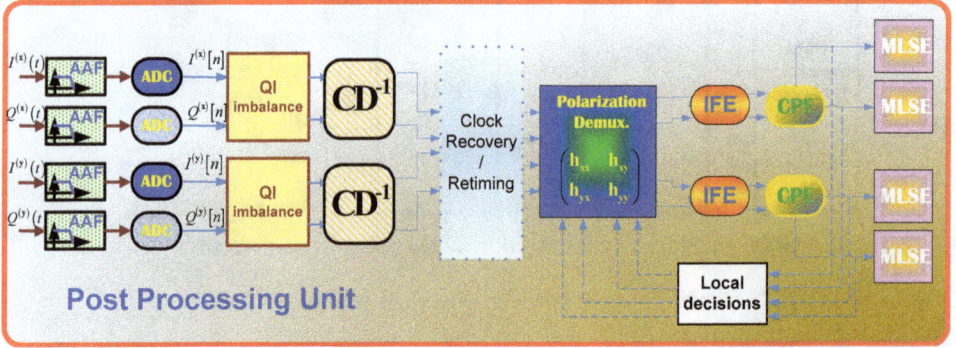

Fig. 9. PPU for 112Gb/sec DP-QPSK coherent detection metro transmission system

At the input of the PPU there are four lanes corresponding to the in-phase and quadrature signals, which can be expressed analytically as follows:

$$
I_{\mathbf{R}}^{(x)}(t) = K \cdot \left(\begin{array}{l} \cos^2 \psi \left[\sum_l r_l \left(t - \dfrac{\tau}{2} \right) e^{j\theta_l^{(x)}} \right] + \sin^2 \psi \left[\sum_l r_l \left(t + \dfrac{\tau}{2} \right) e^{j\theta_l^{(x)}} \right] - \\ -\dfrac{1}{2}\sin 2\psi \left[\sum_l r_l \left(t - \dfrac{\tau}{2} \right) e^{j\theta_l^{(y)}} - \sum_l r_l \left(t + \dfrac{\tau}{2} \right) e^{j\theta_l^{(y)}} \right] \end{array} \right) + I_{\mathbf{N}}^{(x)}(t)
$$

$$
I_{\mathbf{R}}^{(y)}(t) = K \cdot \left(\begin{array}{l} \cos^2 \psi \left[\sum_l r_l \left(t + \dfrac{\tau}{2} \right) e^{j\theta_l^{(y)}} \right] + \sin^2 \psi \left[\sum_l r_l \left(t - \dfrac{\tau}{2} \right) e^{j\theta_l^{(y)}} \right] - \\ -\dfrac{1}{2}\sin 2\psi \left[\sum_n r_l \left(t - \dfrac{\tau}{2} \right) e^{j\theta_l^{(x)}} - \sum_l r_l \left(t + \dfrac{\tau}{2} \right) e^{j\theta_l^{(x)}} \right] \end{array} \right) + I_{\mathbf{N}}^{(y)}(t)
$$

(5.2)

where $I_{\mathbf{R}}^{(x)}(t)$ and $I_{\mathbf{R}}^{(y)}(t)$ represent complex baseband currents in each polarization. Equation (5.2) accounts for a linear channel, including the following effects: CD, first order PMD, laser phase noise at both Tx and Rx (local oscillator) lasers, intradyne reception and additive Gaussian noise formed by optical amplifiers. In Equation (5.2) $e^{j\theta_l^{(x)}}, e^{j\theta_l^{(y)}}$ represent the x- and y-polarization information symbols, and $I_{\mathbf{N}}^{(x)}(t), I_{\mathbf{N}}^{(y)}(t)$ denote the complex noise terms respectively. The effect of PMD in (5.2) is expressed through ψ (the angle between the input state of polarization in the fiber and the principal state of polarization) and τ being the DGD between the polarization modes. The influences of CD, phase noises and intradyne frequency are represented by the shaping pulse:

$$
r_l(t) = \left(h_{CD} * \left[g(t - lT) e^{j\phi_{\mathrm{T}}(t)} \right] \right) \cdot e^{j\left[\omega_{IF} t + \phi_{\mathrm{R}}(t) \right]}
$$

(5.3)

where $\phi_T(t)$ and $\phi_R(t)$ represent the phase noises at the Tx and Rx respectively, $h_{CD}(t)$ being the impulse responses accounting for CD, ω_{IF} denotes the angular intradyne frequency, and $g(t)$ is the shaping pulse at the transmitter which is assumed non-return to zero (NRZ) throughout this manuscript. In Fig. 9, each lane, corresponding to the real and imaginary parts of x- and y- polarization in (5.2), is filtered out by an AAF which is modeled as a 5th order Butterworth LPF. In turn, the signal at each lane is sampled and quantized by an ADC at sampling rate of 28 Gsamples/sec with 5-bit resolution.

The proposed PPU is similar to those presented in (Ip & Kahn, 2007; Fludger et al. 2008; Kuschnerov et al., 2009). The important differences are: (a) the incoming signal bandwidth is intentionally reduced by the AAFs to comply with the Nyquist sampling theorem, (b) sampling and DSP are being done at symbol rate, and (c) the AAF-related introduced ISI is compensated by post processing MLSE.

The lack of orthogonality between I- and Q-signals, stemming mainly from non-perfect 2-pol. hybrid, is compensated by the I-Q imbalance blocks (Savory, 2010). CD compensation is performed by CD^{-1} blocks by sampling (1.3) and using Fast Fourier Transform together with overlap-add/save method. After removing a bulk amount of chromatic dispersion, clock recovery unit corrects the mismatch between the Tx and Rx clocks, and interpolating the input signal to the optimal sampling point. Polarization Demux block in Fig. 9 carries out both polarization demultiplexing and PMD compensation functionalities. IFE (Leven, 2007) and CPE (Viterbi A.J. & Viterbi A.M., 1983) blocks perform intradyne frequency estimation and carrier phase estimation to compensate for frequency mismatch and lasers' phase noises respectively.

5.2.3 Simulation results and discussion

Combined CD and PMD tolerance of 112Gb/sec coherent detection DP-QPSK system is presented on Fig. 10.

Here, polarization and CD effects have been examined, while the same laser serves as the local oscillator and the transmit laser, i.e. no frequency mismatch occurs. Carrier phase estimation and timing recovery were assumed ideal. In fact, the effect of sampling phase is significantly reduced due to the AAF-MLSE combination as explained in (Gorshtein, 2010). Yet, an optimization is required due to the least mean squares (LMS) presence.

Each Monte-Carlo simulation set includes 400,000 random bits. The CD equalizer obeys a zero forcing criterion, and is implemented in the frequency domain, while the CD value is predetermined. The number of states in the MLSE is 16. Histogram estimation method is used for channel estimation where a novel blind equalization scheme is used, as described in section 2 (and in (Gorshtein, 2010)).

Fig. 10 reveals that the proposed system successfully compensates for $20,000 \left[\dfrac{ps}{nm \cdot km} \right]$ of CD (corresponding to 1000-1200km fiber) together with up to 100ps of DGD. The overall penalty of about 2 dB as compared to its back-to-back (b2b) scenario, for worst case scenario is observed. The main reason for this penalty comes from the AAF (Gorshtein, 2010) whereas MLSE engine is used in each lane to struggle this intentionally introduced ISI in the most optimal manner.

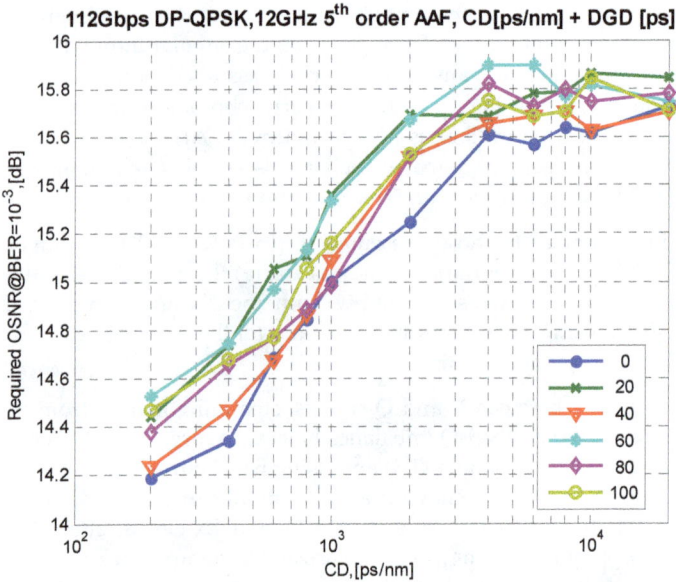

Fig. 10. Combined CD and PMD tolerance of 112Gb/sec DP-QPSK coherent detection metro transmission system

5.2.4 Summary

Sub-section 5.2 presented a "Coherent Metro" MLSE-based solution for 112Gb/sec transmission, which can be used in new deployed systems (green field). The considerations of choosing optimal modulation format were discussed. DP-QPSK modulation format is the most suitable for metro transmission in the proposed system. The DSP that is required in coherent systems, briefly described above, is significantly more complicated than as compared to a direct detection system. Major bottlenecks of a coherent system are requirements for tremendous ADC sampling rate, operation speed of the aforementioned DSP and associated complexity and power dissipation. Significant relaxations of all above can be achieved by introducing AAF, allowing efficient practical implementation, whereas accompanying ISI can be compensated by the MLSE with only 2dB penalty (as compared to b2b) for 100ps DGD in 1200km link.

5.3 Section summary and conclusions

In this section ultra-high speed 112Gb/sec systems for brown and green field applications were proposed. In the brown field, 4-wavelength direct detection is suggested, where reduced bandwidth, already existing and mature, *10G components* are used for 28Gb/sec transmission in each lane. Since most deployed 10G systems use simplest modules at both Tx and Rx sides, new RBW-OOK and RBW-DB modulation formats, which can be transmitted and received in such a network, were proposed and examined. Four MLSE engines are used to compensate for ISI, introduced by the reduced bandwidth components, and for performance improvement. MLSE reception of the proposed

Advanced Modulation Formats and MLSE Based Digital Signal Processing for 100Gbit/sec
Communication Through Optical Fibers

117

modulation formats extends the link tolerance to *residual* CD and first order PMD, as compared to conventional HD receiver. In many brown field 10G *metro* networks, the available OSNR is in the range 18-24dB. RBW-DB requires lower OSNR to achieve the same BER performance as compared to RBW-OOK, due to the *controlled ISI* that is inherent in DB format. Using full bandwidth receiver with RBW-DB transmission can dramatically reduce the OSNR requirement. Yet, for all examined modulation formats, the extended CD tolerance is limited to less than 70km.For full compensation of 80km spans an additional fixed optical compensation module is needed.The proposed solution significantly reduces system cost both due to the use of previous generation 10G components and by replacing expensive tunable dispersion compensation element by a constant fixed dispersion compensation module. This approach transfers the tuneability task to the digital domain, executed by the MLSE engine.

When designing a new green field metro network , coherent detection is preferred. The received photocurrents are proportional to the transmitted optical field components (amplitude and phase) enabling full digital compensation of transmission impairments, e.g., CD, PMD, non-perfect and non-synchronized lasers and so on. DP-QPSK modulation format is examined, due to its higher spectral efficiency as compared to binary formats on one hand, and better performance than modulation formats with richer vocabulary on the other hand. Major bottlenecks of this approach are ultra-high sampling rate, DSP operating speed, complexity and power dissipation. Most coherent 100G systems investigated today employ two-fold oversampling, requiring ultra-fast ADCs. Noting that in metro systems, OSNR of 16-18 dB can be obtained, ADC sampling rate, DSP complexity and power can be dramatically reduced, by employing the baud rate sampling, preceded by the appropriate AAF. MLSE is used to compensate for ISI, which is intentionally introduced by heavy anti-aliasing low pass filtering. It is shown that the proposed system can tolerate CD of about 1200km together with 100ps DGD, when only minor (2dB) penalty is present due to the introduction of the AAF.

6. Long-haul transmission

The link lengths of long-haul transmission systems is 1200km and above. Due to such high distances and attenuation, many optical amplifiers are required. Therefore the available OSNR in the link becomes more deficient. Hence, to achieve the required performance, the system should perform very close to the theoretical optimum. In turn, effective 112Gb/sec transmission with coherent detection would be the most promising solution. Furthermore, since nearly optimal performance is required, all the DSP equalization techniques, sketched in the previous section, must operate on (two-fold) oversampled signals, requiring higher speed ADC (of the order of 64Gsamples/sec for DP-QPSK). In this case, the MLSE is not needed, as such oversampled systems can recover the full performance as indicated both theoretically (IP & Kahn, 2007) and shown experimentally by various groups: (Fludger et al., 2008; Kushnerov et al., 2009; Savory 2010) and references therein. Since in oversampled systems, AAF practically does not introduce such a severe ISI (as compared to the baud rate sampling scenario presented in the previous section), MSLE is not obligate, and even obsolete due to its complexity and power consumption (especially at such a high operating rate).

7. Conclusion

High speed (100G b/sec and beyond) optical communication undergoes a revolution nowadays. Together with increasing the volume of transmitted information per second, currently deployed systems become severely limited by CD and PMD. New cost effective, low power and low complexity solutions are desired. Depending on the transmission range, different system design and equalization schemes are required. In general, digital compensation of transmission impairments is more cost effective as compared to optical compensation, at the expense of power dissipated by the associated DSP ASIC. High speed ADCs and equalization post processing form main bottlenecks for this kind of solutions.

In this chapter novel MLSE-based approaches for 112Gb/sec transmission were proposed for "extended short reach" and metro links. The optical fiber was described from the communication theory system point of view. Various trade-offs between spectral efficiency, complexity, power and cost for each class of the links were presented and discussed in details. The principles of MLSE equalization were outlined and a new blind channel estimation technique was introduced. In addition, a method for upgrading existing 10G links to achieve 112Gb/sec using 4-wavelengths, based on "previous generation" low cost 10G components instead of 28G components was proposed. Two novel modulation formats RBW-OOK and RBW-DB were examined for different short and metro links categories, based on direct detection and MLSE equalization. Coherent detection scheme for green field single carrier 112Gb/sec DP-QPSK metro applications with reduced power consumption and lower complexity was proposed.

For "extended short reach" links coherent detection is not justified. Hence 4-wavelength 112Gb/sec transmission based on direct detection and simplest modulation formats is preferred. The proposed solution benefits from the fact that 10G APDs are available, whereas their 28G counterparts are still not available. Therefore, the proposed method, based on MLSE detection of RBW-OOK and RBW-DB modulation formats, is a very attractive solution possible with appropriate combination of cost, complexity and power for 112Gb/sec transmission in these ranges.

10Gb/sec brown field transmission with optical dispersion compensation and direct detection is very popular in metro links. To achieve 112Gb/sec, currently deployed 10G systems may be used. 4-wavelength slots can be dedicated to carry 112Gb/sec, and existing low cost 10G components can be used for 28Gb/sec transmission in each slot. Depending on the link length, available OSNR and the sensitivity of the photo-detectors several variants of RBW-OOK and RBW-DB can be used.

Extending the reach for 112Gb/sec metro transmission, and later on in green field networks, cost effective solutions can be achieved by employing "Coherent Metro" systems. Coherent detection allows the entire removal of DCFs, effectively increasing the available OSNR in the link. The task of channel equalization is performed in the digital domain, provided that full information of the amplitude and phase of the optical field is recovered during the detection process. The combination of AAF and MLSE is proposed to form power efficient and cost effective solution for single carrier 112Gb/sec DP-QPSK transmission. This method provides significant relaxation on ADC sampling rate and operation speed of the following equalization of transmission impairments, at the expense of additional 2dB of OSNR penalty which seems acceptable in metro links.

Advanced Modulation Formats and MLSE Based Digital Signal Processing for 100Gbit/sec
Communication Through Optical Fibers

119

Since in long-haul transmission every single dB of OSNR is critical, there is no room for OSNR - ADC speed compromise, and the incoming signal must be oversampled. In this case the MLSE is obsolete, due to extra ASIC complexity and power dissipation.

8. References

Agazzi, O. et al. (2005), Maximum-Likelihood Sequence Estimation in Dispersive Optical Channels, *Journal of Lightwave Technology*, Vol. 23, No.2, (February 2005), pp.749 – 763

Agrawal, G. (2002). *Fiber optic communications systems*, John Wiley & Sons, Inc., ISBN 0-471-21571-6, USA

Fludger, C. et al. (2008), Coherent Equalization and POLMUX-RZ-DQPSK for Robust 100-GE Transmission, *Journal of Lightwave Technology.*, Vol. 26, No. 1, (January 2008), pp. 131-141.

Foggi, T. et al. (2006), Maximum-Likelihood Sequence Detection With Closed-Form Metrics in OOK Optical Systems Impaired by GVD and PMD, *Journal of Lightwave Technology*, Vol. 24, No.8, (August 2006), pp.3073 – 3086

Gorshtein, A. et al. (2010), Coherent Compensation for 100G DP-QPSK with One Sample per Symbol Based on Anti-Aliasing Filtering and Blind Equalization MLSE, *Photonic Technology Letters.*, Vol.22, No.16, (August 2010), pp. 1208-1210

Ip, E. & Kahn, J. (2007). Digital Equalization of Chromatic Dispersion and Polarization Mode Dispersion. *Journal of Lightwave Technology.*, Vol.25, No.8, (August 2007), pp. 2033-2043

Kaminov, I. & Li, T. (2001). *Optical Fiber Telecommunications*, Academic Press Elsevier Science, ISBN 0-12-395173-9, San Diego, California, USA

Kuschnerov, M. et al. (2009), DSP for Coherent Single-Carrier Receivers, *Journal of Lightwave Technology.*, Vol. 27, No. 16, (August 2009), pp. 3614-3622.

Leven, A. et al. (2007), Frequency Estimation in Intradyne Reception, *Photonic Technology Letters*, Vol. 19, No. 6, (March 2007) pp.366-368.

Optical Interworking Forum (May, 2010), 100G Forward Error Correction White Paper, In: *OIF-FEC-100G-01.0*, 14.05.2011, Available from http://www.oiforum.com/public/documents/OIF_FEC_100G-01.0.pdf

Proakis, J. (1995). *Digital Communications*, McGraw-Hill, ISBN 0-07-051726-6, Singapore

Roberts, K. et al. (2009), Performance of Dual-Polarization QPSK for Optical Transport Systems, *Journal of Lightwave Technology.*, Vol.27, No.16, (August 2009), pp. 3546-3559

Røyset, A. & Hjelme, D. (1998), Symmetry Requirements for 10-Gb/s Optical Duobinary Transmitters, *Photonic Technology Letters*, Vol.10, No.2, (February 1998), pp. 273-275

Savory, S. (2010), Digital Coherent Optical Receivers: Algorithms and Subsystems, *IEEE Journal of Selected Topics in Quantum Electronics*, Vol. 16, No. 5, (October 2010), pp. 1164-1179

Viterbi, A.J. & Viterbi, A.M. (1983), Nonlinear Estimation of PSK-modulated Carrier Phase with Application to Burst Digital Transmission, *IEEE Transactions on Information Theory*, Vol. IT-29, No. 4, (July 1983) pp. 543–551.

Winzer, P. & Essiambre, R. (2006). Advanced Optical Modulation Formats, *Proceedings of the IEEE*, Vol. 94, No. 5, (May 2006), pp. 952-985

6

2 Terabit Transmission over Installed SMF with Direct Detection Coherent WDM

Paola Frascella and Andrew D. Ellis
Photonics System Group, Tyndall National Institute
& Department of Physics, University College Cork
Ireland

1. Introduction

The way people communicate has continued to evolve in the last decade; information is becoming more visual and digital. Every message exchanged between people is highly likely to be accompanied by high-definition photos or video and transported over long intercity distances. This is the era of Visual Networking (Cisco white paper, 2011a), where social networking websites dominate the market and image based content is increasingly user-generated using advanced personal mobile devices. In 2010, 14.3 petabytes (10^{15} bytes) of mobile/wireless traffic were offloaded onto the fixed network each month. Driven in part by the increase in devices and the capabilities of those devices, there will be two networked devices per capita in 2015, up from one networked device per capita in 2010, resulting in a 32% compound annual growth rate (CAGR) of the total (fixed plus mobile) internet traffic. On a long term scale (e.g. the last ten years), the CAGR has been approximately 19% and the total number of internet users grew from 361 million in Dec 2000 to 2,095 million in March 2011 (www.internetworldstats.com, 2011). Moreover, as telecom technology is deployed in emerging economic powers including Brazil, India and China and is seen by the World Bank as the key to economic independence in sub-Saharan Africa and other areas of the developing world (Reuters, 2010a and 2010b), the exponential growth of global internet traffic will continue, reaching a capacity of one zettabyte (10^{21}) per month shortly after 2015 (Cisco white paper, 2011b).

The **deployed** networks mostly use standard **single-mode fibres** (SMF), which support a single propagating mode, and erbium doped fibre amplifiers (EDFAs) for data transport. The fibre is installed undersea, underground and even sometimes running in the air-suspended from overhead cables. Optical fibre is dominant in submarine, long haul and metropolitan area networks, and is beginning to dominate high-performance access networks. The demand for high-capacity data transmission over the installed fibre networks is evident. Innovative solutions to support the continuing increase in capacity currently falls into two alternative approaches: one focuses on direct physical changes to the network to enable the transport of significantly higher capacities, the second on how to transmit more capacity on the existing deployed networks. The first direction involves the study of *new optical fibres* for more efficient transport of information (Zhu et al., 2011), and new network architectures, essentially allowing the replacement of electrical switches with optical implemented alternatives (Dunne et al., 2009). Such a radical change in the network will be adopted when the proposed upgrade to a new fibre and/or a new architecture will offer the network operator groundbreaking improvements, enabling increases in revenue generation above the upgrade cost. The second approach, a more short-term solution, enables moderate upgrade for an *immediate satisfaction of the capacity demand*, in contrast to the introduction of novel technologies which often require long terms and high investments. In this chapter we focus on a solution within the second approach, providing increased capacity over existing infrastructure at minimum cost and complexity.

In its original form, Ethernet combines low implementation cost, high reliability and relative simplicity of installation to become the *de facto* local area network standard. Ethernet has evolved and adapted to meet the increasing bandwidth demands of end-users. The latest variants, 40 and 100 Gigabit Ethernet (GbE) were recently standardised by the IEEE data transport applications over both copper and optical fibre. Other enhancements, such as the support for operations, administration and maintenance (OAM) functionality, have contributed to the emergence of **Carrier-Class Ethernet** as the dominant transport technology in telecommunication networks. Today it is safe to assume that nearly all internet traffic starts and ends with an Ethernet connection. With zettabyte data volumes the server farms, used to host and distribute Visual Networking services, require low cost ultra high-capacity intra and inter-data centre connections. Indeed recent requests for Terabit Ethernet (Lee, 2011; Lam et al., 2010) have motivated the work that we will present in this chapter.

In parallel, dual polarisation quadrature phase shift keying (DP-QPSK) was developed for telecom applications (ITU-T G.709/Y.1331) for the transport of Ethernet without recourse to inverse multiplexing. The spectral resource (the optical fibre bandwidth) is already highly shared through wavelength division multiplexing (WDM) in current networks, and will need to implement high-spectral efficiency techniques in order to carry Terabit Ethernet data in the future. It is widely accepted that **multicarrier systems**, such as Coherent WDM (CoWDM) and other variants of optical Orthogonal Frequency Division Multiplexing (OFDM), are strong candidates for Terabit Ethernet (TbE) transmission over metro area networks (10-300 km) (Sanjoh et al., 2002; Ellis & Gunning, 2005; Lowery et al., 2006; Shieh & Authaudage, 2006; Djordjevic & Vasic, 2006; Jansen, 2007; Goldfarb et al., 2007; Chen, H. et al., 2009; Hillerkuss et al., 2011; Zhao & Ellis, 2010). With these techniques, in order to achieve Tbit/s capacities individual WDM channels are further expanded into bands, each containing many orthogonal subcarriers. **Orthogonality** opens the possibility to transmit higher capacities with reduced cost per bit, without recourse to disruptive network upgrades. Emerging grid-less reconfigurable add-drop multiplexers (ROADMs) (Poole et al., 2011), which are beginning to dominate the market (www.infonetics.com, 2011), in combination with flexible multicarrier solutions offer high capacities in the Tbit/s region and increase the network efficiency (Thiagarajan et al., 2011; Christodoulopoulos et al., 2011; Takara et al., 2010; Bocoi et al., 2009). For a single carrier m-QAM solution, the required optical signal-to-noise ratio increases more rapidly than the capacity increases. In contrast, multi-carrier solutions, such as all-optical OFDM and CoWDM, do not suffer from this limitation and allow for very flexible and scalable total transmitted capacities.

Multicarrier solutions, which meet growing capacity requirements, must offer compatibility with Ethernet. Moreover cost-effective implementations are essential, especially for short network connections as in financial institutions and data centre providers. CoWDM is a promising candidate for future high-speed Ethernet transport. In this Chapter we transmit Ethernet packets and implement forward error correction (FEC), showing how this determines the system performance. We identified critical clock stability issues unique to multicarrier systems (Frascella et al., 2010b) and demonstrated the impact on the system design of the more stringent BER of an Ethernet client(Frascella et al., 2010a). In segments of the network where high capacity is needed at the lowest cost, **direct detection** could be used to avoid the cost, complexity and power consumption of digital coherent receivers. In this chapter, we consider the field transmission of a 2 Tbit/s multibanded direct detection CoWDM signal over installed SMF, first using EDFA amplification only (Frascella et al., 2010c), and then use Raman amplification to enhance the potential reach (Frascella et al., 2011). Mixed Ethernet (with FEC) and PRBS payloads are used to study both the Ethernet transmission and the performances against fibre impairments of the optical multiplexing format. Fourtynine subcarriers were measured with pre-FEC bit error ratio (BER) performance lower than 10^{-5} and post-FEC frame-loss ratio (FLR) below 10^{-9} for Ethernet transmission over unrepeated 124 km of SMF. Outage probability due to polarisation mode dispersion (PMD) is estimated from BER measurements extended over several hours, showing the robustness of CoWDM format. The reach of direct detected 40 Gbaud Terabit capacities is predicted for single-mode fibre based systems as a function of the amplifier spacing, suggesting that CoWDM is suitable for Terabit Ethernet transport over metropolitan links, reaching 1,400 km at spacing of 80 km and up to 130 km unrepeated transmission.

2. High-capacity transmission over installed SMF

In laboratories, the total capacity and the spectral efficiency have drastically grown thanks to the introduction of higher modulation formats and digital coherent detection. In March 2011, records were achieved of 101 Tbit/s and 11 bit/s/Hz in a single-mode single-core optical fibre using coherent detection by (Qian et al., 2011). However, there are no scientific reports of higher-capacity **field results** than the 3.2 Tb/s demonstrated in early 2001 (Chen D. et al., 2001), which was achieved with 80 standard WDM channels carrying 40 Gbit/s NRZ-OOK spaced at 100 GHz across the L and C-band with Raman amplification and FEC over 3 spans of 82 km long SMF. The highest spectral efficiencies with high-capacity are achieved with orthogonal multiplexing, both in laboratory (Qian et al., 2011) and field experiments (Frascella et al., 2010c; Xia et al., 2011), although other techniques (e.g. based on pre-filtering) also allow high spectral efficiencies (Gavioli et al., 2010; Roberts, 2011). Multi-band transmission with orthogonal multiplexing over field deployed fibre started in 2010 where 759 Gbit/s total capacity was achieved with off-line processed coherently detected DP-QPSK-OFDM and information spectral density (ISD) of 2.35 bit/s/Hz (assuming the use of 7% FEC overhead) over a total of 764 km of SMF (Dischler et al., 2010). 2 Tb/s capacity with orthogonal multiplexing was first achieved in 2010 using real time direct detection (Frascella et al., 2010c) and then in 2011 offline coherent detection (Xia et al., 2011). The reach and the ISD (respectively, 0.7 bit/s/Hz/pol and 3 bit/s/Hz) were determined by the repeater spacing and receiver complexity.

2.1 Coherent WDM (CoWDM)

Coherent WDM is an all-optical implementation of OFDM where phase control of adjacent subcarriers is exploited to minimise inter-subcarrier crosstalk interference arising from non-ideal orthogonally-matched filters (or demultiplexing of orthogonal subcarriers). OFDM itself is a specific implementation of orthogonal systems developed in the 1950s (Mosier & Clabaugh, 1958) and extensively studied in the 1960s (Deman, 1964; Chang, 1966; Ito & Yokoyama, 1967; Zimmerman, 1967), where similarly to orthogonality condition kept in the time domain to avoid inter-symbol interference (ISI) there is an orthogonality condition in the frequency domain to avoid inter-channel crosstalk interference (Proakis & Salehi, 2008). This condition may be expressed as:

$$\int_{-\infty}^{+\infty} X_k(f) X_j^*(f)\, df \overset{Th.}{=} \int_{-\infty}^{+\infty} x_k(t) x_j^*(t)\, dt = \begin{cases} E & k=j \\ 0 & k\neq j \end{cases} \tag{1}$$

where the first equality is Parseval's theorem, $E = \int_{-\infty}^{+\infty} |x_k(t)|^2\, dt$ is the energy of the signal $x_k(t)$ of the k-th channel and $X_k(f)$ its spectrum (and Fourier transform). If we consider the signal waveforms only to differ in frequency, then the orthogonality condition in Eq. 1 introduces a condition on the signal spacing. OFDM is a particular case of orthogonal system, where the spacing between frequencies is equal to the symbol rate (1/T):

$$\omega_k - \omega_j = 2\pi/T \tag{2}$$

Various flavours of orthogonal systems have been proposed, including half of the symbol rate (Chang, 1970; Rodrigues & Darwazeh, 2002; Zhao & Ellis, 2010), or close approximations to Eq. (2) (Yamamoto et al., 2010). Whilst all of these systems satisfy the orthogonality condition and may thus be strictly classified as Orthogonal FDM systems, for the last decade (2002-2011) the terminology "OFDM" has been understood to apply to systems with very low inter-subcarrier crosstalk satisfying Eq. 2, and implemented using Fourier Transforms (Weinstein & Ebert, 1971).

Orthogonally multiplexed multicarrier systems were first proposed for long-haul optical systems in 2002 (Sanjoh et al., 2002), when OFDM was already standardised for DAB HDTV and UMTS. Later on, different varieties were proposed by (Ellis & Gunning, 2005; Feced et al., 2005; Lowery et al., 2006; Djordjevic & Vasic, 2006; Shieh & Authaudage, 2006), and extensively studied in laboratory experiments (Jansen et al., 2008a, 2008b; Shieh et al., 2008; Yonenaga et al., 2009; Sano et al., 2007, 2009; Chandrasekhar et al., 2009; Liu et al., 2009; Schmogrow et al., 2011).

CoWDM derives from the concept that at high symbol rates the orthogonality condition is only maintained if the optical phases ($\phi_{k,j}=d\omega_{k,j}/dt$) of the subcarrier k and j are constant, and aligned to ensure that any residual crosstalk is distributed away from the eye crossing. When CoWDM was first simulated (Ellis & Gunning, 2005), patented (Ellis et al., 2009b) and experimentally verified (Ellis & Gunning, 2005; Gunning et al. 2005; Healy et al., 2006) it was clear that the phase control of each subcarrier, implemented at the transmitter, could ensure orthogonality (reduced BER penalty) by using commercially available modulators and photodiodes with bandwidths comparable to the symbol-rate, rather than the full system capacity, as required for DSP based OFDM (Lowery, 2010). It has been recently been demonstrated that the advantage of phase control is correlated to the transmitter and receiver structure (Ibrahim et al., 2010). Note that the benefit of phase control is negligible in the case of coherently detected dual quadrature signals (Zhao & Ellis, 2011) and maximum in the case of direct detection of single quadrature signals. A signal-to-residual crosstalk ratio (SXR) for OOK signals has been defined (Ibrahim et al., 2010), as the ratio between the signal power in the absence of crosstalk (which corresponds to the signal power in the '1' bits) and the crosstalk, i.e. the sum of the crosstalk in '1' bits and '0' bits. Taking into account only the crosstalk interference coming from the adjacent subcarriers (j+1 and j-1) and assuming no ISI, the SXR for direct detected OOK signals is:

$$SXR_{dd} = \left| x_{j,j}(0) \right|^2 \Big/ \left\{ 2 \left[\left| x_{j-1,j}(0) \right|^2 + \left| x_{j+1,j}(0) \right|^2 + \left| x_{j-1,j}(0) x_{j,j}(0) \cos\left(\phi_j - \phi_{j-1} \right) \right| + \right. \right.$$
$$\left. \left. + \left| x_{j+1,j}(0) x_{j,j}(0) \cos\left(\phi_j - \phi_{j+1} \right) \right| + 2 \left| x_{j-1,j}(0) x_{j+1,j}(0) \cos\left(\phi_{j-1} - \phi_{j+1} \right) \right| \right] \right\}$$

(3)

where $x_{k,j}(0)$ is the baseband representation at the sampling instant of the signal pulse shape for the k-th subcarrier (corresponding to a frequency spectrum H_{Tx}) after optical filtering, demultiplexing and electrical filtering (all included in the frequency response H_{Rx}) targeted to the subcarrier j:

$$x_{k,j}(t) = \frac{1}{2\pi} \int_{-\infty}^{+\infty} H_{Tx}\left(\omega - \omega_k + \omega_j \right) H_{Rx}\left(\omega + \omega_j \right) e^{j\omega t} d\omega$$

(4)

The cosine terms represent the phase of the beats between all the three subcarriers j-1, j and j+1, which pass through the optical filter. From Eq. (3) is evident that by controlling the phase difference between adjacent subcarriers the crosstalk could be minimised (or equivalently the SXR maximised) by setting the CoWDM phase condition:

$$\left| \phi_j - \phi_{j-1} \right| = \frac{\pi}{2}$$

(5)

CoWDM and in general all-optical OFDM are not only robust against dispersion and nonlinearities (Ellis et al., 2009a; Healy et al., 2006; Hillerkuss et al., 2011; Sano et al., 2007; Frascella et al., 2010b), but also remove speed limitations set by electronics as well as linearity issues introduced during EO/OE conversions (Huang et al., 2009). All-optical OFDM promises to achieve Terabit transport in real time, when compared to DSP-based OFDM (Schmogrow et al., 2011; Xia et al., 2011). The ability to operate using direct detection, and over existing dispersions maps (Ellis et al., 2009a), offers the potential for low-cost non-disruptive upgrades making CoWDM an attractive proposition for cost sensitive applications.

2.2 Transmission of 2 TbE with EDFA only

The experimental setup used for the field demonstration of the 2 Tbit/s CoWDM is illustrated in Fig. 1. One 10.7 Gbit/s pseudo-random binary sequence (PRBS) with a $2^{31}-1$ pattern length was aggregated with three forward error correction (FEC) encoded 10 Gigabit Ethernet (10 GbE) WAN PHY (9.953 Gbit/s) streams. The PRBS tributary was synchronised to the FEC encoded Ethernet signal and used to monitor the system performance and identify impairments; it was later on replaced with a fourth Ethernet stream to verify the performance of 2 TbE. Fourtynine subcarriers were generated from seven DFB lasers using sine wave driven amplitude modulators (Healy et al., 2007; Frascella et al., 2010b). The

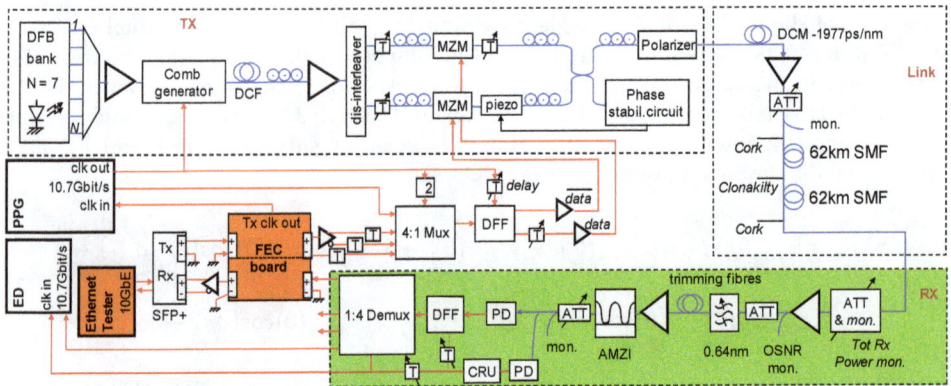

Fig. 1. Experimental setup for 2 Tbit/s transmission over 124 km of field-installed SMF (Frascella et al., 2010b). PPG- pulse pattern generator, DFF- D-flip flop, CRU- clock recovery unit; PD- photodiode, ED- error detector, DCM- dispersion compensating fibre.

electrically multiplexed 42.84 Gbaud signals were used to modulate the 49 CoWDM subcarriers with NRZ-OOK, where odd and even channels modulated by data and delayed inverted data respectively. The transmitted spectrum had a total bandwidth of 2.8 THz (guard-bands of 85.67 GHz were introduced to minimize the interference between bands) and transported a total of 2.1 Tbit/s, giving a spectral efficiency of 0.7 bit/s/Hz in a single polarization after taking into account 7% FEC overhead. Inter-subcarrier phases were controlled using an electrically driven piezo fibre-stretcher, where the optimum condition was established by measuring the 42.6 GHz beat frequency between two adjacent subcarriers. The capacity could have been readily doubled using polarisation multiplexing (Cuenot et al., 2007). The fibre link, pre-compensated for chromatic dispersion, used EDFA amplification only. The field-installed SMF was looped-back at Clonakilty to the measurement laboratory in Cork, giving a total span length of 124 km and an associated loss of 26 dB, which is a challenge in terms of OSNR, as it would be for all unrepeatered links of similar reach. The total signal launch power into the SMF was found to be optimum at +16.9 dBm.

At the receiver, channel selection is performed by a passband filter of 0.64 nm bandwidth, and a one-tap optical FFT is realised with an asymmetric Mach-Zehnder interferometer (AMZI). After direct detection, the BER and FLR of each optical subcarrier were measured. The BER is shown in Fig. 2, where the best (#25, 1552.74 nm) and the worst performing (#48, 1562.77 nm) subcarriers are highlighted with different symbols (Fig. 2(a)). At the maximum received power of –12 dBm, the BER of the worst performing subcarrier (#48) was 1.3×10^{-5}. The band containing subcarrier #48 had an OSNR of 30.8 dB (Fig. 2(b)- the received OSNR is defined per band, as the ratio of the signal power in a band (2.5 nm) over the noise power in 0.1 nm bandwidth). The best performing subcarrier (#25) achieved a BER of 3×10^{-9} and an OSNR about 1 dB higher. Fig. 3(c) shows the received eye-diagrams after optical demultiplexing for the best (#25) and worst (#48) subcarriers with the crosstalk between adjacent subcarriers remaining minimized at the centre of the eye after transmission over fibre.

We believe that the 6 dB difference (at 10^{-5}) observed in the required OSNR between the best and worst subcarriers can be attributed to: a) the wavelength sensitivity within the comb generation module (see optical spectrum in Fig. 3 (left-axis)); b) the residual gain variation in the optical amplifiers; c) phase errors between adjacent comb lines after transmission (Gunning et al., 2005); and also d) polarisation mode dispersion (Frascella et al., 2010b). An improvement of the flatness of the 49 subcarriers and the introduction of phase control for each individual subcarrier would guarantee an equal OSNR for all subcarriers. This enables the launch power of all subcarriers to be increased to the nonlinear threshold, therefore improving the OSNR and BER of the worst subcarriers. The average receiver sensitivity at a BER of 1×10^{-5} across the measured sensitivities for the 49 subcarriers (filled diamonds in Fig. 3) was ~ –15.5dBm. A BER of 2×10^{-15} is required to achieve a FLR of 10^{-12} when transmitting an Ethernet frame (Frascella et al., 2010a). FEC boards employing a simple Reed Solomon RS(255,239) code (BER threshold of 1×10^{-4}, dashed red line #1 in Fig. 2), as from ITU-T G.975.1 Recommendation, will leave a system margin of 3 dB received power/OSNR. Enhanced FEC realised with interleaved RS(1023,1007)/BCH(2047,1952) code (BER threshold of 2×10^{-3}, dashed red line #2 in Fig. 2) can leave a bigger margin, i.e. 7 dB. Moreover, the removal of additional loss from the system (i.e. variable attenuator and power monitor) could improve this margin even further.

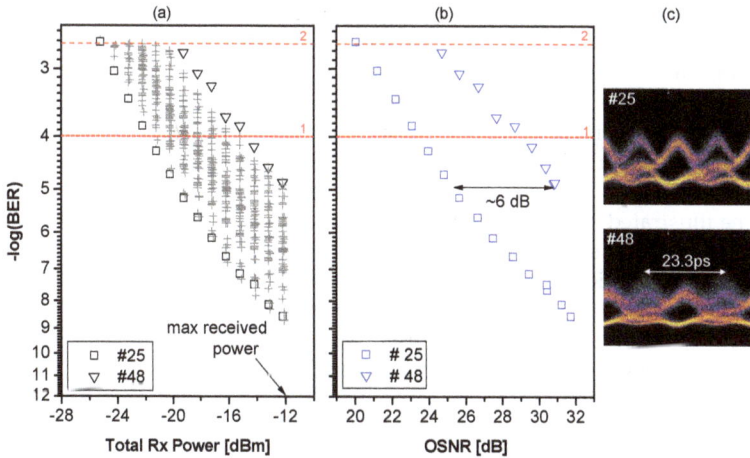

Fig. 2. BER performance after transmission for the best (#25) and worst (#48) subcarriers in terms of (a) total received power, and (b) received OSNR. Grey crosses are all the other 47 subcarriers, represented only on the left graph for clarity. The dashed lines in red represent the threshold for (1) the used FEC board and (2) an enhanced FEC threshold of 2×10^{-3}. (c) Respective eye-diagrams at maximum received power.

Fig. 3 also shows the Ethernet performance of all the 49 subcarriers. No frame-losses were observed for any of the 49 subcarriers for the received total power shown in Fig. 3 (open triangles) when a maximum number of frames (4.3×10^9) allowed in the Ethernet tester for a single run was set, suggesting a frame loss rate of below 2.3×10^{-10}. When replacing the PRBS tributary with an Ethernet stream, no frame losses were observed for all four FEC-encoded Ethernet WAN PHY streams of the optical subcarrier #19. This represented the first attempt of such a high Ethernet capacity, 2 TbE, transmission over an unrepeatered installed fibre of an inter-city distance.

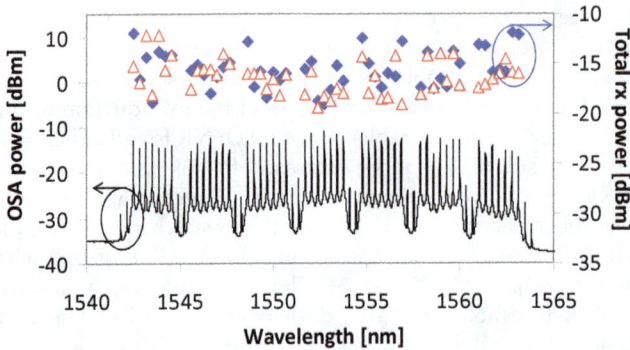

Fig. 3. Left: received optical spectrum after transmission with a resolution bandwidth of 0.02 nm. Right: total received power at BER of 1.0×10^{-5} (filled diamonds) and at FLR of 2.3×10^{-10} (open triangles).

The BER performance of the PRBS tributary for a random subcarrier (#19) at the maximum received power (-12 dBm) was monitored over 6 hours to estimate the impact of dynamic effects in the field-installed fibre. The BER variation against time, plotted in Fig. 4, shows fluctuations in the BER of up to two orders of magnitude with a peak BER of $\sim 5 \times 10^{-5}$. We attribute these variations mainly to PMD, but the features from 5 hours onwards might also be due to a memory overload of the algorithm used. The feedback control was implemented on a PC, which stored all the phase error values; therefore at the end of the measurement cycle, the feedback delay was increased due to increasing memory usage. The results in Fig. 4 can also be illustrated as a probability density function (PDF) of log(BER), as in Fig. 5 (a). In this case, the PDF shows two distinct peaks: the one with the greater amplitude corresponding to a typical Maxwellian distribution associated with PMD, and the other peak can be attributed to either: the proximity of the field-installed fibre cable to a major motor-vehicular transport link, or the response of the phase-stabilization circuit to the intermittent frequency modulation present on the synchronizing clock.

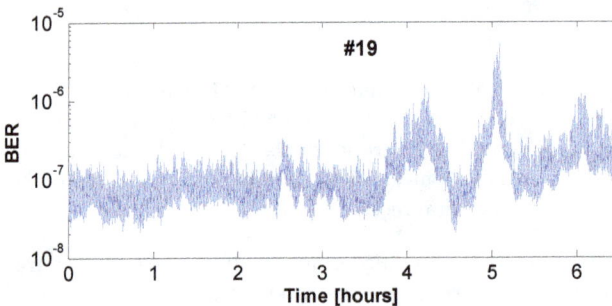

Fig. 4. Long term BER measurement over time for subcarrier #19. The BER gating window was set to 100 ms.

The outage probability, which is defined as the probability that a system outage occurs, could be estimated to understand how the observed effects degraded the system performance. An outage occurs whenever the BER is greater than 10^{-12} after FEC decoding.

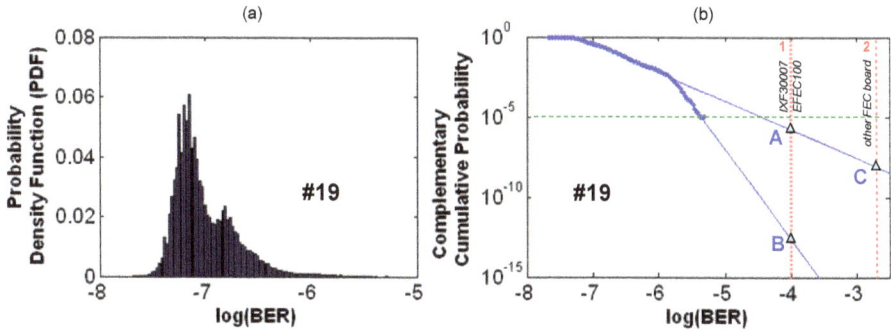

Fig. 5. (a) Probability Density Function (PDF); and (b) complementary cumulative probability calculated from the long term BER measurement relative to an average performance subcarrier (#19). The blue lines are extrapolating slopes from the data, and dashed red lines represent FEC thresholds as from Fig. 2. The green dashed line represents a desirable outage probability (Kaminow et al., 2008).

From the PDF values in Fig. 5(a), one can calculate the complementary cumulative probability which is plotted in Fig. 5(b) (blue dots). The complementary cumulative probability is defined as the probability that the BER is greater than a certain value (x-axis). The outage probability will then correspond to the intercept between the complementary cumulative probability and the FEC threshold. We consider two extrapolations from the complementary cumulative probability, the first omitting the infrequent high BER events giving an upper bound on the outage probability of 2×10^{-6} (point A in Fig. 5(b)) and the second including these events, giving a lower bound of 3×10^{-12} (point B). Consequently, at full received power, this particular subcarrier delivered an outage probability below the widely used specification of 10^{-5} (Kaminow et al., 2008). The outage probability could be substantially improved if an enhanced FEC board, represented as a dashed red line #2 in Fig. 5(b), were used, which would allow outage probabilities below 8×10^{-9} (point C).

2.3 Hybrid EDFA/Raman amplification

2.3.1 Stimulated Raman Scattering

Raman Scattering is a spontaneous isotropic process where a pump wave is inelastically scattered by the fibre material. The process results in the annihilation of the original photon and the generation of a phonon and a photon at a lower frequency. In optical fibres, only forward (FW) and backward (BW) scattered photons are guided. The vibrational energy level of silica dictates the peak value of the Raman shift, $\Omega_R = \omega_p - \omega_s \sim 13$ THz between the pump frequency ω_p and the signal frequency ω_s. If a weak probe is injected into the fibre at an appropriate frequency shift from an intense pump, we obtain Stimulated Raman Scattering (SRS) and the maximum SRS gain g_R is 0.6×10^{-13} m/W at 1.55 µm (Stolen et al., 1972; Pizzinat et al., 2003)). The SRS gain has a wide bandwidth (when pump and signal are co-polarised (Headley & Agrawal, 2005) that exceeds 10 THz, making SRS favourable for WDM amplification.

The power exchange between the pump and the signal is governed by the following set of coupled equations, under the approximation of undepleted pumps (Agrawal, 2001):

$$\frac{dI_s}{dz} = g_R I_s I_p - \alpha_s I_s$$

$$\frac{dI_{FW}}{dz} = -\alpha_p I_{FW} \qquad (6)$$

$$\frac{dI_{BW}}{dz} = +\alpha_p I_{BW}$$

where the intensities of the signal I_s and pumps I_p (sum of FW and BW), I_{FW}, I_{BW}, are all functions of the propagation direction z. Also $I_p(0)=I_{FW}(0)+I_{BW}(0)$ and $I_{FW}(0)=I_{BW}(L)=P_p/A_{eff}$ where P_p is the nominal power of the pumps in Watts and A_{eff} is the effective area of the fibre. α_p (=0.069 km^{-1} for 1427 and 1455 µm) and α_s (=0.046 km^{-1}) are the attenuation coefficients at the pump and signal wavelength respectively. Solutions of Eq. (6) are:

$$I_{FW}(z) = I_{FW}(0)\exp(-\alpha_p z)$$

$$I_{BW}(z) = I_{BW}(0)\exp(\alpha_p z) = I_{BW}(L)\exp(-\alpha_p L)\exp(\alpha_p z) \qquad (7)$$

$$I_s(z) = I_s(0)\exp\left\{g_R\left(\int I_p(z)dz\right) - \alpha_s z\right\}$$

and

$$\int I_p(z)dz = \frac{I_{FW}(0)\exp(-\alpha_p z)}{-\alpha_p} + \frac{I_{BW}(0)\exp(\alpha_p z)}{\alpha_p} + c \qquad (8)$$

where c is a constant. We could determine c by setting $z=0$ in Eq. (4):

$$c = \frac{I_{FW}(0)}{\alpha_p} - \frac{I_{BW}(0)}{\alpha_p} \qquad (9)$$

The net gain that the signal experiences from one end of the fibre to the other end is given by $G_{net}(z)$:

$$G_{net}(z) = \frac{I_s(z)}{I_s(0)} = G_{RA}(z)\exp(-\alpha_s z) \qquad (10)$$

where the Raman gain $G_{RA}(z)$ evolves along the propagation direction z, giving its main contribution within a pump effective length $L_{eff} = \left(1 - e^{-\alpha_p L}\right)/\alpha_p$:

$$G_{RA}(z) = \exp\left\{g_R L_{eff}\left[I_{FW}(0) + I_{BW}(0)e^{\alpha_p z}\right]\right\} \qquad (11)$$

2.3.2 Noise in Distributed Raman Amplification (DRA)

A particular advantage of distributed amplification is the reduced effective noise figure coupled with a more uniform power distribution along the fibre, which enables an improved compromise between nonlinearities and noise performances.

In the case of uni-directional Raman amplification and assuming that the loss coefficients at the pump and signal wavelength are similar ($\alpha_p \approx \alpha_s$), the effective noise figure is given by (Desurvire, 2002):

$$NF_{RA} = 1 + 2n_{eq} \tag{12}$$

where the equivalent input noise factor (Desurvire, 1986 & 2002; Walker et al., 1991; Kikuchi, 1990), n_{eq}, which could be interpreted as the equivalent number of photons per mode at the amplifier input, is given by

$$
\begin{aligned}
n_{eq}^{FW} &= \frac{a_0}{\alpha_s} \exp\left(-\frac{a_0}{\alpha_s}\right) \left\{ Ei\left(\frac{a_0}{\alpha_s}\right) - Ei\left[\frac{a_0}{\alpha_s} \exp(-\alpha_s L)\right] \right\} \\
n_{eq}^{BW} &= 1 - \frac{1}{G_{net}} + \frac{\alpha_s}{a_0} \left\{ \exp(\alpha_s L) - \frac{1}{G_{net}} \right\}
\end{aligned}
\tag{13}
$$

where $Ei(x)$ is the exponential integral function (Abramovitz & Stegun, 1972), $a_0 = (\ln G_{net} + \alpha_s L)/L_{eff} = \ln G_{RA}/L_{eff}$ is the SRS gain coefficient and G_{BW} is the net gain due to BW pumping only.

When BW pumping is the main source of gain, but bidirectional pumping is also employed, we could express the equivalent noise figure for the Raman amplifier as (Bromage et al., 2004) in Eq. (14) where G_{FW} is the increase in optical signal power when FW pumps are switched on:

$$NF_{RA} = \frac{NF_{BW}}{G_{FW}} \tag{14}$$

2.3.3 Double Rayleigh Backscattering (DRB)

Rayleigh backscattering (RB) is light scattered backwards to the direction of propagation by material density imperfections occurring in the fibre induced during manufacture. The backscattered intensity increases with fibre length; in SMF (Rayleigh coefficient R_s~6.164×10^{-5} km^{-1}) it reaches a constant value of 32 dB below the signal power after approximately 20 km.

The backscattered light may itself be backscattered through the Rayleigh backscattering process, resulting in a doubly backscattered signal. In a conventional fibre without distributed amplification, this process has negligible impact on the system performance, however if the system contains bidirectional gain this effect could significantly degrade the equivalent noise figure of the system. This is particularly the case in the presence of Raman gain which provides distributed gain along the fibre length, but may also be observed in systems employing bi-directional lumped amplifiers. Within a Raman amplified fibre, the DRB signal is given by (Nissov, 1999):

$$P_{DRB}(z) = P_s(z) R_s^2 \int_0^z \frac{1}{G_{net}^2(x)} \int_x^L G_{net}^2(y)\,dy\,dx \tag{15}$$

The total equivalent noise figure of a distributed Raman amplifier (DRA), taking into account of amplified spontaneous emission (ASE) and DRB, could be then expressed as (Essiambre et al., 2002):

$$NF_{DRA} = NF_{RA} + \frac{1}{G_{net}} \frac{\frac{5}{9} P_{DRB}}{h\nu \sqrt{\Delta\nu_{el}^2 + \frac{\Delta\nu_{opt}^2}{2}}} \tag{16}$$

where $5/9 P_{DRB}$ is the DRB power copolarised with the signal, $\Delta\nu_{opt}$ and $\Delta\nu_{el}$ are respectively the equivalent double-side bandwidth of the optical signal and the electrical filter at the receiver.

2.3.4 Impact of cross phase modulation

Raman amplification can enhance the performances of the CoWDM system in terms of required OSNR and delivered BER (or Q factor). However, the optimization of a DRA system is complex because the distribution of the gain along the transmission fibre improves the OSNR by keeping the signal power from falling to very low levels (low noise figure), but simultaneously increases the signal distortions that result from Kerr and other signal nonlinearities (higher nonlinear phase shift and effective length). Fig. 6 illustrates our experimental setup, which transmits 2 Tbit/s NRZ-OOK CoWDM with PRBS 2^{31} -1 over the 124 km installed fibre link with distributed amplification. The hybrid EDFA/Raman system consisted of three units: EDFA #1 in Fig. 6, FW and BW Raman pumping units. At the output of the transmitter (point A in Fig. 6) the OSNRs (measured per CoWDM band, as the ratio between the signal power in 2.5 nm and the noise power in 0.1 nm) were between 38 and 39 dB, which dropped to below 37dB at the output of EDFA#1. The first stage EDFA also included optical supervisory channel equipment in order to ensure the integrity of the installed link.

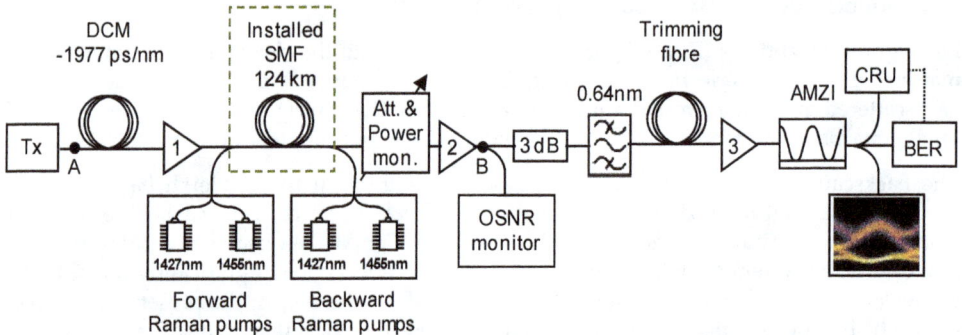

Fig. 6. Experimental setup for hybrid Raman/EDFA transmission over 124 km installed-SMF of 2 Tbit/s OOK CoWDM.

Conventionally, the equivalent noise figure NF_{AB} between the points A and B in Fig. 6 is defined as:

$$NF_{AB} = \frac{NF_1}{e^{-\alpha_{DCM}L_{DCM}}} + \frac{NF_{DRA}-1}{e^{-\alpha_{DCM}L_{DCM}}G_{E1}} + \frac{NF_2-1}{e^{-\alpha_{DCM}L_{DCM}}G_{E1}G_{net}} \tag{17}$$

where NF_1 and G_{E1} are noise figure and gain of EDFA #1, NF_2 is the second EDFA's noise figure, α_{DCM} and L_{DCM} are loss coefficient and length of the DCM, $G_{net} = G_{net}(L)$ is the net gain over the SMF as from Eqs. (10) and (11). NF_{DRA} is the equivalent noise figure of the Raman amplifier which takes into account of DRB and ASE as in Eq. (16), and results in a typical output OSNR in the region of around 34dB after Raman amplification.

In order to calculate the evolution of the signal power, the conventional Raman amplification formulas (Eqs. (10) and (11)) should be modified to take into account the insertion loss of the variable attenuator α_{ATT}, insertion losses of the FW pump coupler α_x, and efficiency parameters of the pump intensities η_1, η_2 which will also take into account the two wavelengths used within each pump module. We therefore have:

$$G_{net}(z) = \alpha_{ATT} e^{-\alpha_s z} G_{RA}(z) \tag{18}$$

$$G_{RA}(z) = \alpha_x \exp\left\{ g_R L_{eff} \left[\eta_1 I_{BW}(L) e^{\alpha_p(z-L)} + \eta_2 I_{FW}(0) \right] \right\} \tag{19}$$

Fig. 7 shows the analytically calculated Raman gain (blue solid line), along with experimental measurements (red squares) to confirm the accuracy of the model. This allows the calculation of the signal and DRB powers, assuming zero pump depletion as shown in Fig. 8. Fig. 8(a) shows the case of backwards pumping only with 2 pump wavelengths both at +27 dBm; Fig. 8(b) is the case of bidirectional pumping each with two pump wavelengths at +27 dBm. Even for bidirectional pumping, the DRB power is very low (maximum of –40 dBm) and does not degrade the equivalent noise figure significantly. The equivalent link noise figures are 22.7 dB using backwards pumping only and 11.7 dB using bidirectional pumping, dominated by ASE generated in the Raman amplifier itself. However, with bidirectional pumping, the signal power excursions are relatively low (increased path averaged power), so for a given launch power, we would anticipate a significant increase in the impact of cross-phase modulation (XPM).

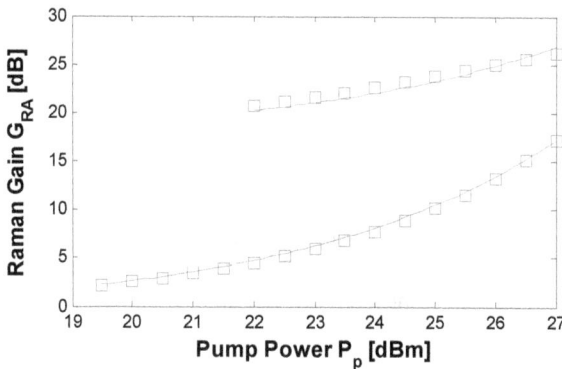

Fig. 7. Raman gain versus pumps' power: experimental data (red squares) and fitting (η_1= 0.68, η_2= 0.36, α_x=1.26 corresponding to ~1 dB extra loss, g_R=0.68×10^{-13} m/W- blue line).

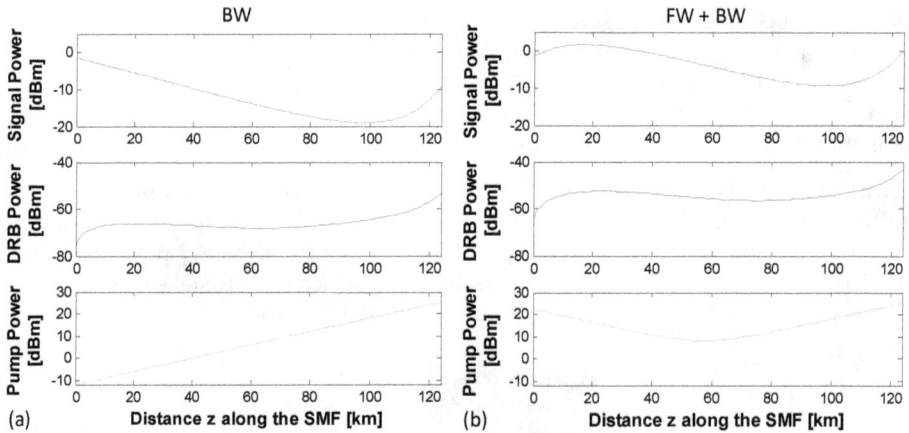

Fig. 8. Trend of the signal (P_s), DRB (P_{DRB}) and pump (P_p) powers along the SMF.

For a well designed optically pre-amplified receiver, the Q factor for NRZ OOK signal is limited by signal-ASE beat noise and is given by:

$$Q = \frac{I_1 - I_0}{\sigma_1 + \sigma_0} \approx \frac{I_1}{\sigma_1} = \sqrt{SNR}$$
$$Q_{dB} = 20\log_{10} Q = 10\log_{10} SNR \tag{20}$$

where I_1 and I_0 are the detected photocurrents for the '1' and '0' (I_0=0 for OOK) and σ_1 and σ_0 are the standard deviations of the noise on the '1' and '0' respectively. On a logarithmic scale the Q factor scales linearly with the SNR (and hence with the OSNR) with a slope of 0.5. This linearly increasing trend is inverted when signal nonlinearities become dominant; as the signal power increases, the Q factor and the BER decrease. Assuming that the system is limited by signal-spontaneous beat noise and XPM, the power dependence of the signal-to-noise ratio may be expressed as (Mitra & Stark, 2001):

$$SNR = \frac{P_{ch}e^{-\left(P_{ch}/P_{XPM}\right)^2}}{P_{ASE} + P_{ch}\left(1 - e^{-\left(P_{ch}/P_{XPM}\right)^2}\right)} \approx \frac{P_{ch}\left[1 - \left(\frac{P_{ch}}{P_{XPM}}\right)^2\right]}{P_{ASE} + \frac{P_{ch}^3}{P_{XPM}^2}} \tag{21}$$

where P_{ch} is the power per channel (or subcarrier), P_{ASE} the optical noise (ASE) per polarisation (see Eq. (22)) and P_{XPM} a nonlinear threshold associated with XPM. Near the peak of this function, the exponential terms may be approximated with the first two terms of the Taylor series expansion, giving the right hand form of Eq. (21). The optical noise contains contributions from the transmitter OSNR, the Raman amplifier and the receiver preamplifier, and is given by:

$$P_{ASE} = \left[n_{eqBW} + n_{sp}\left(\frac{1}{G_{net}} - 1\right)\right]h\nu\Delta\nu_{opt} + P_{ASE,E1} \tag{22}$$

where $\Delta \nu_{opt}$ is the receiver optical bandwidth, transparency is assumed (the input power to the Raman amplifier is restored at the output of the receiver preamplifier) and $P_{ASE,E1}$ is the contribution from the transmitter OSNR. The XPM modelled by (Mitra & Stark, 2001) for a multi-span system assumes decorrelation in space and time, which corresponds to particular ratios of nonlinear, dispersion and walk of lengths. However, for a single span system, the impact of such decorrelation is reduced, and we could anticipate that the nonlinear intensity would be inversely proportional to the Raman effective length. Taking into account this, we find that a good fit to our experimental data is obtained if we multiply the nonlinear threshold from (Mitra & Stark, 2001) by the ratio of effective lengths with and without Raman pumping, as following:

$$P_{XPM} = \sqrt{\frac{BD\Delta\lambda L_{eff,EDFA}}{2\gamma^2 \ln(N_{ch}/2)L_{eff,R}^2}} \tag{23}$$

where γ and D are the nonlinear and the chromatic dispersion coefficients of the SMF, N_{ch} is the total subcarrier number, B is the subcarrier bandwidth and $\Delta\lambda$ is the subcarrier spacing. The effective Raman length $L_{eff,R}$ and the effective EDFA length $L_{eff,EDFA}$ are:

$$L_{eff,R} = \int_0^L G_{net}(z)\,dz \tag{24}$$

$$L_{eff,EDFA} = \frac{1-e^{-\alpha_s L}}{\alpha_s} \tag{25}$$

where L is the SMF span length. From Eq. (21) we observe that the Q-factor (or equivalently the SNR) of the system will vary substantially when varying both the Raman gain and the launched power into the optical fibre, because of the trade-off between OSNR (or NF) and nonlinearities. Assuming dominance of XPM, we can predict the Q-factor of the system when varying both the Raman gain and the launched power into the optical fibre, as in the contour plot depicted in Fig. 9. An optimum operating area is identified were the Q-factor is maximum; the optimum launch power into the SMF with Raman amplification is few dBs lower than the EDFA only system, and it may be varied by around 5 dB for approximately the same delivered BER or Q-factor. An equivalent Raman gain must be provided to counterbalance the lower launch power. Note that for gains below 17.2 dB only the BW pumps were used, whilst for higher gains maximum BW pump power was combined with an appropriate level of forward pumping.

2.3.5 Experimental results

Experimentally two cross-sections (σ_1 and σ_2) of Fig. 9 were studied. For a fixed Raman gain of 17.2 dB (BW only), the gain of the first EDFA was varied in order to evaluate the nonlinear power threshold (corresponding to cross section σ_1 in Fig. 9). Fig. 10 shows the Q-factor in dB, calculated directly from the BER measurements for the worst-performing optical subcarrier #48 (left-axis, blue circles) at the receiver, against the launch power, varied via the gain of the first EDFA. A similar trend was measured on subcarrier #17; hence negligible variation in the nonlinear performance is expected across the 21 nm bandwidth. Fig. 10 also shows the OSNR for the 7th band (containing subcarrier #48) measured at the

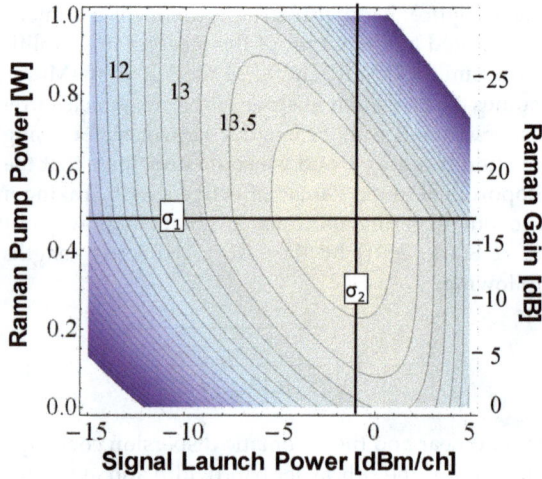

Fig. 9. Predicted Q-factor as a function of power per subcarrier and Raman gain, assuming XPM and limitations, 3.4 dB multiplexing Q-penalty and finite transmitter OSNR (see Fig. 10 for details). See text for details of Raman pump conditions. Contour levels are in steps of 0.25. Operating conditions for BER measurements shown as cross-sections σ_1 and σ_2.

output of the first EDFA (#1 in Fig. 6) showing the small variation in OSNR, as a result of variations in the amplifier population inversion. At low power levels, the degradation in Q-factor (left-axis, blue circles) is caused by a lower OSNR (right-axis, red circles); at high powers, the OSNR increases but the Q-factor starts degrading due to nonlinearities. The optimal operating condition was about –1 dBm per subcarrier launched into the SMF. Fig. 10 also shows the analytical fit for the Q-factor, taking into account the transmitter OSNR and a 3.4 dB multiplexing Q-penalty observed for this subcarrier (which could be found in the experimental measurement (Fig. 12)) showing agreement within 0.5 dB across the entire launch power range studied.

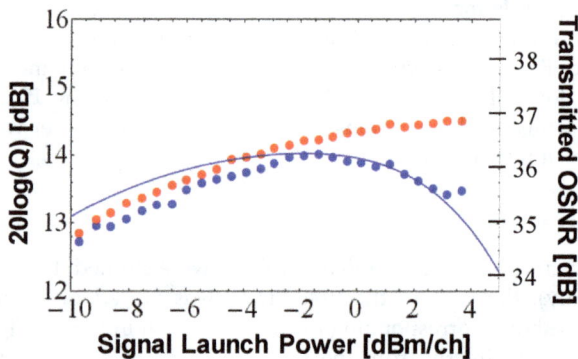

Fig. 10. Q-factor calculated from BER measurements (blue circles, left-y axis) for optical subcarrier #48, and measured OSNR (for the associated band) at the output of the transmitter EDFA (red circles, right-y axis) against power per subcarrier at the input of 124 km installed-SMF. Solid line represents analytically predicted performance (see Fig. 9).

At the optimum launch power, the Raman gain was increased from 17.2 dB by increasing the FW pump power or decreased from this level by reducing the BW pump power (cross-section σ_2 in Fig. 9). In Fig. 11 the Q-factor (calculated from measured BER for optical subcarrier #48 (left-axis)) against the on-off Raman gain is plotted as blue circles. The measured OSNR at the output of the receiver preamplifier is also shown on the right-axis. Experimental optimum working conditions were found to be a Raman gain of 17.2 dB (BW Raman only) and launch power of –1 dBm. This agrees well with the analytical predictions (solid lines in Fig. 10 and 11). At this operating point, all 49 subcarriers were characterized in terms of BER performance, and the corresponding Q-factors are shown in Fig. 12 (right-y axis) along with the transmitted spectrum, in order to identify the 49 wavelengths. An average Q-factor of 15 dB across the 49 subcarriers was observed after transmission (Frascella et al., 2011), which gave a system margin of ~4 dB when related to a BER of 2×10^{-3} (enhanced FEC threshold) for 2 Tbit/s system based on CoWDM. BW Raman amplification induced a 3 dB OSNR improvement when compared to an EDFA only amplification system (Frascella et al., 2010b), but at maximum BW Raman gain only a 1.2 dB improvement in the Q-factor was achieved due to the impact of cross phase modulation.

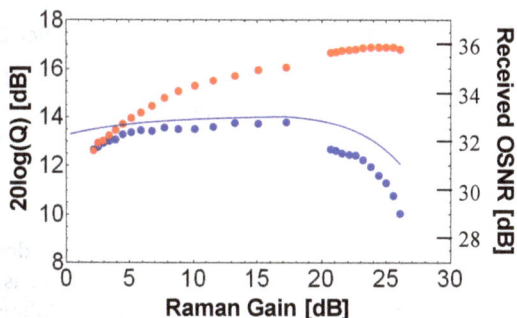

Fig. 11. Q-factor calculated from BER measurements (blue circles, left-y axis) for optical subcarrier #48, and measured OSNR (for the associated band) at the output of the receiver preamplifier (red circles right-y axis) as a function of Raman gain (below 17dB, backwards pumping only). Solid line represents analytically predicted performance (see figure 9).

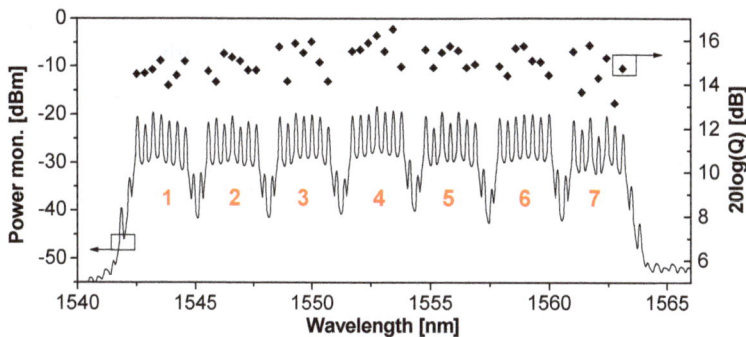

Fig. 12. Spectrum at the transmitter output, with band numeration shown (left-y axis). Calculated Q-factor from BER measurements of all 49 subcarriers with hybrid Raman/EDFA amplification (right y-axis).

3. Discussion

Having established the accuracy of the analytical predictions using both gain measurements and Q-factor analysis, Eq. (21) may be used to estimate the maximum reach of a 40 Gbaud direct detected OOK CoWDM system. Further assuming periodic dispersion compensation such that the signal power and the pulse shape at the input of each span is the same, the nonlinear threshold remains as given in Eq. (23), and the optical noise power needs to be multiplied by the number of fibre spans N_A.

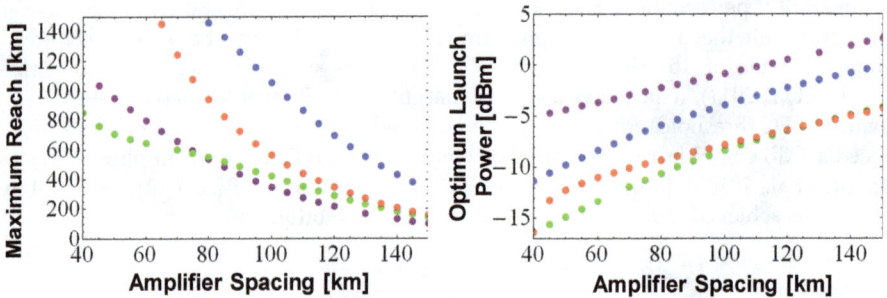

Fig. 13. Reach (left) and optimum per subcarrier launch power (right) of a 40 Gbaud OOK CoWDM system for delivering a BER of 10^{-5} (corresponding to frame loss free Ethernet performance) calculated under the experimental conditions (3.4 dB maximum multiplexing penalty and finite transmitter OSNR). Purple, EDFA only; Blue, Backwards pumping; Red forwards pumping; Green, bidirectional pumping. Pump powers at maximum in each case.

This approach may be used to determine the total reach L_T dependence on the amplifier spacing L_A (implicit in N_A) and the optimum signal launch power P_{in}, as plotted in Fig. 13 for a target SNR of 16.3 dB, corresponding to a worst case BER of 10^{-5} (allowing from frame loss free Ethernet performance after FEC) and a 3.4 dB multiplexing penalty. As expected, the reach is larger for low amplifier spacing, which offers reduced gains and reduced levels of ASE, although this solution tends to result in increased cost due to the increased number of amplifier sites. The EDFA only case offers the lowest reach, whilst the BW Raman amplification alone allows the maximum increase in reach. The forward pumping allows instead for significantly lower optimum signal launch powers (right hand graph in Fig. 13), but the OSNR benefits are reduced by the increased effective length.

At 124 km spacing, it is confirmed that the best reach is achieved at a launch power close to –1 dBm in good agreement with the measured nonlinear threshold. Using reduced amplifier spacing of 80 km or below, the reach is increased to beyond 1,400 km, confirming the suitability of 40 Gbaud direct detected OOK CoWDM system for use in ultra high capacity metro area networks employing dispersion management.

4. Conclusions

This chapter demonstrates that direct detection CoWDM with EDFA amplification only is suitable for Terabit Ethernet transport over unrepeatered spans up to ~130 km. Raman amplification would allow for an increased system margin, where necessary. Experimental demonstration showed that one 124 km span transmission with Raman amplification left a

Q-factor system margin of about 4 dB, which is consistent with theoretical expectations. For a repeated system employing EDFAs, with 80 km amplifier spacing the reach of direct detected 40 Gbaud OOK 2 Tbit/s is expected to be 500 km whilst with Raman amplification, reaches in excess of 1,000km are possible for repeater spacing below 100 km (Healy et al., 2007). Hence orthogonal multiplexing and direct detection constitute a feasible low-cost per bit solution for metropolitan links based on single-mode fibres.

5. Acknowledgments

The authors acknowledge F.C.G. Gunning, C. Antony, N. MacSuibhne, S.K. Ibrahim from the Photonics Systems Group of the Tyndall Institute and P. Gunning from BT Innovate and Design for invaluable assistance with the experimental demonstrations; W. McAuliffe and D. Cassidy from BT Ireland for provision of and access to the installed optical fibre; D. Pearce from Ixia Europe for the loan of Ethernet Protocol Test Equipment. This work was supported in part by Science Foundation Ireland (SFI) under grant number 06/IN/I969

6. References

Abramovitz, M. & Stegun, I.A. (1972). Handbook of Mathematical Functions. Dover Publications, New York, 1972, chapters 15 and 22.

Agrawal, G.P. (2001). Nonlinear Fiber Optics. 3rd ed., Academic Press, San Diego, CA, 2001.

Agrawal, G.P., (2002). Fiber-Optic Communication Systems. 3rd edition, Wiley Inter-Science, New York, 2002.

Bocoi, A., Schuster, M., Rambach, F. & Kiese, M., Bunge, C.-A. & Spinnler, B. (2009). Reach-Dependent Capacity in Optical Networks Enabled by OFDM. Proceedings of Optical Fiber Communication (OFC) Conference 2009, OMQ4.

Bromage, J., Bouteiller, J.-C., Thiele, J., Brar, K., Nelson, L.E., Stulz, S., Headley, C., Boncek, R., Kim, J., Klein, A., Baynham, G., Jørgensen, L.V., Grüner-Nielsen, L., Lingle, R.L.Jr. & DiGiovanni, D.J. (2004). WDM Transmission Over Multiple Long Spans With Bidirectional Raman Pumping. IEEE Journal of Lightwave Technology, Vol.22, No.1, pp.225-232.

Chandrasekhar, S., Liu, X., Zhu, B. & Peckham, D.W. (2009). Transmission of a 1.2-Tb/s 24-Carrier No-Guard-Interval Coherent OFDM Superchannel over 7200-km of Ultra-Large-Area Fiber. European Conference on Optical Communications (Sep 2009), PD2.6.

Chang, R. W. (1966). Synthesis of band-limited orthogonal signals for multichannel data transmission. Bell System Technological Journal, Vol. 45, pp. 1775-1796 (Dec. 1966).

Chang, R. (1970). Orthogonal frequency multiplex data transmission system. US Patent 3,488,445, filed Nov 1966, published Jan 1970.

Chen, H., Chen, M. & Xie, S. (2009). All-optical sampling orthogonal frequency-division multiplexing scheme for high-speed transmission system. IEEE Journal of Lightwave Technology, Vol.27, No.21, pp. 4848-4854.

Chen, D., Wheeler, S., Nguyen, D., Färbert, A., Schöpflin, A., Richter, A., Weiske, C.-J., Kotten, K., Krummrich, P.M., Schex, A. & Glingener, C. (2001). 3.2 Tb/s field trial (80 x 40 Gb/s) over 3 x 82 km SSMF using FEC, Raman and tunable dispersion compensation. Optical Fiber Communication Conference (2001), PD36.

Christodoulopoulos, K., Tomkos, I. & Varvarigos, E. (2011). Dynamic Bandwidth Allocation in Flexible OFDM-based Networks. Proceedings of *Optical Fiber Communication (OFC) Conference* 2011, OTuI5.

Cisco white paper (Feb 2011). *Cisco Visual Networking Index: Global Mobile Data Traffic Forecast Update, 2010–2015,* available from http://www.cisco.com

Cisco white paper (June 2011). *Entering the Zettabyte Era, 2010–2015,* available from http://www.cisco.com

Cuenot, B., Gunning, F.C.G., McCarthy, M., Healy, T. & Ellis, A.D. (2007). 0.6 Tbit/s capacity and 2 bit/s/Hz spectral efficiency at 42.6 Gsymbol/s using a single DFB laser with NRZ coherent WDM and polarization multiplexing. Proceedings of CLEO-Europe (2007), CI8-5-FRI.

Deman, P. (1964). Frequency and Time Allocation Multiplex System. US Patent 3163718, filed Jun 1962, published Dec 1964.

Desurvire, E. (2002). Erbium Doped Fiber Amplifiers. Principles and Applications. John Wiley & Sons, New Jersey 2002, pp.121-136.

Dischler, R., Klekamp, A., Buchali, F., Idler, W., Lach, E., Schippel, A., Schneiders, M., Vorbeck, S. & Braun, R.-P. (2010). Transmission of 3 x 253-Gb/s OFDM-Superchannels over 764 km field deployed single mode fibers. *Optical Fiber Communication Conference* (2010), PDPD2.

Djordjevic, I.B. & Vasic, B. (2006). Orthogonal Frequency Division Multiplexing for High-Speed Optical Transmission. *Optics Express,* Vol.14, pp. 3767-3775, May 1, 2006.

Dunne, J., Farrell, T. & Shields, J. (2009). Optical Packet Switch and Transport: A New Metro Platform to Reduce Costs and Power by 50% to 75% While Simultaneously Increasing Deterministic Performance Levels. *International Conference on Tranpsarent Optical Networks* (July 2009), Mo.D4.4.

Ellis, A.D. & Gunning, F.C.G. (2005). Spectral density enhancement using Coherent WDM. *IEEE Photonics Technology Letters,* Vol.17, No.2, pp. 504-506.

Ellis, A.D., Tomkos, I., Mishra, A.K., Zhao, J., Ibrahim, S.K., Frascella, P. & Gunning, F.C.G. (2009a). Adaptive Modulation Schemes. *Digest of LEOS Summer Topical Meetings* (2009), TuD3.2.

Ellis, A.D., Gunning, F.C.G. & Healy, T.C. (2009b). Communication Systems. US patent filed Oct 2006, published Mar 2009, US 2009/0074416 A1.

Ellis, A.D., Zhao, J. & Cotter, D. (2010). Approaching the Non-Linear Shannon Limit. *IEEE Journal of Lightwave Technology,* Vol.28, No.4, pp.423-433.

Essiambre, R.-J., Winzer, P.J. & Grosz, D.F. (2006). Impact of DCF properties on system design. *Journal of Optical Fiber Communications* Reports 3, pp.221-291.

Essiambre, R.-J., Kramer, G., Winzer, P.J., Foschini, G.J. & Goebel, B. (2010). Capacity Limits of Optical Fiber Networks. *IEEE Journal of Lightwave Technology,* Vol.28, No.4, pp.662-701, Feb 2010.

Feced, R., Rickard, R. & Richard, E. (2005). Reference Phase and Amplitude Estimation for Coherent Optical Receiver. US Patent 2005/0180760, filed Feb 2004, published Aug 2005.

Frascella, P., Mac Suibhne, N., Gunning, F.C.G., Ibrahim, S.K., Gunning, P. & Ellis, A.D. (2010c). Unrepeatered field transmission of 2 Tbit/s multi-banded coherent WDM over 124km of installed SMF. *Optics Express,* Vol.18, No.24, pp. 24745-24752 (Nov 2010).

Frascella, P., Gunning, F.C.G., Ibrahim, S.K., Gunning, P. & Ellis, A.D. (2010b). PMD tolerance of 288 Gbit/s Coherent WDM and transmission over unrepeated 124 km of field-installed single mode optical fiber. *Optics Express*, Vol.18, No.13, pp. 13908-13914, June 2010.

Frascella, P., Ibrahim, S.K., Gunning, F.C.G., Gunning, P. & Ellis, A.D. (2010a). Transmission of a 288Gbit/s Ethernet Superchannel over 124km un-repeated field-installed SMF. *Optical Fiber Communication Conference* (Mar 2010), OThD2.

Gavioli, G., Torrengo, E., Bosco, G., Carena, A., Curri, V., Miot, V., Poggiolini, P., Belmonte, M., Guglierame, A., Brinciotti, A., La Porta, A., Forghieri, F., Muzio, C., Osnago, G., Piciaccia, S., Lezzi, C., Molle, L. & Freund, R. (2009). 100 Gb/s WDM NRZ-PM-QPSK Long-Haul Transmission Experiment over Installed Fiber Probing Non-Linear Reach with and without DCUs. *European Conference on Optical Communications* (Sep 2009), Tu3.4.2.

Goldfarb, G., Li, G. & Taylor, M.G. (2007). Orthogonal Wavelength-Division Multiplexing using Coherent Detection. *IEEE Photonics Technology Letters*, Vol.19, No.24, pp. 2015-2017.

Gunning, F.C.G., Healy, T., Manning, R.J. & Ellis, A.D. (2005). Multi-banded Coherent WDM Transmission. *European Conference on Optical Communications*, Proceedings Vol.6, Paper Th4.2.6 (Sep 2005).

Headley, C. & Agrawal, G.P., (2005). Raman Amplification in Fiber Optical Communications Systems. Elsevier Academic Press, ISBN: 0-12-044506-9.

Healy, T., Gunning, F.C.G. & Ellis, A.D., (2006). Phase Stabilisation of Coherent WDM Modulator Array. *Optical Fiber Communication Conference* (Mar 2006), OTuI5.

Healy, T., Ellis, A.D., Gunning, F.C.G., Cuenot, B. & Rukosueva, M. (2006). 1 b/s/Hz Coherent WDM Transmission over 112 km of Dispersion Managed Optical Fiber. *Optical Fiber Communication Conference* (Mar 2006), JThB10.

Healy, T., Gunning, F.C.G., Ellis, A.D. & J.D. Bull (2007). Multi-wavelength source using low drive-voltage amplitude modulators for optical communications. Optics Express, Vol.16, No.6, pp.2981-2986, March 2007.

Healy, T., Gunning, F.C.G., Pincemin, E., Cuenot, B. & Ellis, A.D. (2007). 1,200 km SMF (100 km spans) 280 Gbit/s Coherent WDM Transmission using Hybrid Raman/EDFA Amplification. *European Conference on Optical Communications* 2007, Mo.1.3.5.

Hillerkuss, D., Schmogrow, R., Schellinger, T., Jordan, M., Winter, M., Huber, G., Vallaitis, T., Bonk, R., Kleinow, P., Frey, F., Roeger, M., Koenig, S., Ludwig, A., Marculescu, A., Li, J., Hoh, M., Dreschmann, M., Meyer, J., Ben Ezra, S., Narkiss, N., Nebendahl, B., Parmigiani, F., Petropoulos, P., Resan, B., Oehler, A., Weingarten, K., Ellermeyer, T., Lutz, J., Moeller, M., Huebner, M., Becker, J., Koos, C., Freude, W. & Leuthold, J. (2011). 26 Tbit s[-1] line-rate super-channel transmission utilizing all-optical fast Fourier transform processing. *Nature Photonics*, Vol.74, pp.1-8, DOI:10.1038, May 2011.

Huang, Y.-K., Qian, D., Saperstein, R.E., Ji, P.N., Cvijetic, N., Xu, L. & Wang, T. (2009). Dual-Polarization 2x2 IFFT/FFT Optical Signal Processing for 100-Gb/s QPSK-PDM All-Optical OFDM. *Optical Fiber Communication Conference* (Mar 2009), OTuM4.

Ibrahim, S.K., Zhao, J., Gunning, F.C.G., Frascella, P., Peters, F.H. & Ellis, A.D. Towards a Practical Implementation of Coherent WDM: Analytical, Numerical, and Experimental Studies. *IEEE Photonics Journal*, Vol.2, No.5, Oct 2010.

Ito, S. & Yokoyama, S. (1967). Phase-modulated frequency division multiplex system. US Patent 3,349,182, filed June 1964, published Oct 1967.

Jansen, S.L. et al. (2008a). OSA-JON, Vol.7, pp. 173-182, 2008.

Jansen, S.L., Morita, I. & Tanaka, H. (2008b). 10 x 121.9-Gb/s PDM-OFDM transmission with 2-b/s/Hz spectral efficiency over 1000 km of SMF. *Optical Fiber Communication Conference* (Feb 2008), PDP2.

Jansen, S.L., Morita, I., Takeda, N. & Tanaka, H. (2007). 20-Gb/s OFDM Transmission over 4,160-km SSMF Enabled by RF-Pilot Tone Phase Noise Compensation. Proceedings of *Optical Fiber Communication (OFC) Conference* 2007, PDP 15.

Kaminow, I.P., Li, T. & Willner, A.E., (2008). Optical Fiber Telecommunications V A: Components and Subsystems, Elsevier 2008.

Kikuchi, K. (1990). Generalized formula for optical amplifier noise and its application to erbium-doped fibre amplifiers. Electronics Letters, Vol.26, No.22, p.1851 (1990).

Lam, C.F., Liu, H., Koley, B., Zhao, X., Kamalov, V. & Gill, V. (2010). Fiber Optic Communication Technologies: What's Needed for Datacenter Network Operations. IEEE Communications Magazine, Vol.48, No.7 (2010).

Lee D. (2011). Scaling Networks in Large Data Centers. *Optical Fiber Communication Conference* (Mar 2011), OWU1.

Liu, X., Buchali, F. & Tkach, R.W. (2009). Improving the nonlinear tolerance of polarization-division-multiplexed CO-OFDM in long-haul fiber transmission. *IEEE Journal of Lightwave Technology*, Vol.27, No.16, pp. 3632-3640, Aug. 2009.

Lowery, A.J., Du L. & Armstrong, J. (2006). Orthogonal Frequency Division Multiplexing for Adaptive Dispersion Compensation in Long Haul WDM Systems. *Optical Fiber Communication Conference* (2006) PDP39.

Lowery, A.J. (2010). Design of arrayed-waveguide grating routers for use as optical OFDM demultiplexers. *Optics Express*, Vol.18, No.13, pp. 14129-14143 (Jun 2010).

Mitra, P.P. & Stark, J.B. (2001). Nonlinear limits to the information capacity of optical fibre communications. Nature, Vol.411, pp. 1027-1030, 2001.

Mosier, R.R. & Clabaugh, R.G. (1958). Kineplex, a bandwidth efficient binary transmission system. *AIEE Transactions*, Vol.76, pp. 723-728 (Jan 1958).

Nissov, M., Rotwitt, K., Kidorf, H.D. &Ma, M.X. (1999). Rayleigh crosstalk in long cascades of distributed unsaturated Raman amplifiers. *Electronics Letters*, Vol.35, pp. 997-998.

Pizzinat, A., Santagiustina, M. & Schivo, C., (2003). Impact of Hybrid EDFA-Distributed Raman Amplification on a 4 x 40-Gb/s WDM Optical Cmmunication System. *IEEE Photonics Technology Letters*, Vol.15, No.2, pp. 341-343.

Poole, S., Frisken, S., Roelens, M. & Cameron C. (2011). Bandwidth-flexible ROADMs as Network Elements. Proceedings of *Optical Fiber Communication (OFC) Conference* 2011, OTuE1.

Proakis, J.G. & Salehi, M., (2008). Digital Communications. 5th edition, Mc Graw Hill, New York, 2008.

Qian, D., Huang, M.-F., Ip, E., Huang, Y.-K., Shao, Y., Hu, J., & Wang, T. (2011). 101.7-Tb/s (370 x 294-Gb/s) PDM-128QAM-OFDM transmission over 3 x 55-km SSMF using Pilot-based phase noise mitigation. *Optical Fiber Communication Conference* (2011), PDPB5.

Reuters, (2010a) "World Bank Sees African Economies Rebounding in 2010," March 18, 2010.

Reuters, (2010b) "Davos Special Report: Africa Rising," January 26, 2010.

Roberts, K. (2011). Digital Signal Processing for Coherent Optical Communications: Current State of the Art and Future Challenges. *OSA Annual Meeting, Advanced Photonics Congress*, 12-15 June 2011, Toronto, Canada, SPWC1.

Rodrigues, M.R.D. & Darwazeh, I (2002). Fast OFDM: A proposal for doubling the data rate of OFDM schemes. Proceedings of IEEE/IEE International Conference on Telecommunications (ICT 2002), Beijing, China. (pp. 484 - 487).

Sanjoh, H., Yamada, E. & Yoshikuni, Y. (2002). Optical orthogonal frequency division multiplexing using frequency/time domain filtering for high spectral efficiency up to 1 bit/s/Hz. *Optical Fiber Communication Conference* (2002), ThD1.

Sano, A., Yamada, E., Masuda, H., Yamazaki, E., Kobayashi, T., Yoshida, E., Miyamoto, Y., Kudo, R., Ishihara, K. & Takatori, Y. (2009). No-Guard-Interval Coherent Optical OFDM for 100-Gb/s Long-Haul WDM Transmission. *IEEE Journal of Lightwave Technology*, Vol.27, No.16, Aug 2009.

Sano, A., Masuda, H., Kobayashi, T., Yamada, E., Miyamoto, Y., Hibino, Y., Ishihara, K., Takatori, Y., Hagimoto, K., Yamada, T. & Sakamaki, Y. (2007). 30 x 100 Gb/s all-optical OFDM transmission over 1300 km SMF with 10 ROADM nodes. *European Conference on Optical Communications* (Sep 2007), PD1.7.

Schmogrow R. et al. (2011). 101.5 Gbit/s Real-time OFDM transmitter with 16QAM modulated subcarriers. *Optical Fiber Communication Conference* (Mar 2011), OWE5.

Shieh, W. & Authaudage, C. (2006). Coherent optical orthogonal frequency division multiplexing. *Electronics Letters*, Vol.42, No.10, p. 587.

Shieh, W., Yang, Q. & Ma, Y. (2008). 107 Gb/s coherent optical OFDM transmission over 1000-km SSMF using orthogonal band multiplexing. *Optics Express*, Vol.16, No.9, pp. 6378-6386.

Stolen, R.H., Ippen, E.P. & Tynes, A.R., (1972). Raman Oscillation in Glass Optical Waveguides. *Applied Physics Letters* 20 (2), pp. 62-64, 1972.

Takara, H., Kozicki, B., Sone, Y., Tanaka, T., Watanabe, A., Hirano, A., Yonenaga, K. & Jinno, M. (2010). Distance-Adaptive Super-Wavelength Routing in Elastic Optical Path Network (SLICE) with Optical OFDM. Proceedings of *European Conference on Optical Communication (ECOC)*, Sep 2010, Paper We.8.D.2.

Thiagarajan, S., Frankel, M. & Boertjes, D. (2011). Spectrum Efficient Super-Channels in Dynamic Flexible Grid Networks – A Blocking Analysis. Proceedings of *Optical Fiber Communication (OFC) Conference* 2011, OTuI6.

Walker, G.R., Spirit, D.M., Williams, D.L. & Davey, S.T. (1991). Noise performance of distributed fiber amplifiers. Electronics Letters, Vol.27, No.15, p.1390 (1991).

Weinstein, S. & Ebert, P. (1971). Data transmission by frequency division multiplexing the discrete Fourier transform. IEEE Transactions on Communication Technology, Vol.COM-19, pp. 628-634, October 1971.

www.infonetics.com (2011). ROADM WSS component market on pace for 20% annual growth through 2015. available at www.infonetics.com , May 2011.

Xia, T.J., Wellbrock, G.A., Huang, Y.-K., Ip, E., Huang, M.-F., Shao, Y., Wang, T., Aono, Y., Tajima, T., Murakami, S. & Cvijetic, M. (2011). Field experiment with mixed line-rate transmission (112-Gb/s, 450-Gb/s, and 1.15-Tb/s) over 3,560 km of installed fiber using filterless coherent receiver and EDFAs only. *Optical Fiber Communication Conference* (2011), PDPA3.

Yamamoto, Y. & Inoue, K. (2003). Noise in amplifiers. *IEEE Journal of Lightwave Technology*, Vol.21, No.11.

Yamamoto, S., Yonenaga, K., Sahara, A., Inuzuka, F. & Takada, A. (2010). Achievement of sub-channel frequency spacing less than symbol rate and improvement of dispersion tolerance in optical OFDM transmission. IEEE Journal of Lightwave Technology, Vol.28, pp.157-163 (2010).

Yonenaga, K., Inuzuka, F., Yamamoto, S., Takara, H., Kozichi, B., Yoshimatsu, T., Takada, A. & Jinno, M. (2009). Bit-rate-flexible all-optical OFDM transceiver using variable multi-carrier source and DQPSK/DPSK mixed multiplexing. *Optical Fiber Communication Conference* (Mar 2009), pp. 1-3, OWM1.

Zhao, J. & Ellis, A.D., (2010). A Novel Optical Fast OFDM with Reduced Channel Spacing Equal to Half of the Symbol Rate Per Carrier. *Optical Fiber Communication Conference* (Mar 2010), OMR1.

Zhao, J. & Ellis, A.D. (2011). Electronic Impairment Mitigation in Optically Multiplexed Multicarrier Systems. IEEE Journal of Lightwave Technology, Vol.29, No.3, pp.278-290, (2011).

Zhu, B., Taunay, T.F., Fishteyn, M., Liu, X., Chadrasekhar, S., Yan, M.F., Fini, J.M, Monberg, E.M., Dimarcello, F.V., Abedin, K., Wisk, P.W., Peckham, D.W. & Dziedzic, P. (2011). Space-, Wavelength-, Polarisation-Division Multiplexed transmission of 56-Tb/s over a 76.8-km seven-core fiber. *Optical Fiber Communication Conference* (Mar 2011), PDPB7.

Zimmerman, M.S. & Kirsch, A.L. (1967). The AN/GSC-10 (KATHRYN) Variable Rate Data Modem for HF Radio. IEEE Transactions on Communication Technology, Vol.COM-15, No.2, pp. 197-204 (1967).

7

Integration of Eco-Friendly POF Based Splitter and Optical Filter for Low-Cost WDM Network Solutions

Mohammad Syuhaimi Ab-Rahman, Hadi Guna,
Mohd Hazwan Harun, Latifah Supian and Kasmiran Jumari
Universiti Kebangsaan Malaysia, Selangor Darul Ehsan
Malaysia

1. Introduction

The field of *'green technology'* encompasses a continuously evolving group of methods and materials, from techniques for generating energy to non-toxic cleaning products. The present expectation is that this field will bring innovation and changes in daily life of similar magnitude to the *'information technology'* explosion over the last two decades. In these early stages, it is impossible to predict what *'green technology'* may eventually encompass (Gupta and Khurana 2010).

Nowadays, the whole world of telecommunications and information communities is facing a more and more serious challenge, namely on one side the transmitted multimedia-rich data are exploding at an astonishing speed and on the other side the total energy consumption by the communication and networking devices and the relevant global CO_2 emission are increasing terribly. It has been pointed out that 'currently 3% of the world-wide energy is consumed by the information & communications technology (ICT) infrastructure that causes about 2% of the world-wide CO_2 emissions, which is comparable to the world-wide CO_2 emissions by airplanes or one quarter of the world-wide CO_2 emissions by cars' (Janota and Hrbček 2011).

According to the recent research report of Ericsson Media Relations, energy costs account for as much as half of a mobile operator's operating expenses. Therefore, telecommunications applications can have a direct, tangible impact on lowering greenhouse gas emissions, power consumption, and achieve efficient recycling of equipment waste. Moreover, to find radio networking solutions that can greatly improve energy-efficiency as well as resource-efficiency is not only benefit for the global environment but also makes commercial sense for telecommunication operators supporting sustainable and profitable business. Within the framework of *'green communications'*, a number of paradigm-shifting technical approaches can be expected, including but not limited to energy-efficient network architecture & protocols, energy-efficient wireless transmission techniques (*e.g.*, reduced transmission power & reduced radiation), cross-layer optimization methods, and opportunistic spectrum sharing without causing harmful interference pollution (Ericsson 2008).

The green technology wavelength division multiplexing based on polymer optical fiber (GT-WDM-POF) network solution is presented. Green technology polymer optical fiber (GT-POF) splitter has been fabricated by environmental friendly fusion technique, as an effective transmission media to split and recombine a number of different wavelengths. Two different wavelengths from ecologically friendly light emitting diode (LED) were fully utilized to transmit two different sources of systems; Ethernet connection and video transmission system. Red LED which in 650nm wavelength capable to download and upload data through Ethernet cable while green LED in 520nm wavelength transmits a video signal. Special filter has been placed between the splitter and receiver-end to ensure GT-WDM-POF network system can select and generate a single signal as desired. Efficiency of both devices and network were observed. The material, fabrication method, system & application approach in this chapter are based on the environment friendly solution to reduce the power consumption & wastage without affecting the system performance. Our GT-POF splitter and GT-WDM-POF network solution proposed in this paper are the first reported up to this time.

1.1 Eco-friendly sources

LEDs, or light emitting diodes, are the light source in solar powered products. This solid-state product is composed of a semiconductor diode. A semiconductor is a material that can conduct electricity. A semiconductor diode is composed of a semi-conductive crystal that has added impurities in order to create a positive and negative side; since current flows in one direction through the diode. A region is then created in between the positive and negative zones, called a *PN*-junction, which is where the action takes place within the diode; in our case it emits light. LEDs are preferably used due to the many advantages can be offered: *Efficiency*: An incandescent light requires much more energy to properly heat the filament in order to generate light. The light produced by an LED is a cool light. More light is produced per watt in an LED than an incandescent. Even more energy can be saved if the light is solar powered.

LED bulbs are widely used in visible light communication which replacing *Wi-Fi* and offer much higher capacity. *Color*: LEDs do not require filters, like colored bulbs, in order to create a specific colored light. Color is produced based on the material of the semiconductor. The different color represents the frequency of carrier signal that can be used to increase the data capacity by introducing Wavelength Division Multiplexing (WDM). *Size*: LEDs come in many different sizes since they are not constrained to creating a vacuum in which to house the filament. LEDs can be smaller than 2 mm. *On/Off Time*: An LED takes only microseconds to achieve its full brightness. This is ideal in a solar powered light that is running off of a battery that has determinate energy life. On/Off time or blinking of Led also represent the capacity of data that can be modulated. *Cycling*: In applications that are cycled between on and off frequently, like an outdoor solar light, LEDs are ideal since they won't burn out quickly. *Lifetime*: The lifetime of the LED greatly exceeds its incandescent, and even its fluorescent counterparts. An average lifetime of an incandescent light is 1,000-2,000 hours and a fluorescent bulb is 10,000-15,000 hours. The LED, on the other hand, has a typical lifetime of 35,000-50,000 hours. *Light Dispersement*: An LED is designed to focus its light, so where an incandescent or fluorescent may seem brighter since the light radiates in all directions, the LED light can be directed to a specific location without the use of an external reflector.

The smaller beam size enable the light be injected to the multimode fiber for communication application. *Ecologically Friendly*: LEDs are more efficient than others, as stated above; so they conserve electricity, especially if they are solar powered. LEDs do not contain toxic chemicals like fluorescent bulbs do. Several incandescent and fluorescent bulbs will be used during the lifetime of a single LED. If your desire is to light a space and save the environment, then the clear choice is the LED. Solar powered LEDs are an additional benefit in that they require no additional energy costs.

Three major benefits of this solid-state lighting technology, shown in Fig. 1, can be summarized as follows: *firstly*, the inherent capability of solid-state sources to generate light with high efficiency is resulting in giant energy savings. *Secondly*, potentially huge environmental benefits are a result of the efficiency and durability of solid-state emitters, particularly light-emitting diodes based on inorganic semiconductors. *Thirdly*, solid-state emitters allow one to control the emission properties with much greater precision, thereby allowing one to custom-tailor the emission properties for specific applications (Schubert, Kim et al. 2006).

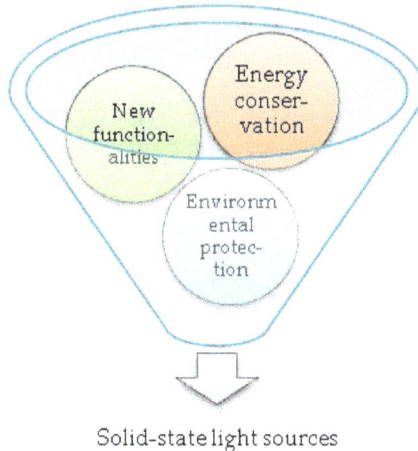

Fig. 1. Benefits enabled by solid-state light sources

1.2 Green technology medium

Nowadays, polymer optical fiber (POF) become alternative transmission media replacing copper or even glass fiber for short-haul communication. POF links are becoming increasingly popular for applications such as computer or peripheral connections, control and monitoring, board interconnects and even domestic *Wi-Fi* systems. Unlike glass fiber, POF remains flexible while having a large diameter core and high numerical aperture (Keene and Selli 1989), lead to high capacity they can bring along the fiber. Moreover, the fiber is easy to handle with the potential for constructing networks using simple conductor and easy installation procedures while retaining some of the advantages of optical fiber such as electromagnetic interference (EMI) immunity, non-conducting cable, small size and security. Another feature is the use of visible light to transmit information (Kuzyk 2007).

Due to wide advantages of POF over copper or even glass fiber, POF are used widely in various optical networks. Recent communication system over POF desires increasingly more bandwidth and therefore the wavelength division multiplexing (WDM) system is the solution that allows the transmission of information over more than just a single wavelength (color) and thus greatly increases the POF's bandwidth. WDM is a technique that multiple signals are carried together as separate wavelengths (color) of light in a multiplexed signal (Kuzyk 2007; Grzemba 2008; Ziemann, Zamzow et al. 2008).

1.3 Multipurpose applications

Video Communications efficiently utilizes energy, reduces unnecessary travel, and significantly shrinks carbon emissions. Video Communication allows our organization to reduce carbon footprint, since one of the most effective ways to reduce carbon dioxide emissions is to reduce unnecessary travel. It provides measureable result to help us achieve more with less. Video Communication is the green way to communicate. Through the introduction of Video Conferencing your activities can; a) communicate better and optimize work-life-balance, b) become more environmentally responsible and c) reduce the organization carbon footprint. The International environmental organizations specify video communications as an effective green-technology to reduce carbon footprint, thereby reducing global warming effects. The Nature Conservancy lists video conferencing as one of its 'Easy Things You can Do to Help Our Climate.' The World Wildlife Fund has released reports that demonstrate how video conferencing dramatically reduces carbon emissions.

Communication using video conferencing technology offer an ideal solution to enable us to reduce the amount of time you spend in your car or travelling, while allowing us to visually connected to anyone, anywhere, at any time. It improves the environment with lower hydrocarbon emissions, and a reduction of fuel consumption and pollution.

1.4 Proposed network

In wavelength division multiplexing based on polymer optical fiber (WDM-POF) system, many transmitters with different lights color to carry single information. For example, red light with 650nm wavelength modulated with Ethernet signal while blue, green, and yellow lights carry image information, radio frequency (RF), and television signal, respectively (Ericsson 2008). As shown is Fig. 2, Wavelength Division Multiplexer is the first passive device required in WDM-POF system and it functions to combines optical signals from multiple different single-wavelength end devices onto a single fiber.

Conceptually, the same device can also perform the reverse process with the same WDM techniques, in which the data stream with multiple wavelengths decomposed into multiple single wavelength data streams. The reverse process is called as de-multiplexing.

In general, POF splitter Conceptually, POF splitter has similar function, operates to couple or combine several optical data pulse as a single coupled signal. Hence, the development of wavelength division multiplexer based on POF splitter is possible.

Typically, the commercial POF splitter that manufactured commercially by some manufacturer priced expensively at approximately more than US$250 in global market. There have been many techniques of fabricating POF splitter. These techniques include

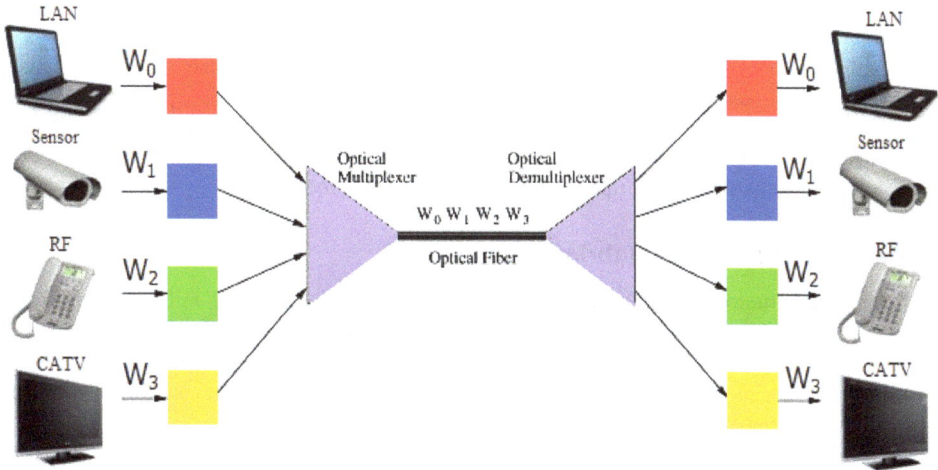

Fig. 2. Wavelength Division Multiplex (WDM) with an optical multiplexer and demultiplexer

twisting and fusion, side polishing, chemical etching, cutting and gluing, thermal deformation, moulding, biconical body and reflective body (Ehsan, Shaari et al. 2009).

For this chapter, fusion technique is practically proposed to fabricate POF splitter. Essentially, the term of *'fusion'* defines the act or procedure of liquefying or melting by the application of heat (Imoto, Maeda et al. 1987). In order to develop the economical POF splitter, this study is undertaken to modify the typical fusion technique, whereby the technique is fully implemented by handwork. The heating elements and immune-to-heat tube (from the previous fusion technique) are changed in terms of availability and the appropriate twisting and pulling strengths are tuned specifically for the modified fusion technique (Kelly, May et al. 1995; Ab-Rahman, Guna et al. 2008). In this chapter, the characterization of the *GT-POF splitter* is carried out in order to determine the performance of device. Besides, study on how far the WDM-POF system can go, and how far color filter influences the output power of the system also reported.

2. The WDM technology

Three major benefits of this solid-state lighting technology can be summarized as follows: *firstly*, the inherent capability of solid-state sources to generate light with high efficiency is resulting in giant energy savings. *Secondly*, potentially huge environmental benefits are a result of the efficiency and durability of solid-state emitters, particularly light-emitting diodes based on inorganic semiconductors. *Thirdly*, solid-state emitters allow one to control the emission properties with much greater precision, thereby allowing one to custom-tailor the emission properties for specific applications (Schubert, Kim et al. 2006).

In wavelength division multiplexing based on polymer optical fiber (WDM-POF) system, many transmitters with different lights color to carry single information. For example, red light with 650nm wavelength modulated with Ethernet signal while blue, green, and yellow lights carry image information, radio frequency (RF), and television signal, respectively

(Ericsson 2008). WDM is the first passive device required in WDM-POF system and it functions to combines optical signals from multiple different single-wavelength end devices onto a single fiber. Conceptually, the same device can also perform the reverse process with the same WDM techniques, in which the data stream with multiple wavelengths decomposed into multiple single wavelength data streams. The reverse process is called as de-multiplexing. As we know, the most essential devices needed in common WDM-POF technology are transmitter, multiplexer, demultiplexer and receiver.

3. Environmental friendly fabrication method

For this chapter, fusion technique is practically applied to fabricate POF splitter. Essentially, the term of *'fusion'* defines the act or procedure of liquefying or melting by the application of heat (Imoto, Maeda et al. 1987). In order to develop the economical POF splitter, this chapter is undertaken to modify the typical fusion technique, whereby the technique is fully implemented by handwork.

The heating elements and immune-to-heat tube (from the previous fusion technique) are changed in terms of availability and the appropriate twisting and pulling strengths are tuned specifically for the modified fusion technique (Kelly, May et al. 1995; Ab-Rahman, Guna et al. 2008). In this chapter, the characterization of the GT-POF splitter is carried out in order to determine the performance of device. Besides, chapter on how far the WDM-POF system can go, and how far color filter influences the output power of the system also reported.

The $1 \times N$ *GT-POF splitter* is an optical device, which ended by N number of POF ports, while the other side ended by one POF port. Like other typical splitter, it is also possible to work bidirectional, whereby it works from the N ports into 1 port (for coupling signal purpose), or *vice versa* (for splitting signals purpose). As an example, 1×4 *GT-POF splitter* developed by the jointing of four *Polymethylmethacrylate* (PMMA) POFs. Each inputs and output is connected with POF connecter as shown in Fig. 3.

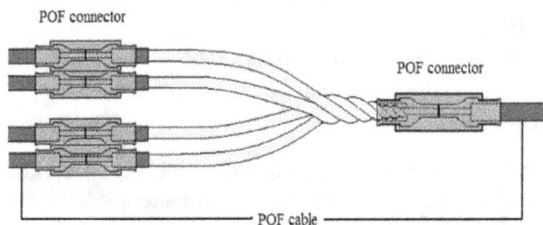

Fig. 3. The design of the 1×4 GT-POF splitter

For the filter design which able to eliminate unwanted signal and select the wavelength of the system as desired as shown in Fig. 4.

In development process of $1 \times N$ *GT-POF splitter* based on POF technology, multimode SI-POF type made of PMMA 1 mm core size fully utilized in this paper, as PMMA is one of the most commonly used optical materials, Due to its intrinsic absorption loss mainly contributed by carbon–hydrogen stretching vibration in PMMA core POF (Kuzyk 2007).

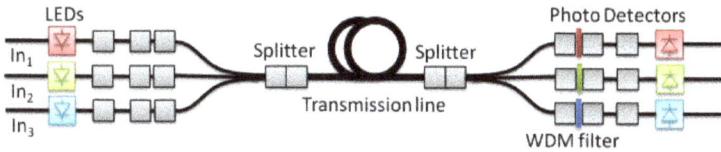

Fig. 4. GT-WDM-POF network architecture using 1 × 3 GT-POF splitter and color filters. The splitter can also be used as multiplexer and the splitter-filter combination can be performed as demultiplexer

To fabricate the final product of optical 1 × N GT-POF splitter, some stages has to be done, start from fiber fusion, bundle formation and finalized with cable jointing. Fabricated through fusion method by fuses and combine N number POFs (in bundle arrangement) and fabricate it ends part in a shape of fused-taper-twisted fibers (diameter 1 mm). POFs will be twisted and pulled down while it is fused in a heat of flame. Heating process was done indirectly, while POFs covered by metal tube. Thus, heat was provided for POFs through metal tube heating.

For characterization process, here we choose a number of samples of 1 × 4 GT-POF splitter to measure the efficiency of the GT-POF splitter. The developed splitter must be able to properly coupling an optical signal to generate a single coupled signal efficiently, with low power loss. Optical power meter has been used to measure the optical power from POFs.

Bidirectional optical loss measurement is carried out in order to determine either side of the 4 × 4 GT-POF splitter with lower optical loss as final product of 1 × 4 GT-POF splitter before cutting the middle of the 4 × 4 GT-POF splitter. Red LED injected through each of inputs individually and separately from the right side (lights propagate leftward) in order to measures output powers and calculates the optical loss. Then, similar procedure is repeated for rightward measurement. Finally, optical loss for fused bundle in both directions analytically compared. The procedure explained above visualized as in Fig. 5.

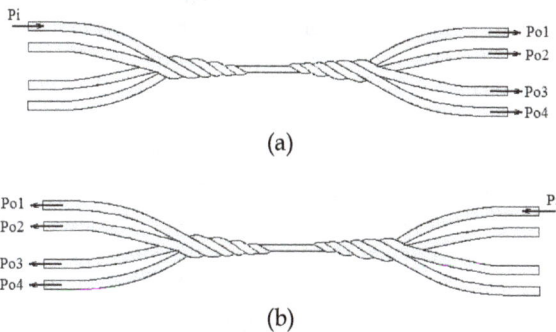

(a)

(b)

Fig. 5. Measurement process for bi-directional 4 × 4 GT-POF splitter

In order to measure the power efficiency of splitter, at first, red LED injected from transmitter pit into single POF cable (1 mm of core diameter) and obtained power defined as input power while output power obtained by injecting LED into POF splitter (through single POF port) and each POF ports (on cascaded side) measured by optical power meter. The procedure of input/output power measurement depicted in Fig. 6.

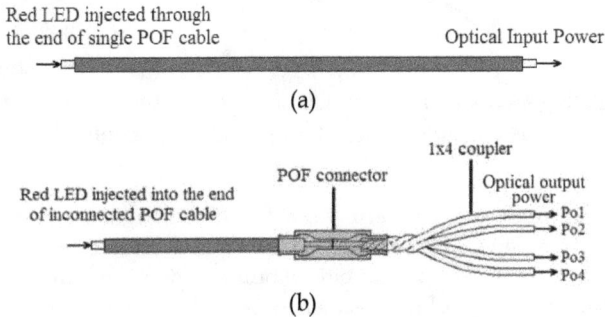

Red LED injected through
the end of single POF cable   Optical Input Power

(a)

1x4 coupler

POF connector   Optical output
Red LED injected into the end   power
of inconnected POF cable   Po1
                                                                 Po2
                                                                 Po3
                                                                 Po4

(b)

Fig. 6. The illustration of (a) input and (b) output power measurement

4. Results

In order to investigate precisely the exact value of power intensity for each POF outputs of fused bundle, bidirectional optical loss measurement has been carried out whereby injecting red LED through each of POF inputs on both sides of fused bundle separately. The average optical loss for fused POF bundle has been yet calculated for both directions (leftward and rightward) and then analytically compared. According to the observation above, it is revealed that the optical loss for fused bundle in different direction was not identical. Analytically, fused POF bundle has lower optical loss in rightward direction. Thus, right side of fused POF bundle selected as POF splitter because it might couple multiple optical signals and produce single optical signal with lower attenuation and higher efficiency compared to the other side. Indeed, optical loss for fused bundle mainly caused by physical changes on POF especially on fused taper twisted in which POFs in bundle arrangement were all fused, twisted and merged.

Through modified fusion technique (see Fig. 7), the ideal sample of fused-taper-twisted POFs successfully produced which the diameter of fused taper-twisted POF approaching 1 mm. As the result of injecting red LED through the fused POF bundle, it is observed that red light pass smoothly through fused taper-twisted POFs. It is observed that all POF outputs did illuminate red light individually with different intensity power. So this means that no deformation occurred along fused taper-twisted POFs and this sample has ability of signal coupling.

For commercial purpose, research also produces a *GT-POF splitter* come out with a proper housing with concept of '*1Malaysia*' (called *one Malaysia*) as shown in Fig. 8, the housing wrapped by a beautiful Malaysian's attire sticker to promote the specialty of this local product. The efficiency of the signal transmitted by this splitter can be seen in Fig. 9.

By utilizing this *GT-POF splitter*, research also conduct a second project called *GT-WDM-POF network* by integrating a color filter inside the POF connection as shown in Fig.10. A certain information or data are carried by the transmitters where each transmitter carries signals of different wavelengths specified by the LED. The specialized designed color films are used to filter out any other wavelength that is not within the range. It will only allow one wavelength to get through the film and thus conveyed the data carried at the receiver.

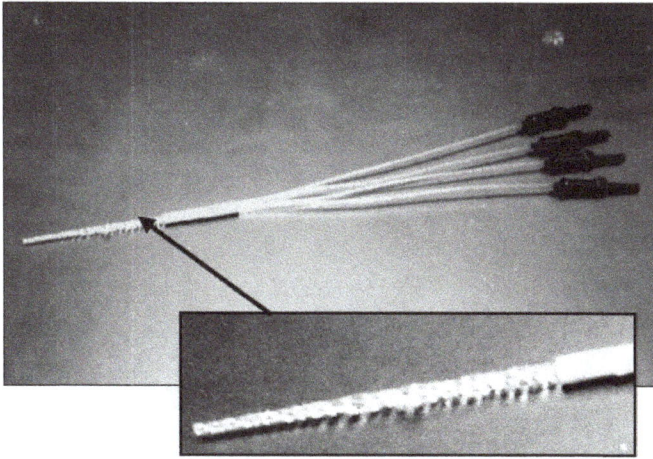

Fig. 7. Prototype of the novel 1 × N GT-POF splitter; zoomed in picture visualize the fused-taper-twisted POF

Fig. 8. Packed Prototype of the 1 × N POF Splitter & Demultiplexer called 1Malaysia ™ splitter

Back to the GT-POF splitter, the fused taper-twisted part (refer to Fig. 7), where every four POFs were fused or combined becoming as so-called single POF, play major role in coupling four individual optical signals. The fused taper-twisted POFs should be fabricated as well as all fibers in bundle arrangement fused (combine one another via heat exposure) completely. Otherwise, the POF splitter would probably fail to work according to its main role, coupling the numbers of individual optical signal (Ab-Rahman, Guna et al. 2009; Ab-Rahman, Guna et al. 2009; Ab-Rahman, Guna et al. 2009).

The error could be occurred on it either while fabrication process or characterization test stages imposed on them. Irregularities of controlled heat while heating process exposed on the POFs become one of the major problem (Ab-Rahman, Guna et al. 2008; Ab-Rahman, Guna et al. 2009; Ab-Rahman, Harun et al. 2009), due to it lower melting point makes core structure of POF could be more sensitive on heating process. Once it is damaged, it is hard

Fig. 9. Prototype of 1 × N GT-POF splitter. The input signal is split into four channels which are suitable to distribute the application equally to the number of destination

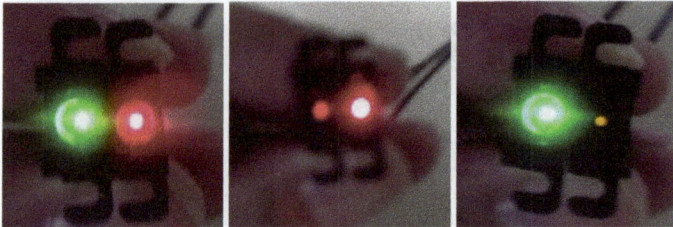

Fig. 10. Prototype of the 1 × N POF Demultiplexer. The device is used to split the signal to different frequency (color). The multiplexed signal is separated according to the application (data & video signal) respectively

to let a light pass through the core, or even not pass at all. It is so important to stop twisting and pulling POFs while the POF was getting frozen in order to prevent micro-scaled crack on core. That is why we use the metal tube while we conduct the indirect heating to fiber, it is to reduce this kind of damaged.

Furthermore, comparing with Imoto (1987) works, this fabrication method are produce less harmful gases such as nitrogen, sulfur or Carbon monoxide as produced by reaction of burning process using oxyhydrogen burner. The change of original diameter of POF considerably led to the change on optical properties including numerical aperture and maximum acceptance angle. All of these changes spoil light propagation principle based on total reflection; there would be much more rays of light refracted and propagate beyond cladding to atmosphere (Held 2002).

Indirect heating was used to minimize the undesirable deformation in the fused fiber bundle. This allows us easier fabrication and accurate control of the biconical taper. Furthermore, the continuous processing capability leads us to the reduction in fabrication time and improved yield. This method is expected to drastically reduce coupler fabrication costs (Imoto, Maeda et al. 1987).

The change of original diameter of POF considerably led to the change on optical properties including numerical aperture and maximum acceptance angle. All of these changes spoil light propagation principle based on total reflection; there would be much more rays of light refracted and propagate beyond cladding to atmosphere (Held 2002).

After characterization had been done, if the LED injected directly through single POF cable, the obtained input power was 12 μW. As expected, output will be obtained minimum is 3 μW (as the power expected to be divided into 4 POFs equally). From the observation of power output measurement, the final result was obtained as shown in Fig. 11.

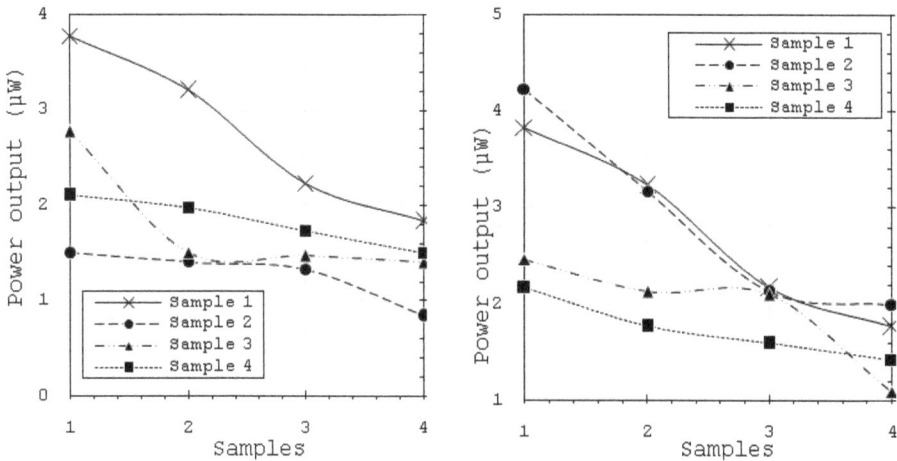

Fig. 11. The analysis of optical loss for fused bundle in both directions

Furthermore, it is stated that splitting ratio of the developed 1 × N GT-POF *splitter* was not totally homogenous. In *GT-WDM-POF* network, the homogeneity concerned as essential criteria for POF splitter as optical multiplexer in coupling multiple optical signals. For example, if red, blue, green, and yellow lights injected through Port 1, Port 2, Port 3 and Port 4, respectively, the blue light with 470 nm wavelengths will experience the greatest optical loss after all different wavelengths multiplexed. This would result in much more power dissipation in de-multiplexing mixed wavelengths into single blue wavelength whereby the optical power output of de-composed blue wavelength would be far less.

These happened because all POFs in bundle arrangement were not fused completely. Physically, the twisted-effect still could be seen on fused POFs and this indicates that all POFs not fully fused. In other word, the fused-taper-twisted POF was not formed well. Since the fusion technique was done by handwork, we found hard to fabricate and ensure all POF were fully fused. Obtained data was manipulated to determine the attenuation of each POF outputs as depicted in Fig. 12.

From analytical observation above, it is indicated that attenuation of each output is 0.25 dB, 2.93 dB, - 0.63 dB and – 0.14 dB, respectively. Different values show that each POF output has different characteristic of attenuation. After calculating, the minimum insertion loss for prototype is 2.41 dB. We found that standard 1 × N GT-POF *splitter* has 2.1 dB of minimum

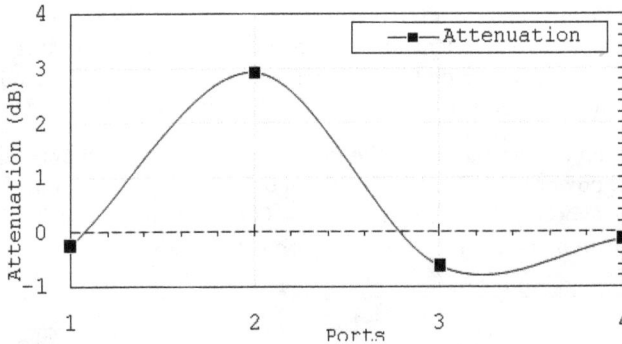

Fig. 12. Different attenuation on each POF output

insertion loss. So, the deviation of insertion loss between our novel and commercial splitter is not so large.

Comparison of hand-made and commercial splitter have been observed, in term of market price. Overall price for 1 × 4 *GT-POF splitter* cost is less than US$3, but for the commercial one which available in market is cost not less than US$250. Nowadays, many technology have been provided to coupling a signal, Low Cost 1 × 2 Acrylic-based Plastic Optical Fiber Coupler (Ehsan, Shaari et al. 2009) for example, but knowing that the fabrication techniques was very complicated, not to mention about their massive equipment needed, here *GT-POF splitter* can be seen as one of the most promising solution to face this kind of problem.

In this chapter, the optical loss is categorized as extrinsic loss due to the physical change of POF, LED projection to POF and the core-to-core connection and (Appajaiah, Wachtendorf et al. 2007; Kuzyk 2007). It is learned that the physical change of POF caused by fabrication process, where by diameter of POFs increasingly decrease to approach 1 mm and the POFs finally has taper-twisted shape. In characterization process, optical loss may present through the direct LED projection to POF surface. Besides, optical loss may also present through the connection between the fused taper-twisted POFs and POF cable (Appajaiah, Wachtendorf et al. 2007). The other aspect that playing an important role to transmit two different signal represented by different color on transmitter devices is the filter which is placed between the *GT-POF splitter* and the receiver section. In this research, two different LED was utilized; red LED (650nm) transmit an internet line through LAN connection and green LED (520nm) to deliver a high quality video signal to be displayed on a monitor screen. Analysis on the effectiveness of the filter itself also carried out. Here the comparison result of the efficiency of both green and red LED on their way to deliver a different signal to be split by *GT-POF splitter*, and optical power meter was placed in the output port right before the receiver port, as shown in Fig. 13.

The deviation between both signals was reach 3dB, while video transmission system showed a better quality of transmission system in low-cost WDM-POF system. The image quality of the video through WDM-POF method can be seen in Fig. 14.

Comparison for the optical line either using the filter or not, has been analyzed. The insertion loss of the cable with or without red filter is visualized in Fig. 15, also with it logarithm and linear function of the data.

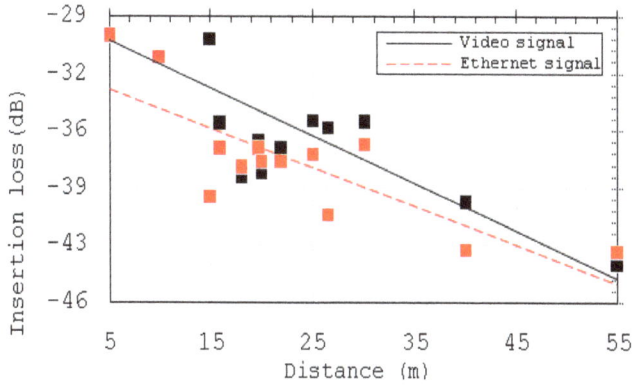

Fig. 13. Power loss comparison between green and red LED (and in linear function), green LED represent the video quality of the system while red LED represent the rate of download and upload through internet line

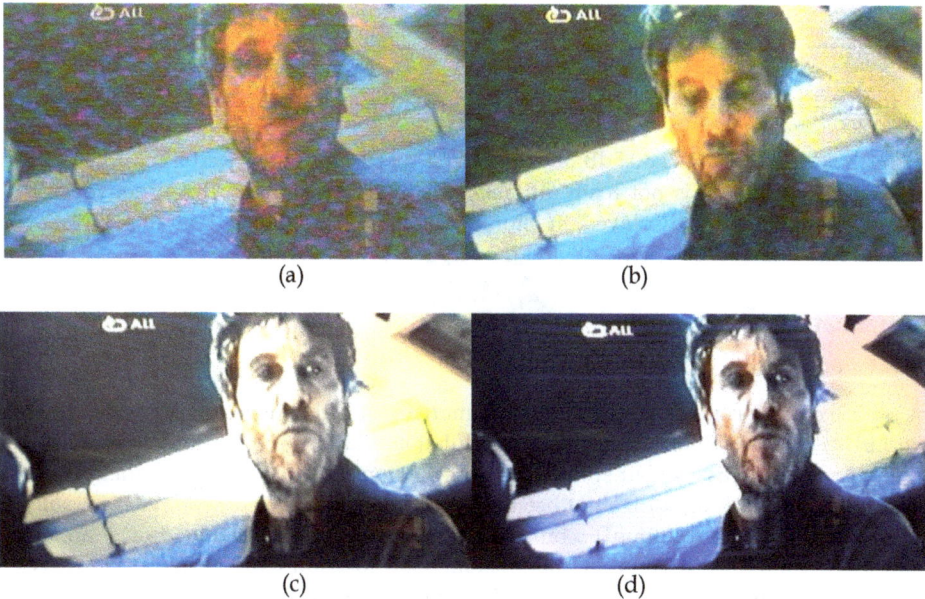

Fig. 14. Video quality of WDM-POF system of (a) 50m, (b) 30m, (c) 20m and (D) 10m of optical transmission line

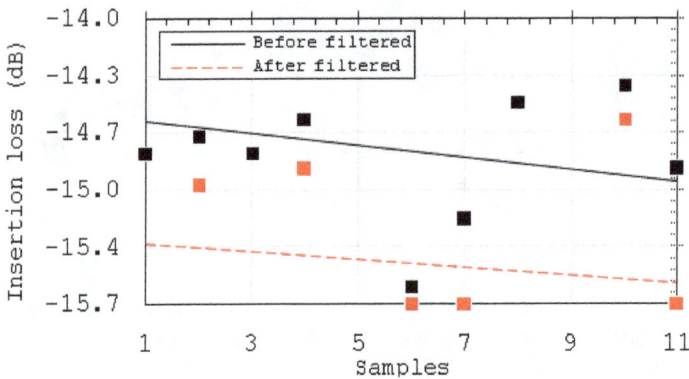

Fig. 15. Insertion loss of the cable (in logarithm function) before and after we place the red filter.

Figure 15 shows that almost 0.5dB breakdown occurred once we placed a red filter into the line. But this deviation is not really influenced either speed rate on LAN network or the video quality which is displayed in monitor. Hence, it is shown that the deviation of insertion loss is about less then -3dB and the highest is reach -7dB.

5. Applications

5.1 In-vehicle entertainment networking

The optical splitter is then applied in the automotive test bed to develop in-car infotainment. Two LEDs are used to perform intensity modulation with the internet data (red) and video signal (green) respectively. Our splitter has going advancement to function as demultiplexer to enable the multiplexed LED light been separated and interpreted next. As the result, the capacity of the communication become has double (due to use of two LEDs) and more application can be embedded in the system. Fig. 16 shows the result of LED WDM communication in the in-car application. The small diagram below shows the LED WDM communication configuration. Our proposed system is the lowest cost of WDM system that is reported up to this time.

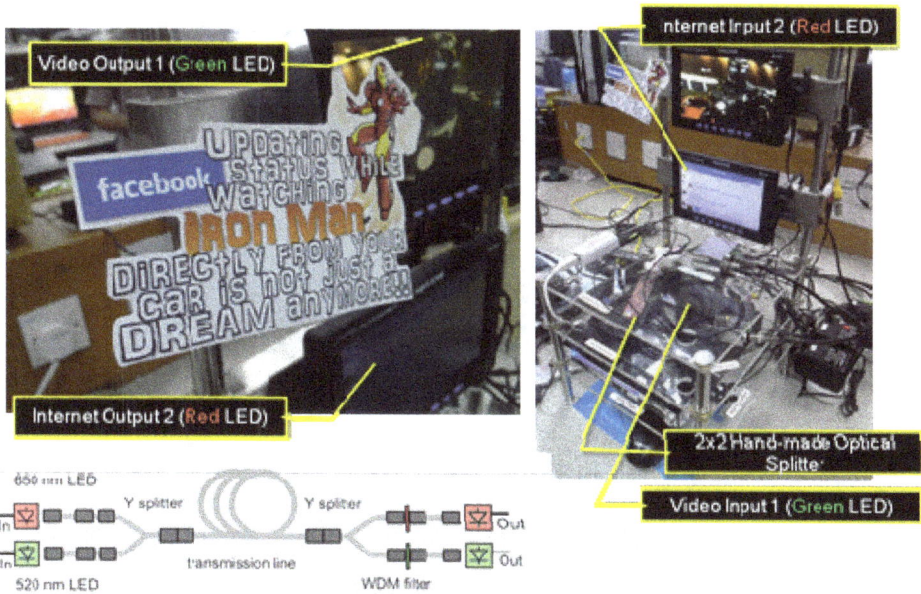

Fig. 16. *GT-POF splitter* and demultiplexer play an important role in WDM communication in-vehicle entertainment networking (protected by patent no. PI2010700001)

5.2 Low cost video signal splitting solution

Closed-circuit television (CCTV) is not a cheap and simple system. The most powerful it is the more money we spend to built the overall system. But now, with a novel POF-based device, video signal splitting process will never be face pricing constraints anymore. So here, we can see that this device able to become the next video signal splitting solution (see Fig. 17). While some other CCTV and network supplier competing each other to provide the most powerful for surveillance system which come out with an expensive budget, now with combination between *GT-POF splitter* and an inexpensive *GT-WDM-POF network*, price will never be a main problem to install CCTV system anymore.

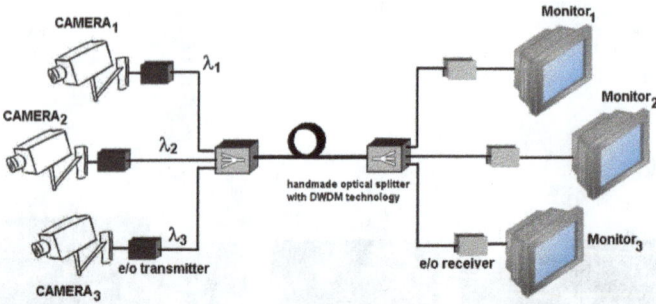

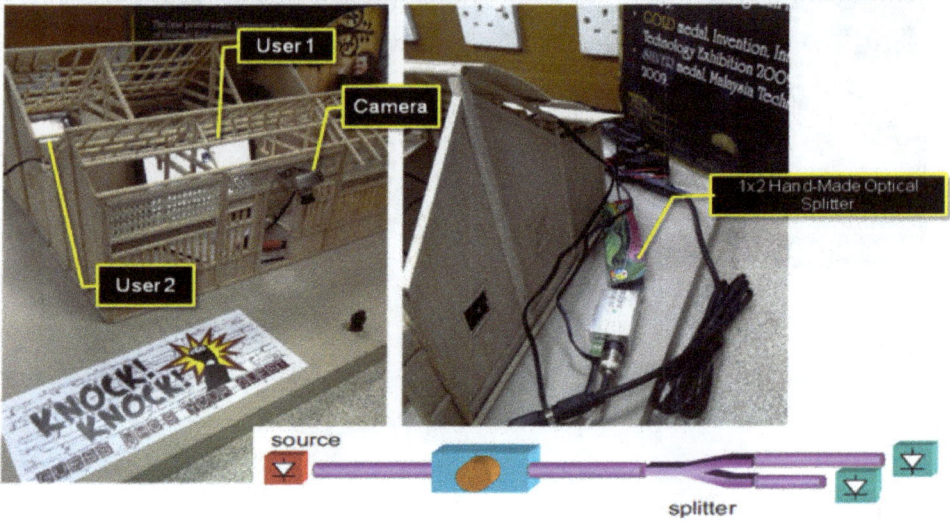

Fig. 17. Application for home networking and surveillance system in-house (protected by patent no. PI2010700001)

'Knock.. knock.. whose guest stands in front of the main door?', is one of the most powerful campaign regarding to this surveillance camera system. We can place a mini video camera in front of our main door of the house. Connecting the camera with some monitors display which is placed into a number of rooms inside the house. While we connect into some nods using a *GT-POF splitter* to split a video signal into a number of rooms, using a simple image processing devices, image from the main door will be transmitted into the house. Now, the members of a family inside the house, will never fighting anymore, ask one to another to open the main door while all of the members are busy with their own activity inside their own room. Because now, *GT-POF splitter* able to transmit an image from the main door into the rooms, and they will accurately noticed whose guest is standing in front of the door.

6. Conclusion

Fusion technique has been successfully practiced to fabricate optical $1 \times N$ GT-POF splitter. Which N represents numbers of output channel have been fabricated. Final analysis shows that efficiency of 1×4 GT-POF splitter output has splitting ratio 25:16:31:28 % and 2.41 dB of minimum optical loss. Hence, the obtained result reveals that a novel $1 \times N$ GT-POF splitter has great potential to be employed as economical wavelength divisions multiplexer because it able to couple several different wavelengths with few main advantages that are low optical loss and low-cost. An intensive study suggested in order improving the homogeneity of this prototype. This device is highly recommended for GT-WDM-POF system as it is not as costly as other commercial POF splitter. Filters play an important role in giving a higher insertion loss from the GT-WDM-POF system, but the quality of a number of output ports is not badly destructed due to the color band gap from the filter itself, speed rate of the internet still stable and the resolution of the video image is quite good.

7. Acknowledgment

This research has been conducted in Computer& Network Security Laboratory, Universiti Kebangsaan Malaysia (UKM). This project is supported by Ministry of Science, technology and Environment, Government of Malaysia, 01-01-02-SF0493 and Research University Grant fund UKM-GUP-TMK-07-02-108. All of our novel fabrication method of POF splitter, 1Malaysia™ splitter, $1 \times N$ GT-POF splitter and also the GT-WDM-POF network solution were protected by patent numbered PI2010700001.

8. References

Ab-Rahman, M. S., H. Guna, et al. (2009). 1xN Self-Made Polymer Optical Fiber Based Splitter for POF-650nm-LED based Application. 2009 International Conference on Electrical Engineering and Informatics, Selangor, Malaysia.

Ab-Rahman, M. S., H. Guna, et al. (2009). "Cost-effective 1x12 POF-Based Optical Splitters as an Alternative Optical Transmission Media for Multi-Purpose Application." IJCSNS International Journal of Computer Science and Network Security 9(3): 72-78.

Ab-Rahman, M. S., H. Guna, et al. (2009). "Fabrication and Characterization of Optical 1x12 Fused-Taper-Twisted Polymer Optical Fiber Splitters." Journal of Optical Communications 30(1): 16-19.

Ab-Rahman, M. S., H. Guna, et al. (2008). "Fabrication and Characterization of Customer-Made 1x3 POFBased Optical Coupler for Home Networking." IJCSNS International Journal of Computer Science and Network Security 8(12): 43-48.

Ab-Rahman, M. S., H. Guna, et al. (2009). "Bidirectional Optical Power Measurement for High Performance Polymer Optical Fiber-based Splitter for Home Networking." Australian Journal of Basic and Applied Sciences 3(3): 1661-1669.

Ab-Rahman, M. S., M. H. Harun, et al. (2009). Comparative Analysis of Power Efficiency of Handmade 1x12 Polymer Optical Fiber-Based Optical Splitter. 2009 International Conference on Electrical Engineering and Informatics, Selangor, Malaysia.

Appajaiah, A., V. Wachtendorf, et al. (2007). "Climatic exposure of polymer optical fibers: Thermooxidative stability characterization by chemiluminescence." Journal of Applied Polymer Science 103(3): 1593-1601.

Ehsan, A. A., S. Shaari, et al. (2009). Low Cost 1x2 Acrylic-based Plastic Optical Fiber Coupler with Hollow Taper Waveguide. 25TH of Progress in Electromagnetics Research Symposium, Beijing, China, Progress in Electromagnetics Research Symposium, PIERS.

Ericsson (2008) "Green Power to Bring Mobile Telephony to Billions of People."

Grzemba, A. (2008). MOST: the automotive multimedia network, Franzis.

Gupta, P. and H. Khurana (2010). Public Entrepreneurship: A Dynamic Strength For Budding Green Technology. Proceedings of the 4th National Conference; INDIACom-2010, New Delhi.

Held, G. (2002). Understandign Data Communications, Boston, EUA : Addison-Wesley.

Imoto, K., M. Maeda, et al. (1987). "New biconically tapered fiber star coupler fabricated by indirect heating method." Lightwave Technology, Journal of 5(5): 694-699.

Janota, A. and J. Hrbček (2011). Slovak ETC System Implemented – What Next? Transport Systems Telematics. J. Mikulski, Springer Berlin Heidelberg. 104: 30-37.

Keene, I. W. and R. K. Selli (1989). Passive components for plastic optical fibre transmission links. Plastics Materials for Optical Transmission, IEE Colloquium on.

Kelly, C., G. May, et al. (1995) "WDM Technologies in Telecommunications."

Kuzyk, M. (2007). Polymer fiber optics: materials, physics, and applications, CRC/Taylor & Francis.

Schubert, E. F., J. K. Kim, et al. (2006). Solid-state lighting- : a benevolent technology. Bristol, ROYAUME-UNI, Institute of Physics.

Ziemann, O., P. Zamzow, et al. (2008). POF handbook: optical short range transmission systems, Springer.

8

Secure Long-Distance Quantum Communication over Optical Fiber Quantum Channels

Laszlo Gyongyosi and Sandor Imre
Budapest University of Technology and Economics,
Department of Telecommunications
Hungary

1. Introduction

In today's communication networks, the widespread use of optical fiber and passive optical elements allows to use quantum key distribution (QKD) in the current standard optical network infrastructure. In the past few years, quantum key distribution schemes have attracted much study. The security of modern cryptographic methods, like asymmetric cryptography, relies heavily on the problem of factoring large integers (Rivest et al., 1978), (Schneier, 1996). In the future, if quantum computers become reality, any information exchange using current classical cryptographic schemes will be immediately insecure (Shor, 1994), (Shor, 1997). Current classical cryptographic methods are not able to guarantee long-term security. Other cryptographic methods, with absolute security must be applied in the future.

Cryptography based on the principles of quantum theory is known as *quantum cryptography* (Bennett et al., 1982), (Bennett & Brassards, 1984), (Bennett, 1992), (Imre & Balázs, 2005). Using current network technology, in order to spread quantum cryptography, interfaces must be implemented that are able to manage together the quantum and classical channels. The information-theoretic security of optical-fiber based quantum communication is the fundamental question of quantum cryptography. Quantum cryptographic schemes use photons as information carriers. The physical properties of photons make it possible to use quantum bits to realize unconditionally secure quantum communication over *long distances* (Duan et al., 2001) using the current standard optical fiber network. On the other hand, the success of secure long-distance quantum communications and global quantum key distribution systems depends strongly on the development of efficient quantum *repeaters* (Van Meter et al., 2009).

This chapter is organized as follows. First is a brief overview of the optical-fiber based QKD protocols. Then, we give a description of a QKD protocol designed for long-distance quantum communications between the quantum repeater nodes - called the DPS (Differential Phase Shift) QKD protocol. Next we show the results on the information-theoretic security analysis of DPS QKD protocol. Finally, we give an introduction to the quantum repeaters, then we summarize the results.

1.1 QKD for optical fibers

The safety of quantum cryptography relies on the no-cloning theorem. According to no-cloning theorem, any eavesdropping activity on the quantum channel necessarily perturbs the state of the qubits, thus Alice and Bob can detect the presence of Eve in the communication. In quantum cryptography Eve cannot clone the sent qubits perfectly, thus she has to use an ancilla quantum state, interact with the sent quantum state. This chapter will analyze the DPS QKD protocol, using efficient computational information geometric algorithms. The DPS QKD protocol was introduced for practical reasons, since the earlier QKD schemes were too complicated to implement in practice. The DPS QKD protocol can be an integrated part of current network security applications, hence it's practical implementation is much easier with the current optical devices and optical networks. Moreover, the DPS QKD protocol can be implemented in long-distance quantum communications, between the *quantum repeater* nodes. As follows, we will focus on this QKD scheme, however there are many other QKD schemes available, see (Branciard et al., 2005), (Dušek et al., 2006), (Hübel et al., 2007), (Gomez-Sousa & Curty, 2009), (Kwiat et al., 2001), (Niederberger et al., 2005), (Renner et al., 2005).

The DPS QKD scheme was designed to offer a well-implementable and more efficient practical solution with better key generation rates to realize quantum cryptography, than classical QKD approaches. As follows, it provides the best way to achieve long-distance QKD over optical-fiber quantum channels. In the DPS quantum cryptography protocol, the sender and the receiver use weak coherent state pulses, and logical bits are encoded in the relative phase of the pulses. The sender encodes every logical bit in two signals, and at the receiver's side, Bob use the two signals to decode the sent logical bit. The relevance of the DPS QKD protocol could have been increased dramatically in practical applications, since the differential phase shift QKD protocol is much more simpler in hardware design than the well known QKD protocols, such as BB84 or the Six-state QKD protocols. On the other side, contrary to it's easy implementation and it's much simpler working mechanism, the DPS QKD's protocol unconditional security is still not proven. The proposed geometrical analysis shows a method to quantify the secure key generation rate of the DPS QKD protocol, which is still missing from the literature. The possible attacks against the DPS protocol have been studied deeply. In this section we analyze the information-theoretical impacts of quantum cloner based attacks against the DPS QKD protocol.

As the most general attack against the protocol, we analyze coherent attacks, based on two different types of quantum cloner machines. The first section is organized as follows. First is a short brief on the DPS QKD protocol, and then we show the results on the information-theoretic security analysis of DPS QKD protocol. Finally, we summarize the results. In the second part of the chapter we discuss long-distance optical-fiber based quantum communications.

2. The DPS QKD protocol

In practical implementations of QKD protocols, Alice, the sender, uses weak coherent pulses (WCP) instead of a single photon source. As has been shown, WCP based protocols have a security threat, since an eavesdropper can perform a photon number splitting attack against the protocol (Inoue et al., 2003), (Honjo et al., 2004). These kinds of attacks are based on the fact that some weak coherent pulses contain more than one photon in the same polarization state, which provides information to the eavesdropper without any disturbance. The DPS protocol is robust against such photon number splitting attacks in practice, however a theoretical lower bound on the security of the protocol is still missing from the literature (Inoue et al., 2003), (Honjo et al., 2004). The working mechanism of the DPS QKD protocol is based on the same idea as the B92 protocol (Bennett, 1992): even two non-orthogonal quantum states are sufficient to perform a secure quantum key distribution. In the DPS protocol, Alice encodes the logical bits in the phase of the pulses. If the phases are modulated by 0, then Alice sends a logical zero, and if the phase between the two pulses is π, then she encodes a logical one. If the relative phase between two pulses is 0, then Bob will detect 0, and similarly, if the phase between the two pulses is π, then he will obtain a logical 1.

In the sending process, Alice generates coherent states of the same intensity μ, and from these states she forms a sequence, as follows:

$$\Psi = \ldots \left| e^{i\varphi_{k-1}} \sqrt{\mu} \right\rangle \left| e^{i\varphi_k} \sqrt{\mu} \right\rangle \left| e^{i\varphi_{k+1}} \sqrt{\mu} \right\rangle \ldots = \ldots \left| \psi(k-1) \right\rangle \left| \psi(k) \right\rangle \left| \psi(k+1) \right\rangle \ldots, \tag{1}$$

where the phases can be set at 0 or π, hence for a logical zero we have $e^{i\varphi_k} = e^{i\varphi_{k+1}}$, and for a logical one, the difference between the two phases is π, and $e^{i\varphi_k} \neq e^{i\varphi_{k+1}}$. Since the logical bits are encoded in the phases between the signals, the k-th signal has relevance in the determination of both the k-th and $(k+1)$-th logical bits, hence $\Psi \neq \ldots \left| \psi(k-1) \right\rangle \otimes \left| \psi(k) \right\rangle \otimes \left| \psi(k+1) \right\rangle \ldots$, and this fact increases the complexity of any security analysis (Inoue et al., 2003), (Honjo et al., 2004). From this viewpoint, the DPS protocol has been analyzed by Takesue *et al.* (Takesue et al., 2005). In this section we show that the complexity of the DPS QKD protocol's security analysis can be decreased dramatically, using fast computational information geometric methods (Gyongyosi & Imre, 2010), (Gyongyosi & Imre, 2010a), (Nielsen et al., 2007), (Nielsen & Nock, 2008), (Nielsen & Nock, 2008a), (Nielsen & Nock, 2009).

The security of the DPS QKD protocol lies in the fact that the sender randomly prepares and sends to Bob two non-orthogonal quantum states, similarly to the B92 protocol (Bennett, 1992). The DPS QKD protocol geometrically can be modeled in the same way as the B92 protocol, however its implementation is much easier in practice, since the DPS QKD protocol does not require a bright reference pulse as does the B92 protocol (Bennett, 1992), (Inoue et al., 2003), (Honjo et al., 2004). In practical implementations the DPS scheme can be realized by WCP pulses with an average photon number less than 1, and the sent WCP pulse can be described by

$$\left\langle \alpha \middle| -\alpha \right\rangle = e^{-2|\alpha|^2}, \tag{2}$$

where $|\alpha|^2 = \mu \ll 1$ is the average photon number per pulse (Inoue et al., 2003), (Honjo et al., 2004). The B92 protocol uses the same $0, \pi$ modulation-scheme, however the B92 protocol practical implementation is more complicated than the DPS QKD's scheme (Inoue et al., 2003), (Honjo et al., 2004).

The general setup of the DPS QKD protocol is illustrated in Fig. 1. The time difference between the pulses is known at the receiver's device.

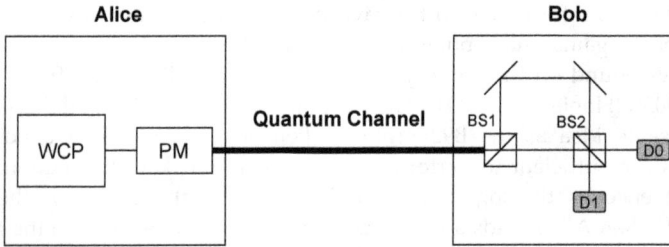

Fig. 1. A schematic view of DPS QKD protocol. Alice uses Weak Coherent Pulses (WCP) and a Phase Modulator (PM) to generate the signals. Bob decodes them with Beam Splitters (BS1, BS2) and photon detectors (D0, D1)

The optimal secure key rate of the protocol has been guaranteed only for individual attacks where the eavesdropper acts on the photons individually (Inoue et al., 2003), (Honjo et al., 2004). Here, we will analyze the most general collective attacker model, since the security of DSP QKD protocol against this general attack still remains an open question. In this attacker model, we analyze only the cloned photons from the given pulse, and we will give an approximation on the information obtainable by the eavesdropper. The analysis focuses on the eavesdropper's information about the given key, and the eavesdropper's cloned quantum states.

In the experimental realization of DPS QKD protocol, the signal consists of a weak coherent state and a strong phase reference. The relative optical phase between weak coherent state and reference pulse is either 0 or π, and these kinds of signals were already used in classical QKD schemes. In the attacker model, for simplicity we model these signals as two non-orthogonal quantum states, and we will analyze the still open questions related to the lower bounds on eavesdropper's obtainable information. As has been shown by Inoue (Agrawal, 1997) and Honjo (Honjo et al., 2004), the photon number splitting attacks can not be realized only with zero-error.

2.1 Practical quantum cryptography

Quantum cryptography uses the fundamental principles of quantum mechanics, and provides unconditionally secure communication. Optical channels have a fundamental role in quantum cryptography, since the information is encoded in the polarization states of photons.

The implementation of these quantum key distribution (QKD) schemes is much simpler than other approaches of quantum information processing, since these QKD schemes can be realized using the current optical network architecture. Quantum cryptography requires only the isolation of quantum states, unlike quantum computers, where the controlling of the interaction between quantum states is also a required task. In the case of quantum cryptography, the parties would like to communicate over macroscopic distances, hence optical fiber is a very practical choice to propagate and preserve the physical properties of photonic qubits (Paterson et al., 2004).

The experimental optical QKD schemes were intended to use single-photon sources and single-photon detectors, – but for practical reasons, many advanced classical optical communication techniques have been integrated into these systems. Several experimental QKD solutions with different quantum information encoding approaches have been proposed since Bennett and Brassard introduced their scheme (Bennett & Brassard, 1984). The practical applications have brought out many question as to the security of different QKD schemes. By using optical fiber, it is possible to send the quantum states without decoherence. On the other hand, photon losses are still a significant issue in optical-fiber based quantum communications.

The decoherence of optical channels is often negligible in practice, however it may be critical in some practical implementations. The sent quantum states can easily be lost in the quantum channel, hence the sender has to resend the qubits. The losses in the optical fiber determine the speed of secret key generation and the information which can be leaked to an eavesdropper. The unconditional security of most optical-fiber based QKD protocols against sophisticated attacks is unquestionable; however the theoretical proof of this for all the QKD schemes is still missing. To develop a practical and unconditionally secure optical QKD scheme, it is necessary to maximize the maximal bridgeable distance and the speed of key generation (Duan et al., 2001).

One of the most important properties of all QKD schemes is that these protocols use standard telecommunication components and optical networks for their implementation. On the other hand, many attacks against the protocol can be achieved by these simple components, such as beamsplitter attacks, intercept-resend or photon number splitting attacks. A strict analysis of the efficiency of these methods is required to prove the security of practical quantum cryptography. The proposed method bridges the gap between information-theoretic proofs and the open questions regarding the security of practical QKD implementations. We can analyze the correlation between the information-theoretic security of the protocol and the length of the optical-fiber or the speed of secret key generation. The ideas and results that we present in this chapter can be useful for the analysis of the information-theoretic security of QKD and future quantum communication protocols and for studying the information-theoretic aspects of the security of quantum networks.

2.2 Demonstration of QKD

The first implementations of quantum cryptography were based on polarization encoding, hence the logical values of the qubits were encoded in the polarization angles of the photons and the communication distance was only a few tens of centimeters (Agrawal, 1997), (Townsend, 1997).

The optical-fiber channel based QKD was first demonstrated in 1993 by Townsend, Rarity and Tapster (Townsend, 1997). Their method was based on phase-coding, and the length of the optical fiber was 10 km. Later, optical-fiber based QKD was extended from phase-coding to polarization encoding (Galtarossa & Menyuk, 2005), however the communication was implemented only over a distance of 1.1 km. Later, many optical-fiber based QKD systems were demonstrated, such as the scheme presented by Dynes, (Dynes et al., 2007), Rosenberg (Rosenberg et al., 2007), and Villoresi (Villoresi et al., 2004). As a result of these demonstrations, with the help of optical quantum channels, quantum communication can now be implemented over large distances. In the following few years, many valuable research and results were demonstrated, and the speed of key generation and the modulation techniques of the applied tools were increased dramatically. However, there are still many factors in real-life QKD implementations that add several imperfections to the working mechanism of the system (Kwiat et al., 2001).

However, as an important result of optical-fiber based QKD schemes, the photons which realize quantum states can be transmitted over long distances. The optical fibers make it possible to preserve the quantum states against the noise of the environment. In many practical long distance optical-fiber QKD systems, the information is rather encoded in the relative phase of two pulses, with very short time separation, since these encoding schemes can be applied more reliably for long-distance optical communications.

Currently, the longest transmission distance of optical-fiber based BB84 protocol is about 200 km, however advanced phase-modulation techniques make it possible to use the protocol over longer distances (Stucki et al., 2009), (Duan et al., 2001). QKD schemes were implemented over a 250 km long optical fiber, however these distances are still small compared to the distances in standard optical communication networks (Curty et al., 2008), (Inoue et al., 2003), (Honjo et al., 2004), (Kwiat et al., 2001), (Manderbach et al., 2007), (Takesue et al., 2005), (Stucki et al., 2009). The Six-state quantum cryptography protocol uses six polarization states to encode the logical bits in the qubits. The optical-fiber based Six-state QKD protocol tolerates more noise than the classical BB84 protocol (Kwiat et al., 2001), but its practical implementation requires more optical elements, which increases the noise in the system. Because of the greater noise level, the secret key generation rate of an optical-fiber based Six-state protocol is below that of the standard BB84 protocol. In the last few years many new optical-fiber based QKD approaches have been developed, such as the SARG04 protocol (Branciard et al., 2005), the Gaussian QKD scheme (Cerf et al., 2001), and the discrete-modulation QKD protocols (Inoue et al., 2003), (Honjo et al., 2004).

The DPS QKD scheme was designed to offer a more efficient practical solution with better key generation rates than classical QKD approaches (Honjo et al., 2004), (Inoue et al., 2003). In the DPS quantum cryptography protocol, the sender and receiver use weak coherent state pulses, and the logical bits are encoded in the relative phase of the pulses. The sender encodes every logical bit in two signals, and at the receiver's side, Bob use the two signals to decode the sent logical bit. The DPS QKD protocol has deep relevance to practical applications, since the differential phase shift QKD protocol is much simpler in hardware design than the well known QKD protocols, such as BB84 or the Six-state QKD protocols. Results on the security of DPS QKD protocol was published by Assche (Van Assche et al., 2004), Branciard, Gisin, and Scarani (Branciard et al., 2008), and several other attacks against various QKD schemes have been studied by Acín04 (Acín et al., 2004), Curty, Tamaki, and Moroder (Curty et al., 2008), Takesue et al. (Takesue et al., 2005), and Branciard et al. (Branciard et al., 2005), Curty et al. (Curty & Lütkenhaus, 2004), Fuchs et al. (Fuchs et al., 1997), Branciard, Devetak (Devetak & Winter, 2005), Cerf (Cerf et al., 2002), D'Ariano (D'Ariano & Macchiavello, 2003), Dušek (Dušek et al., 2006), Fasel (Fasel et al., 2004), Branciard (Branciard et al., 2005), and Gomez-Sousa and Curty (Gomez-Sousa & Curty, 2009), Hübel (Hübel et al., 2007), Niederberger (Niederberger et al., 2005), Renner05 (Renner et al., 2005).

The standard practical implementations of optical-fiber based QKD are bi-directional schemes, which means that the signal sent from Alice to Bob uses a bright carrier pulse, which was sent previously from Bob. The pulse travels from Bob to Alice, hence it can be easily manipulated by an eavesdropper, who can perform an arbitrary operation on the pulse using her standard optical elements. Moreover, it is also possible to use her optical signals to replace the original signals sent by the legal parties.

Before we start to discuss the information-theoretic aspects of secure communication over optical-fiber based quantum channels, we give a short description of the physical properties of optical quantum channels.

3. Physical properties of optical quantum channels

In practical optical-fiber based quantum communication, the unitary transformations of the qubits are made by standard optical components such as beamsplitters, wave plates and phase shifters. These devices are sufficient for realizing all unitary transformations. Moreover, to send a qubit over long distances without the destruction and noise of the environment, the *current standard optical infrastructure* can be used.

In practical quantum cryptography the propagation of single photons over long-distance optical channels is still an exciting question, since the classical optical relay devices cannot be used anymore. In contrary to classical optical communications, in a quantum communication system the signal consists of individual quantum states, which, according to the no-cloning theorem, cannot be copied and repeated (Duan et al., 2001). The development of quantum repeaters that can handle quantum states is still under research, and it is impossible with current technology to amplify photons on the level of single quantum states. To describe the properties of optical quantum channels, we introduce the α db/km loss coefficient of the optical fiber channel, and the length L of the channel (Agrawal, 1997), (Townsend, 1997). For a fixed value of the loss coefficient α db/km, the communication rate of the optical-fiber based QKD implementations can be analyzed as functions of the optical fiber length L. Many techniques have been developed to increase the efficiency of optical-fiber based quantum communication, such as increasing of photons per emitted pulse, but there are still many error correction and privacy amplification steps, which reduce further the speed of key generation. Currently, the fastest key generation speed in the standard BB84 QKD scheme was measured to be about 1.02 Mbps over a 20 km length optical fiber quantum channel (Niederberger et al., 2005). As we will show in Section 4.3, there is a connection between the radius of the smallest quantum informational ball and the fiber length (Gyongyosi & Imre, 2010).

In this section we discuss the optical fiber quantum channels. One of the most important questions from the security aspect of optical quantum channels is the loss of the optical channels. According to the no-cloning theorem, the quantum states sent cannot be repeated by a quantum repeater, hence the losses of the optical channel could cause many problems in practical communications. The level of loss determines the secret key generation rate and the maximal achievable transmission distance. From the viewpoint of the security of optical-based QKD, there is no difference between the quantum states lost in the quantum channel and the eavesdropped photons. In the security analysis, we have to count all the lost photons as eavesdropped qubits.

The relevance of optical fiber links for quantum communication was discovered by Agrawal et al (Agrawal, 1997). As has been shown, the loss in the optical fiber link mostly depends on the length L of the channel (Fasel et al., 2004), (Galtarossa & Menyuk, 2005), (Gomez-Sousa

& Curty, 2009), (Hübel et al., 2007). This length parameter affects the value of the transmission parameter t exponentially:

$$t = 10^{\frac{-\alpha L}{10}},$$ (3)

where the parameter α determines the attenuation in dB/km. According to experimental results, this parameter depends on the wavelength used within the optical fiber, for a 1330 nm optical channel $\alpha \approx 0.34\ dB/km$, and for 1550 nm it is about $\alpha \approx 0.2\ dB/km$ (Fasel et al., 2004), (Galtarossa & Menyuk, 2005), (Hübel et al., 2007).

As has been shown by Fasel, Gisin, Ribordy, Zbinden (Fasel et al., 2004), there are two unavoidable effects in optical-fiber based communications, chromatic dispersion and polarization mode dispersion (Agrawal, 1997). Chromatic dispersion can be handled by optical elements, however the second phenomena cannot be compensated for, hence it can cause decoherence in polarization encoding based schemes. In recent optical fiber implementations, all of these effects can be suppressed and stable quantum communication can be implemented in practice (Gyongyosi & Imre, 2010).

Free-space optical quantum channels are mostly used in short-distance and ground-space links, and several free-space quantum channel implementations have been demonstrated in the past few years (Manderbach et al., 2007), (Villoresi et al., 2004). The attenuation of free-space optical channels is $\alpha < 0.1\ db/km$, and for an L length optical channel the transmission can be expressed in terms of the length and the attenuation of the channel as follows:

$$t \approx \left[\frac{d_r}{d_S + DL} \right]^2 \cdot 10^{\frac{-\alpha L}{10}},$$ (4)

where the parameters d_r and d_S are the apertures of the sending and receiving telescopes, and D is the divergence of the beam (Agrawal, 1997), (Galtarossa & Menyuk, 2005), (Hübel et al., 2007).

3.1 The effect of noise of quantum channels

Besides the fact that the Bloch sphere provides a very useful geometrical approach to describe the density matrices, it also can be used to analyze the noise of the optical fiber quantum channel models. From algebraic point of view, quantum channels are linear trace-preserving completely positive maps, while from a geometrical viewpoint, the quantum channel is an affine transformation. While, from the algebraic view the transformations are defined on density matrices, in the geometrical approach, the transformations are interpreted as Bloch vectors.

The image of the quantum channel's linear transform is an *ellipsoid* on the Bloch sphere (see Fig. 2). To preserve the condition for a density matrix ρ, the noise on the quantum channel $\mathcal{N}$ must be trace-preserving, i.e. $Tr\mathcal{N}(\rho) = Tr(\rho)$, and it must be completely positive, i.e. for any identity map I, the map $\mathcal{N} \otimes I$ maps a semi-positive Hermitian matrix to a semi-positive Hermitian matrix (Hayashi et al., 2005).

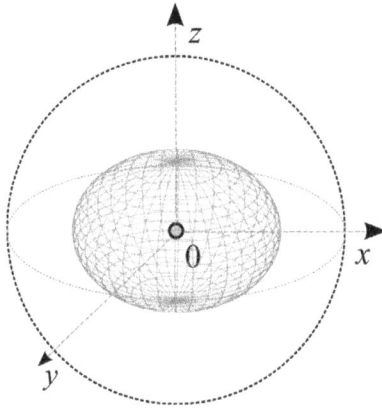

Fig. 2. Geometrically the image of the noisy quantum channel is an ellipsoid

We will use the terms *"unital"* and *"non-unital"* quantum channels. This distinction means the following thing: for a unital quantum channel $\mathcal{N}$, the channel map transforms the I identity transformation to the I identity transformation, while this condition does not hold for a non-unital channel. The optical fiber quantum channel belongs to the "non-unital" family (Hayashi et al., 2005).

To express it, for a unital quantum channel, we have $\mathcal{N}(I) = I$, while for a non-unital quantum channel, $\mathcal{N}(I) \neq I$. As we will see in Section 3.2, this difference can be rephrased in a geometrical interpretation, and the properties of the channel maps of the quantum channels can be analyzed using informational geometry. For a unital quantum channel, the center of the geometrical interpretation of the channel ellipsoid is equal to the center of the Bloch sphere. This means that a unital quantum channel preserves the average of the system states.

On the other hand, for a non-unital quantum channel, the center of the channel ellipsoid will differ from the center of the Bloch sphere. For an ideal enclosing scheme, the average of the pure orthogonal input states is equal to the center of the Bloch sphere. The main difference between unital and non-unital channels is that the non-unital channels do not preserve the average state. It follows from this that the numerical and algebraic analysis of non-unital quantum channels is more complicated than in the case of unital ones. While unital channels shrink the Bloch sphere in different directions with the center preserved, non-unital quantum channels shrink both the original Bloch sphere and move the center of the ball from the origin of the Bloch sphere. This fact makes our analysis more complex, however, in many cases, the physical systems cannot be described with unital quantum channel maps.

One of the most important quantum channels describing the transmission of information through optical-fibers, is *also a non-unital*.

Unital channel maps can be expressed as convex combinations of the four unitary Pauli operators (X, Y, Z and I), hence unital quantum maps are also called Pauli channels. Since the unital channel maps can be expressed as the convex combination of the basic unitary transformations, the unital channel maps can be represented in the Bloch sphere as different rotations with shrinking parameters. On the other hand, for a non-unital quantum map, the map cannot be decomposed into a convex combination of unitary rotations and the transformation not just shrinks the ball, but also moves its center from the origin of the Bloch sphere.

The geometrical interpretations of a unital and a non-unital quantum channels are illustrated in Fig. 3.

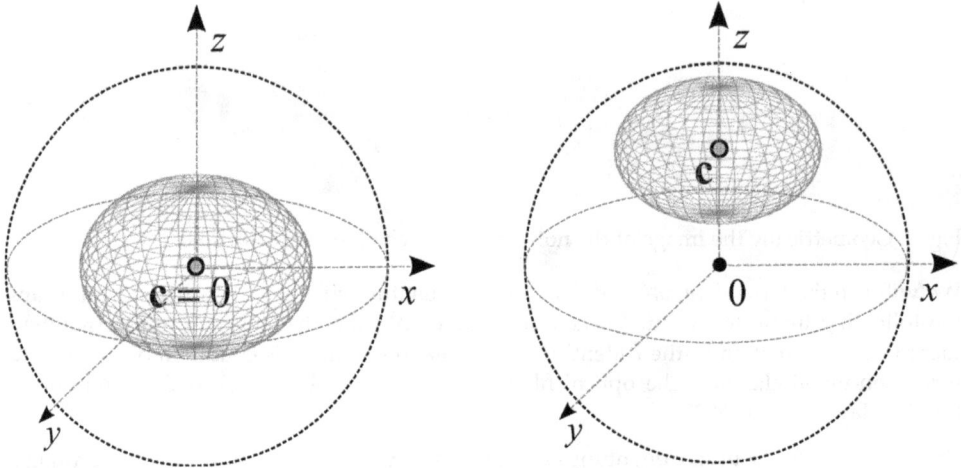

Fig. 3. The geometrical interpretation of a unital (a) and a non-unital (b) quantum channels

The unital channel maps can be expressed as convex combinations of the basic unitary transformations, while non-unital quantum maps cannot be decomposed into a convex combination of unitary rotations, because of the geometrical differences between the two kinds of maps. The geometrical approaches can help to reduce the complexity of the analysis of the different quantum channel models, and as we will show in Section 5, the problem of secure quantum communication over optical fiber channels can be converted into geometrical problems. The connection between the channel maps and their geometrical interpretation on the Bloch sphere makes it possible to give a simpler and more elegant solution for several hard, and still unsolved problems.

3.2 The noisy optical-fiber quantum channel

In this section, we introduce the general discussion of optical-fiber quantum channel - the amplitude damping channel. The effect of amplitude damping has great importance in optical communications, since this channel model describes energy dissipation. In practical optical or quantum communications, where quantum states or quantum bits are used, the loss of energy from the quantum system causes amplitude damping. In many practical applications, energy dissipation is an unavoidable phenomenon, and analysis of the amplitude damping quantum channel is therefore a relevant issue.

A non-unital amplitude damping quantum channel $\mathcal{N}$ can be described in the Kraus representation (Nielsen & Chuang, 2000), using a set of Kraus matrices $\mathcal{A} = \{A_i\}$, in the following form

$$\mathcal{N}(\rho) = \sum_i A_i \rho A_i^\dagger, \tag{5}$$

where $\sum_i A_i^\dagger \rho A_i = I$, and

$$A_1 = \begin{bmatrix} \sqrt{p} & 0 \\ 0 & 1 \end{bmatrix}, \text{ and } A_2 = \begin{bmatrix} 0 & 0 \\ \sqrt{1-p} & 0 \end{bmatrix}, \tag{6}$$

where p represents the probability that the channel leaves the $|0\rangle$ input state unchanged. In practical optical-fiber based applications, this parameter represents the probability of energy loss from losing a particle. The channel flips the input state from $|0\rangle$ to $|1\rangle$ with probability $1-p$. If $p=0$, then the output of the channel is $|1\rangle$, with probability 1. For $|1\rangle$ input states, the channel leaves the input qubit untouched, and the output of the channel is $|1\rangle$ with probability 1. As can be concluded, the output of a non-unital amplitude damping quantum channel depends on the state of input qubit, and for, $p=0$, the channel output is $|1\rangle$ with probability 1.

For an optical quantum channel, the set of Kraus operators $\mathcal{A} = \{A_i\}$ can be transformed to the King-Ruskai-Szarek-Werner (KRSW) ellipsoid channel model (King & Ruskai, 2001), (Ruskai et al., 2001). In the KRSW channel model, the ellipsoid channel parameters are $\{t_k, \lambda_k\}$, where $k = 1,2,3$. The analysis of non-unital channels is a more complicated task than for unital quantum channels since, in the KRSW representation, one or more parameters $\{t_k\}$ can be non-zero, which results in a more complicated calculation. The effect of $\{t_k \neq 0\}$ is that the average output $\rho = \sum_i p_i \rho_i$ of the channel moves away from the origin of the Bloch sphere, meaning that the center of the smallest enclosing quantum informational ball is not equal to the origin of the Bloch sphere $\frac{1}{2}I$. In the Bloch sphere representation, the effect of the amplitude damping channel on the initial input state $\rho = \frac{1}{2}(1+|\mathbf{r}_{in}|)$, where $|\mathbf{r}_{in}|$ is the length of the initial Bloch vector, can be analyzed. The output state is denoted by $\mathcal{N}(\rho) = \frac{1}{2}(1+|\mathbf{r}_{out}|)$, hence the amplitude damping channel can be expressed using Bloch vectors $\mathbf{r}_{in}$ and $\mathbf{r}_{out}$ in the following way:

$$\mathbf{r}_{out} = \begin{pmatrix} \mathbf{r}_{out}^{(x)} \\ \mathbf{r}_{out}^{(y)} \\ \mathbf{r}_{out}^{(z)} \end{pmatrix} = \begin{pmatrix} \sqrt{1-p} & 0 & 0 \\ 0 & \sqrt{1-p} & 0 \\ 0 & 0 & 1-\frac{p}{2} \end{pmatrix} \begin{pmatrix} \mathbf{r}_{in}^{(x)} \\ \mathbf{r}_{in}^{(y)} \\ \mathbf{r}_{in}^{(z)} \end{pmatrix} + \begin{pmatrix} 0 \\ 0 \\ \frac{p}{2} \end{pmatrix}. \tag{7}$$

The amplitude damping channel performs an affine map on the input state and the effect of the channel can be visualized in the Bloch sphere representation. The optical-fiber quantum channel can described in the KRSW ellipsoid channel model, with the following channel parameters (King & Ruskai, 2001), (Ruskai et al., 2001):

$$t_x = 0, \ t_y = 0, \ t_z = 1 - p,$$
$$\lambda_x = \sqrt{p}, \ \lambda_y = \sqrt{p}, \ \lambda_z = p, \tag{8}$$

where $p \in [0,1]$ is the channel parameter. In the Bloch sphere representation, the smallest value of $D(\rho \| \sigma)$ corresponds to the contour closest to the location of the density matrix.

In Figure 4(a), the Euclidean distances from the origin of the Bloch sphere to center $\mathbf{c}^*$ and to point ρ are denoted by m_σ and m_ρ, respectively. To determine the optimal length of vector $\mathbf{r}_\sigma$, the algorithm moves point σ. As we move vector $\mathbf{r}_\sigma$ from the optimum position, the larger contour corresponding to a larger value of quantum relative entropy D will intersect the channel ellipsoid surface, thereby increasing $\max_{\mathbf{r}_\rho} D(\mathbf{r}_\rho \| \mathbf{r}_\sigma)$. The optimal quantum informational ball is illustrated in light-grey in Figure 4(b). The first vector, m_σ, measures the Euclidean distance between the average and the center of the Bloch ball, while the second one, m_ρ, gives us the Euclidean distance from the center to the optimal channel output state.

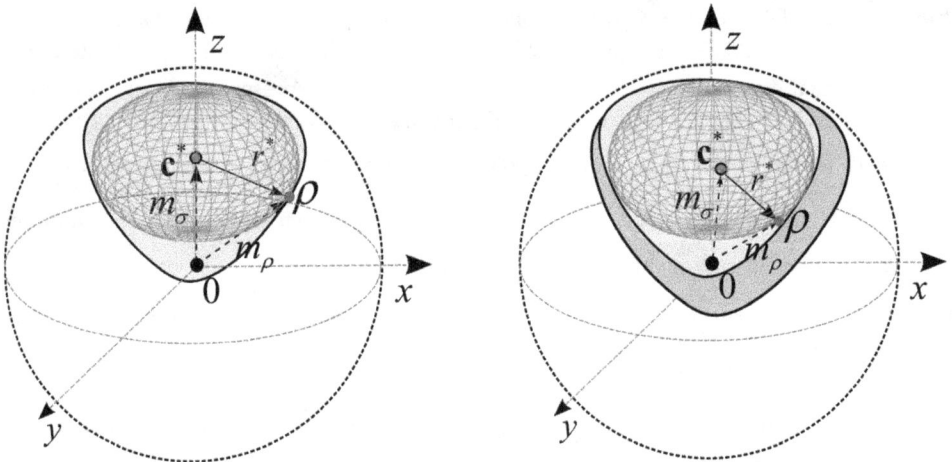

Fig. 4. Intersection of quantum informational ball and channel ellipsoid of optical-fiber quantum channel (a). As the vector moves from the optimal position, the quantum ball cannot be used to analyze the optical fiber quantum channel (b)

From a geometrical analysis, it can be concluded that the optimum input states for an optical-fiber quantum channel are unentangled, non-orthogonal quantum states (Gyongyosi & Imre, 2010a), (King & Ruskai, 2001), (Ruskai et al., 2001). The proposed geometrical approach - based on the quantum relative entropy function as distance measure between quantum states - has symmetries with the King-Ruskai-Szarek-Werner ellipsoid model (King & Ruskai, 2001), (Ruskai et al., 2001).

3.3 About the additivity of optical-fiber quantum channels

The additivity property of optical-fiber quantum channels is still an exciting subject of current research. There are some non-unital channels for which strict additivity is known, however the general rule for non-unital quantum channels is still not proven. The additivity of non-unital quantum channels is still an open question and currently under research. The additivity of optical quantum channels is still a remarkable and valuable research field in quantum information theory and it could have deep relevance to future quantum communications. The additivity of unital quantum channels is known and it has been proven that strict additivity holds for all unital quantum channels (Cortese, 2002). The fact that the optimum input states for an optical-fiber quantum channel $\mathcal{N}$ are unentangled, non-orthogonal quantum states, can be confirmed (King & Ruskai, 2001), (Ruskai et al., 2001). The geometrical analysis has shown that, for an optical-fiber quantum channel $\mathcal{N}$, there is no advantage in putting entangled quantum states to the input, and optimal results can be achieved by using non-orthogonal quantum states (Gyongyosi & Imre, 2010a). The average channel output state is denoted by σ, the center of the smallest quantum informational ball is denoted by c^*. The geometrical analysis has shown that, for an optical-fiber quantum channel $\mathcal{N}$, there is no advantage in putting entangled quantum states to the input, and optimal results can be achieved by using non-orthogonal quantum states.

In Fig. 5, we show the smallest quantum informational balls with their radii vectors of amplitude damping channel $\mathcal{N}$ for orthogonal and non-orthogonal inputs. As it can be confirmed geometrically, the optimal channel capacity can be achieved by non-orthogonal input states.

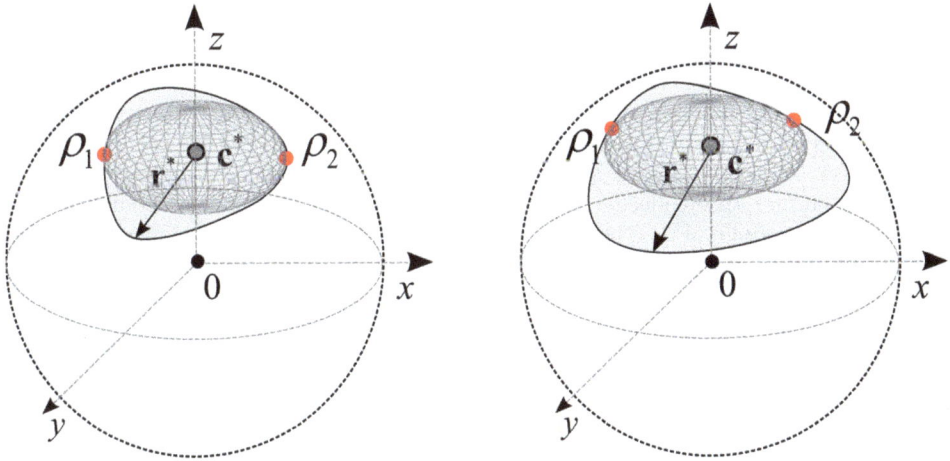

Fig. 5. The smallest quantum informational balls for amplitude damping channel model using orthogonal and non-orthogonal inputs

The optimal joint capacity of optical-fiber quantum channel $\mathcal{N}_{12} = \mathcal{N}_1 \otimes \mathcal{N}_2$, and the largest possible radius of the smallest enclosing quantum informational superball can be obtained by using non-orthogonal input states. The average channel state σ of the amplitude damping channel $\mathcal{N}$ is located on the horizontal line between the optimal output states ρ_1 and ρ_2. As a conclusion, the average state σ of the channel ellipsoid is the average of the two optimal states ρ_1 and ρ_2. As follows, the optimal input states are not symmetric, hence these states are cannot be fitted to a horizontal line. The smallest quantum informational superball can be used to determine the optimal channel input states for which input states super-additivity holds. These states are asymmetric and hard to find them without this geometrical approach, using just a numerical analysis. The computation of the smallest quantum informational ball is based on quantum relative entropy function $D(\cdot \| \cdot)$, as distance measure.

4. Attacker models

This section analyzes the information-theoretic security of the most important practical optical-fiber based QKD protocols, such as BB84, the Six-state QKD protocol and the DPS QKD scheme. We analyze deeper the collective attacker model, since this attack can be considered as the most general attacker model. However, its physical realization requires many advanced devices, which are still not accessible for an eavesdropper, for future applications, a deep analysis of this attacker model would be appropriate and useful.

In the *individual* attacker model, the eavesdropper uses a probe and she entangles the sent quantum state and her probe state independently for every qubits. This model allows an eavesdropper to store her probes in quantum memory (Shuai et al., 2006) and she is able to measure the stored qubits independently, using the measurement information stolen from the steps of post-processing. Eve can combine the measurement strategies, she can apply POVM or standard von-Neumann measurements.

The *collective* attacker model, which is the subject of this section, is very similar to the individual attacker model. The eavesdropper is also able to use quantum memory to store the quantum states, however in this attacker model she is able to use a more advanced measurement strategy. In this case, Eve has the ability to use global generalized measurement on all the stored qubits as a single quantum state, using her advanced quantum computer. The collective attack is a more sophisticated and more surreptitious attack than the individual type of attack (as remarked already, the technological requisites are still missing for this). On the other hand, the theoretical analysis of this model would be very useful for the future.

In the attacker model analyzed here, the eavesdropper performs her attacks collectively on the qubits, and measures the stored quantum states using advanced measurement techniques and quantum computers. The general model of collective attack is illustrated in Fig. 6, the currently unavailable devices are colored in gray.

Finally, the third attacker model is the *coherent* attacker model. This model can be considered as the most general type of attack, since Eve can entangle the whole transmission with her probe of arbitrary dimensionality (Inoue et al., 2003), (Honjo et al., 2004).

For both type of attacks many security proofs exist, however the security of these QKD schemes in practice is still not shown, and is still an open question (Inoue et al., 2003), (Honjo et al., 2004). We give a very efficient information geometric approach to practically analyze the security of these protocols.

Fiber-Optical Quantum Channel

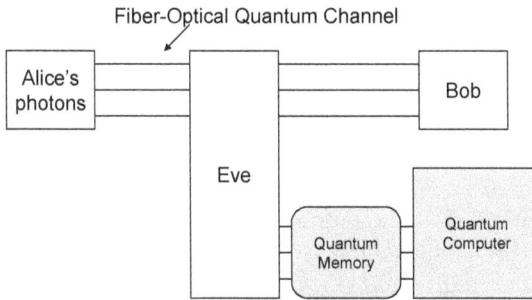

Fig. 6. The collective attacker model. The eavesdropper performs her attacks collectively on the qubits, and she measures the stored quantum states using advanced measurement techniques and quantum computers

4.1 Physically allowed cloning attacks for quantum cryptography

In secret quantum communications the best eavesdropping attacks on quantum cryptography are based on imperfect cloning machines. Using a probe, the eavesdropper imperfectly clones the sender's quantum state, keeps one copy, and sends the other. The physically allowed transformations of Eve's quantum cloner on Bob's qubit can be described in terms of Completely Positive (CP) trace preserving maps, which are affine map. The effects of a quantum cloner can be given in the tetrahedron representation (Gyongyosi & Imre, 2010a), (Hayashi, 2006).

Quantum cryptography is an emerging technology that offers new forms of security protection, however the quantum cloning based attacks against the protocol will play a crucial role in the future (Branciard et al., 2008), (Cerf et al., 2002), (Niederberger et al., 2005), (Townsend, 1997). We identify the quantum cloning based attacks in the quantum channel, and find potential and efficient solutions for their detection in secret quantum communications. The collective and coherent attacks against quantum cryptography are based on imperfect quantum cloners. The type of quantum cloner used depends on the quantum cryptography protocol. Against the Four-state (BB84) protocol, Eve, the eavesdropper, uses a phase-covariant cloner, while for the Six-state protocol, the optimal results can be achieved by the universal quantum cloner (UCM) (Biham & Mor, 1997), (Cerf et al., 2002), (Cerf, 2000), (Gyongyosi & Imre, 2010b), (Gyongyosi & Imre, 2010c), (Gyongyosi & Imre, 2010d). We use an efficient computational geometric method to analyze the quantum information theoretical impacts of physically allowed attacks on the quantum channel.

4.1.1 Preliminaries

In quantum cryptography the best eavesdropping attacks use the quantum cloning machines (Biham & Mor, 1997), (Cerf et al., 2002), (Cerf, 2000), (Gisin et al., 2001), (Gyongyosi & Imre, 2010b), (Gyongyosi & Imre, 2010c), (Gyongyosi & Imre, 2010d), (Nielsen & Chuang, 2000), (Yao, 1995). However, an eavesdropper can not measure the state $|\psi\rangle$ of a single quantum bit, since the result of her measurement is one of the single quantum system's eigenstates. The measured eigenstate gives only very poor information to the eavesdropper about the original state $|\psi\rangle$. The process of cloning of *pure* states can be generalized as

$$|\psi\rangle_a \otimes |\Sigma\rangle_b \otimes |A\rangle_x \rightarrow |\Psi\rangle_{abx}, \tag{9}$$

where $|\psi\rangle$ is the state in the *Hilbert space* to be copied, $|\Sigma\rangle$ is a *reference* state, and $|A\rangle$ is the *ancilla* state (Bhar et al., 2007), (Gisin et al., 2001), (Nielsen & Chuang, 2000). A cloning machine is called *symmetric* if at the output all the clones have the same fidelity, and asymmetric if the clones have different fidelities.

The no-cloning theorem has important role in quantum cryptography, since it makes no possible to copy a quantum state perfectly. In 1996 Bužek and Hillery published the method of imperfect cloning, while the original no-cloning theorem was applied only to perfect cloning (Bužek & Hillery, 1996). The asymmetric cloning machines have been discussed for eavesdropping of quantum cryptography in (Cerf, 2000). For attacks on some quantum cryptography protocol, it has been proven that the best strategy uses quantum cloning machines (Cerf et al., 2002), (D'Ariano & Macchiavello, 2003). In this section we characterize the cloning machines by the *informational theoretical* meaning of quantum cloning activity in the quantum channel.

Alice's side is modeled by random variable $X = \{p_i = P(x_i)\}, i = 1, \dots N$. Bob's side can be modeled by an other random variable Y. The Shannon entropy for the discrete random variable X is denoted by $H(X)$, which can be defined as $H(X) = -\sum_{i=1}^{N} p_i \log(p_i)$, for conditional random variables, the probability of the random variable X given Y is denoted by $p(X|Y)$. Alice sends a random variable to Bob, who produce an output signal with a given probability. Eve's cloner in the quantum channel increases the uncertainty in X, given Bob's output Y.

The general model for the quantum cloner based attack is illustrated in Fig. 7.

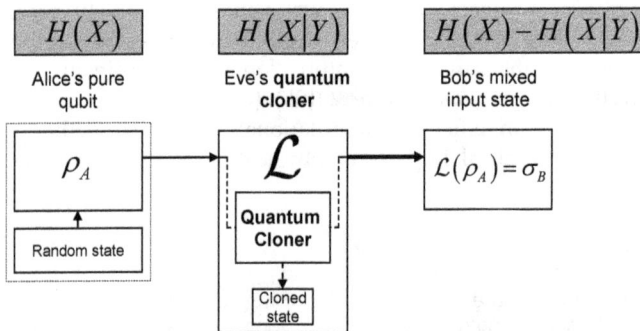

Fig. 7. The attacker model and the entropies

The informational theoretical noise of Eve's quantum cloner increases conditional Shannon entropy $H(X|Y)$, where

$$H(X|Y) = \sum_{i=1}^{N_X} \sum_{j=1}^{N_Y} p(x_i, y_j) \log p(x_i|y_j), \tag{10}$$

The security analysis is focused on the cloned mixed quantum state, received by Bob. The type of the quantum cloner machine depends on the actual protocol. For BB84, Eve chooses the phase covariant cloner (Rezakhani et al., 2005), while for the Six-state protocol she uses the universal quantum cloner (UCM) machine. Alice's pure state is denoted by ρ_A, Eve's cloner modeled by an affine map $\mathcal{L}$, and Bob's mixed input state is denoted by $\mathcal{L}(\rho_A) = \sigma_B$. We can use the fact, that for random variables X and Y, $H(X,Y) = H(X) + H(Y|X)$, where $H(X)$, $H(X,Y)$ and $H(Y|X)$ are defined by probability distributions.

We measure in a geometrical representation the information which can be transmitted in a presence of an eavesdropper on the quantum channel. The security of the quantum channel can be analyzed by *radius* r^* of the smallest enclosing ball of Bob, which describes the maximal transmittable information from Alice to Bob in the *attacked* quantum channel:

$$r^* = max_{\{all\ possible\ x_i\}} H(X) - H(X|Y). \tag{11}$$

To compute the radius r^* of the smallest informational ball of quantum states and the entropies between the cloned quantum states, instead of classical Shannon entropy, we will use von-Neumann entropy $S(\cdot)$ and quantum *relative entropy* $D(\cdot\|\cdot)$ functions (Gyongyosi & Imre, 2010a). Geometrically, the presence of an eavesdropper causes a detectable mapping to change from a noiseless one-to-one relationship, to a stochastic map. If there is no cloning activity on the channel, then $H(X|Y) = 0$, and the radius of the smallest enclosing quantum informational ball on Bob's side will be maximal (Gyongyosi & Imre, 2010).

4.1.2 Quantum cloning and QKD security

The security of QKD schemes relies on the *no-cloning* theorem (Bennett et al., 1982), (Wootters & Zurek, 1982). Contrary to classical information, in a quantum communication system the quantum information cannot be copied perfectly. If Alice sends a number of photons $|\psi_1\rangle, |\psi_2\rangle, ..., |\psi_N\rangle$ through the quantum channel, an eavesdropper is not interested in copying an arbitrary state, only the possible polarization states of the attacked QKD scheme. To copy the sent quantum state, an eavesdropper has to use a quantum cloner machine, and a known *"blank"* state $|0\rangle$, onto which the eavesdropper would like to copy Alice's quantum state. If Eve wants to copy the i-th sent photon $|\psi_i\rangle$, she has to apply a unitary transformation U, which gives the following result:

$$U(|\psi_i\rangle \otimes |0\rangle) = |\psi_i\rangle \otimes |\psi_i\rangle, \tag{12}$$

for each polarization states of qubit $|\psi_i\rangle$. A photon chosen from a given set of polarization states can only be perfectly cloned, if the polarization angles in the set are distinct, and are all mutually orthogonal (Cerf, 2000), (Wootters & Zurek, 1982). The unknown non-orthogonal states cannot be cloned perfectly, the cloning process of the quantum states is possible only if the information being cloned is classical. The polarization states in the QKD protocols are not all orthogonal states, which makes it impossible an eavesdropper to copy the sender's quantum states (Imre & Balázs, 2005), (Cerf, 2000), (Hayden et al., 2003),

(Wootters & Zurek, 1982). In the collective-type attacks, Eve imperfectly clones the sender's quantum state using her quantum state probe, she sends one copy to Bob and keeps the other copy. The effects of Eve's quantum cloner on Bob's qubit can be described in the terms of CP, trace preserving maps. The map of the quantum cloner compresses the Bloch-ball, as an affine map. This affine map has to be a complete positive, trace preserving map, which shrinks the Bloch ball along the x, y and z directions.

4.2 Inside a quantum cloner

This chapter strongly emphasizes the applicability of quantum cloners in secret quantum communications. In this section we see inside the quantum cloner using one of the most general quantum cloner models: the *universal* quantum cloner (Bhar et al., 2007), (Bužek & Hillery, 1996), (Hillery et al., 1999). The universal cloner produces two identical faithful copies of the input quantum state, and the quality of the output states is *independent* from the fidelity of the input state – hence it is really universal, as follows from its name. To describe the working mechanism of the universal quantum cloner, first we have to specific the inputs of the cloner machine. The input states of the quantum cloners are:

-   the *original* unknown input quantum state,
-   a *blank* quantum state onto which the unknown input state is to be cloned,
-   and the state of the quantum cloner, which also can be referred to as the *ancilla state* or the *environment*.

The process of quantum cloning can be divided into two important parts:

-   first part: *preparation* state,
-   second part: *cloning* state.

In the first phase, the quantum cloner uses elementary single-qubit rotations, and CNOT transformations, while in the second phase only CNOT transformations are applied. It follows from this that the quantum circuit of a quantum cloner can be constructed from elementary quantum circuits, using simple quantum gates (Bhar et al., 2007).

The quantum circuit of an universal quantum cloner machine is illustrated in Fig. 8.

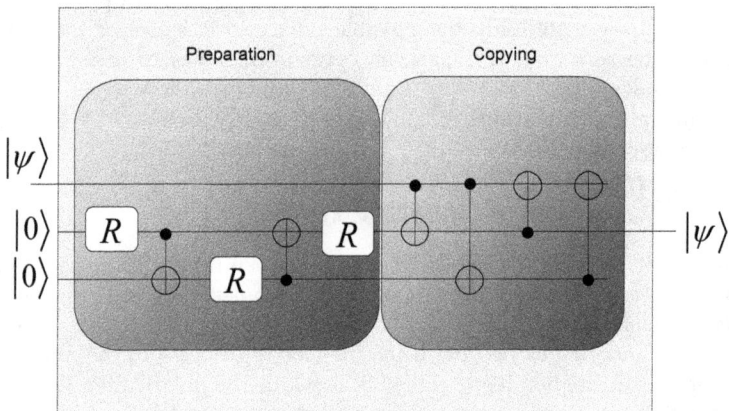

Fig. 8. Inside the universal quantum cloning circuit

The inputs of the quantum cloner are the unknown quantum state to be cloned, and the ancilla state: in this case, we have four dimensional ancilla, which represents the blank state for the copying and the environment of the quantum cloner machine. The blank state and the second ancilla state are known states – they are in the pure $|0\rangle$ state. The output of the quantum cloner machine consists of two clone quantum states and an ancilla state (Bhar et al., 2007).

4.3 Attacks against the DPS QKD protocol

To analyze the possible effects of an eavesdropper, - as in the previous case - we will use two kinds of quantum cloners, the most general universal quantum cloner and the phase covariant cloner. Other possible attacks against the protocol, such as sequential attacks, unambiguous state discrimination attacks or minimum error discrimination methods have been analyzed, and upper bounds have been obtained (Inoue et al., 2003), (Honjo et al., 2004). The security bounds for this type of attack were analyzed by Biham and Mor (Biham & Mor, 1997), and they have concluded the same bound holds for the protocol. In this attacker model, the eavesdropper tries to clone each of the quantum states sent by Alice, following an independent cloning strategy. The eavesdropper can change her strategy in a probabilistic way, hence in practical QKD applications, Eve can stop her cloning activity for a while, and then later, she can attack again. By changing the used strategies, she can decrease the probability of detection of her activity in the quantum channel.

In a collective attack, an eavesdropper can use a quantum memory to store her quantum states, and she can delay the whole measurement process. She can collect the required information from the steps of key agreement between Alice and Bob, which can be used to choose the best measurement strategy on the collected quantum states. As has been shown by Devetak and Winter (Devetak & Winter, 2005), the generic security bound for an collective attack can be given by the Csiszár–Körner bound (Csiszár & Körner, 1978) for one-way postprocessing as $I(A:B) - \min(I_{AE}, I_{BE})$ with $I_{AE} = \max_{Eve} \chi(A:E)$, where I_E is the eavesdropper's information about the raw key of Alice and Bob, and $I(A:B)$ can be expressed as

$$H(A) + H(B) - H(AB).\tag{13}$$

To compute the χ *Holevo quantity*, we will use the fact, that this quantity can be expressed as the radius of the smallest enclosing quantum informational ball, hence

$$
\begin{aligned}
r^{*} &= I(A:E) = S(E) - S(E|A) \\
&= \max \chi(A:E) = \max\left(S(\rho_E) - \sum_a p(a) S(\rho_{E|a}) \right),
\end{aligned}\tag{14}
$$

where a is Alice's output with probability distribution $p(a)$, and $\rho_{E|a}$ is Eve's ancilla and $\rho_E = \sum_a p(a)\rho_{E|a}$ is the partial state of the eavesdropper. We note, that the same equation can be applied between Alice and Bob, when Bob is also able to store the quantum states.

The eavesdropper's most general strategy can include many possible variations which cannot be parameterized efficiently. However the security bounds for general or coherent attacks are the same as for collective attacks, hence the geometrical approach can be used to analyze both collective and coherent attacks. As has been shown by Branciard, Gisin, and Scarani (Branciard et al., 2008), the simplest realization of a collective attack against the DPS QKD protocol is the beam-splitting attack, hence here we use this type of attack to describe the informational-theoretic security of the DPS QKD protocol.

To describe the coherent beam-splitting attack, we model Alice's sent states as a sequence of coherent states $\otimes_i |\psi(i)\rangle$, where each $\psi(i)$ is chosen from the set $\{+\psi, -\psi\}$, and the logical value of the bit is 0 if $\psi(i-1) = \psi(i)$, and 1 if $\psi(i-1) = -\psi(i)$. In the collective beam-splitting attack, the eavesdropper uses a beamsplitter to get a fraction of the signal. The remaining fraction of the signal, denoted by τ, is sent directly to Bob, hence Bob will receive the state $\otimes_i |\psi(i)\sqrt{\tau}\rangle$, and similarly, the eavesdropper's state can be described as $\otimes_i |\psi(i)\sqrt{1-\tau}\rangle$. The eavesdropper's information can be given by using the von Neumann entropy, as

$$I_E^{DPS} = S(\rho_E) - \frac{1}{2}S(\rho_{E|0}) - \frac{1}{2}S(\rho_{E|1}), \tag{15}$$

where it is assumed that the probability of each logical bit value is equal, hence

$$\rho_E = \frac{1}{2}\rho_{E|0} + \frac{1}{2}\rho_{E|1}. \tag{16}$$

Using the coding scheme $\psi(i-1) = \psi(i)$ for 0, and $\psi(i-1) = -\psi(i)$ for 1, the state of $\rho_{E|0}$ and $\rho_{E|1}$ can be expressed as

$$\rho_{E|0} = \frac{1}{2}P_{+\psi_E, +\psi_E} + \frac{1}{2}P_{-\psi_E, -\psi_E} \text{ and } \rho_{E|1} = \frac{1}{2}P_{+\psi_E, -\psi_E} + \frac{1}{2}P_{-\psi_E, +\psi_E}, \tag{17}$$

where $\psi_E = \psi\sqrt{1-\tau}$ and P_{ψ_E} is the projector. Using ψ_E, we can introduce a new parameter (Inoue et al., 2003), (Honjo et al., 2004)

$$\gamma = e^{-|\psi_E|^2} = e^{-\mu(1-\tau)}, \tag{18}$$

where μ is the intensity of the sent weak coherent pulse. Using this parameter, the inner product between $|\langle +\psi_E | -\psi_E \rangle| = \gamma^2$, for given μ intensity, the eavesdropper's information can be expressed as

$$I_E^{DPS}(\mu) = 2H\left[\frac{(1-\gamma^2)}{2}\right] - H\left[\frac{(1-\gamma^4)}{2}\right] =$$

$$2H\left[\frac{(1-|\langle +\psi_E | -\psi_E \rangle|)}{2}\right] - H\left[\frac{(1-|\langle +\psi_E | -\psi_E \rangle|^2)}{2}\right], \tag{19}$$

where H is the Shannon entropy function.

The connection between the practically achievable secret key rate K of the protocol and the radius r^* of the smallest enclosing quantum informational ball of the eavesdropper can be given by (Inoue et al., 2003), (Honjo et al., 2004):

$$K(\mu) = \left[I(A:B) - r^*\right]R = \left[I(A:B) - \max_{Eve}\left(S(\rho_E) - \sum_a p(a)S(\rho_{E|a})\right)\right]R \tag{20}$$
$$= v\left(1 - e^{-\mu\tau}\right)\left(1 - I_E^{DPS}(\mu)\right),$$

where R is the raw key rate and v is the repetition rate.

5. Geometrical description of DPS QKD protocol

In phase-coding QKD schemes, a signal consists of a superposition of two time-separated pulses. These methods, instead of polarization encoding, encode the information in the relative phase between two pulses. However, the polarization and phase encoding schemes are equivalent mathematically (Inoue et al., 2003), (Honjo et al., 2004), hence in the information geometrical security analysis, we can use the Bloch-ball representation to study the security of the protocol. We can use the following translation between the basis states $|0\rangle$, $|1\rangle$ on the Bloch-ball, and the relative phases of the first signal $|S_1\rangle$, and the second signal $|S_2\rangle$:

$$\{|0\rangle, |1\rangle\} = \left\{\frac{1}{\sqrt{2}}(|S_1\rangle + |S_2\rangle), \frac{1}{\sqrt{2}}(|S_1\rangle - |S_2\rangle)\right\}. \tag{21}$$

Hence for example, the $|\nearrow\rangle$ polarization state on the Bloch-ball can be rewritten in the following form:

$$|\nearrow\rangle = \frac{1}{\sqrt{2}}(|0\rangle + |1\rangle) = \frac{1}{\sqrt{2}}(|S_1\rangle + i|S_2\rangle). \tag{22}$$

In the case of $|\nearrow\rangle$, the relative phase between signals $|S_1\rangle$ and $|S_2\rangle$ is π. As we can conclude, the information encoded in the polarization and in the relative phases are equivalent.

In the analysis of the DPS QKD protocol, we can use the following conventions of the relative phases of the signals and the polarization states on the Bloch-ball:

$$\frac{1}{\sqrt{2}}(|S_1\rangle + |S_2\rangle) = |0\rangle = |\leftrightarrow\rangle,$$
$$\frac{1}{\sqrt{2}}(|S_1\rangle + i|S_2\rangle) = \frac{1}{\sqrt{2}}(|0\rangle + |1\rangle) = |\nearrow\rangle. \tag{23}$$

In Fig. 9, we have illustrated these conventions in the notations of the security analysis. The first and the second signals are denoted by $|S_1\rangle$ and $|S_2\rangle$.

The soundness of these notations of the proposed geometric analysis is based on the fact that the relative phases between the pulses can be represented by polarization angles on the

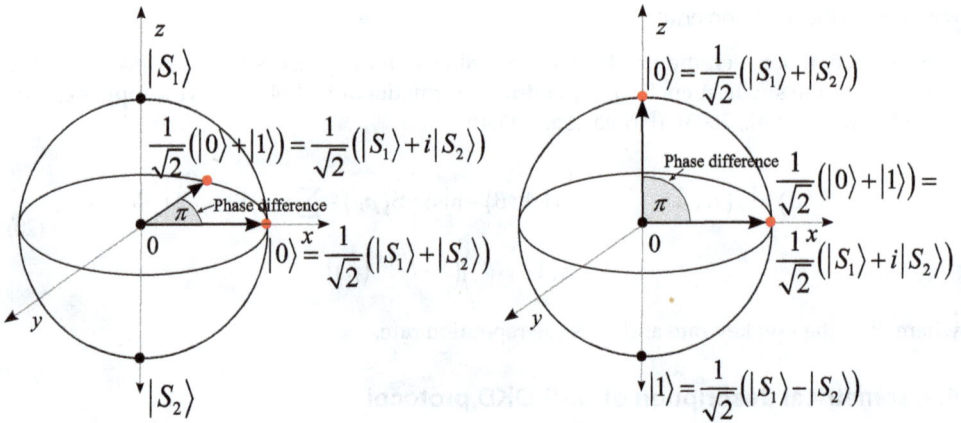

Fig. 9. The notations used in the geometrical analysis of DPS QKD protocol. The relative phases between the pulses can be represented by polarization angles on the Bloch-ball. The phase encoding and polarization encoding schemes are equivalent mathematically and in the geometrical security analysis

Bloch-ball, since the phase encoding scheme and the polarization encoding scheme are mathematically the same (Agrawal, 1997), (Inoue et al., 2003), (Gyongyosi & Imre, 2010), (Honjo et al., 2004). Hence, the DPS QKD protocol can be modeled in terms of the polarization states of the B92 protocol. It uses only two polarization states, and the key can be described by a random sequence $B = (b_1, b_2, \ldots b_N)$ of logical bits, and the generated N-length qubit string is:

$$|\psi\rangle = |\psi_{b_1}\rangle \otimes |\psi_{b_2}\rangle \otimes \ldots \otimes |\psi_{b_N}\rangle = \bigotimes_{i=1}^{N} |\psi_{b_i}\rangle, \tag{24}$$

where b_i is the basis of the i-th qubit. The i-th qubit $|\psi_{b_i}\rangle$ in the string is generated according to the B92 coding convention (Bennett, 1992), as $|\psi_0\rangle = |\leftrightarrow\rangle$ and $|\psi_1\rangle = |\nearrow\rangle$.

In Fig. 10, we illustrated the two polarization states of the DPS QKD protocol, used in the information geometric security analysis.

In general, these polarization states can be expressed by means of some orthogonal basis $\{|0\rangle, |1\rangle\}$ as follows:

$$|\pm\alpha\rangle = a|0\rangle \pm b|1\rangle, \tag{25}$$

where the coefficients can be given by

$$a = \sqrt{\frac{1}{2}\left(1 + e^{-2\mu_\alpha}\right)} \quad \text{and} \quad b = \sqrt{\frac{1}{2}\left(1 - e^{-2\mu_\alpha}\right)}, \tag{26}$$

and $a, b \in \mathbb{R}$ and $a^2 + b^2 = 1$, and $a > b$ if $\mu_\alpha \neq 0$.

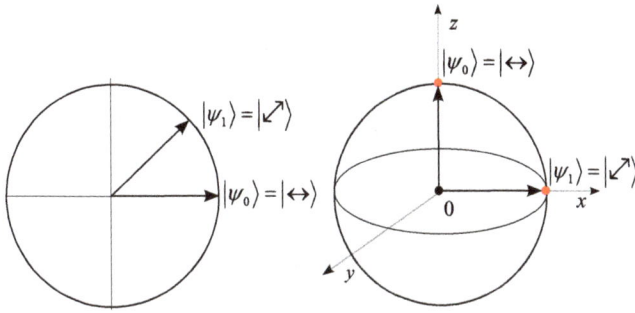

Fig. 10. The polarization states of the protocol

In Fig. 11, we show the polarization states of $|\pm\alpha\rangle = a|0\rangle \pm b|1\rangle$ in the Bloch-ball representation, and we depict the π phase difference between quantum states $\{|\rho_1\rangle,|\rho_2\rangle\}$ and $\{|\rho_3\rangle,|\rho_4\rangle\}$. In practice, Alice sends a WCP signal, whose phases are randomly modulated by 0 or π. These WCP signals are modeled by the quantum states $\{|\rho_1\rangle,|\rho_2\rangle\}$ and $\{|\rho_3\rangle,|\rho_4\rangle\}$ on the Bloch-ball.

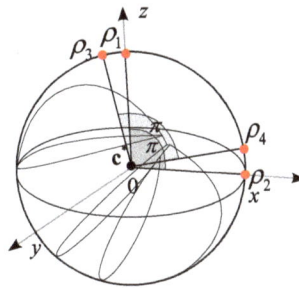

Fig. 11. The WCP pulses of DPS QKD protocol are modeled by different polarization states

In Fig. 12, we show the result of the eavesdropper's attack. For the best results, Eve uses the universal cloner for non-equatorial states ρ_1, ρ_2 and ρ_3, and the phase-covariant cloner for equatorial states (Bechmann-Pasquinucci & N. Gisin, 1999), (Cirac & Gisin, 1997), (Gyongyosi & Imre, 2010).

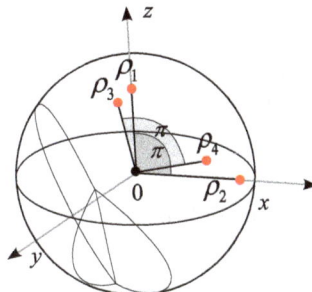

Fig. 12. The tessellation of the Bloch-ball for cloned quantum states differs from the diagram of the pure states originally sent

As can be concluded, the smallest enclosing quantum informational ball contains all the cloned states. The length of the radius of the smallest quantum informational ball describes the eavesdropper's maximally obtainable information.

In Fig. 13 we show the smallest enclosing quantum informational ball and the convex hull of the quantum states.

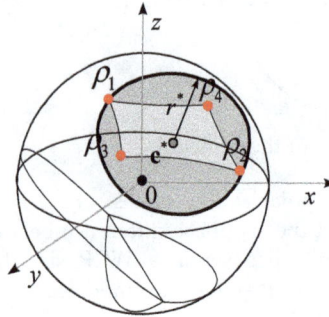

Fig. 13. The radius of the smallest quantum informational ball describes the eavesdropper's maximum obtainable information. The algorithm computes the length of the information-theoretical radius by the determination of the convex hull of mixed quantum states

We have computed the radius of the smallest quantum informational ball, which measures the information obtained by the attacker.

5.1 Coherent attack against the DPS QKD aprotocol

The DPS QKD protocol was introduced for practical reasons, since the earlier QKD schemes were too complicated to implement in practice. The DPS QKD protocol has high relevance to practice. The DPS QKD protocol can be integrated into current network security applications, hence its practical implementation is much easier with the current optical devices and optical networks. We introduce a fundamentally new method to analyze the information-theoretic security of the DPS QKD protocol.

To study the security of the protocol, we will analyze the collective attacker model against the optical-fiber based DPS QKD scheme. In this type of attack, Eve is equipped with beamsplitters, optical switch, detectors, quantum memory and a quantum computer (Biham & Mor, 1997), (Chen et al., 2006), (Gyongyosi & Imre, 2010a). As has been shown by Branciard, Gisin, and Scarani (Branciard et al., 2008), the simplest realization of a collective attack against the DPS QKD protocol is the beam-splitting attack, hence in this section we use this type of attack to describe the informational-theoretic security of the DPS QKD protocol.

The optical implementation and the optical devices of the eavesdropper's coherent attack are illustrated in Fig. 14.

Before we start to analyze the information-theoretic aspects of collective attacks against the optical-fiber based DPS QKD protocol, we give a short account of the parameters of photon detecting probabilities and the total efficiency of the quantum channel. As we have explained previously, the single photon detection probability can be expressed as $p_{true} \approx \mu t$, where μ is the average number of photons per pulse and t is the total transmission efficiency of the optical fiber. We also use the quantum efficiency η of the receiver, and the loss of the detector of the receiver by L_B db/km. The optical-fiber is characterized by its length L and its loss coefficient α db/km (Branciard et al., 2005), (Duan et al., 2001).

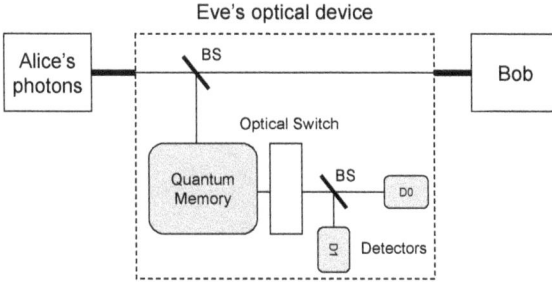

Fig. 14. Eve's optical device for coherent attack (BS-Beam Splitter)

Eve would like to calibrate her beamsplittered transmission to be equal to this t, however the sent beam with n_p pulses and with average photon number $n_p \mu t$ will be used by the eavesdropper. According to her strategy, she will use another beam with average photon number $n_p \mu(1-t)$, and the probability that the eavesdropper obtains the value of a logical bit at a certain time and Bob has also detected the photon, is $\mu(1-t)$.

In the collective attack, the eavesdropper can use a quantum memory, hence she is able to change her strategy, and she can store the pulses. However, the legal parties can delay the public announcement for an arbitrarily long time, hence the decoherence of the eavesdropper's quantum register makes it impossible to use the stored states (Biham & Mor, 1997). Moreover, as has been shown (Inoue et al., 2003), (Honjo et al., 2004), the eavesdropper can use an optical interferometer with an optical switch instead of a beamsplitter, hence the success probability of Eve can be increased to $2(\mu(1-t))$. As a conclusion, using a beamsplitter, the eavesdropper is able to exactly determine the values of the bits for a fraction of the pulse which is $2(\mu(1-t))$, and for the remaining $1-2(\mu(1-t))$ fraction of the states, the eavesdropper enjoys only a 50% chance of getting the correct result. As has been shown (Agrawal, 1997), if the total transmission efficiency of the optical fiber is $t \ll 1$, then the mutual information between Eve and Bob is independent of this parameter, hence in this case it is independent of the transmission properties of the quantum channel. As one possible solution, the efficiency of the eavesdropper attack can be decreased, if the legal parties choose the average photon number μ to be small, independently of the total transmission efficiency of the optical fiber (Inoue et al., 2003), (Honjo et al., 2004).

5.2 Numerical results for the attacked DPS QKD protocol

In Fig. 15 we summarize the results for the proposed information geometric analysis of the security of the DPS QKD protocol against collective attacks. In Fig. 15(a), we introduced a new radius $r_{secure} = I(A:B) - r^*$ derived from Eq. (20) and analyzed it in the function of the length of the optical fiber.

In Fig. 15(b), the secret key generation rate of the attacked quantum channel is derived from 15(a). The results are based on the analysis made above, and the properties of optical-fiber based quantum communication. In this model, Eve is equipped with a quantum memory, hence she is able to store the quantum states, which introduces a time delay in the communication (Chen et al., 2006), (Inoue et al., 2003), (Honjo et al., 2004). The analysis shows the correlation between the length of the optical fiber and the maximal secure key generation rate. The secret key generation rates are computed from the radius r_{secure} as a function of the fiber-length. The secure key generation rate was derived from the radius of the smallest quantum informational ball (Gyongyosi & Imre, 2010).

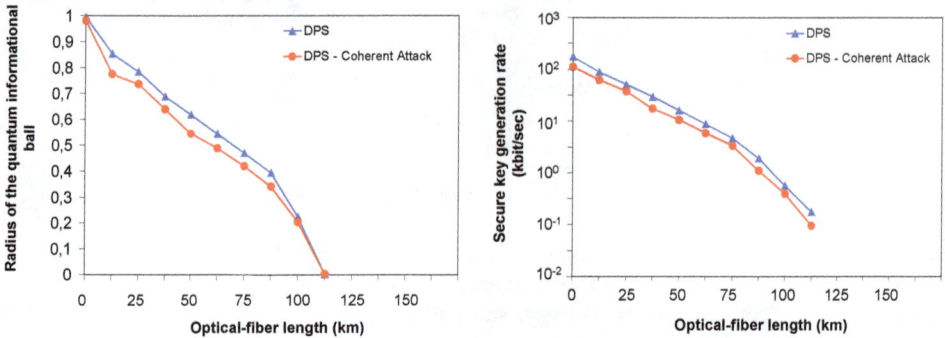

Fig. 15. Radii of the information-theoretic ball (a) and secure key generation rates (b) as a function of optical fiber length for DPS QKD protocols for an eavesdropper-free channel and an eavesdropped channel. The secret key rates are derived by the proposed information geometric algorithm

The secure key generation rate of the protocol for a coherent attack is computed from the radius of the quantum informational ball. It is slightly different from the result of the eavesdropper-free protocol, however there is no significant decrease in the radius of the smallest quantum informational ball (Gyongyosi & Imre, 2010).

In Fig. 16 we show the results of our analysis for a collective attack. The effects of the disturbance caused by the eavesdropper is analyzed in the range of $[0,0.5]$. The upper and lower bounds of radii of the eavesdropper's smallest quantum ball are shown as the function of the disturbance, using UCM and phase-covariant quantum cloners.

We have used the mutual information analysis to show the security of the DPS QKD protocol against coherent attacks. The radius of the smallest enclosing informational ball, hence the maximal obtainable information of the eavesdropper, increases with the level of disturbance. However, in the tolerated range of the disturbance level of the DPS QKD protocol, the analyzed quantum cloners make no possible for an eavesdropper to realize a successful attack in practice. As it is well described by the radii of the smallest quantum informational balls, the UCM based attack allows Eve less information than the phase-covariant based attack, which result confirms the mutual information analysis of UCM and phase-covariant cloner based attacks (Bechmann-Pasquinucci & N. Gisin, 1999), (Cirac & Gisin, 1997), (Gyongyosi & Imre, 2010).

The results of the information-theoretic based analysis confirmed the fact, that coherent attack does not help to Eve to increase her information about the key.

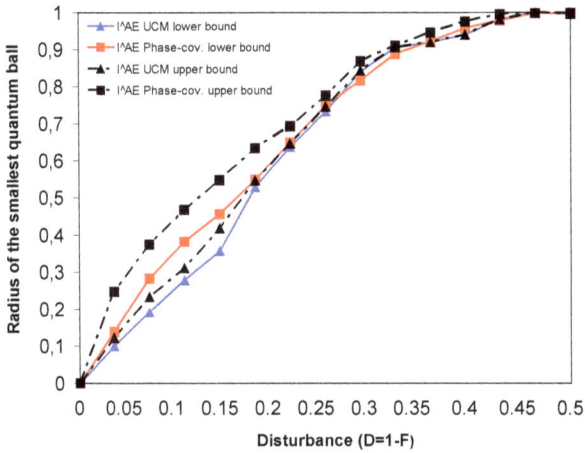

Fig. 16. Results of information geometrical security analysis of DPS QKD protocol for collective attack. The eavesdropper's obtainable information is described by the radius of the smallest enclosing quantum informational ball (lower bound - solid lines, upper bound - dashed lines)

5.2.1 Numerical results for the attacked DPS QKD protocol

The DPS QKD protocol offers information-theoretically secure quantum communication over optical fiber quantum channels, and the practical implementations of the protocol make it possible to use quantum cryptography in a simple and efficient way, with relatively high secret key generation rates over long distances. The protocol can be implemented easily with current optical network structure and optical fibers, and makes it possible for quantum cryptography to become a popular and easy-implementable practical cryptographic system in the future.

Practical QKD schemes require a reliable medium, which can transmit the photons with reasonable losses. The optical-fiber based QKD approaches seem to be the most appropriate choice for many practical reasons. This section analyzed the information-theoretic security of actual optical-fiber based QKD schemes using efficient information geometric approaches. In the proposed security analysis we introduced the smallest enclosing ball representation, which is used to describe the information-theoretic security of the QKD scheme. We analyzed the information-theoretic impacts of the most general eavesdropping attacks against the protocols, and we discovered the connection between the length of the optical fiber and the radius of the smallest enclosing quantum informational ball. Using the quantum informational ball representation, we derived the connection between the secret key generation rate over optical channels. The secure key rate is calculated from the information-theoretic radii of the smallest enclosing quantum informational balls, as a function of the length of the optical-fiber.

The proposed security analysis was focused on the most general coherent attack. Although sufficiently advanced technical devices — such as quantum memory or quantum computers — are still not available, the security of quantum communication against these attacks will be a very important issue in the future. To demonstrate the applicability of the presented information geometric algorithm, we analyzed the information-theoretic security of the DPS QKD protocol.

6. Long-distance quantum communications with optical fiber channels

This chapter discusses secure long-distance quantum communications. As we have concluded in Section 5.2.1, the DPS QKD protocol can be the key protocol to achieve secure long-distance quantum communication over the optical-fiber based infrastructure. However, the secure long distance quantum communication would not be possible without the quantum repeater. In this section, we describe the working mechanism of quantum repeater, which is the key in the implementation of DPS QKD scheme over long distances with the help of the optical fiber network.

The success of future long-distance quantum communications and global quantum key distribution systems depends on the development of efficient quantum repeaters. A quantum repeater is not simply a signal amplifier, in contrast to classical repeaters. The quantum repeater is based on the transmission of entangled quantum states between the repeater nodes. As we will show in Section 6.1, there are several differences between a classical and a quantum repeater. In the quantum communication networks of the future, besides long distance communication, other networks structures could be implemented, such as self-organizing, truly probabilistic quantum networks, see (Gyongyosi & Imre, 2010e).

6.1 The quantum repeater

The quantum repeater nodes create highly entangled EPR (Einstein-Podolsky-Rosen) states with high fidelity of entanglement. The entangled quantum states can be sent through the quantum channel as single quantum states or as multiple photons. In the first case the fidelity of the shared entanglement could be higher, however it has lower probability of success in practice, since these quantum states can be lost easily on the noisy quantum channel (Duan et al., 2001). In order to recover fidelity of entanglement from noisy quantum states purification is needed. If the quantum repeaters could communicate with each other through idealistic quantum channels, the fidelity of the shared pairs would be nearly maximal, which could decrease dramatically the purification steps required.

Sharing of quantum entanglement plays critical role in quantum repeaters. The fidelity of the entanglement decreases during the transmission through the noisy quantum channel (Ladd et al., 2006), (Van Loock et al., 2008). Therefore, in practical implementations, the quantum entanglement cannot be distributed over very long distances; instead, the EPR states are generated and distributed between smaller segments (Van Meter et al., 2009).

A practical approach of the quantum repeater is called the "hybrid quantum repeater" (Van Meter et al., 2009), (Munro et al., 2008), (Jiang et al., 2008). The hybrid quantum repeater uses atomic-qubit entanglement and optical coherent state communication (Munro et al., 2010). In practice, the *base stations* of the quantum repeaters are connected by optical fibers, the entangled quantum states are sent through these fibers (Louis et al., 2008), (Sangouard et al., 2009).

Quantum repeaters use the purification protocol to increase the fidelity of transmission (Sangouard et al., 2009), (Stephens et al., 2008). The rate of entanglement purification depends on the fidelity of the shared quantum states, since the purification step is a probabilistic process. Moreover, the success probability of the purification of the entangled quantum states depends on the fidelity of the entangled states – if the fidelity of entanglement of the shared state is low, then the success probability of its purification will be also low. Another important disadvantage of the purification algorithm is that it requires a lot of classical information exchange between the quantum nodes (Devitt et al., 2008).

The quantum repeater itself can be regarded as a quantum computer, which can realize the quantum teleportation algorithm and the purification steps. The quantum transformations at the receiver side of teleportation require classical inputs, since the quantum teleportation protocol uses them to recover the unknown quantum state from the entangled quantum state (Munro et al., 2010), (Louis et al., 2008), (Sangouard et al., 2009). In the sharing process, the sender base station entangles the quantum state with another separate physical qubit, and then it is multiplexed into the quantum channel (Van Meter et al., 2009). At the receiver's side, the multiplexed pulses are demultiplexed, and the receiver entangles each pulse with a free quantum state. If the entangling operation is successful, then the sender and the receiver share an EPR state (Duan et al., 2001), (Munro et al., 2010), (World Wide Science, 2011).

The entanglement creation uses the quantum communication channel; hence some noise is added to the transmitted states. As follows, in the next step, the created entanglement has to be purified (Bernardes et al., 2010), (Munro et al., 2010), (Van Meter et al., 2009). The purification is an error-correcting scheme, and it uses local quantum operations only – hence these operations can be realized in the separated base stations locally (Van Meter et al., 2009). The purification step takes two EPR pairs and by the usage of local quantum transformation and classical communication, it combines the two EPR states into one, higher-fidelity EPR pair (Munro et al., 2010).

In the next step, the unknown quantum state can be teleported by the quantum teleportation scheme, furthermore using entanglement swapping it can be transmitted to the final destination. The entanglement swapping (Munro et al., 2010), (Van Meter et al., 2009) is equal to a set of quantum teleportation steps – hence the entanglement swapping is an "extended teleportation protocol", which is able to bridge the gap between the physically separated stations in long distances (Duan et al., 2001), (Van Meter et al., 2009). During quantum teleportation, the sender's input quantum state is destroyed and recovered at the receiver's side, using shared entanglement between the parties. The receiver needs two classical bits to recover the unknown quantum state. In the protocol, both the sender and the receiver have to use local quantum operations only, which are based on classical information.

The purification process destroys the Bell pair, hence two quantum states in the middle station have been freed, and they can be reused in the next teleportation. The chain of repeaters has been constructed to extend of the distance between those nodes which share an EPR state. The purification scheme is able to correct the errors of the transmission which occurs at the node-switching and the fiber-based communication devices. The entanglement swapping can be extended to long distances, and by means of the node-to-node quantum communications, a global-scale quantum network can be constructed in the future (Van Meter et al., 2009).

In practical implementations, many EPR states can be shared between two nodes, and from these imperfect EPR pairs, the selection of the most appropriate pairs could be a complex algorithmic problem. In the literature several purification algorithms have been discussed, such as the symmetric purification, the pumping method, greedy scheduling or the banded purification method (Van Meter et al., 2008). The purification process is a probabilistic process, which means, that it can fail in some cases, but this probability decreases as the fidelity of the input EPR states grows (Van Meter et al., 2009).

6.2 The implementation of quantum repeaters

The design of a quantum repeater has been studied by Van Meter et al. (Van Meter et al., 2009), and in 2008 they presented a system design for a practical quantum repeater (Munro et al., 2008). Quantum repeaters will be very important for long-distance quantum communications, distributed quantum systems, and secure quantum communications. On the other hand, there are several differences between a classical and a quantum repeater. Due to the fundamental differences between the classical states and the quantum states, the quantum repeater works in a completely different way. The quantum repeater is not a signal amplifier, the way a classical repeater is.

The working mechanism of the quantum repeater is based on the following quantum protocols (Munro et al., 2008), (Van Meter et al., 2009):

- *purification of quantum states.*
- *entanglement swapping (quantum teleportation).*

The purification protocol is used to increase the fidelity of the distributed quantum states. Quantum teleportation is used to transport unknown quantum states using EPR states. Basically, these two quantum protocols represent the fundamental basis of a quantum repeater.

From an engineering point of view, the development of quantum repeaters is one of the biggest and most important challenges. As follows from the theoretical background of the quantum repeater, it requires a classical communication channel to assist the quantum communication. The quantum teleportation protocol requires classical information, otherwise the unknown quantum state cannot be recovered in a physically separated, distant location.

The quantum repeater uses quantum entanglement for the "retransmission" of a quantum state. We would like to guarantee the successful transfer of the quantum state, hence the implementation of an error correcting scheme is required. The quantum repeater will use the purification protocol, which can be used to increase the fidelity of the transmission.

In the work of Van Meter et al. (Van Meter et al., 2009), a new algorithm has been introduced for this purpose, they called it *banded purification* (Van Meter et al., 2009). The banded purification scheme could improve the utilization of the resources and the "repeating" of the quantum states.

The fidelity of the transmitted quantum states depends on the fidelity of the distributed EPR states. The entangled Bell states have to be shared among the stations in an initial step — these EPR states will be used in the teleportation protocol. After the EPR states have been distributed and the teleportation of the unknown quantum state has been finished, an error-correction scheme has to be used to correct any possible errors produced during the transformation.

The quantum repeater itself is a quantum computer, which can realize the quantum teleportation algorithm and the purification steps. These quantum transformations require classical inputs, since the quantum teleportation protocol requires them.

Entanglement swapping is the process in which the quantum state repeats from *A* to *B*, through several intermediate base stations, using quantum teleportation and purification.

The working mechanism of the quantum repeater can be divided into the following main steps:

1. Creation of Bell states distributed among the quantum base stations,
2. Purification of the shared EPR states,
3. Teleportation between the nodes,
4. Entanglement swapping.

In a quantum communication network, the quantum repeater works differently from the classical repeater. The whole transmission rather could be called "entanglement swapping," since it is this step which realizes the transmission. Now, let's see the steps of the working mechanism of the quantum repeater.

In the first step, high-quality entangled pairs are shared between the base stations. These base stations could be a few tens of kilometers from each other. As has been shown by Van Meter et al. (Van Meter et al., 2009), using optical devices for quantum communication, the entangled quantum states can be created with a fidelity of 0.63 for 20 kilometers, which can even be increased by the purification step.

In the sharing process, the sender base station entangles a quantum state with another separate physical qubit, then it is multiplexed into the optical-fiber - or the quantum channel. At the receiver side, the multiplexed pulses are demultiplexed, and the receiver entangles each pulse with a free quantum state. If the entangling operation succeeds, then the sender and the receiver share an EPR state.

The process of entanglement creation between the base stations is illustrated in Fig. 17.

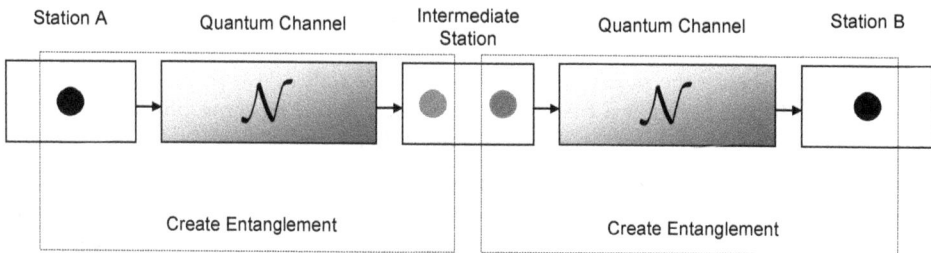

Fig. 17. The background of the quantum repeater is entanglement creation between the adjacent nodes. The quantum states are transmitted by quantum teleportation and local operations in the local nodes

The entanglement creation uses the quantum communication channel, hence some noise is added to the process. As follows, in the next step, the created entanglement has to be purified. The next step is purification, which can help to increase the fidelity of the shared EPR states before the unknown quantum state is teleported. The purification is an error-correcting scheme, and it uses local quantum operations only: hence these operations can be realized locally in the separated base stations (Van Meter et al., 2009).

Now, let's see what this purification does. The *purification* step takes two EPR pairs and by the use of local quantum transformation and classical communication, it combines the two EPR states into one EPR pair, which has greater fidelity.

The theoretical working mechanism of the purification step is illustrated in Fig. 18.

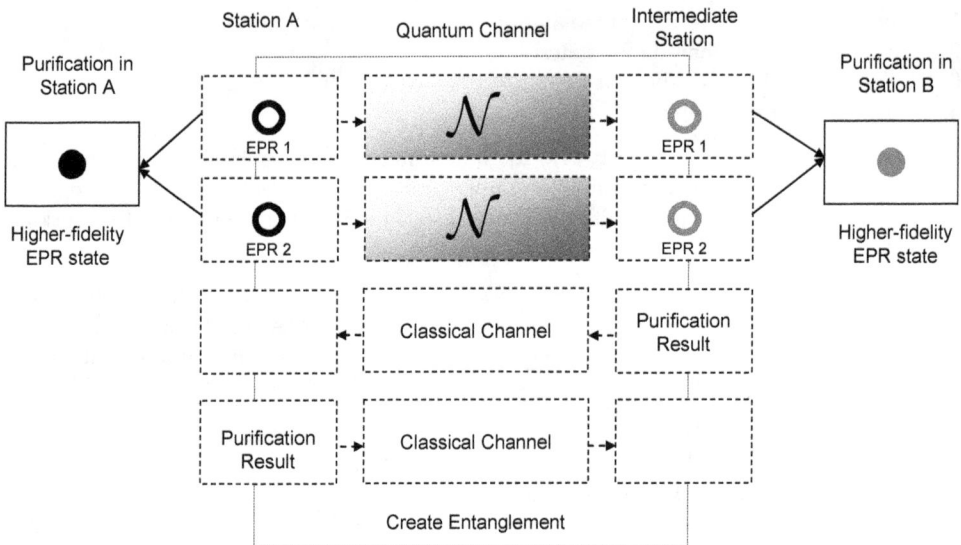

Fig. 18. The noisy quantum states can be purified by entanglement purification. This step requires a lot of resources and classical communication

The purification step has great relevance to the fidelity of the working mechanism of the quantum repeater, however the enhancement of the efficiency of this scheme is still under research (Munro et al., 2009). In the purification step, from two input EPR pairs one EPR pair is generated, hence the purification of the two EPR states destroys one EPR state. As follows, one EPR pair becomes a free qubit, hence this step frees some physical resources (Munro et al., 2009). The purification step could result in two possible outcomes:

1. the purification operation *fails*, and both EPR pairs are then freed;
2. the purification operation *succeeds*: the result is one EPR pair, with higher fidelity. If the fidelity is still low, this state can be used in the next purification step, otherwise it can be used for teleporting and entanglement swapping.

This purification can be done between several other EPR pairs, hence an efficient algorithm is required to choose the Bell pairs to be purified. This method is called *scheduling*, and it has great importance from the viewpoint of the physical resources and the rate at which the fidelity of the shared EPR pairs can be increased (Bernardes et al., 2010).

After the purification step has been finished, in the next step the unknown quantum state can be teleported by the quantum teleporting scheme and by entanglement swapping it can be transmitted to its final destination (Bernardes et al., 2010).

The next step is teleportation and entanglement swapping. Entanglement swapping is equal to a set of quantum teleportation steps: hence entanglement swapping can be viewed as an "extended teleportation protocol," which is able to bridge the large distances between physically separated stations. The quantum teleportation protocol is a component of entanglement swapping, and while quantum teleportation is realized between base stations at short distances, entanglement swapping is an "extended teleportation" which is realized between the sender and the receiver stations. In the quantum teleportation scheme, Alice's input quantum state is destroyed and recovered at Bob's side, using shared entanglement between the parties. Bob needs two classical bits to recover the unknown quantum state. In this protocol, both Alice and Bob have to use local quantum operations only, which are based on classical information.

If we apply quantum teleportation between nodes at short distances, the final result will be referred as entanglement swapping. The Bell states are shared between the adjacent base stations, hence entanglement swapping has the following steps (Van Meter et al., 2009):

- entanglement sharing between Station 1 and Station 2,
- entanglement sharing between Station 2 and Station 3,
- teleportation from Station 1 to Station 2,
- local operations at Station 2, measurement of both quantum states: quantum bits are freed at Station 2,
- classical communication from Station 2 to Station 3,
- local operations at the Station 3.

This process destroys a Bell pair, hence two quantum states in the middle station have been freed, and they can be reused in the next teleportation. The result of the whole process is a swapped entanglement, which connects Station A and Station B.

As the result of the swapped entanglement, the unknown quantum state has been transferred from the sender base station to the receiver base station, through an arbitrary number of intermediate base stations. The structure of this quantum repeating consists of a chain of base stations—or quantum repeaters—and the information is transmitted via quantum teleportation. The chain of repeaters was constructed for the purpose of extending the distance between those nodes that share an EPR state. In the proposed construction this means that in the swapping process, two n-hop Bell pairs are combined into one $2n$ EPR state. This architecture is called the "doubling architecture" and, as has been shown by Briegel et al. (Briegel et al., 1998), the performance of the system declines polynomially rather than exponentially as the distance increases.

In the zero level of entanglement swapping, zero swaps has been made, in the first level one, while in the second level, two swap transformations have been realized. After the transformations have been finished, the initial EPR pair is stretched to reach all the hops, the EPR states of the intermediate hops—which mean three EPR states—have been destroyed (Van Meter et al., 2009).

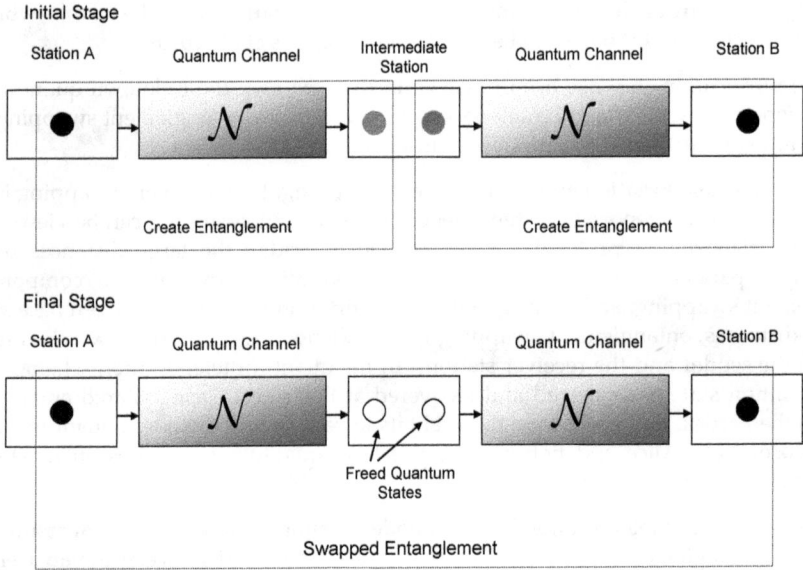

Fig. 19. The entanglement swapping realized by the local transformations made in the Intermediate Node. The swapping operation frees up the "intermediate" EPR states

The zero level of swapping with n base stations is illustrated in Fig. 20.

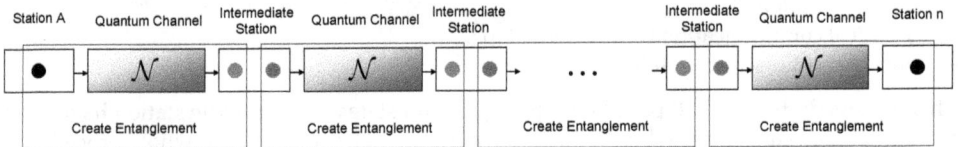

Fig. 20. In the first phase, the adjacent repeater nodes shares entanglement with each other

In this phase, each pair of adjacent base stations share an EPR pair with each other. In the next step, the adjacent base stations start to communicate with each other using quantum teleportation and classical communication. As a result of this step, the EPR state will span two base stations. The entanglement swapping will free up four quantum states.

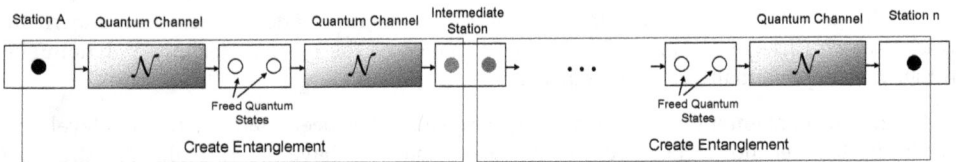

Fig. 21. After the EPR states have been shared, local transformations are made. These transformations free up quantum states in the nodes

In the second level, the entanglement swapping is realized between three base stations, which will result in four spanned base stations, and in six freed quantum states (Van Meter et al., 2009). The second level of entanglement swapping is illustrated in Fig. 22.

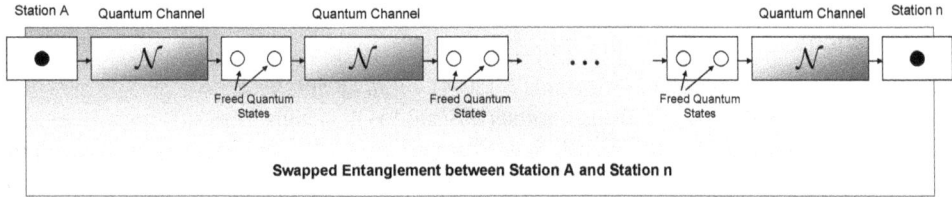

Fig. 22. After the local transformations have been realized in the intermediate stations, the result is an entanglement between node A and node B

The proposed method can be extended to n levels of entanglement swapping and the number of base stations spanned to 2^n. The architecture can be used for long-distance communication and in the quantum networks of the future (Munro et al., 2010), (Van Meter et al., 2009). Entanglement swapping is realized by the purification step and the quantum teleportation protocol combined with classical communications. The purification scheme is able to correct the errors of the transmission which occurs in the node-switching and the fiber-based communication devices. Entanglement swapping can be extended to long distances, and with the help of node-to-node quantum communications, a global scale quantum network can be constructed in the future.

7. Conclusion

In the first part of the chapter, we analyzed the DPS QKD scheme. The DPS QKD protocol has become more popular among quantum cryptographic protocols, since it offers higher key rates and easier implementation. The DPS protocol is tailored for practical applications – such as long-distance quantum communication over optical fiber quantum channels – and represents a more applicable protocol than other discrete- and continuous-variable QKD protocols, which were invented by theorists. The differential phase-shift protocol has better rates and its practical realizations are much simpler. However the security of the DPS QKD protocol is still an open question, since it unconditional security has not been fully approved yet. In this chapter, we analyzed the most general collective attack against the protocol by information geometrical methods, and for different attacker strategies. The proposed information geometrical method analyses the information-theoretic security of the DPS QKD protocol, and it could offer a very useful practical algorithmical solution to solve the still open and unknown questions related to the information-theoretic security of quantum cryptographic protocols. In the second part we have summarized the working mechanism of quantum repeater which will have great importance in optical fiber based secure long-distance quantum communications. The most ideal QKD protocol between the quantum repeater nodes should be the DPS protocol in the future, according to the flexibility, efficiency, easy implementability and low communication complexity.

8. Acknowledgment

The authors would like to thank the financial support from „Sandor Csibi" Ph.D. Researcher Scholarship at the Budapest University of Technology, Faculty of Electrical Engineering and Informatics, Hungary.

9. References

Acín, A.; Gisin, N.; Masanes, L. & Scarani, V. (2004). *Int. J. Quant. Inf.* 2, 23.

Agrawal, G. (1997). *Fiber-Optic Communication Systems* "Wiley, New York".

Bennett, C.H.; Brassard, G.; Breidbard, S. & Wiesner, S. (1982). Quantum cryptography, or unforgeable subway tokens. In D. Chaum, R. Rivest, and A. T. Sherman, eds., *Advances in Cryptology – Proc. CRYPTO '82* . Plenum Press.

Bennett, C.H. & Brassard, G. (1984). Quantum cryptography: public key distribution and coin tossing, Int. conf. Computers, Systems Signal Processing, Bangalore, India, December 10-12,175-179.

Bennett, C.H; (1992). Quantum cryptography using any two non orthogonal states, Phys. Rev. Lett. 68, 3121-3124.

Bernardes, N.K.; Praxmeyer, L. & van Loock, P. (2010). Rate analysis for a hybrid quantum repeater, arXiv:1010.0106v1.

Bhar, A.; Chattopadhyay, I. & Sarkar, D. (2007). No-Cloning and No-Deleting Theorems through the Existence of Incomparable States Under LOCC, *QUANTUM INFORMATION PROCESSING*, Volume 6, Number 2, 93-99, DOI: 10.1007/s11128-006-0041-2.

Biham, E. & Mor, T. (1997). Security of Quantum Cryptography against collective attacks, *Phys. Rev. Lett. 78* , 2256-1159.

Branciard, C.; Gisin, N. & Scarani, V. (2008). *New J. Phys.* 10, 013031.

Branciard, C.; Gisin, N.; Kraus, B. & Scarani, V. (2005). *Phys. Rev. A* 72, 032301.

Briegel, H.J.; Dür, W.; Cirac, I. & Zoller, P. (1998). Quantum repeaters: the role of imperfect local operations in quantum communication. *Physical Review Letters*, 81:5932 5935.

Bužek, V. & Hillery, M. (1996). Quantum copying: Beyond the no-cloning theorem, *Phys. Rev. A* 54, 1844–1852.

Cerf, N, Lévy. M. & Van Assche, G. (2001). *Phys. Rev. A* 63, 052311.

Cerf, N. (2000). Asymmetric quantum cloning machines in any dimension, *J.Mod.Opt.* 47 187, http://arxiv.org/abs/quant-ph/9805024.

Cerf, N.; Bourennane, M.; Karlsson, A. & Gisin, N. (2002). *Phys. Rev. Lett.* 88, 127902.

Cortese, J. (2002) "The Holevo-Schumacher-Westmoreland Channel Capacity for a Class of Qudit Unital Channels", LANL ArXiV e-print quant-ph/0211093.

Cirac, I. & Gisin, N. (1997) *Phys. Lett. A*, 229, 1.

Curty, M. & Lütkenhaus, N. (2004). *Phys. Rev. A* 69, 042321.

Curty, M.; Tamaki, K. & Moroder, T. (2008). *Phys. Rev. A* 77, 052321.

Csiszár, I. & Körner, J. (1978). *IEEE Trans. Inf. Theory* 24, 339.

D'Ariano, G. & Macchiavello, C. (2003). *Phys. Rev. A* 67, 042306.

Devetak, I. & Winter, A. (2005). Distillation of secret key and entanglement from quantum states. *Proceedings of the Royal Society A*, 461:207-235.

Devitt, S.J.; Munro, W.J. & Nemoto, K. (2008). High Performance Quantum Computing, arXiv:0810.2444.

Duan, L.; Lukin, M. D.; Cirac, J, I. & Zoller, P. (2001). "Long-distance quantum communication with atomic ensembles and linear optics," *Nature 414, 413.*

Dušek, M.; Lütkenhaus, N. & Hendrych, M. (2006). in *Progress in Optics*, edited by E. Wolf, "Elsevier, New York", Vol. 49, p. 381.

Dynes, J.; Yuan, Z. L.; Sharpe, A. W & Shields, A. J (2007) "Practical quantum key distribution over 60 hours at an optical fiber distance of 20km using weak and vacuum decoy pulses for enhanced security," *Opt. Express* 15, 8465.

Fannes, M. (1973). A continuity property of the entropy density for spin lattices. *Communications in Mathematical Physics*, 31:291.

Fasel, S.; Gisin, N.; Ribordy, G. & Zbinden, H. (2004). *Eur. Phys. J. D* 30, 143.

Fuchs, C.; Gisin, N.; Griffiths, R. B.; Niu, C.-S. & Peres, A. (1997). *Phys. Rev. A* 56, 1163.

Galtarossa, A., & Menyuk, C. R. (2005). Polarization Mode Dispersion (Springer, Berlin).

Gisin, N.; Ribordy, G.; Tittel, W. & Zbinden, H. (2001). Quantum cryptography. Quantph/0101098.

Gomez-Sousa, H. & Curty, M. (2009). *Quant. Inf. Comput.* 9, 62.

Gyongyosi, L. & Imre, S. (2010): Algorithmical Analysis of Information-Theoretic Aspects of Secure Communication over Optical-Fiber Quantum Channels, *Journal of Optical and Fiber Communications Research*, Springer New York, ISSN 1867-3007 (Print) 1619-8638 (Online).

Gyongyosi, L. & Imre, S. (2010): Information Geometrical Analysis of Additivity of Optical Quantum Channels, *IEEE/OSA Journal of Optical Communications and Networking (JOCN)*, IEEE Photonics Society & Optical Society of America, ISSN: 1943-0620; 2010.

Gyongyosi, L. & Imre, S. (2010): Algorithmic Superactivation of Asymptotic Quantum Capacity of Zero-Capacity Quantum Channels, *Information Sciences, Informatics and Computer Science Intelligent Systems Applications*, ELSEVIER, ISSN: 0020 0255; accepted. (In Press, 2011.).

Gyongyosi, L. & Imre, S. (2010): Information Geometrical Approximation of Quantum Channel Security, *International Journal On Advances in Security*, Published by: International Academy, Research and Industry Association, ISSN: 1942-2636.

Gyongyosi, L. & Imre, S. (2010): Quantum Singular Value Decomposition Based Approximation Algorithm, *Journal of Circuits, Systems, and Computers (JCSC)*, World Scientific, Print ISSN: 0218-1266, Online ISSN: 1793-6454.

Gyongyosi, L. & Imre, S. (2011). Quantum Cellular Automata Controlled Self-Organizing Networks, in *"Cellular Automata"*, INTECH, ISBN 978-953-7619-X-X.

Bechmann-Pasquinucci, H. & Gisin, N. (1999). *Phys. Rev. A* 59, 4238.

Hayashi, M.; Imai, H.; Matsumoto, K.; Ruskai, M.B. & Shimono, T. (2005). Qubit channels which require four inputs to achieve capacity. *QUANTUM INF.COMPUT.*, 5:13.

Hayashi, M. (2006). *Quantum Information: An Introduction*. Springer-Verlag.

Hayden, P.; Leung, D.; Shor, P. & Winter, A. (2003). Randomizing quantum states: Constructions and applications. quant-ph/0307104.

Hillery, M.; Bužek, V. & Berthiaume, A. (1999). Quantum secret sharing. *Phys. Rev. A*,59:1829.

Honjo, T., Inoue, K. & Takahashi, H. (2004) Differential-phase-shift quantum key distribution experiment with a planar light-wave circuit Mach-Zehnder interferometer, *Opt. Lett. 29*, 2797.

Hübel, H.; Vanner, R.; Lederer, T.; Blauensteiner, B.; Lorünser, T.; Poppe, A. & Zeilinger, A. (2007). *Opt. Express 15*, 7853.

Imre, S. & Balázs, F. (2005*): Quantum Computing and Communications – An Engineering Approach*, Published by John Wiley and Sons Ltd.

Inoue, K.; Waks, E. & Yamamoto, Y. (2003). Differential-phase-shift quantum key distribution using coherent light, *Phys. Rev. A 68*, 022317.

Jiang, L.; Taylor, J.M, Nemoto, K.; Munro, W.J.; Van Meter, R. & Lukin, M.D. (2008). *Quantum Repeater with Encoding*, arXiv:0809.3629.

King, C. & Ruskai, M. B. (2001). „Minimal entropy of states emerging from noisy quantum channels", *IEEE Trans. Info. Theory 47*, 192 - 209.

Kwiat, P.; Enzer, D. G.; Hadley, P. G. & Peterson, C. G. (2001). "Experimental Six-state quantum cryptography," in International Conference on Quantum Information, *2001 OSA Technical Digest Series* (Optical Society of America), paper FQIPB4.

Ladd. T.; van Loock, P.; Nemoto, K.; Munro, W.J. & Yamamoto, Y. (2005) .*New J. Phys. 8*, 184.

Louis, S.G.R.; Munro, W.J.; Spiller, T.P. & Nemoto, K. (2008). *Phys. Rev. A 78*, 022326.

Munro, W.J.; Harrison, K.A.; Stephens, A.M.; Devitt, S.J. & Nemoto, K. (2010). *Nature Photonics*, 10.1038/nphoton.2010.213.

Munro, W.J.; Van Meter, R.; Louis, S.G.R. & Nemoto, K. (2008). *Phys. Rev. Lett.* 101, 040502.

Niederberger, A.; Scarani, V. & Gisin, N. (2005). *Phys. Rev. A 71*, 042316.

Nielsen, M. & Chuang, I. L. (2000). *Quantum Computation and Quantum Information*, Cambridge University Press.

Nielsen, F.; Boissonnat, J-D. & Nock, R. (2007) On Bregman Voronoi diagrams. *In Proceedings of the 18th Annual ACM-SIAM Symposium on Discrete Algorithms (SODA'07)*, pages 746–755, Philadelphia, PA, USA, Society for Industrial and Applied Mathematics.

Nielsen, F. & Nock, R. (2008). Bregman Sided and Symmetrized Centroids. *ICPR 2008*, ICPR'08, (arXiv:0711.3242).

Nielsen, F. & Nock, R. (2008). On the smallest enclosing information disk. *Inf. Process. Lett. IPL'08*, 105(3): 93-97.

Nielsen, F. & Nock, R. (2009). Approximating Smallest Enclosing Balls with Application to Machine Learning, International Journal on Computational Geometry and Applications (IJCGA'09).

Paterson, K.; Piper, F. & Schack, R. (2004). Why quantum cryptography?, eprint arXiv:quant ph/0406147.

Renner, R.; Gisin, N. & Kraus, B. (2005). *Phys. Rev. A 72*, 012332.

Rezakhani, A.; Siadatnejad, S.; Ghaderi, A. H. (2005). Separability in Asymmetric Phase Covariant Cloning, *Phys. Lett. A* 336, 278, 10.1016/j.physleta.2004.12.015, arXiv:quant ph/0312024v2.

Rivest, R.; Shamir, A. & Adleman, L. (1978). "A method for obtaining digital signatures and public-key cryptosystems." *Communications of the ACM* 21(2): 120-126.

Rosenberg, D.; Harrington, J. W.; Rice, P. R.; Hiskett, P. A.; Peterson, C. G.; Hughes, R. J. & Nordholt, J. E. (2007). "Long-Distance Decoy-State Quantum Key Distribution in Optical Fiber," *Phys. Rev. Lett.* 98, 010 503.

Ruskai, M.; Szarek, S. & Werner, E. (2001). "An Analysis of Completely-Positive Trace Preserving Maps on 2 by 2 Matrices", LANL ArXiV e-print quant-ph/0101003.

Sangouard, N.; Simon, C.; de Riedmatten, H. & Gisin, N. (2009). Quantum repeaters based on atomic ensembles and linear optics arXiv:0906.2699.

Schmitt-Manderbach, T.; Weier, H.; FÄurst, M.; Ursin, R.; Tiefenbacher, F.; Scheidl, T.; Perdigues, J.; Sodnik, Z.; Kurtsiefer, C.; Rarity, J. G.; Zeilinger, A. & Weinfurter, H. (2007). "Experimental Demonstration of Free-Space Decoy-State Quantum Key Distribution over 144 km," *Phys. Rev. Lett.* 98, 010 504.

Schneier, B. (1996). *Applied Cryptography*. John Wiley & Sons.

Shor, P. (1994). Algorithms for quantum computation: discrete logarithms and factoring. In *Proc. 35th Ann. IEEE Symp. Foundations of Comp. Sci.*, pp. 124–134. IEEE Press, doi:10.1109/SFCS.1994.365700. eprint arXiv:quant-ph/9508027.

Shor, P. (1997). Polynomial time algorithms for prime factorization and discrete logarithms on a quantum computer. *SIAMJ. Comp.,26(5):*1484-1509.

Shuai, C.; Yu-Ao, C.; Thorsten, S.; Zhen-Sheng, Y.; Bo, Z.; Jorg, S. & Jian-Wei, P. (2006). "Deterministic and Storable Single-Photon Source Based on a Quantum Memory." *Physical Review Letters 97*, 173004.

Stephens, A.M.; Evans, Z.W.; Devitt, S.J.; Greentree, A.D.; Fowler, A.G..; Munro, W.J.; O'Brien, J.L.; Nemoto, K. & Hollenberg, L.C.L. (2008). A Deterministic optical quantum computer using photonic modules, *Phys. Rev. A.* 78, 032318.

Stucki, D.; Walenta, N.; Vannel, F.; Thew, R. T.; Gisin, N.; Zbinden, H.; Gray, S.; Towery, C. R. & Ten, S. (2009). "High rate, long-distance quantum key distribution over 250km of ultra low loss fibers," *New J. of Phys.* 11, 075 003.

Takesue, H.; Diamanti, E.; Honjo, T.; Langrock, C.; Fejer, M.M.; Inoue, K.; & Yamamoto, Y. (2005). *New J. Phys.* 7, 232.

Townsend, P. (1997). Quantum cryptography on multiuser optical fiber networks," *Nature* 385, 47.

Van Assche, G.; Cardinal, J. & Cerf, N. J. (2004). *IEEE Trans. Inf. Theory* 50, 394.

Van Loock, P.; Lütkenhaus, N.; Munro, W.J. & Nemoto, K. (2008). *Phys. Rev. A* 78, 062319.

Van Meter, R.; Ladd, T.; Munro, W.J. & Nemoto, K. (2008). System Design for a Long-Line Quantum Repeater, arXiv:0705.4128v2 [quant-ph].

Van Meter, R.; Ladd, T.D.; Munro, W.J. & Nemoto, K. (2009). *IEEE/ACM Transactions on Networking* 17, 1002.

Villoresi, P.; Tamburini, F.; Aspelmeyer, M.; Jennewein, T.; Ursin, R.; Pernechele, C.; Bianco, G.; Zeilinger, A. & Barbieri, C. (2004). "Space-to-ground quantum-communication using an optical ground station: a feasibility study," arXiv:quant-ph/0408067v1.

Wootters, W. & Zurek, W. H. (1982). A single quantum cannot be cloned. *Nature*, 299:802 803, doi:10.1038/299802a0.

WorldWideScience.org (2011), Sample records for quantum computer development from *WorldWideScience.org*.

Yao, A (1995). Security of quantum protocols against coherent measurements. In *Proceedings of the 1995 ACM Symposium on Theory of Computing*, pages 67-75.

9

Nonlinear Compensation Using Multi-Subband Frequency-Shaped Digital Backpropagation

Ezra Ip[1] and Neng Bai[2]
[1]NEC Labs America, Princeton, NJ
[2]University of Central Florida, Orlando, FL
USA

1. Introduction

As dense wavelength-division multiplexed (DWDM) systems push towards ever higher spectral capacities, the nonlinear Shannon's limit is rapidly being approached (Essiambre et al., 2008). Much research has been devoted to nonlinear compensation (NLC) algorithms that can undo deterministic nonlinear impairments to increase nonlinear capacity. All NLC algorithms are ultimately based approximate solutions to the inverse nonlinear Schrödinger equation (NLSE) that describes signal propagation in fiber. In forward propagation, assuming the signal is sufficiently narrowband where the slowly varying envelope approximation holds, a signal evolves as a Manakov equation (Marcuse et al. 1997):

$$\frac{\partial \mathbf{u}}{\partial z} = \left(\hat{\mathbf{D}} + \hat{\mathbf{N}} \right) \mathbf{u} , \tag{1}$$

where $\mathbf{u}(z,t) = \begin{bmatrix} u_x(z,t) & u_y(z,t) \end{bmatrix}^T$ is the Jone's representation of the slowly-varying baseband electric field envelope, $u_x(z,t)$ and $u_y(z,t)$ are the two polarization components.

$\hat{\mathbf{D}} = -\frac{1}{2}\alpha - \beta_1 \frac{\partial}{\partial t} - j\beta_2 \frac{1}{2!}\frac{\partial^2}{\partial t^2} + \beta_3 \frac{1}{3!}\frac{\partial^3}{\partial t^3}$ and $\hat{\mathbf{N}} = j\frac{8}{9}\gamma|\mathbf{u}|^2$ are the linear and nonlinear operators, with α, β_1, β_2, and β_3 being 2×2 matrices representing attenuation, polarization-mode dispersion, group velocity dispersion and dispersion slope of the fiber, and γ being the fiber's nonlinear parameter. In the absence of noise, the forward propagation equation in (1) can be inverted via (Li et al. 2008):

$$\frac{\partial \mathbf{u}}{\partial(-z)} = \left(\hat{\mathbf{D}} + \hat{\mathbf{N}} \right) \mathbf{u} . \tag{2}$$

This operation is analogous to passing the received signal through a fictitious channel where each element in the fictitious channel exactly inverts the real elements in the forward-propagating channel (Fig. 1). In the presence of optical noise arising from optical amplifiers, the inverse NLSE is inexact. As the behavior of the NLSE becomes chaotic at high power, relative signal distortion increases, causing the nonlinear Shannon's limit to arise. All

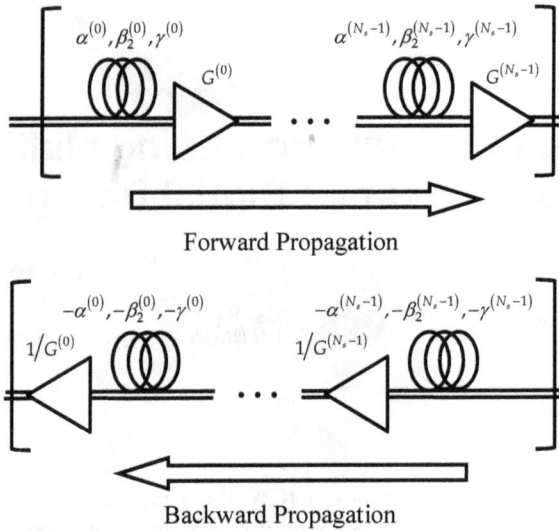

Fig. 1. Channel inversion via backpropagation.

nonlinear compensation (NLC) methods can be shown to be approximate solutions of the inverse NLSE.

The most advanced nonlinear compensation method is "digital backpropagation" (DBP), where the electric field $\mathbf{u}(z,t)$ is recovered by an optical-to-electrical downconverter and then sampled at the Nyquist rate (Ip & Kahn, 2010). The channel impairments are then inverted by numerically solving (2), usually via the split-step Fourier method (SSFM). The SSFM is an iterative algorithm that divides the fiber channel into small steps and then successively passing the signal through the linear and nonlinear operators at each step. For the SSFM algorithm to be accurate, the step size has to be small enough so that the phase rotation in time due to application of $\hat{\mathbf{N}}$ and the phase rotation in frequency due to application of $\hat{\mathbf{D}}$ are both sufficiently small. The step size requirement has been studied in (Sinkin et al. 2003, Zhang & Hayee 2008).

To date, digital backpropagation has not been demonstrated in real-time due to its high algorithmic complexity. Much recent research effort has focused on finding approximate algorithms that can approach DBP in performance, but has significantly lower algorithmic complexity. Some promising results have recently been reported, such as by lowpass filtering the nonlinear operator (Du & Lowery, 2010, Li, et al. 2011). In comparison with the traditional SSFM, which is "frequency-flat," filtered backpropagation exploits chromatic dispersion in fiber that causes different frequency components of a signal to propagate at different speeds. From the point of view of a given frequency component within a signal, the other frequency components "walkoff," causing an averaging effect on their resulting nonlinear interaction thus reducing the variance of their nonlinear distortion. As walkoff increases with frequency separation, the frequencies that lie closest to the frequency component of interest will contribute greater nonlinear distortion than those frequencies that are far away. This effect can be fully exploited by using pre- and post-filters in the calculation of the nonlinear operator.

We will explore an enhanced DBP algorithm, where the nonlinear perturbation $\hat{N} \cdot u$ at each step is computed via multiple subbands: for each subband, different pre- and post-filters are used. Although the complexity required to compute $\hat{N} \cdot u$ is increased, the greater accuracy of the estimate enables larger step sizes. Furthermore, performance can be traded off against complexity by varying either the number of steps or number of subbands.

The outline of this chapter will be as follows. In Section 2, the theory of filtered DBP will be introduced from the point of view of casting the NLSE as a third-order Volterra series. The equations for computing pre-filter will be derived. In Section 3, the DSP architecture needed to implement filtered DBP will be given, and we will revisit the physical intuition for FS-BP. Simulation results will be presented in Section 4.

2. Theory

2.1 Single-polarization

2.1.1 Volterra series model

We begin by considering the NLSE for a single-polarization signal. Let $u(z,t)$ be the scalar electric field in the signal polarization of interest. Ignoring pulse polarization-mode dispersion (PMD) in fiber, the inverse scalar NLSE is given by (Agrawal, 2001):

$$\frac{\partial u}{\partial(-z)} = -\frac{\alpha}{2}u - j\frac{\beta_2}{2}\frac{\partial^2 u}{\partial t^2} + j\gamma|u|^2 u \tag{3}$$

Resolving $u(z,t) \sim \sum_k u_k(z)e^{j\omega_k t}$ in terms of its spectral components $u_k(z)$, where the frequencies are evenly spaced $\omega_k = k\Delta\omega$, we obtain a set of coupled equations:

$$\frac{du_k}{d(-z)} = \left(-\frac{\alpha}{2} + j\frac{\beta_2\omega_k^2}{2}\right)u_k + j\gamma\sum_{\substack{j,l,m \\ \in S[k]}} u_j u_l u_m^* . \tag{4}$$

In the absence of nonlinearity ($\gamma = 0$), the right-hand side of (4) gives the chromatic dispersion compensation filter as $H^-(\omega_k) = \int_0^L \frac{\alpha(z)}{2} - j\frac{\beta_2(z)}{2}\omega_k^2 \, dz$. When $\gamma \neq 0$, the set $S[k] \in \{j,l,m : j+l-m=k\}$ denote the frequencies that interact through the Kerr nonlinearity to produce a polarization at ω_k. The term for which $j = m = k$ denotes intra-channel self-phase modulation (ISPM); the terms for which $j = m \neq k$ denotes intra-channel cross-phase modulation (IXPM); while the remaining terms are intra-channel four-wave mixing (IFWM). Using a third-order perturbation technique developed by (Nazarathy 2008), the signal can be expanded as $u_k(z) = u_k^{(1)}(z) + u_k^{(3)}(z)$, where the first- and third-order terms satisfy:

$$\frac{du_k^{(1)}}{d(-z)} = \left(-\frac{\alpha}{2} + j\frac{\beta_2\omega_k^2}{2}\right)u_k^{(1)}, \text{ and} \tag{5}$$

$$\frac{du_k^{(3)}}{d(-z)} = \left(-\frac{\alpha}{2} + j\frac{\beta_2\omega_k^2}{2}\right)u_k^{(3)} + j\gamma \sum_{\substack{j,l,m \\ \in S[k]}} u_j^{(1)}u_l^{(1)}u_m^{(1)*} \,. \tag{6}$$

Let $v_k(z) = u_k(z)\exp\left(\int\limits_z^L -\left(\frac{\alpha(z')}{2} - j\frac{\beta_2(z')\omega_k^2}{2}\right)dz'\right)$ be a distanced-normalized spectral

components of the signal, where L is the step size. We can similarly decompose $v_k(z)$ as

$v_k(z) = v_k^{(1)}(z) + v_k^{(3)}(z)$, from which:

$$\frac{dv_k^{(1)}}{d(-z)} = 0 \,, \text{ and} \tag{7}$$

$$\frac{dv_k^{(3)}}{d(-z)} = j \sum_{\substack{j,l,m \\ \in S[k]}} v_j^{(1)}v_l^{(1)}\left(v_m^{(1)}\right)^* \times \gamma(z)\exp\left(\int\limits_z^L \alpha(z') + j\Delta\beta_{jlm}(z')dz'\right). \tag{8}$$

where $\Delta\beta_{jlm}(z) = \beta_2(z)\Delta\omega^2(l-k)(j-k)$.

We first assume the fiber parameters remain constant throughout this step, and there are no optical amplifiers in between. The nonlinear perturbation is then given by a third-order Volterra series:

$$v_k^{(3)}(0) = -j \sum_{\substack{j,l,m \\ \in S[k]}} D_{jlm}^{FWM} v_j^{(1)}v_l^{(1)}\left(v_m^{(1)}\right)^* \,, \tag{9}$$

where,

$$D_{jlm}^{FWM} = \gamma\exp\left(\left(\alpha + j\Delta\beta_{jlm}\right)L\right) \cdot \frac{1-\exp\left(-\left(\alpha + j\Delta\beta_{jlm}\right)L\right)}{\alpha + j\Delta\beta_{jlm}} \,, \tag{10}$$

We can define $L_{jlm}^{FWM} = \dfrac{1-\exp\left(-\left(\alpha + j\Delta\beta_{jlm}\right)L\right)}{\alpha + j\Delta\beta_{jlm}}$ to be the "effective length" of the FWM

process involving frequencies ω_j, ω_l and ω_m. We then have $D_{jlm}^{FWM} = \gamma e^{(\alpha + j\Delta\beta_{jlm})L} \cdot L_{jlm}^{FWM}$.

Clearly, as $|l-k|$ and $|j-k|$ increases (i.e., frequencies far from ω_k), $\Delta\beta_{jlm}$ increases, and

hence the strength of their nonlinear coupling on ω_k (i.e., D_{jlm}^{FWM}) decreases.

More generally, we can also consider that in the integration from $z = L$ to $z = 0$, there are N_s spans of fiber, with an optical amplifier after each span (Fig. 2). Let $\alpha^{(n)}$, $\beta_2^{(n)}$ and $L_f^{(n)}$ be the attenuation, dispersion and length of the n-th fiber, and let $G^{(n)}$ be the gain of the n-th amplifier. Carrying out the integration inside the exponent in (8) we get:

$$\frac{dv_k^{(3)}}{d(-z)} = j \sum_{\substack{j,l,m \\ \in S[k]}} v_j^{(1)} v_l^{(1)} \left(v_m^{(1)}\right)^* \times \frac{\gamma^{(n)}}{G^{(n)}} \exp\left(\left(\alpha^{(n)} + j\Delta\beta_{jlm}^{(n)}\right)(z_n - z)\right) \times$$
$$\left[\prod_{n'=n+1}^{N_s-1} \frac{1}{G^{(n')}} \exp\left(\left(\alpha^{(n')} + j\Delta\beta_{jlm}^{(n')}\right)L_f^{(n')}\right)\right] \tag{11}$$

for $z_{n-1} \le z < z_n$.

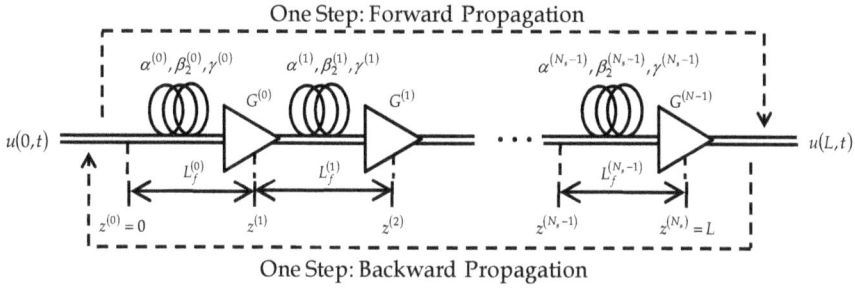

Fig. 2. Solving the NLSE with step size equal to N_s heterogeneous spans of fiber.

At the output of the step, the perturbation term have a third-order Volterra series:

$$v_k^{(3)}(0) = -j \sum_{\substack{j,l,m \\ \in S[k]}} D_{jlm}^{FWM} v_j^{(1)} v_l^{(1)} \left(v_m^{(1)}\right)^*, \tag{12}$$

where

$$D_{jlm}^{FWM} = \sum_{n=0}^{N_s-1} \frac{\gamma^{(n)}}{G^{(n)}} \exp\left(\left(\alpha^{(n)} + j\Delta\beta_{jlm}^{(n)}\right)L_f^{(n)}\right) L_{jlm}^{FWM(n)} \times$$
$$\left[\prod_{n'=n+1}^{N_s-1} \frac{1}{G^{(n')}} \exp\left(\left(\alpha^{(n')} + j\Delta\beta_{jlm}^{(n')}\right)L_f^{(n')}\right)\right], \text{ and} \tag{13}$$

$L_{jlm}^{FWM(n)} = \dfrac{1 - \exp\left(-\left(\alpha^{(n)} + j\Delta\beta_{jlm}^{(n)}\right)L_f^{(n)}\right)}{\alpha^n + j\Delta\beta_{jlm}^{(n)}}$ is the "effective length" of the FWM process

involving frequencies ω_j, ω_l and ω_m in the n-th fiber.

For the special case where all N_s spans are identical and the amplifier gains $G^{(n)} = \exp\left(\alpha^{(n)}L^{(n)}\right)$ exactly equalize the loss of the previous span, (13) can be simplified to:

$$D_{jlm}^{FWM} = \gamma L_{jlm}^{FWM} \exp\left(j\Delta\beta_{jlm}L_f \frac{N_s+1}{2}\right) \frac{\sin\left(\Delta\beta_{jlm}L_f N_s/2\right)}{\sin\left(\Delta\beta_{jlm}L_f/2\right)}. \tag{14}$$

2.1.2 Time-domain approximation

The numerical complexity of evaluating the Volterra series in (9) or (12) is $O(M^3)$, where M is the total number of spectral components in (4). However, suppose it is possible to define an approximate set of Volterra coefficients $\tilde{D}_{jlm}^{FWM} \approx D_{jlm}^{FWM}$:

$$v_k'^{(3)}(0) \approx -j \sum_{\substack{j,l,m \\ \in S[k]}} \tilde{D}_{jlm}^{FWM} v_j^{(1)} v_l^{(1)} \left(v_m^{(1)}\right)^*, \tag{15}$$

where $v_m'^{(1)} = W_k(\omega_m) v_m^{(1)}$ is obtained by passing the spectral components $v_m^{(1)}$ through a carefully designed frequency-shaping pre-filter $W_k(\omega_m)$ whose coefficients are such that:

$$W_k(\omega_j) W_k(\omega_l) W_k^*(\omega_m) = \tilde{D}_{jlm}^{FWM}, \text{ for } (j,l,m) \in S[k] \tag{16}$$

It can be shown that:

$$v'^{(3)}(0,t) = \sum_k v_k'^{(3)}(0) e^{j\omega_k t} = -j \left| \sum_k v_k'^{(1)} e^{j\omega_k t} \right|^2 \left(\sum_k v_k'^{(1)} e^{j\omega_k t} \right) = -j \left| v'^{(1)}(t) \right|^2 v'^{(1)}(t), \tag{17}$$

where $v'^{(3)}(0,t) \approx v^{(3)}(0,t)$ approximates the actual nonlinear perturbation. As according to (16), the filter coefficients are chosen to approximate D_{jlm}^{FWM} at the frequency index k, $v'^{(3)}(0,t) \approx v^{(3)}(0,t)$ therefore has the highest accuracy around ω_k. The procedure indicated by (17) is as follows:—

1. Multiply the input signal by the pre-filter $v_m'^{(1)} = W_k(\omega_m) v_m^{(1)}$.

2. Take the inverse Fourier transform of $v_k'^{(1)}$ to obtain $v'^{(1)}(t)$.

3. Compute $v'^{(3)}(0,t)$ using (17)

4. Take the Fourier transform $v'^{(3)}(0,t)$ to obtain the coefficient $v_k'^{(3)}$.

5. Repeat steps #1 to #4, using different pre-filters for each frequency index k, until the nonlinear perturbation at all frequency indices are obtained.

Since the fast Fourier transform can be solved efficiently in $O(M \log M)$ operations, the above procedure has lower numerical complexity $O(M^2 \log M)$ than the corresponding Volterra series in (9) or (12), which have complexity $O(M^3)$.

It is possible to obtain even greater complexity savings – at the expense of accuracy – if instead of using a different pre-filter each time for #1 to #4, the N frequency indices are partitioned into "subbands" (Weidenfeld et al. 2011). Suppose we modify (16) to:

$$W_b(\omega_j) W_b(\omega_l) W_b^*(\omega_m) = \tilde{D}_{jlm}^{FWM}, \text{ for } (j,l,m) \in \{S[k]: \omega_k \in \Omega_b\}. \tag{16b}$$

i.e., we design the frequency-shaping $W_b(\omega_j)W_b(\omega_l)W_b^*(\omega_m)$ that approximates D_{jlm}^{FWM} at all frequency indices k where ω_k lies within the subband Ω_b. We invoke the same steps #1 to #3 above replacing W_k with W_b, and for step #4, we use: —

4b. Take the Fourier transform $v'^{(3)}(0,t)$ to obtain the coefficients $v_k'^{(3)} : \omega_k \in \Omega_b$.

In other words, since $v'^{(3)}(0,t)$ has good accuracy around the center frequency of the subband for which W_b was designed, we keep multiple frequency indices at a time. The algorithmic complexity then scales as $O(BM\log M)$, where B is the total number of subbands used.

With $v'^{(3)}(0,t)$ computed, the spectral components of the backpropagated signal can now be found by:

$$u'_k(0) = \left(v_k^{(1)} + \xi v_k'^{(3)}(0)\right)\exp\left(\int_0^L \left(\frac{\alpha(z')}{2} - j\frac{\beta_2(z')\omega_k^2}{2}\right)dz'\right),\qquad(18)$$

where $v_k^{(1)} = u_k(L)$. The parameter ξ denotes the overall nonlinear perturbation is scaled, and its value should be optimized for a given launch power level and system dispersion map. It is noted that (18) represents an asymmetric split-step solution, since the nonlinear operator (addition by $v'^{(3)}(0,t)$) is computed first, followed by the linear operator (multiplication by dispersion compensation filter).

2.1.3 Calculating the pre-filter

To find an appropriate frequency-shaping pre-filter, we decompose (16b) into amplitude and phase equations:

$$\log|W_b(\omega_j)| + \log|W_b(\omega_l)| + \log|W_b(\omega_m)| \approx \log|D_{jlm}^{FWM}|, \text{ and}\qquad(19)$$

$$\angle W_b(\omega_j) + \angle W_b(\omega_l) - \angle W_b(\omega_m) \approx \angle D_{jlm}^{FWM},\qquad(20)$$

where $(j,l,m) \in \{S[k]:\omega_k \in \Omega_b\}$. These are linear systems of $O(M^3/B)$ equations and M unknowns, and are thus highly over-determined. However, (18) and (19) can be solved via a "best fit" method such as the Moore-Penrose pseudoinverse.[1]

If N is large, the systems of equations may be intractable to solve. Typically, a large N is required when the frequency response $H^+(\omega_k) = \int_0^L -\frac{\alpha(z)}{2} + j\frac{\beta_2(z)}{2}\omega_k^2 \, dz$ has large

---

[1] Care should be taken to ensure that the phase of D_{jlm}^{FWM} is properly unwrapped over the indices of interest, before matrix inversion. Note this is a three-dimensional phase-unwrapping procedure.

dispersion over the step size taken. This may be encountered in dispersion-unmanaged transmission for example: if $\varphi_{cd}(\omega) = \beta_2 L \omega^2$ changes rapidly with frequency, the signal must be decomposed with fine spectral resolution in order for each component to remain "frequency-flat" over the step size considered. If (19) and (20) has too many equations to be solved numerically, it is possible to derive the pre-filter heuristically. Consider (14), which gives the Volterra coefficients for a step size equal to N identical fiber spans. Assume the spans are sufficiently long $(L_f \gg 1/\alpha)$. Then, it can be shown that for realistic fiber parameters, $L_{jlm}^{FWM} \approx 1/\alpha$ in the neighborhood of $l \approx m \approx k_b$, where k_b is the frequency index of the center of the subband. This neighborhood also contains the strongest components of $D_{jlm}^{FWM} : (j, l, m) \in \{S[k] : \omega_k \in \Omega_b\}$. Hence,

$$\angle D_{jlm}^{FWM} \approx \Delta\beta_{jlm} L_f \frac{N_s + 1}{2} = -\frac{\beta_2}{2} \Delta\omega^2 L_f \left(j^2 + l^2 - m^2 - k^2\right) \frac{N_s + 1}{2}. \tag{21}$$

Noting that phase is quadratic due to dispersion, suppose we choose:

$$\angle W_b(\omega_k) \approx -\frac{\beta_2}{2}(k - k_b)^2 \Delta\omega^2 L_f \frac{N_s + 1}{2}, \tag{22}$$

which centers the quadratic characteristic about ω_{k_b} in the middle of the subband. Similarly for the amplitude equation:

$$
\begin{aligned}
\left|D_{jlm}^{FWM}\right| &\approx \frac{\gamma N_s}{\alpha} \frac{1 - \frac{1}{3}\left(\Delta\beta_{jlm} N_s L_f / 2\right)^2}{1 - \frac{1}{3}\left(\Delta\beta_{jlm} L_f / 2\right)^2} \\
&\approx \frac{\gamma N_s}{\alpha} \left[1 - \frac{1}{3}\left(\Delta\beta_{jlm} N_s L_f / 2\right)^2\right] \left[1 - \frac{1}{3}\left(\Delta\beta_{jlm} L_f / 2\right)^2\right] \\
&= \frac{\gamma N_s}{\alpha} \left[1 - \frac{1}{12}\Delta\beta_{jlm}^2 \left(N_s^2 - 1\right) L_f^2\right] \\
&\approx \frac{\gamma N_s}{\alpha} \exp\left(-\frac{1}{12}\Delta\beta_{jlm}^2 \left(N_s^2 - 1\right) L_f^2\right) \\
&= \frac{\gamma N_s}{\alpha} \exp\left(-\frac{1}{48}\beta_2^2 \Delta\omega^4 \left(j^2 + l^2 - m^2 - k^2\right)^2 \left(N_s^2 - 1\right) L_f^2\right)
\end{aligned} \tag{23}
$$

We thus pick a fourth-order amplitude characteristic about ω_{k_b}:

$$\left|W_b(\omega_k)\right| \approx \left(\frac{\gamma N_s}{\alpha}\right)^{1/3} \exp\left(-\frac{1}{48}\beta_2^2 (k - k_b)^4 \Delta\omega^4 \left(N_s^2 - 1\right) L_f^2\right). \tag{24}$$

The pre-filter W_b can now be found by combining (22) and (24):

$W_b(\omega) = \left|W_b(\omega)\right| \exp\left(j\angle W_b(\omega)\right)$.

2.2 Dual-polarization

For dual-polarization systems, an identical derivation can be carried replacing (3) with the

Manakov equation in (2). As before, let $v_k(z) = u_k(z)\exp\left(\int_z^L -\left(\frac{\alpha(z')}{2} - j\frac{\beta_2(z')\omega_k^2}{2}\right)dz'\right)$ be the

distanced-normalized spectral components of the signal, and let $v_k(z) = v_k^{(1)}(z) + v_k^{(3)}(z)$. It

can be shown that the nonlinear perturbation term $v_k^{(3)}(z) = \left[v_{x,k}^{(3)}(z) \quad v_{y,k}^{(3)}(z)\right]^{\mathrm{T}}$ satisfies:

$$v_{x,k}^2(L) = -j \sum_{\substack{j,l,m \\ \in S[k]}} D_{jlm}^{FWM}\left(v_{x,j}^{(1)}v_{x,l}^{(1)}\left(v_{x,m}^{(1)}\right)^* + v_{x,j}^{(1)}v_{y,l}^{(1)}\left(v_{y,m}^{(1)}\right)^*\right), \text{ and} \tag{25a}$$

$$v_{y,k}^2(L) = -j \sum_{\substack{j,l,m \\ \in S[k]}} D_{jlm}^{FWM}\left(v_{y,j}^{(1)}v_{y,l}^{(1)}\left(v_{y,m}^{(1)}\right)^* + v_{y,j}^{(1)}v_{x,l}^{(1)}\left(v_{x,m}^{(1)}\right)^*\right). \tag{25b}$$

where the Volterra coefficients D_{jlm}^{FWM} are almost identical to those derived for the single-polarization case, with a scaling factor of $(8/9)^{1/3}$ to absorb the factor of $8/9$ in the Manakov equation. Hence for step sizes (i) less than one fiber span, (ii) equal to N_s heterogeneous fibers spans, and (iii) equal to N_s identical fiber spans, D_{jlm}^{FWM} is given by (10), (13) or (14). Thus, the reduced-complexity procedure outlined previously can also be used to find $v_k^{\prime(3)}(z) \approx v_k^{(3)}(z)$, with the subband filters given by (19) and (20) using the analytical method, or (22) and (24) using the heuristic method:

$$v_x^{\prime(3)}(0,t) = \sum_k v_{x,k}^{\prime(3)}(0)e^{j\omega_k t} = -j\left(\left|\sum_k v_{x,k}^{\prime(1)}e^{j\omega_k t}\right|^2 + \left|\sum_k v_{y,k}^{\prime(1)}e^{j\omega_k t}\right|^2\right)\left(\sum_k v_{x,k}^{\prime(1)}e^{j\omega_k t}\right), \tag{26a}$$

$$v_y^{\prime(3)}(0,t) = \sum_k v_{y,k}^{\prime(3)}(0)e^{j\omega_k t} = -j\left(\left|\sum_k v_{x,k}^{\prime(1)}e^{j\omega_k t}\right|^2 + \left|\sum_k v_{y,k}^{\prime(1)}e^{j\omega_k t}\right|^2\right)\left(\sum_k v_{y,k}^{\prime(1)}e^{j\omega_k t}\right), \tag{26b}$$

which can be simplified as:

$$\mathbf{v}^{\prime(3)}(0,t) = -j\left|\mathbf{v}^{\prime(1)}(t)\right|^2 \mathbf{v}^{\prime(1)}(t). \tag{27}$$

3. Digital signal processing architecture

The digital signal processing architecture that implements multi-subband, frequency-shaped backpropagation (FS-BP) is shown in Fig. 3(a) (Ip & Bai 2011). It is assumed that a coherent receiver recovers the in-phase (I) and quadrature (Q) components of the electric field in the two signal polarizations, which are synchronously sampled with digital-to-analog converters (DAC). Shown on the left-hand side of Fig. 3(a) are the received samples $\mathbf{u}(z = L_{tot}, t)\big|_{t=nT}$, where L_{tot} is the total length of the system, and T is the sampling interval. The input signal is processed in block sizes of M samples. For single-carrier (SC) transmission, overlap-and-save is used (Oppenheim & Schafer 2009); for orthogonal frequency-division multiplexing (OFDM) transmission, non-overlapping blocks are taken at the input, with the cyclic prefix of the OFDM symbol (block) stripped at the output. In both cases, the signal processing has the canonical model shown in Fig. 3(a). In the absence of nonlinearity, the operations enclosed between the fast Fourier transform (FFT) and inverse FFT (IFFT) performs frequency-domain linear equalization (LE) of chromatic dispersion. In digital backpropagation, the single "linear step" is replaced with a concatenation of nonlinear and linear steps, as outlined in Section 2.

Fig. 3(b) shows a generalized model of the nonlinear step. The input signal is multiplied by the pre-filter derived in Section 2. To avoid aliasing, we upsample the input signal by padding with zeros in the frequency domain, before taking a $2M$–point inverse FFT (IFFT). The nonlinear perturbation is then computed in the time-domain. After taking a $2M$–point FFT to recover the frequency components, we downsample by stripping the high-frequency components, followed by multiplication by a "post-filter." These operations are repeated for the B subbands. Summing their outputs yield an overall nonlinear perturbation, which is then scaled and summed with the through signal.

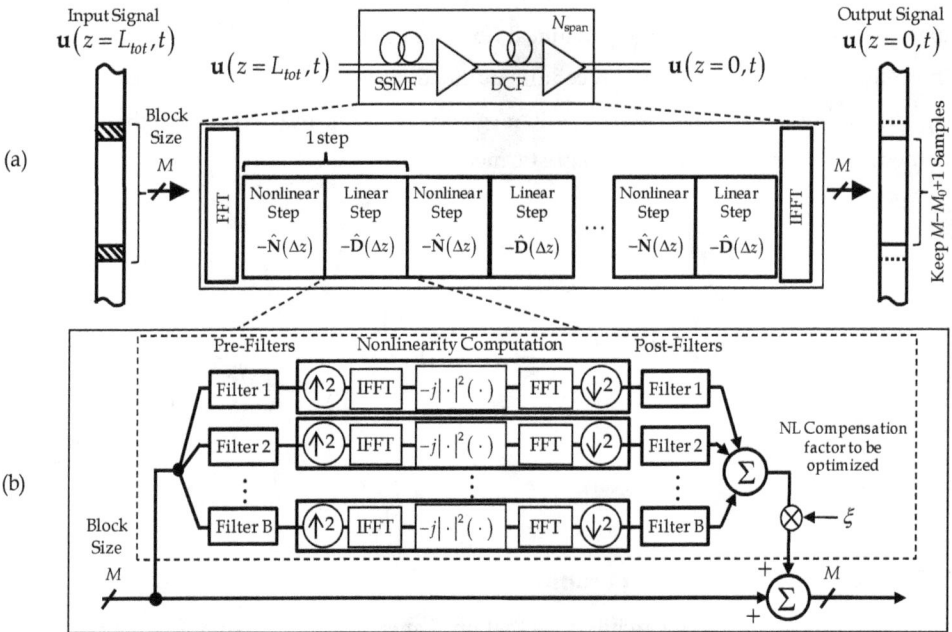

Fig. 3. (a) Digital Backpropagation using overlap-and-save, (b) Enhanced nonlinearity computation using subbanding and frequency shaping by pre- and post-filters.

Note that we have not discussed a method of calculating the post-filter. In Section 2, it was assumed that after computing $\mathbf{v}_k^{\prime(3)}$ in (17) or (27), we keep only the spectral components $\{k : \omega_k \in \Omega_b\}$ in the neighborhood of ω_{k_b}. This corresponds to multiplying $\mathbf{v}_k^{\prime(3)}$ by a rectangular filter whose support spans the bandwidth Ω_b. It is possible to use other filter shapes for the post-filters, which may have better performance than simple rectangular filters. However, this is not covered by the current work.

Algorithmic complexity can be determined directly from Fig. 3(b). Assume the use of overlap-and-save in Fig. 3(a) where adjacent blocks overlap by M_0 samples. We assume a K steps in the backpropagation, and that the post-filters are rectangular filters (no multiplications required). It can be shown that the complexity for FS-BP for single-polarization signals is:

$$C_{FS-BP,1pol} = \left(\left[\left(4(2M)\log_2(2M) + 12M \right)B + 4M \right]K + 4M\log_2 M \right) \big/ (M - M_0 + 1). \quad (28)$$

For dual-polarization signals, the complexity of FS-BP is:

$$C_{FS-BP,2pol} = \left(\left[\left(8(2M)\log_2(2M) + 24M \right)B + 8M \right]K + 8M\log_2 M \right) \big/ (M - M_0 + 1). \quad (29)$$

We note that standard backpropagation is merely a special case of FS-BP where the nonlinearity perturbation is computed without pre- and post-filters (frequency-flat), and only $B = 1$ subband is used. Hence the complexity of standard backpropagation (Std.BP) for single- and dual-polarization signals are:

$$C_{Std.BP,1pol} = \left(\left[4(2M)\log_2(2M) + 8M \right]K + 4M\log_2 M \right) \big/ (M - M_0 + 1), \text{ and} \quad (30)$$

$$C_{Std.BP,2pol} = \left(\left[8(2M)\log_2(2M) + 16M \right]K + 8M\log_2 M \right) \big/ (M - M_0 + 1). \quad (31)$$

Finally, for linear equalization (LE), the complexities for single- and dual-polarization signals are:

$$C_{LE,1pol} = \left(4M + 4M\log_2 M \right) \big/ (M - M_0 + 1), \text{ and} \quad (32)$$

$$C_{LE,2pol} = \left(8M + 8M\log_2 M \right) \big/ (M - M_0 + 1). \quad (33)$$

The motivation for the use of sub-banding, pre-filtering and post-filtering can be understood as follows: in a dispersive fiber, different frequency components of a signal propagate at different speeds. From the point of view of a particular frequency component, the other frequencies walk off. Walk-off has an averaging effect on their mutual nonlinear interaction. Hence, a frequency component of interest experiences stronger nonlinear effects from frequencies closer to it than frequencies further away. To compute nonlinear perturbation at a particular frequency ω_{k_b}, therefore, one should weight the input signal with a pre-filter that emphasizes the frequencies close to it, while suppressing frequencies farther away (Fig. 4). The nonlinearity computed using this method will be accurate around ω_{k_b}. To compute nonlinearity accurately around other frequencies, different pre- and post-filters are needed.

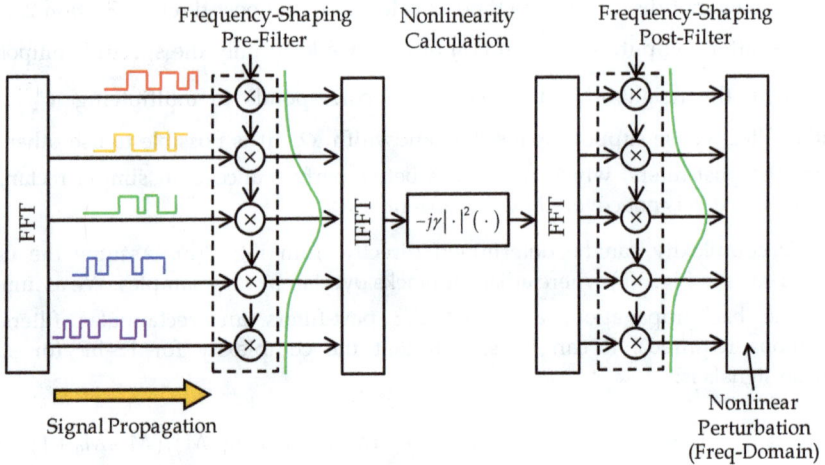

Fig. 4. Physical interpretation of enhanced nonlinearity computation.

4. Results

In this section, we will investigate the efficacy of FS-BP by numerical simulation for two systems (Fig. 5):—

a. OFDM transmission over 12×80-km spans of single-mode fiber (SSMF) (Attenuation = 0.2 dB/km, Dispersion = 17 ps/nm/km and Nonlinear parameter 0.0013 W^{-1}), with full dispersion management after each span using dispersion compensation fiber (DCF) (Attenuation = 0.5 dB/km, Dispersion = −85 ps/nm/km and Nonlinear parameter 0.0053 W^{-1}). The launch power into DCF is assumed to be 6 dB lower than that into SSMF. For the signal, 112-Gb/s OFDM is assumed. The total number of subcarriers (FFT size) is 128. Of these, 102 are modulated with dual-polarization 16-QAM (DP-QAM). A cyclic prefix of 20 is appended for each block.
b. SC transmission over 24×80-km spans of low-dispersion fiber (Attenuation = 0.2 dB/km, Dispersion = 2 ps/nm/km and Nonlinear parameter 0.0013 W^{-1}), with no inline dispersion management. The signal is assumed to be 112-Gb/s DP-16QAM.

For both systems and for all algorithms to be compared, it is assumed that the received signal $\mathbf{u}(L,t)$ is oversampled by a factor of two relative to the chip rate. i.e., For System A, $T = 1/[(112\times10^9/8)\times((128+20)/102)\times2] = 24.6$ ps; for System B, $T = 1/[(112\times10^9/8)\times2] = 35.7$ ps. Overlap-and-save is assumed (Fig. 3(a)). The block size used varies depending on the amount of dispersion to be compensated per step, but the smallest power-of-two is chosen subject to the condition that there be minimal loss of performance due to frequency discretization.

Fig. 6 shows the amplitude and phase of the Volterra coefficients D_{jlm}^{FWM} for System A at DC ($k=0$), assuming a step size equal to four spans (i.e., $K=3$ steps for the entire link). As expected, the largest coefficients are found around $l=0$ and $m=0$, as the corresponding frequencies ω_l and ω_m are closest to ω_k of interest, therefore contributing to the most nonlinear effects.

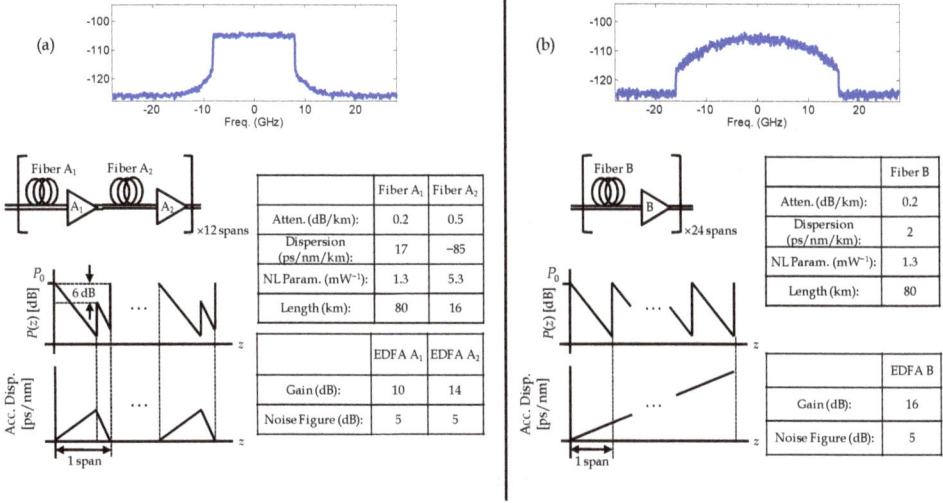

Fig. 5. Simulation setup for (a) System A and (b) System B, showing the signal spectrum, dispersion map, power profile, and fiber & EDFA parameters.

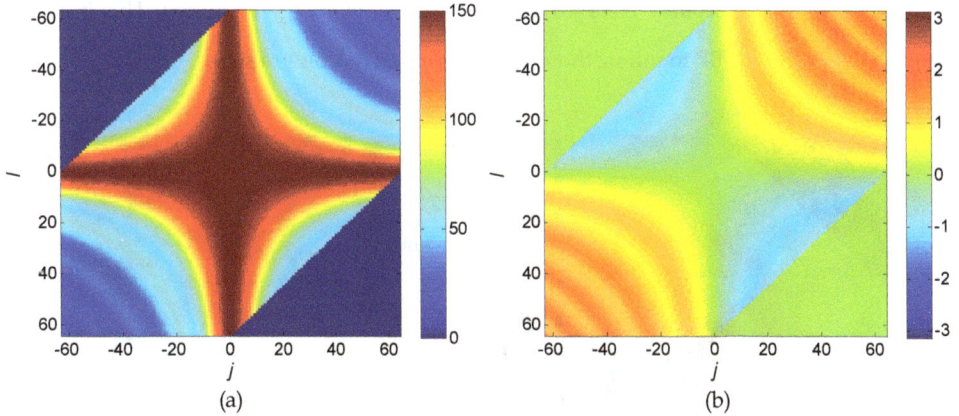

Fig. 6. (a) Amplitude and (b) Phase of D_{jlm}^{FWM} at $k=0$ for System A.

We assume that a block size of $M=128$ is used, and the signal is partitioned into $B=3$ subbands of equal bandwidth. Using (13), (19), and (20), the pre-filter for each subband is computed, and their amplitude and phase responses are shown in Fig. 7. Subband filter #1 is used to compute the nonlinear perturbation at the lowest frequencies (frequency indices $-64 \leq k \leq -22$); subband filter #2 is use for the frequencies near DC ($-21 \leq k \leq 21$), and subband filter #3 for the highest frequencies ($22 \leq k \leq 63$). As expected, the amplitude response of each subband filter has a maximum at the center of each subband. In addition, the phases of the pre-filters are approximately quadratic, accounting for the effect of fiber dispersion.

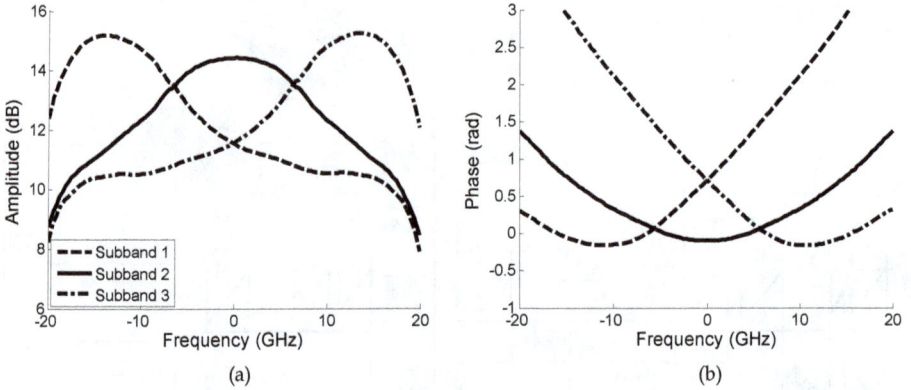

Fig. 7. (a) Amplitude and (b) Phase responses of subband pre-filters for System A.

To see how well the FS-BP technique emulates the actual Volterra coefficients, we evaluate $\tilde{D}_{jlm}^{FWM} = W_k(j)W_k(l)W_k^*(m)$ at $k=0$, since these are the "approximate" Volterra coefficients that will result from the application of (17). Fig. 8 shows the amplitude and phase of $\tilde{D}_{jlm}^{FWM}$ at $k=0$. Compared with Fig. 6, we observe good match between the amplitude and phase of D_{jlm}^{FWM} and $\tilde{D}_{jlm}^{FWM}$ around $l=m=0$ where most of the energy is located. This indicates the effectiveness of FS-BP to mimic the $O(M^3)$ complexity Volterra series.

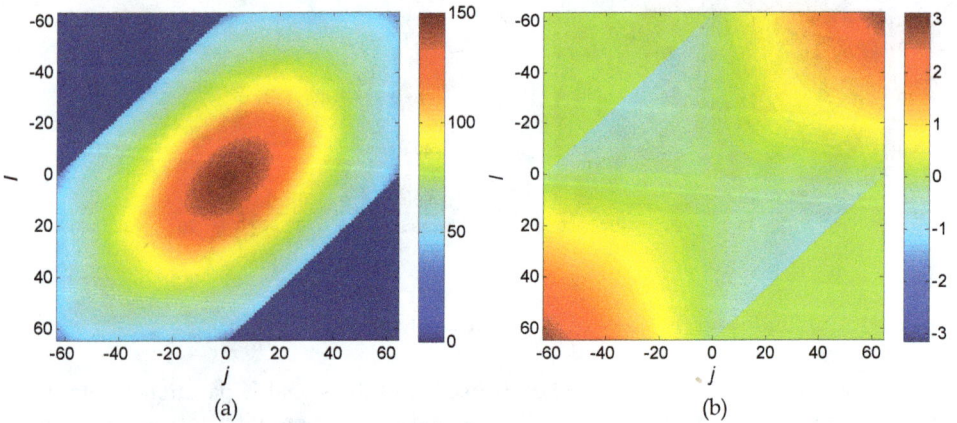

Fig. 8. (a) Amplitude and (b) Phase of $\tilde{D}_{jlm}^{FWM}$ at $k=0$ for System A.

We repeated the results for System B in Figs. 7–9. The block size and number of subband are again assumed to be $M=128$ and $B=3$, and the step size is set to four spans (i.e., $K=6$ for the entire link). The amplitude and phase of the actual Volterra coefficients D_{jlm}^{FWM} at $k=0$ are shown in Fig. 9. We use the "heuristic" model (22) and (24) to compute the pre- filters.

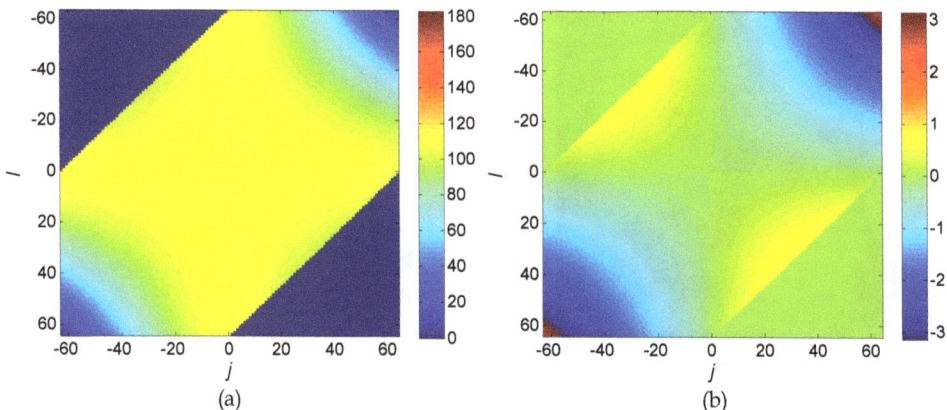

Fig. 9. (a) Amplitude and (b) Phase of D_{jlm}^{FWM} at $k = 0$ for System B.

The phase and amplitude responses of the pre- filters are shown in Fig. 10, with their corresponding $\tilde{D}_{jlm}^{FWM}$ at $k = 0$ shown in Fig. 11. While excellent match between the phases of Fig. 9(b) and Fig. 11(b) is observed, the amplitudes in Fig. 11(a) of the "approximate" Volterra coefficients are larger than the actual Volterra coefficients. Thus, using the pre-filters in Fig. 10 will overestimate the nonlinearity perturbation at each step. In the nonlinearity computation block shown in Fig. 3(b) (also equation (18)), ξ will need to be adjusted to optimize system performance for each power level.

To see how much FS-BP improves system performance across the frequencies, Fig. 12 compares FS-BP with LE for System A, assuming a launch power of $P_0 = -2$ dBm. We use the notation FS-BP[K:B] to denote that K backpropagation steps are used, and in each step, the signal is partitioned into B equal subbands for nonlinearity computation. We evaluate system performance by the Q-factor, which is defined as the mean signal power divided by the mean signal distortion after compensation, i.e., Q is the inverse of error vector

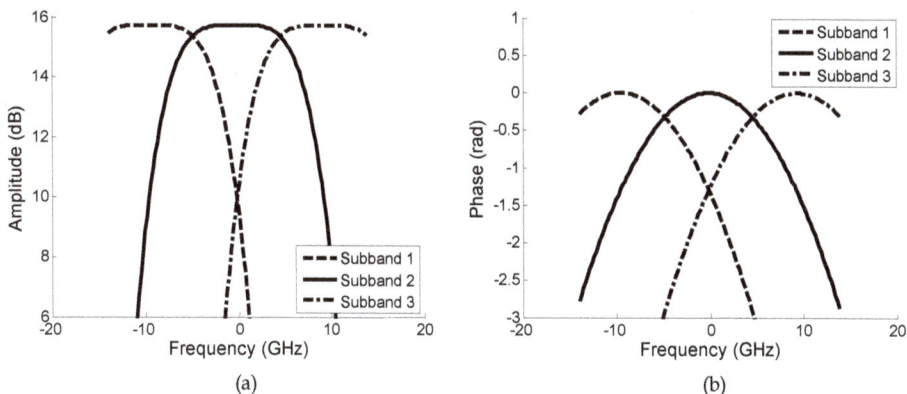

Fig. 10. (a) Amplitude and (b) Phase responses of subband pre-filters for System B.

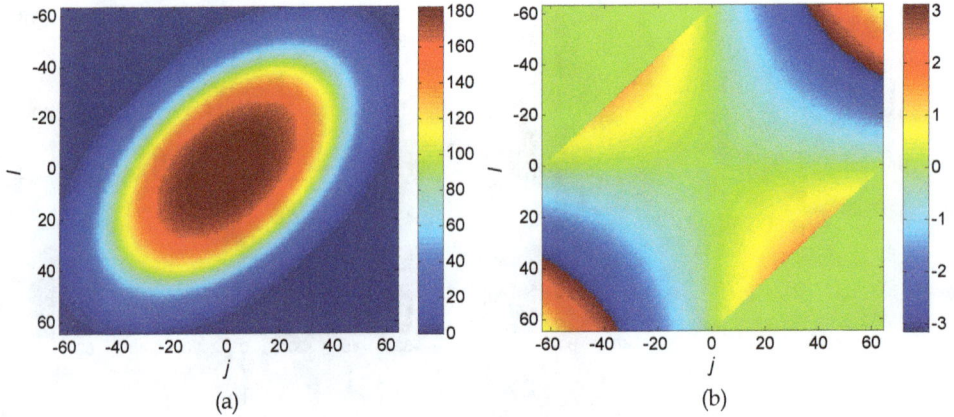

Fig. 11. (a) Amplitude and (b) Phase of $\tilde{D}_{jlm}^{FWM}$ at $k = 0$ for System B.

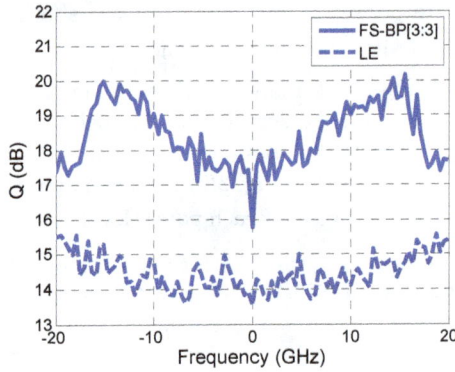

Fig. 12. Q vs. Frequency for System A, comparing different impairment compensation algorithms.

magnitude (EVM). For both LE and FS-BP, it is observed that the outer subcarriers have higher Q than the inner subcarriers, due to reduced number of neighboring subcarriers carrying data contributing to nonlinearity. The use of FS-BP improves the performance for all of the subcarriers. However, the improvement is not uniform. It is observed that the subcarriers around ±15 GHz have the best performance because they correspond to the center of subbands 1 and 3, about which the pre-filters #1 and #3 were optimized.

Next, we investigate system performance versus launch power, and how algorithmic complexity trades off with system performance. We compare (i) linear equalization only (LE), (ii) FS-BP, and (iii) standard backpropagation (Std. BP) with one or more steps per fiber span.

Figs. 13 and 14 show the results for System A and System B. In Figs. 11(a) and 12(a), it is observed that as the number subbands and/or BP sections used is increased, system performance improves as expected. Standard BP with one step per span outperforms the FS-BP algorithms with multi-span step sizes, but has significantly higher algorithmic

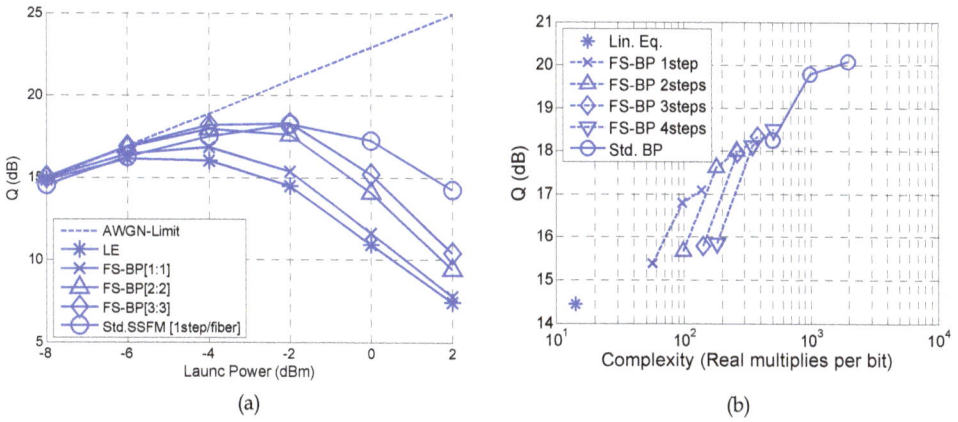

Fig. 13. (a) Performance vs. Launch Power (b) Performance vs. Complexity for System A, comparing different impairment compensation algorithms.

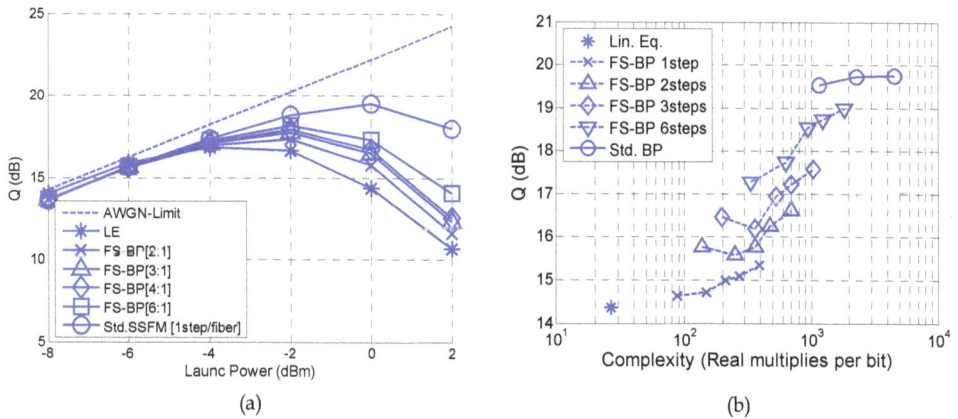

Fig. 14. (a) Performance vs. Launch Power (b) Performance vs. Complexity for System B, comparing different impairment compensation algorithms.

complexity. Figs. 11(b) and 12(b), shows Q versus algorithmic complexity as defined by (28)–(33). The curves labeled "FS-BP K steps" denotes using FS-BP with K DBP steps and varying number of subbands. As the number of subbands is increased from left to right, complexity increases and performance improves. The curve labeled Std. BP denotes standard DBP (frequency-flat) with varying number of steps per span. As the total number of steps increases from left to right, complexity increases and performance improves. Both Systems A and B confirm the performance vs. complexity tradeoff, but they differ on the optimal nonlinearity compensation strategy for a given complexity. For System A, it is observed that for a given complexity, better performance is obtained using fewer steps and more subbands. Conversely, for System B, it is better to employ more steps, but to use only one subband for nonlinearity calculation at each step. This result is intuitively meaningful. In dispersion managed transmission, the signal will have the same amplitude profile after every span (except for noise and nonlinearity, which are small). Hence, dividing the link into larger number of steps will have little performance improvement. By contrast, dividing the signal into subbands will improve the accuracy of the nonlinearity computation, hence better performance. In dispersion unmanaged transmission, the signal profile change rapidly after each step. It is therefore better to use larger number of steps, with strong filtering of the nonlinear perturbation since these will experience strong averaging effect with dispersion. Fig. 10(a) (and equation (24)) confirmed the second-order Gaussian steepness of the amplitude pre-filter. The small dip in Q in Fig. 14(b) at two subbands for FS-BP with 2 and 3 steps is due to the nature of single-carrier signals, where frequencies near DC are the most important. When using one subband, nonlinearity is well compensated near DC, but when using two subbands, the center of the subbands will straddle DC, so nonlinearity is slightly less-well compensated at the most critical frequency.

Finally, Figs. 13(b) and 14(b) indicates that by selecting an optimal number of steps and subbands, FS-BP can provide around 2.5 dB improvement over LE at ten times the algorithmic complexity for both Systems A and B, which is significant savings compared with frequency-flat BP at one step per span. This may make FS-BP an attractive candidate for real-time implementation.

5. Conclusions

In dispersive optical fiber, a given frequency component of a signal experiences stronger nonlinear interactions from frequencies closer to it than frequencies far away. This walkoff effect can be exploited by multiplying the signal with a set of pre-filters, each designed to enable nonlinear perturbation be calculated accurately around a design frequency. By combining the different estimates together, nonlinear perturbation can be calculated accurately across the entire signal bandwidth, allowing backpropagation to use larger step sizes. This muti-subband frequency-shaped backpropagation (FS-BP) approach allows flexible tradeoff between performance and complexity as the number of steps and the number of subbands can be independently varied. We simulated FS-BP for two systems: OFDM transmission over a dispersion-managed link, and single-carrier transmission over a dispersion-unmanaged link. It was found that a dispersion-managed link favors using fewer steps but larger number of subbands; whereas a dispersion-unmanaged link favors using more steps at one subband per step. For both systems, it was found that FS-BP can improve system performance by as much as 2.5 dB at a computational cost ten times that of linear equalization only. This makes FS-BP a potentially candidate for real-time implementation where low algorithmic complexity is essential.

6. References

Agrawal, G. P. (2001). *Nonlinear Fiber Optics 3rd Edition,* Academic Press, ISBN 0120451433, San Diego, CA, U.S.A.

Du, L. B. & Lowery, A. J. (2010). Improved single channel backpropagation for intra-channel fiber nonlinearity compensation in long-haul optical communications systems, *Optics Express*. Vol. 18, No. 16, (August 2010), pp. 17075–17088, ISSN 1094-4087.

Essiambre, R.-J.; Foschini, G. J.; Winzer, P. J.; Kramer, G. & Burrows, E. C. (2008). The capacity of fiber-optic communication systems, *Proceedings of the Optical Fiber Communications Conference (OFC 2008)*, Paper OTuE1, San Diego, CA, U.S.A., March 2008.

Ip, E. & Kahn, J. M. (2010). Fiber impairment compensation using coherent detection and digital signal processing, *Journal of Lightwave Technology*, Vol. 28 No. 4, (February 2010) pp. 502–519, ISSN 0733-8724.

Ip, E. & Kahn, J. M. (2008). Compensation of dispersion and nonlinear impairments using digital backpropagation, *Journal of Lightwave Technology*, Vol. 26, No. 20, (October 2008) pp. 3416–3425, ISSN 0733-8724.

Ip, E. & Bai, N. (2011). The Nonlinear compensation using frequency-shaped multi-subband backpropagation, *Proceedings of the Optical Fiber Communications Conference (OFC 2011)*, Paper OThF4, Los Angeles, CA, U.S.A., March 2011.

Li, L.; Tao, Z.; Dou, L.; Yan, W.; Oda, S.; Tanimura, T.; Hoshida, T. & Rasmussen, J. (2011). Implementation efficient nonlinear equalizer based on correlated digital backpropagation, *Proceedings of the Optical Fiber Communications Conference (OFC 2011)*, Paper OWW3, Los Angeles, CA, U.S.A., March 2011.

Li, X.; Chen, G.; Goldfarb, G.; Mateo, E.; Kim, I.; Yaman, F. & Li, G. (2008). Electronic post-compensation of WDM transmission impairments using coherent detection and digital signal processing, *Optics Express*. Vol. 16, No. 2, (January 2008), pp. 880–888, ISSN 1094-4087.

Marcuse, D., Menyuk, C. R. & Wai, P. K. A. (1997). Applications of the Manakov-PMD equation to studies of signal propagation in optical fibers with randomly varying birefringence, *Journal of Lightwave Technology*, Vol. 15, No. 9, (September 1997) pp. 1735–1746, ISSN 0733-8724.

Nazarathy, M.; Khurgin, J.; Weidenfeld, R.; Meiman, Y.; Cho, P.; Noé, R., Shpantzer, I. & Karagodsky, V. (2008). Phased-array cancellation of nonlinear FWM in coherent OFDM dispersive multi-span links, *Optics Express*. Vol. 16, No. 20, (September 2008), pp. 15777–15810, ISSN 1094-4087.

Oppenheim, A. V. & Schafer, R. W. (2009). *Discrete-Time Signal Processing 3rd Edition*, Prentice Hall, ISBN 0131988425, Upper Saddle River, NJ, U.S.A.

Sinkin, O. V.; Holzlöhner, R.; Zweck, J. & Menyuk, C. (2003). Optimization of the split-step Fourier method in modelling optical-fiber communication systems, *Journal of Lightwave Technology*, Vol. 21, No. 1, (January 2003) pp. 61–68, ISSN 0733-8724.

Weidenfeld, R.; Nazarathy, M.; Noé, R. & Shpantzer, I. (2010). Volterra nonlinear compensation of 100G coherent OFDM with baud-rate ADC, tolerable complexity and low intra-channel FWM/XPM error propagation, *Proceedings of the Optical Fiber Communications Conference (OFC 2010)*, Paper OTuE3, Los Angeles, CA, U.S.A., March 2011.

Zhang, Q. & Hayee, M. I. (2008). Symmetrized split-step Fourier scheme to control global simulation accuracy in fiber-optic communication systems, *Journal of Lightwave Technology*, Vol. 26, No. 2, (January 2008) pp. 302–316, ISSN 0733-8724.

10

Optical Performance Analysis of Single-Mode Fiber Connections

Mitsuru Kihara

Technical Assistance and Support Center,
NTT East Corporation
Japan

1. Introduction

Many single-mode optical fiber (SMF) connection techniques, such as fusion splicing, mechanical splicing, and use of optical connectors, are currently used in fiber-to-the-home (FTTH) systems (Keck et al., 1989; Shinohara, 2005). A fusion splice is fabricated by a fusion splice machine (splicer), which is a precision machine containing fiber alignment, video monitor, and arc discharge functions; it has the highest and stablest performance of all the connections. A mechanical splice is a simple and cost-effective connection that requires no electricity. An optical connector is capable of frequent reconnections. Many kinds of connectors have been developed and used in optical networks. Each connection technique determines how or where it should be used. Fiber connections, except fusion splices, are classified into two types of connection states. One is a connection with physical contact (PC), and the other uses refractive-index matching material. PC-type connectors are mostly used for intra-office fiber connections and on premises where frequent reconnections are required. In contrast, connectors and mechanical splices with refractive-index matching material are mostly used in outside facilities, where frequent reconnections are unnecessary but low cost connections are needed. Field-installable connectors using both PC connectors and refractive-index matching material have recently been developed and used in FTTH systems (Nakajima et al., 2007; Hogari et al., 2010).

The optical performances of these fiber connections have been analyzed and reported (Marcuse, 1976; Young, 1991; Kihara et al., 1996), but some points remain unclear. Unexpected faults occurring during and after installation of these fiber connections might detrimentally affect performance. For instance, when an air gap occurs unexpectedly at the contact point with PC-type connectors or connectors using index matching material, the return loss becomes noticeably worse. In addition, contamination and scratches on an optical connector end surface may cause significant performance deterioration of the mated connectors (Albeanu et al., 2003). Understanding the worst possible optical performance of these fiber connections would make it possible to guarantee the overall performance of a system.

In this chapter, the optical performance of SMF connections is reported, various cases of which have been experimentally investigated: connections with (1) air-filled gaps, (2) a mixture of refractive-index matching material and air-filled gaps, and (3) unexpected use of an incorrectly cleaved fiber end. The cases were assumed to occur accidentally as the result of unexpected failure during and after installation of fiber connections using PC or refractive-index matching material in the field. The various connection cases, classified in their normal and abnormal states, are shown in Fig. 1. In the normal state, the polished fiber ends of a PC connection touch, with no air-filled gap between the ends. A connection using refractive-index matching material has a very small gap between the polished or correctly cleaved fiber ends, and the gap is filled with that material. This chapter details the abnormal connection states of the three connection cases. In section 2, the conventional optical performance analyses of SMF connections based on the D. Marcuse analysis for insertion loss and the W. C. Young et al. analyses for return loss are explained. In section 3, the performance of fiber connections with air-filled gaps is revealed. This case might occur when a fiber connection using PC experiences an unexpected failure, resulting in imperfect PC. In section 4, a loss analysis is reported for fiber connections with a mixture of refractive-index matching material and air-filled gaps. This case might occur when an optical connector or a mechanical splice using refractive-index matching material experiences an unexpected failure. The performance deterioration of fiber connections using an incorrectly cleaved fiber end is demonstrated in section 5. This case might occur when a field-assembly connector or a mechanical splice experiences an unexpected failure. Finally, this chapter is summarized in section 6.

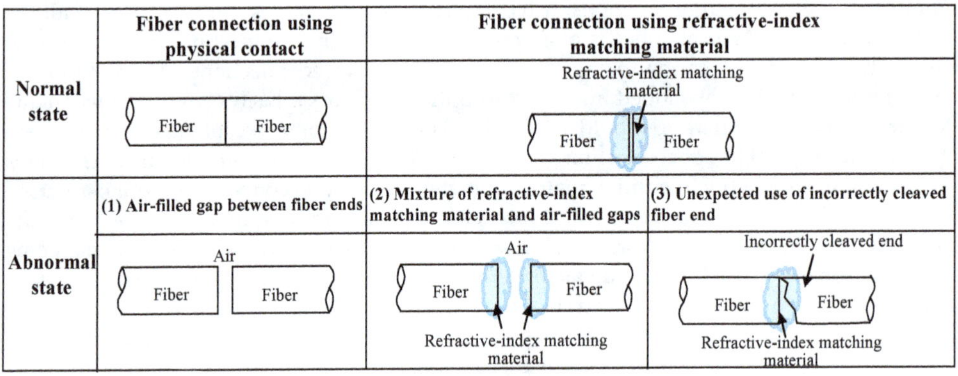

	Fiber connection using physical contact	Fiber connection using refractive-index matching material	
Normal state	Fiber — Fiber	Refractive-index matching material Fiber — Fiber	
	(1) Air-filled gap between fiber ends	**(2) Mixture of refractive-index matching material and air-filled gaps**	**(3) Unexpected use of incorrectly cleaved fiber end**
Abnormal state	Air Fiber — Fiber	Air Fiber — Fiber Refractive-index matching material	Incorrectly cleaved end Fiber — Fiber Refractive-index matching material

Fig. 1. Various states of fiber connections.

2. Overview of conventional analyses of SMF connections

This section explains the conventional optical performance analyses of SMF connections. The two important parameters for the optical performance of fiber connections are insertion loss and return loss. The insertion loss in dB is derived by multiplying -10 by the log of the transmission coefficient T, i.e., $-10 \log(T)$. Here, T denotes the ratio of transmitted light power to incident light power at the fiber connection. Similarly, the return loss in dB is derived by multiplying -10 by the log of the reflection coefficient R, i.e., $-10 \log(R)$. Here, R denotes the ratio of returned light power to incident light power at the fiber connection. In this section, the conventional insertion loss analysis of SMF connections based on that by D. Marcuse is first explained. Then, the W. C. Young et al. analyses for return loss are reported.

2.1 Insertion loss

The insertion loss of SMF connections has been analyzed by D. Marcuse (Marcuse, 1976). According to the analysis, when the fundamental mode of SMF is assumed to be approximately expressed by the Gaussian function, the transmission coefficient T can be calculated for the four major factors shown in Fig. 2. The calculation equations are shown below.

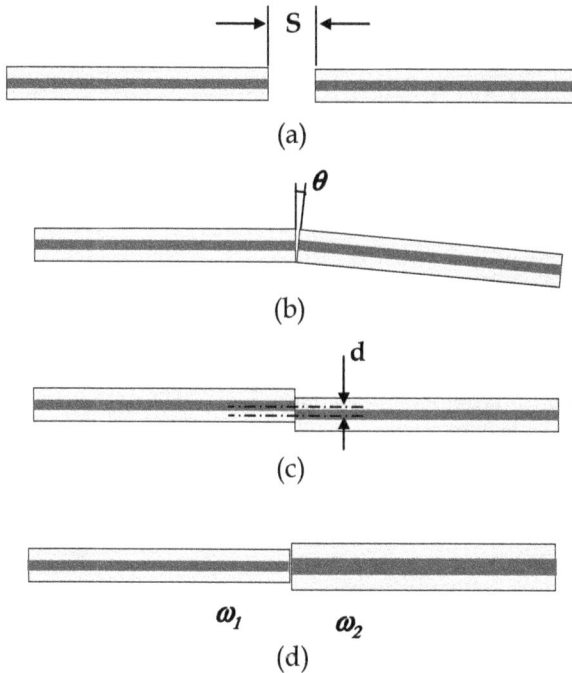

Fig. 2. Four types of insertion loss factors. (a) Gap between fiber ends, (b) misalignment of tilt, (c) misalignment of offset, (d) mode field mismatch.

(a) Gap between fiber ends (when the gap is much larger than the wavelength-order length of the transmitted light)

$$T = \frac{1}{Z^2 + 1} \tag{1a}$$

$$Z = \frac{\lambda S}{2\pi n \omega^2} \tag{1b}$$

(b) Misalignment of tilt

$$T = \exp\left[-\frac{(\pi n \omega \theta)^2}{\lambda^2} \right] \tag{2}$$

(c) Misalignment of fiber offset

$$T = \exp\left[-\frac{d^2}{\omega^2}\right] \tag{3}$$

(d) Mode field mismatch

$$T = \left(\frac{2\omega_1\omega_2}{\omega_1^2 + \omega_2^2}\right)^2 \tag{4}$$

Here, S, θ, d, n, λ, ω, ω_1, and ω_2 are the gap size, tilt, offset, refractive index of the medium between two fibers, wavelength, and the three mode field radii of transmitted light, respectively. These equations are generally and widely used to analyze the insertion loss of an SMF connection.

2.2 Return loss

Return loss is also an important parameter for fiber connections (Young, 1991). A reflection occurs at the boundary between two media with different refractive indices, named a Fresnel reflection (Born & Wolf, 1964). The Fresnel reflection R_0 at the fiber end in a medium is defined by the following equation.

$$R_0 = \left(\frac{n_1 - n}{n_1 + n}\right)^2 \tag{5}$$

Here, n_1 and n denote the refractive indices of the fiber core and the medium, respectively. For instance, when a cleaved fiber end is in air, the refractive indices of the fiber core and air are 1.454 and 1.0, respectively, and the reflection coefficient R_0 is 0.034 (the return loss is 14.7 dB). In this case, the reflected light power is about 3.4 % of the incident power at the fiber end in air, but the value is very large in optical transmission characteristics.

The return loss for a fiber connection without a gap is thought to be negligible. However, we have to consider the return loss for optical fiber connections with a gap between the fiber ends. An analysis of the reflection coefficient caused by a gap between fiber ends is based on multiple reflections behaving like a Fabry-Perot interferometer (Yariv, 1985; Kashima, 1995), which is shown in Fig. 3. In Fig. 3 (a), a flat board with thickness S and refractive index n is placed in a medium with refractive index n_1. Figure 3 (b) shows a fiber connection with a small gap. Here, small means a length of wavelength order. The incident light I_i, transmitted light I_t, and returned light I_r in both figures is considered to behave identically. In Fig. 3 (b), Fresnel reflections occur at the fiber ends because of refractive discontinuity, and some of the incident light is multiply reflected in the small gap. As the phase of the multiply reflected light changes whenever it is reflected, this interferes with the transmitted and reflected lights at the small gap. These multiple reflections between fiber ends are considered to behave like a Fabry-Perot interferometer. The two fiber ends make up the Fabry-Perot resonator. On the basis of the analysis, the reflection coefficient R of optical fiber connections with a gap is defined by the following equations.

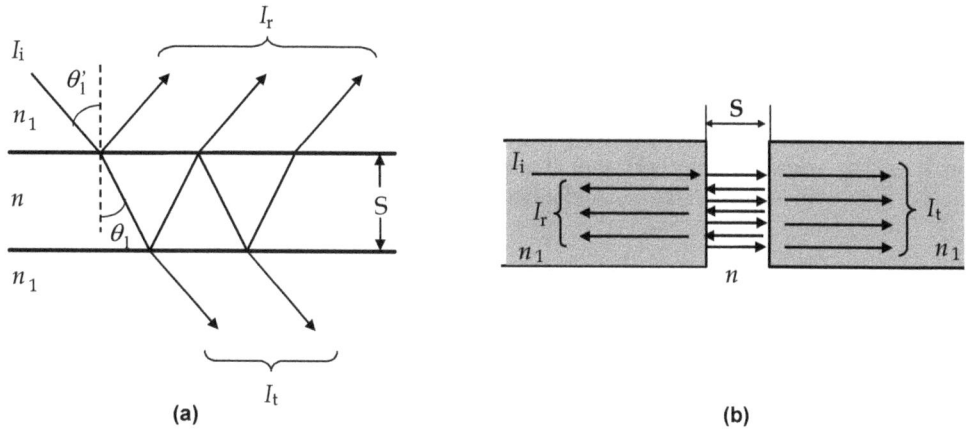

Fig. 3. (a) Fabry-Perot interference model, (b) model of fiber connection with small gap.

$$R = \frac{I_r}{I_i} = \frac{4R_0 \sin^2(\delta/2)}{(1-R_0)^2 + 4R_0 \sin^2(\delta/2)} \tag{6}$$

$$\delta = \frac{4\pi n S \cos\theta_1}{\lambda}$$

Here, δ, n, S, and R_0 are the phase difference, refractive index of the medium, gap size between fiber ends, and reflection coefficient at the fiber core and the medium (Eq. (5)), respectively. If $R_0 \ll 1$, Eq. (6) can be transformed to the following equation.

$$R = 2R_0(1 - \cos\delta) \tag{7}$$

When the fiber ends for the connection are flat, smooth, and perpendicular to the fiber axis, the incident angle $\theta_1{}'$ and the angle θ_1 can be 0 rad. Therefore, Eq. (7) can be transformed to the following equation.

$$R = 2R_0\left(1 - \cos\left(\left(\frac{4\pi n}{\lambda}\right)S\right)\right) \tag{8}$$

This equation is generally used to analyze the return loss of a SMF connection. If more detailed analyses on return loss, such as for polished fiber end connections (a fiber connection whose ends have a high-refractive-index layer) are needed, the work by Young (1991) and Kihara et al. (1996) is recommended.

3. Air-filled gap

This section reveals the performance of fiber connections with air-filled gaps. This case might occur when a fiber connection using PC experiences an unexpected failure, resulting in imperfect PC.

3.1 Wavelength dependence

We focus our investigation on the characteristics of optical fiber connections caused by the gap between the fiber ends. Misalignments of the offset and tilt between the fibers, and the mode field mismatch are not considered. Analysis of optical performance affected by a small gap between fiber ends is based on multiple reflections behaving like a Fabry-Perot interferometer. Here, a small gap means a length of wavelength order. On the basis of the analysis, the transmission coefficient T and the reflection coefficient R of optical fiber connections with an air-filled gap are defined by the following equations.

$$T = \frac{(1 - R_0)^2}{(1 - R_0)^2 + 4R_0 \sin^2(2\pi nS / \lambda)} \tag{9}$$

$$R = \frac{4R_0 \sin^2(2\pi nS / \lambda)}{(1 - R_0)^2 + 4R_0 \sin^2(2\pi nS / \lambda)} \tag{10}$$

The insertion and return losses in dB are derived by multiplying -10 by the log of the transmission and reflection coefficient functions. Here, n_1, n, S, and λ are the refractive indices of the fiber core and of air, and the gap size and wavelength, respectively. R_0 is the reflection coefficient defined by Eq. (5). According to Eqs. (9) and (10), the insertion and return losses depend on wavelength λ and gap size S. The wavelength dependence of the insertion and return losses over a wide wavelength range was experimentally investigated by using mechanically transferable (MT) connectors (Satake et al., 1986). MT connectors without refractive-index matching material generally have small air-filled gaps between their fiber ends (Kihara et al., 2006). The insertion and return losses of MT connectors with an air-filled gap were measured over a wide wavelength range using halogen-lamp or supercontinuum light sources, an optical spectral analyzer, and an optical coupler. The supercontinuum light source can output over +20 dBm/nm more power than the halogen-lamp light source. Two sets of results for MT connectors with air-filled gaps are shown in Figs. 4 (a) and (b), respectively. The circles and lines represent the measured results and the calculations based on Eqs. (9) and (10), respectively. The refractive indices n_1 and n were 1.454 and 1.0, and the gap size S for calculations was 1.13 μm in (a) and 1.3 μm in (b). The calculated and measured data for insertion loss varied between 0.0 and 0.6 dB over a wide wavelength range. The data for return loss varied greatly and resulted in a worst value of 8.7 dB. These two sets of measured results are in good agreement with the calculations. They showed that the insertion and return losses for fiber connections with small air-filled gaps vary greatly and periodically depending on wavelength.

3.2 Gap size dependence

The gap size dependence of the optical performance of fiber connections with an air-filled gap was also investigated. If the gap size between fiber ends is small, the performance could be determined based on the analysis in section 3.1. However, if the gap is larger than a length of wavelength order, radiation loss could occur in it. The attenuation ratio A is defined using the Marcuse equation (1) in terms of the gap between the fiber ends as follows:

$$A = \left[\left(\frac{\lambda S}{2\pi n \omega^2} \right)^2 + 1 \right]^{-1} \tag{11}$$

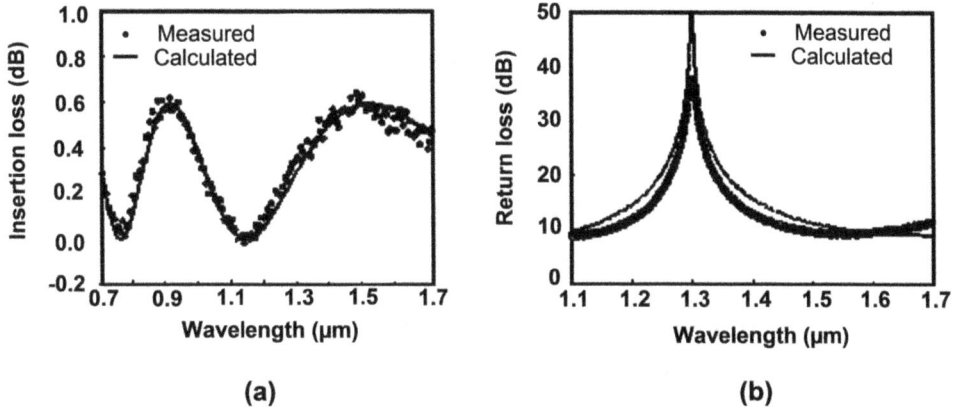

Fig. 4. Wavelength dependence of fiber connections with air-filled gap. (a) Insertion loss results, (b) return loss results.

Here, ω is the mode field radius of the transmitted light. Considering the attenuation in the gap between fiber ends, the transmission coefficient T and the reflection coefficient R are derived from Eqs. (9) and (10) as

$$T = \frac{A(1-R_0)^2}{(1-AR_0)^2 + 4AR_0 \sin^2(2\pi nS / \lambda)} \tag{12}$$

$$R = \frac{[\{1 + A(1-2R_0)\}^2 - 4A(1-2R_0)\sin^2(2\pi nS / \lambda)]R_0}{(1-AR_0)^2 + 4AR_0 \sin^2(2\pi nS / \lambda)} \tag{13}$$

T and R are dependent on gap size S according to Eqs. (12) and (13), which are more complicated than Eqs. (9) and (10), respectively. To demonstrate these dependences, another experiment using an MT connector was performed (Kihara et al., 2010). A feeler gauge (thickness gauge tape) was set and fixed between the two MT ferrules of a connector with a certain gap size by using a clamp spring. By changing the thickness of the feeler gauge, various sizes of gaps were obtained. An air-filled gap was obtained without using refractive-index matching material. The insertion and return losses for the fiber connections with various air-filled gap sizes are shown in Figs. 5(a) and (b), respectively. The circles and lines represent the measured results and the calculations based on Eqs. (12) and (13), respectively. The refractive indices n_1 and n were 1.454 and 1.0, and the wavelength λ for calculations was 1.31 μm in both (a) and (b). The calculated values for insertion and return losses oscillated. This oscillation is caused by the multiple reflection interference in an air-filled gap, which was described earlier. The range of oscillation changed with the gap size. When the gap size was as small as a length of wavelength order, the range of oscillation was large. When the gap size was much larger, the range of oscillation was smaller. This suggests that the insertion and return losses when the gap is small mainly depend on the multiple reflection interference, and that those when the gap is much larger are affected by the radiation loss in an air-filled gap. The measured insertion loss increased with the gap

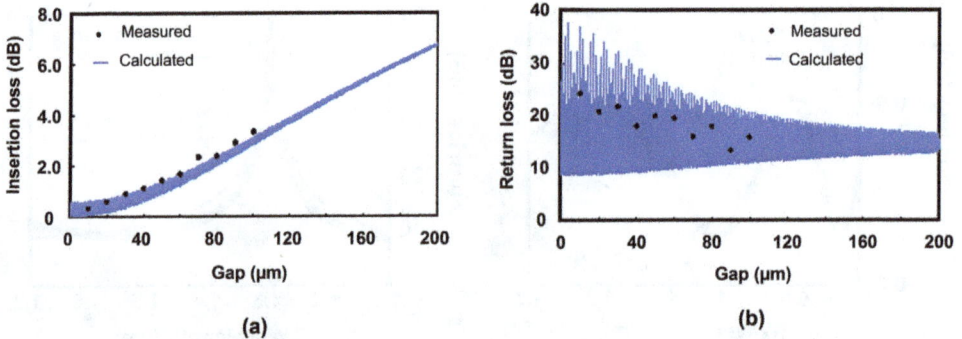

Fig. 5. Gap-size dependence of fiber connections with air-filled gap. (a) Insertion loss results, (b) return loss results.

size as well as the calculated values. The measured return loss varied greatly, but the values were within the oscillation range of the calculated results. The calculated return loss when the gap was much larger was close to 14.7 dB, which is a value of Fresnel reflection at a cleaved fiber end in air. These two sets of measured results are in good agreement with the calculations. Consequently, we theoretically and experimentally revealed the optical performance of fiber connections with various air-filled gap sizes.

3.3 Optical performance of fiber connections with imperfect physical contact

The optical-performance deterioration of a PC-type connector with an imperfect physical contact, i.e., when an air gap occurs unexpectedly at the contact point was also investigated. The experiments using a single-fiber coupling optical fiber (SC) connector (Sugita et al., 1989) were performed. An SC connector is a push-on-type connector and is composed of two plugs and an adaptor. The plug and adaptor are engaged by fitting a pair of elastic hooks into corresponding grooves. Failure to connect the mated connector, such as an incorrect hooking or an existing contamination on a connector end surface, leads to imperfect physical contact and the occurrence of an air-filled gap at the contact point of the connector. An incorrect hooking was intentionally created and 140 SC connector fault samples that had imperfect physical contact were fabricated as investigation samples. The insertion and return losses at a wavelength of 1.3 μm of the fabricated SC connector fault samples are shown in Figs. 6 (a) and (b). The insertion and return losses for SC connectors that maintain physical contact generally are under 0.5 dB and over 40 dB, respectively. In contrast, for the SC connector fault samples with imperfect physical contact, the minimum, maximum, and mean insertion losses were 0.0, 18.1, and 8.7 dB, respectively. The return loss varied between 9.4 and 23.1 dB, and the mean value was 14.6 dB. The results revealed that the optical performance of fiber connections with imperfect physical contact could deteriorate greatly.

Consequently, the optical performances of fiber connections with an air-filled gap are extremely unstable and vary widely. At worst, the insertion and return losses might deteriorate to ~18 and 9.4 dB, respectively. Therefore, air-filled gaps between fiber ends must be prevented from occurring in PC-type connectors.

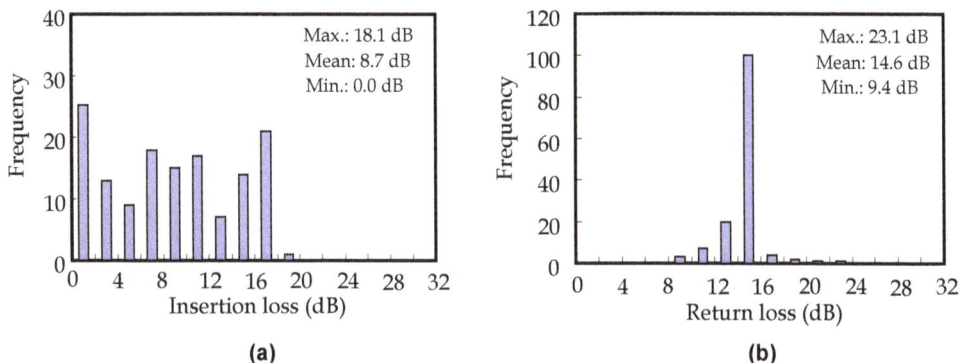

Fig. 6. Optical performance of SC connector fault samples with air-filled gap. (a) Insertion loss results, (b) return loss results.

4. Mixture of refractive-index matching material and air-filled gaps

This section reports a loss analysis for fiber connections with a mixture of refractive-index matching material and air-filled gaps. This case might occur when an optical connector or a mechanical splice using refractive-index matching material experiences an unexpected failure.

4.1 Optical fiber connection with gap

We first focus our investigation on the insertion loss of optical fiber connections caused by the gap between fiber ends. The misalignments of the offset and tilt between the fibers and the mode field mismatch were not taken into account.

There are two analysis techniques for insertion losses caused by these gaps. One is based on multiple reflection analyses, such as that using a Fabry-Perot interferometer, when the gap is small (i.e., of wavelength order). This is expressed by Eq. (9). The other is the Marcuse analysis, which is used when the gap is much longer than the wavelength. This is expressed by Eq. (1). The typical insertion loss results for fiber connections with a small air-filled gap and with refractive-index matching material between the fiber ends are shown in Fig. 7(a). The insertion loss results for fiber connections with long gaps are shown in Fig. 7(b). The measured data were obtained using MT connectors such as described in the previous section. Silicone oil was used as the refractive-index matching material. The circles and squares represent measured results obtained with air-filled and refractive-index matching-material-filled gaps, respectively. The solid and dashed lines indicate the respective calculated results using the above equations. When the gap is small, insertion losses for the air-filled gap vary between 0.0 and 0.6 dB over a wide wavelength range, as shown in Fig. 7 (a). In contrast, the losses for the refractive-index matching-material-filled gap are negligible. According to the multiple reflection analysis, the losses vary between 0.0 and 0.6 dB depending on the gap length if the wavelength is constant. In contrast, when the gap is much longer than the wavelength, the insertion loss worsens and becomes much larger, as shown in Fig. 7 (b). The loss increases with gap length. For instance, the insertion loss for an air-filled gap increases to ~0.8 dB when the gap is 50 μm. These two sets of results are in

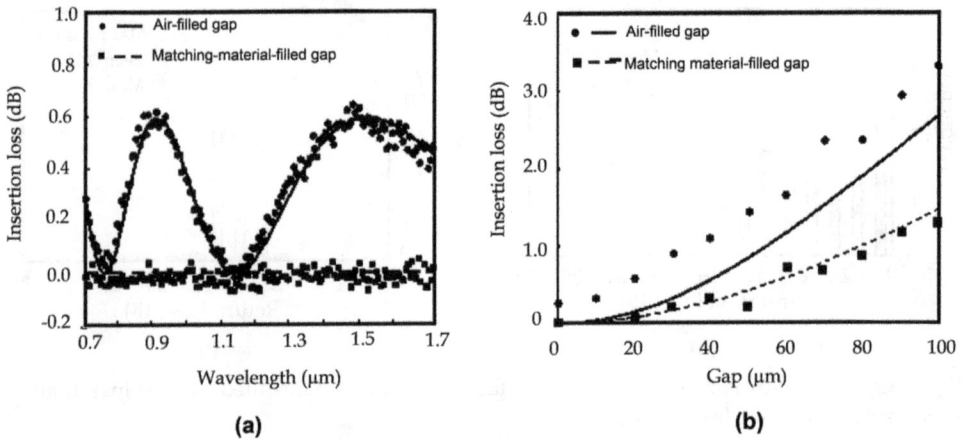

Fig. 7. Optical performance of fiber connections with various gaps. (a) Small gap: 1.1 μm over wide wavelength range of 0.7–1.7 μm. (b) Large gaps: 10 to 100 μm at wavelength of 1.3 μm.

good agreement with the calculations based on the multiple reflection and Marcuse analyses. This indicates that an experiment using feeler gauges is effective for analyzing fiber connections with various gaps.

S. Yoshino et al. reported the results of a mechanical splice fault (Yoshino et al., 2008). The maximum insertion loss change of the mechanical splice with a large gap of less than 50 μm was more than 10 dB during a heat-cycle test. This loss is much larger than the values obtained by the above two analyses. Thus the factors leading to the difference between these results and the conventional theory were experimentally investigated.

4.2 Mixture of refractive-index matching material and air-filled gaps

This section describes the experimental results for fiber connections with a mixture of refractive-index matching material and air-filled gaps. The following experiments using MT connectors with a feeler gauge were performed (Kihara et al., 2009) . MT ferrules without using a feeler gauge (conventionally) were first connected, where refractive-index matching material was used between the ferrule ends. Next, one ferrule pair was disconnected and only one of the ferrule ends was cleaned with alcohol. Then, the cleaned ferrule and the ferrule with refractive-index matching material were connected to a 50-μm feeler gauge. A schematic and photographs of the connected MT ferrules are shown in Fig. 8, and the insertion and return losses of the four fibers in the MT connector are listed in Table 1. The direction of light input to the MT connector changed. The results of the two directions, a and b, are also listed. Every return loss in the same direction was almost equal. The return loss values in the two directions indicated that a mixture of refractive-index matching material and air-filled gaps existed between the fiber ends. In contrast, there was little difference between the insertion losses for different directions within the same fiber, but the insertion loss of each of the four fibers was different. The lowest insertion loss was 3 dB, and the highest was about 40 dB. These results reveal that the insertion loss of fiber connections with a mixture of matching material and air-filled gaps might increase to more than 10 dB.

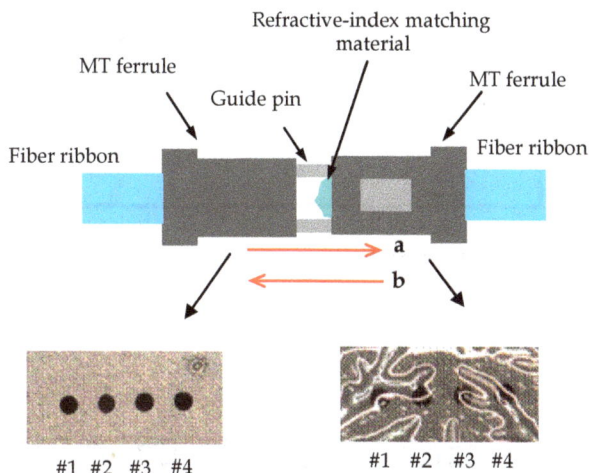

Fig. 8. Schematic view and photographs of MT ferrule samples with mixture of refractive-index matching material and air-filled gaps (50 μm).

Direction a					Direction b				
Fiber number	#1	#2	#3	#4	Fiber number	#1	#2	#3	#4
Insertion loss (dB)	3.2	9.8	20.4	41.7	Insertion loss (dB)	3.1	9.8	21.8	39.5
Return loss (dB)	15.8	16.0	15.9	16.7	Return loss (dB)	24.5	46.2	40.2	47.3

Table 1. Insertion and return loss results.

Another experiment with various gaps: an air-filled gap, a refractive-index matching-material-filled gap, and a mixture of refractive-index matching material and air-filled gaps was conducted. The procedure for creating connections with a mixture of refractive-index matching material and air-filled gaps was described above. The results are shown in Fig. 9. All data are results for a gap of 50 μm. We used 20 individual fiber samples. For fiber connections with air-filled gaps, the minimum, maximum, and mean insertion losses were 0.8, 4.0, and 1.2 dB, respectively. The return losses varied between 15 and 26 dB. With refractive-index matching-material-filled gaps, the insertion losses were less than 0.4 dB, the mean value was 0.25 dB, and the return losses were more than 50 dB. With a mixture of refractive-index matching material and air-filled gaps, the insertion losses on one side attachment were from 1.1 to 42 dB (mean value of 12.0 dB) and the return losses varied between 13 and 47 dB. The insertion losses on the other side attachment were from 0.7 to 35 dB (mean value of 11.3 dB), and the return losses varied between 13 and 18 dB. These results indicate that the insertion and return losses with a mixture of refractive-index matching material and air-filled gaps vary greatly and are unstable.

An MT connector sample with a 50-μm gap containing a mixture of refractive-index matching material and air was made. Then a heat-cycle test in accordance with IEC 61300-2-22 (-40 to 70°C, 10 cycles, 6 h/cycle) on the sample was performed. The insertion and return losses of the sample are shown in Fig. 10. The optical performances changed and were

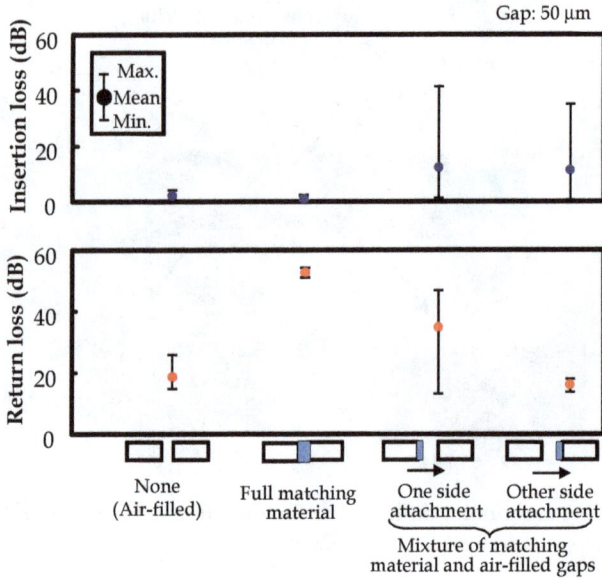

Fig. 9. Relation between gap states of fiber connections and optical performance.

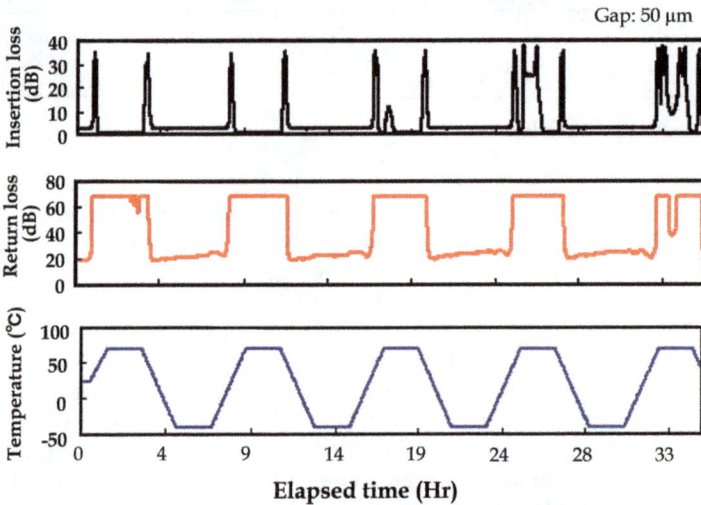

Fig. 10. Heat-cycle test results for fiber connection with mixture of refractive-index matching material and air-filled gaps.

unstable. The insertion loss was initially 2.7 dB and then varied when the temperature changed. The maximum insertion loss was more than 30 dB. The return losses also varied from 20 dB to more than 60 dB. This performance deterioration is thought to be caused by the mixture of refractive-index matching material and air-filled gaps between the fiber ends in the MT connector sample. Refractive-index matching material moved in the gap when the temperature changed, and the mixed-state change of the refractive-index matching material and the air between the fiber ends resulted in the change in optical performance. In a mixed state of refractive-index matching material and air between fiber ends, the boundary between the refractive-index matching material and air could be uneven. In this state, the transmitted light spread randomly in every direction at the boundary, which is similar to an incorrectly cleaved fiber end (uneven end perpendicular to the fiber axis). Therefore, the insertion loss increased to more than 30 dB.

Consequently, the optical performances of fiber connections with a mixture of refractive-index matching material and air-filled gaps are extremely unstable and vary widely. At worst, the insertion loss is more than 30 dB because the light spreads in every direction in the gap between fiber ends. Therefore, it is important to prevent the gap from becoming larger and avoid mixing air into the refractive-index matching material in the gap between fiber ends for these fiber connections.

5. Unexpected use of incorrectly cleaved fiber ends

In this section, the performance deterioration of fiber connections using an incorrectly cleaved fiber end is discussed. This case might occur when a field-assembly connector or a mechanical splice experiences an unexpected failure.

5.1 Incorrectly cleaved fiber end and fiber connection

An incorrectly cleaved fiber end is caused by problems with the fiber cleaver, such as a dropped cleaver or one that has struck something, because a fiber cleaver is a precisely fabricated and sensitive tool. If there are no problems with the fiber cleaver, the fiber will be cleaved correctly and have an ideal flat and smooth end perpendicular to the fiber axis. However, if there are problems, the fiber will be cleaved incorrectly and have an uneven end (NTT East, 2011). The mechanism of correctly and incorrectly cleaving fiber ends is shown in Fig. 11. The procedure of cleaving optical fibers is as follows. A scratch (origin of fracture) is first made on the fiber by the blade of the cleaver. The fiber is then bent at the origin of fracture at an appropriate radius and pushed from the opposite side of the origin of fracture until the fiber is eventually cleaved. Incorrectly cleaved fiber ends are reported to result from an incorrect bend radius during cleaving (Glode et al., 1973; Haibara et al., 1986). If the bend radius is too small, the cleaved fiber end will have a lip. If the bend radius is too large, the cleaved fiber end will have a hackle. A cleaved fiber end with a lip or hackle is not flat or smooth. Therefore, it is important for fiber cleavers to maintain the correct radius when bending fibers.

The normal connection state of correctly cleaved fiber ends and the abnormal state of an incorrectly cleaved fiber ends are shown in Fig. 12. The normal connection state (Fig. 12(a)) has a very small gap between the correctly cleaved fiber ends, and the gap is filled with refractive-index matching material (matching material). The abnormal connection state (Fig. 12(b)) has a large gap between the correctly cleaved (flat) fiber end and incorrectly cleaved (uneven with a lip) fiber end. The gap is filled with refractive-index matching material, but it is so large that it may affect the optical performance. This may be similar to the performance deterioration caused by a large gap between flat fiber ends (Kihara et al., 2009).

Too small radius   **Appropriate radius**   **Too large radius**

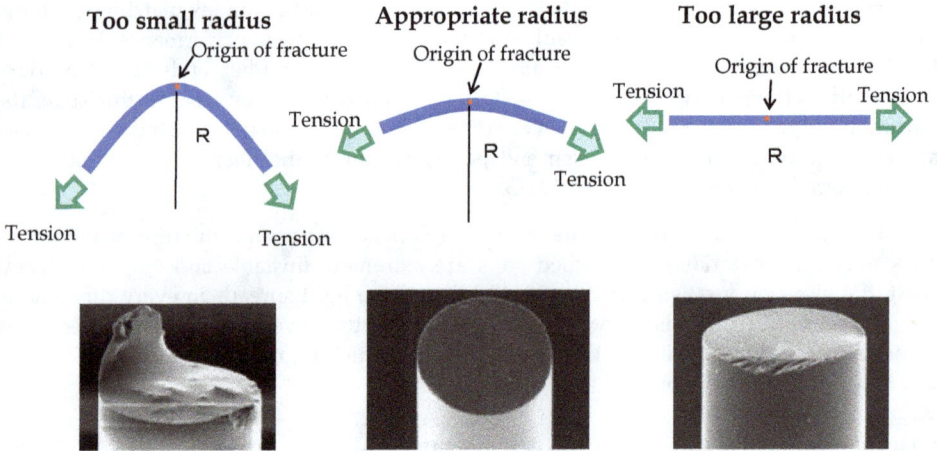

Fig. 11. Mechanism of correctly and incorrectly cleaving fiber ends.

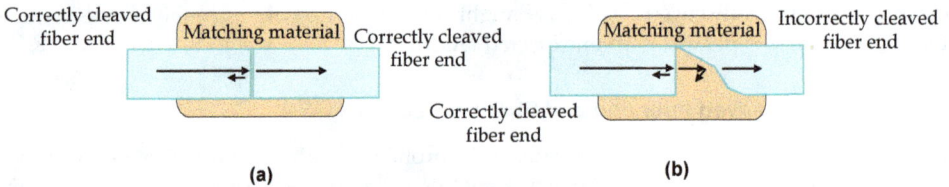

(a)   (b)

Fig. 12. Fiber connection states using correctly and incorrectly cleaved fiber ends.

5.2 Experiments and results

The optical performances of fiber connections using an incorrectly cleaved fiber end were experimentally investigated (Yajima et al., 2011). As investigation samples, 25 field assembly connectors using incorrectly cleaved fiber ends were fabricated. These incorrectly cleaved fiber ends were intentionally made by adjusting the fiber cleaver so that the bend radius would be too small. The cracks of these incorrectly cleaved fiber ends were from 30 to 200 µm in the axial direction. Five connectors using correctly cleaved flat fiber ends were also made for comparison. All 30 connectors were subjected to a heat-cycle test in accordance with IEC 61300-2-22 (-40 to 70°C, 10 cycles, 6 h/cycle) to simulate conditions in the field. We measured the insertion and return losses at a 1.55-µm wavelength.

The results for a sample using correctly cleaved fiber ends (sample 1) are shown in Fig. 13. The initial insertion loss was less than 0.5 dB, and the maximum insertion loss during the test was less than 0.8 dB. The return loss varied from 46 to 62 dB. This variance is thought to have been caused by a tiny refractive-index change in the matching material along with the change in temperature (Kihara et al., 1995). The insertion losses were very low and the return losses were very high. The performance was very stable.

The results for two samples fabricated using incorrectly cleaved fiber ends varied greatly (Figs. 14(a) and (b)). The insertion loss of sample 2 (Fig. 14(a)) varied from 0.4 to 0.9 dB. This variance is thought to have been caused by a tiny offset or tilt between the fiber ends. The

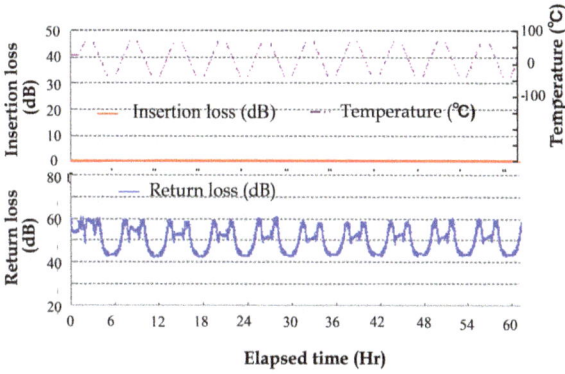

Fig. 13. Heat-cycle test result for connector using correctly cleaved fiber ends (sample 1).

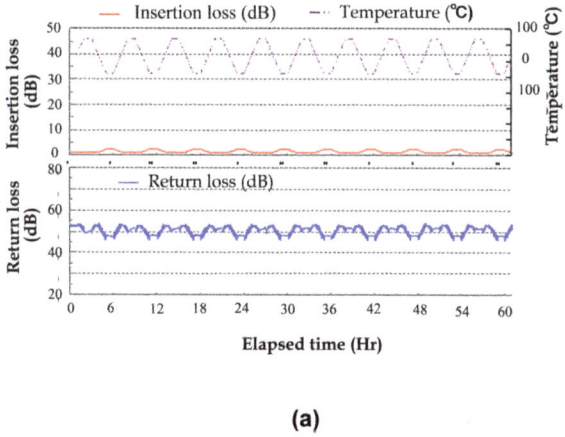

(a)

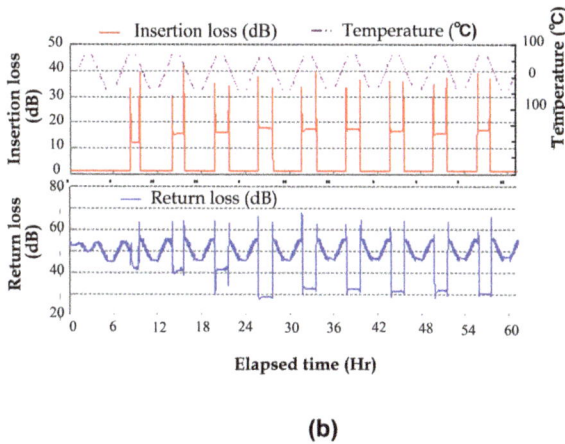

(b)

Fig. 14. Heat-cycle test results for connectors using incorrectly cleaved fiber ends. (a) Sample 2, (b) sample 3.

return loss was always over 40 dB. This performance is very stable and almost the same as that of correctly cleaved fiber ends (sample 1). The initial insertion loss of sample 3 (Fig. 14(b)) was 1.0 dB. The insertion loss of this sample changed from about 1.0 dB to more than 40 dB, according to the change in temperature. The return loss was temporally less than 30 dB. The change of the insertion and return losses is attributed to the partially air-filled gap. The gap was not completely filled with matching material; it was partially filled with air because of the incorrectly cleaved fiber end. The return loss in sample 3 is considered to be smaller than that in sample 2 due to this air gap. These results suggest that the insertion and return losses of fiber connections using incorrectly cleaved fiber ends might change to, at worst, more than 40 dB and less than 30 dB, respectively. This performance is very unstable. Such a substantial increase in insertion loss can affect the quality of an optical network service. Therefore, incorrectly cleaved fiber ends must not be used. A countermeasure is to inspect the cleaved fiber end before installing fiber connections such as a mechanical splice or a field assembly connector (Kihara et al., 2011; Okada et al., 2011).

Overall, the optical performances of fiber connections using an incorrectly cleaved fiber end vary widely. At worst, the insertion and return losses might deteriorate to more than 40 dB and less than 30 dB, respectively.

6. Conclusion

The performances of SMF connections were experimentally investigated for various cases: connections with (1) air-filled gaps, (2) a mixture of refractive-index matching material and air-filled gaps, and (3) unexpected use of an incorrectly cleaved fiber end. The cases were assumed to occur accidentally as the result of unexpected failure during and after installation of fiber connections in the field. In this chapter, the optical performance deterioration of these fiber connections was detailed.

After a brief introduction (section 1), the conventional loss characteristics of SMF connections based on the D. Marcuse and W. C. Young analyses were explained in section 2. From section 3 onward, the insertion and return losses of the fiber connections were reported.

In section 3, the performance of fiber connections with air-filled gaps was explained. This case might occur when a fiber connection using PC experiences an unexpected failure. Generally, the optical performance of a connector that maintains perfect PC will remain environmentally stable. However, when there is an air-filled gap between fiber ends with PC-type connections, the optical performance worsens noticeably. In particular, the return loss varies greatly and might deteriorate to 8.7 dB for one fiber connection at worst.

In section 4, a loss analysis for fiber connections with a mixture of refractive-index matching material and air-filled gaps was reported. This case might occur when an optical connector or a mechanical splice using refractive-index matching material experiences an unexpected failure. The connections normally have a small gap between the fiber ends, and the gap is filled with refractive-index matching material to reduce Fresnel reflection; the optical performance will remain environmentally stable. However, when there is a large gap between fiber ends and the connections use refractive-index matching material, the gap might change to a mixture of refractive-index matching material and air-filled gaps. The optical performance worsens noticeably in this case. In particular, the insertion loss varies greatly and might deteriorate by more than 30 dB for one fiber connection.

In section 5, the performance deterioration of fiber connections using an incorrectly cleaved fiber end was shown. This case might occur when a field-assembly connector or a mechanical splice experiences an unexpected failure. For these fiber connections, it is necessary to strip the fiber coating, clean the stripped fiber with alcohol, and cut the fiber with a cleaver while in the field. If the optical fiber of the connection is not cut correctly, the insertion loss of the fiber connection might deteriorate by more than 30 dB in the same way as that for a fiber connection with a mixture of refractive-index matching material and air-filled gaps.

This chapter discussed the characteristics based on the results obtained for MT, SC, and field assembly connectors. However, the results can be applied to other connectors or fiber connections. These results are considered to be useful for practical construction and operation of optical fiber network systems.

7. Acknowledgments

The author thanks Drs. S. Tomita and H. Izumita at NTT Access Network Service Systems Laboratories (ANSL) and M. Toyonaga at Technical Assistance and Support Center (TASC), NTT East Corporation for their encouragement and helpful suggestions. The author is deeply grateful to M. Uchino at TASC for discussing the air-filled gap experiments, R. Nagano at ANSL for discussing the mixture of refractive-index matching material and air-filled gaps experiments, and H. Watanabe, Y. Yajima, and M. Tanaka at TASC for helpful discussions regarding the incorrectly cleaved fiber end experiments.

8. References

Albeanu, N., Aseere, L., Berdinskikh, T., Nguyen, J., Pradieu, Y., Silmser, D., Tkalec, H., and Tse, E. (2003). Optical connector contamination and its influence on optical signal performance, J SMT, vol. 16, issue 3, (2003) 40–49.

Born, M. and Wolf, E. (1964). Principles of Optics, New York: Macmillan, (1964).

Glode, D., Smith, P. W., Bisbee, D. L., Chinock, E. L. (1973). Optical fibre end preparation for low-loss splices, Bell Sys. Tech. J, vol. 52, (1973) 1579–1587.

Haibara, T., Matsumoto, M., and Miyauchi, M. (1986). Design and development of an automatic cutting tool for optical fibres, IEEE/OSA JLT, 1986, vol. LT-4, No. 9, (1986) 1434–1439.

Hogari, K., Nagase, R., and Takamizawa, K. (2010). Optical connector technologies for optical access networks, IEICE Trans. Electron., Vol. E93-C, No. 7, (2010) 1172–1179.

Kashima, N. (1995). Passive optical components for optical fiber transmission, Norwood, MA: Artech House, (1995).

Keck, D. B., Morrow, A. J., Nolan, D. A., and Thompson, D. A. (1989). Passive components in the subscriber loop, J. Lightwave Technol., vol. 7, (1989) 1623–1633.

Kihara, M., Nagasawa, S., and Tanifuji, T. (1995). Temperature dependence of return loss for optical fiber connectors with refractive index-matching material, IEEE Photon. Tech. Lett., vol. 7, no. 7, (1995) 795–797.

Kihara, M., Nagasawa, S., and Tanifuji, T. (1996). Return loss characteristics of optical fiber connectors, J. Lightwave Technol., vol. 14, Sep. (1996) 1986–1991.

Kihara, M., Tomita, S., and Haibara, T. (2006). Influence of wavelength and temperature changes on optical performance of fiber connections with small gap, IEEE Photon. Tech. Lett., vol. 18, no. 20, (2006) 2120–2122.

Kihara, M., Nagano, R., Uchino, M., Yuki, Y., Sonoda, H., Onose, H., Izumita, H., and Kuwaki, N. (2009). Analysis on performance deterioration of optical fiber connections with mixture of refractive index matching material and air-filled gaps, in Proceedings of the OFC/NFOEC, (2009) JWA4.

Kihara, M., Uchino, M., Omachi, M., and Izumita, H. (2010). Analyzing return loss deterioration of optical fiber connections with various air-filled gaps over a wide wavelength range, in Proceedings of the OFC/NFOEC, (2010) NWE4.

Kihara, M., Watanabe, H., Yajima, Y., and Toyonaga, M. (2011). Inspection technique for cleaved optical fiber ends based on Fabry-Perot resonator, in Proceedings of the OFS-21, (2011), 7753-215.

Marcuse, D. (1976). Loss analysis of optical fiber splice, Bell Sys. Tech. J, vol. 56, (1976) 703–718.

Nakajima, T., Terakawa, K., Toyonaga, M., and Kama, M. (2006). Development of optical connector to achieve large-scale optical network construction, in Proceeding of the 55th IWCS/Focus, (2006) 439–443.

NTT East (2011). Fault cases and countermeasures for field assembly connectors in optical access facilities, NTT Technical Review, Vol. 9, No. 7, (2011).

Okada, M., Kihara, M., Hosoda, M., and Toyonaga, M. (2011). Simple inspection tool for cleaved optical fiber ends and optical fiber connector end surface, in Proceedings of the IWCS, (2011), to be presented.

Satake, T., Nagasawa, S., and Arioka, R. (1986). A new type of a demountable plastic molded single mode multifiber connector, IEEE J. Lightwave Technol., vol. LT-4, (1986) 1232–1236.

Shinohara, H. (2005). Broadband access in Japan: rapidly growing FTTH market, IEEE Com. Mag., vol. 43, (2005) 72–78.

Sugita, E., Nagase, R., Kanayama, K., and Shintaku, T. (1989). SC-type single-mode optical fiber connectors, IEEE/OSA J. Lightwave Technol., vol. 7, (1989) 1689-1696.

Yajima, Y., Watanabe, H., Kihara, M., and Toyonaga, M. (2011). Optical performance of field assembly connectors using incorrectly cleaved fiber ends, in Proceedings of the OECC, (2011) 7P3_053.

Yariv, A. (1985). Introduction to optical electronics, New York: Holt, Rinehart, and Winstone, (1985).

Yoshino, S., Takaya, M., Sonoda, H., Uchino, M., Yuki, Y., Nagano, R., Izumita, H., and Kuwaki, N. (2008). Analysis of mechanical splicing faults in FTTH trial, in Proceeding of the OFC/NFOEC, (2008) NThC4.

Young, W. C. (1991). Optical fiber connectors and splices, Short Course Notes in OFC'91, San Diego (1991).

11

In-Service Line Monitoring for Passive Optical Networks

Nazuki Honda

NTT Access Service System Laboratories, NTT Corp.

Japan

1. Introduction

Broadband optical access services are already being deployed to realize fiber to the home (FTTH) networks that support high speed internet and video on demand services. The number of FTTH/B (building) subscribers is increasing rapidly throughout the world. It was about 61 million by the end of 2010 and is expected to reach 227 million by 2015 (IDATE, 2011). Broadband access network provision currently requires thousands of optical fibers to be accommodated in a central office (CO) for an optical access network. As we must respond to the huge demand for FTTH, we can no longer ignore the operation, administration, and maintenance (OAM) costs in addition to the construction cost (ECOC work shop 8, 2007). Optical fiber maintenance for outside plant is discussed in ITU-T SG6 and seven recommendations, L.25, L.40, L.41, L.53, L.66, L.68, and L.85 have been published. Recommendations L.25 and L.53 describe the fundamental need for optical fiber maintenance functions for preventive and post-fault maintenance. Fiber monitoring and fiber fault location are required as fundamental functions for optical fiber cable network maintenance. These functions make it possible to reduce both troubleshooting time and human resources when repairing damaged cable.

When we monitor an optical fiber network, the network configuration is a critical factor as regards the monitoring method. Passive optical networks (PONs) provide the main FTTH service based on GE-PON (Gigabit Ethernet PON) or G-PON (Gigabit-PON), where several customers' optical network terminals (ONTs) share an optical line terminal (OLT) and an optical fiber (IEEE Std 802.3ah; ITU-T G984.1). The versatile optical fiber monitoring system uses optical time domain reflectometry (OTDR) to locate a fault by measuring an optical pulse response (e.g. Nakao, 2001). However, these PON systems use a simple power splitter near the customer's home to provide an economical IP service, therefore a conventional testing system, where an OTDR is installed in a central office, is unstable because Rayleigh backscattered signals from all the branched optical fibers accumulate in the OTDR trace. Several fiber fault location techniques have been proposed for use with the branched optical fibers of PONs. One example is the proposal by Sankawa et al. (Sankawa, et al., 1990). However, this approach is difficult to employ practically because the change in the loss value becomes small as the number of branches increases. The multi-wavelength OTDR technique uses an expensive arrayed waveguide grating to assign an individual testing wavelength to each branched fiber (Tanaka, et al., 1996). This method requires us to design the characteristics of the waveguide for a specific communication light wavelength. A monitoring method that embeds the OTDR functions in each ONT (Schmuck, et al., 2006) cannot transmit a measured OTDR trace when the optical network has a fault.

This chapter is devoted to an in-service line monitoring system for branched PON fibers. Section 2 presents the basic concepts of an in-service line monitoring system for PON fibers. By employing individually assigned Brillouin frequency shifts (BFS) for the branching region and using a 1650 nm Brillouin OTDR (B-OTDR), we can detect the profile of each branching fiber. Section 3 discusses the design for the BFS separation of branching fibers taking account of the outside plant environment. BFS can be controlled by controlling the amount of dopant in the optical fiber core. The system dynamic range of in-service line PON fiber monitoring is the focus of Section 4 where we discuss 1650 nm B-OTDR and the transmission system power budget in GE-PON. Section 5 describes the measurement of a fiber with an 8-branched optical splitter using a 1650 nm B-OTDR and BFS assigned fiber in each branching region. The in-service line testing of a GE-PON transmission is demonstrated in Section 6.

2. In-service line monitoring system in PONs

2.1 In-service line monitoring technique using U-band test light for access network

An optical fiber line testing system is essential for reducing maintenance costs and improving service reliability in optical fiber networks. To perform maintenance system testing on in-service fibers without causing any degradation in the transmission quality, we use a test light whose wavelength is different from that of the communication light, and install a filter in front of the ONT that cuts off only the test light. Figure 1 shows the wavelength allocation. The ITU-T Supplement on the distributed service network defines the spectrum band for a wavelength division multiplexing (WDM) transmission system. A long wavelength band (L-band) that extends to 1625 nm is used for DWDM/CWDM transmission, and Recommendation G. 983 defines the 1480-1580 nm band (in the S-band) for downstream G-PON and additional service signals to be distributed simultaneously. An ultra long wavelength band (U-band) of 1625-1675 nm can be used as the maintenance band for various services. When monitoring branched PON fiber, the technique must allow in-service line monitoring because other branched fibers accommodated in the same splitter may still be active.

Figure 2 shows the configuration of an optical fiber line testing system for PONs with an optical splitter installed in a central office (Sp-C) and an optical splitter installed in outside plant (Sp-O) near the customer's home (Enomoto, et al., 2005). This testing system comprises an optical testing module (OTM) that contains a B-OTDR, optical fiber selectors (FS) that select the fibers to be tested, an optical coupler to introduce test lights into an optical fiber line, and termination cables with an optical filter that allows a communication light to pass but not a test light. Termination filters are installed in front of the ONTs to cut off the test light. The pass band of the termination filter is designed to be 1260-1625 nm in accordance with the ITU-T L.41 recommendation. The control terminal orders the OTM to perform various optical fiber tests through a data communication network (DCN). This system automatically and remotely measures the optical fiber characteristics, and detects and locates faults in optical fiber networks.

G.Sup39 and L.41

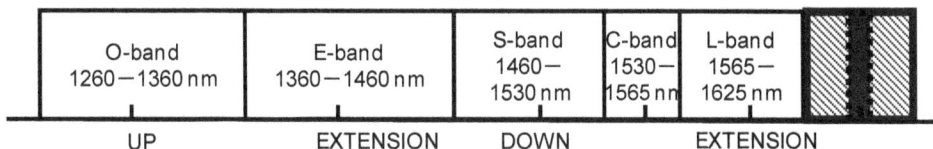

G983.3 (PON)

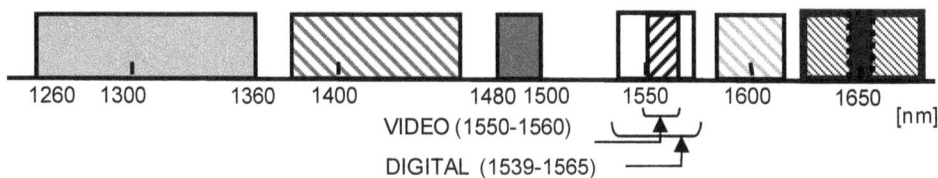

G694.2 (CWDM)

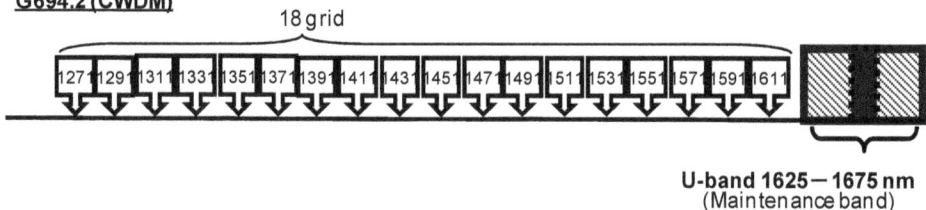

U-band 1625 – 1675 nm
(Maintenance band)

Fig. 1. Wavelength allocation

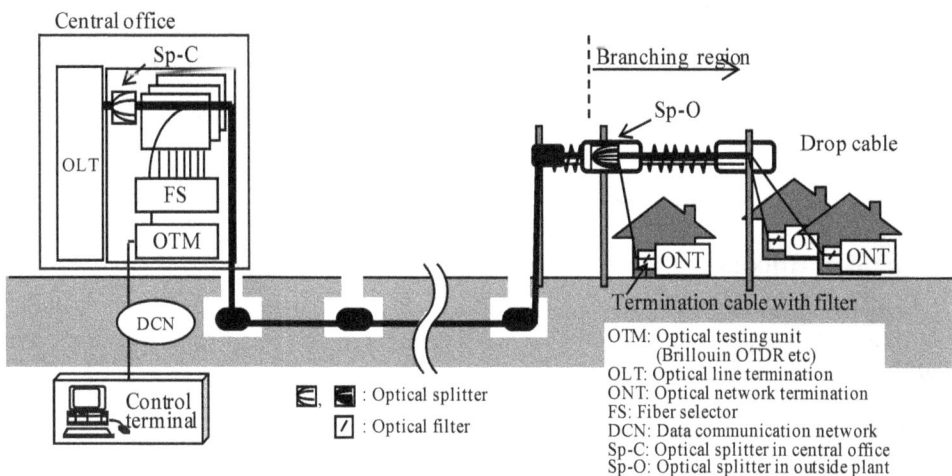

Fig. 2. Configuration of optical fiber line testing system

2.2 PON fiber monitoring using fibers with individually assigned Brillouin frequencies

To enable us to monitor individual branched fibers, we install fibers with individually assigned BFSs in the branched region(Shimizu, et al., 1995; Honda, et al., 2006a; Honda, et al. 2006b). Figure 3 shows the basic configuration of a PON monitoring technique that uses 1650 nm B-OTDR for in-service line monitoring, and the Brillouin scattering light spectra in the branched region. When we input a test light, we can distinguish the Brillouin spectrum peak frequency $v1$, $v2$, ...,vn from each branched fiber. If the peak power of a Brillouin spectrum changes from its initial level, we can determine that the optical fiber with the BFS is faulty. Thus, we can measure the fault location in a branched region from the BFS peak power distribution of the test fibers.

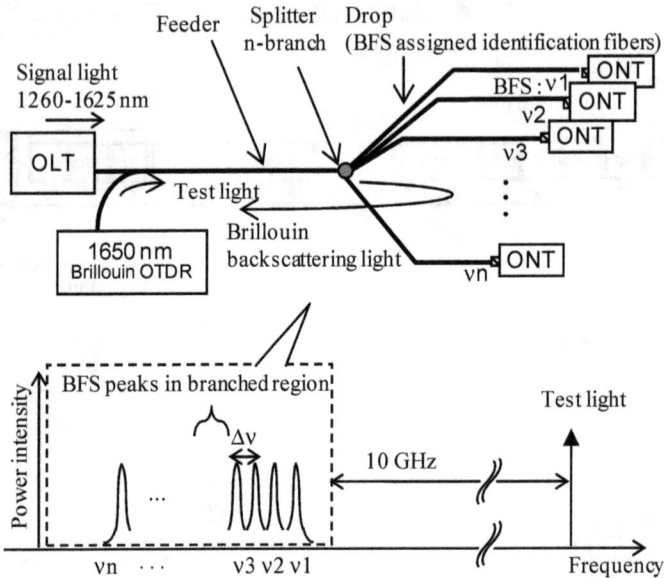

Fig. 3. Fiber monitoring technique for PONs with branched fibers assigned with individual BFSs and BFS spectra in branched region

3. Design of Brillouin frequency shift assigned fiber

To clarify the operating area of the monitoring method proposed in section 2, we discuss the applicable PON branching number derived from the design of the BFS assigned identification fiber in the branched region.

3.1 Design of BFS separation of branched PON fibers for outside environment

The design of BFS assigned fiber is important if we are to realize a method for monitoring a branched PON. The BFS separation Δv of the fibers shown in Fig. 3 must be designed taking account of the temperature-induced fluctuation in the BFS, and strain dependent changes that occur in an outside environment. Δv can be described as

$$\Delta \nu \geq w + \Delta S \cdot C_S + \Delta T \cdot C_T \tag{1}$$

Here w, ΔS, C_S, ΔT, and C_T are the full width at half-maximum (FWHM) of the Brillouin gain spectrum, the fluctuation strain and temperature, and their BFS coefficients, respectively. The Brillouin frequency shift depends on temperature and tension, and the values are 1.08 MHz/°C and 500 MHz/1% strain at 1650 nm, respectively (Izumita, et al., 1997). Since the residual strain change in an optical fiber cable installed as a feeder section is less than 0.01% (Kawataka, et al., 2003), for simplicity we assume that the BFS change has negligible strain dependence. An outside plant is designed to operate in a -40 to 75 °C temperature range (ITU-T L37).

Figure 4 shows the BFS spectra of branching PON fibers (BFSs = $\nu 1, \nu 2, \nu 3, \ldots$ and νn) at temperatures of -40 and 75 °C, and the BFS spectrum of fiber 5 at a typical room temperature of 25 °C. Since the drop cables are accommodated in a customer's residence, we must consider a relative maximum temperature change of 65 °C. Using a typical w value of 50 MHz (Azuma, et al., 1988) and a C_T of 1.08 MHz/°C, we find that the BFS separation must be more than 120 MHz. To monitor the n-branched PON fiber, we need 120 × (n-1) MHz BFS separation.

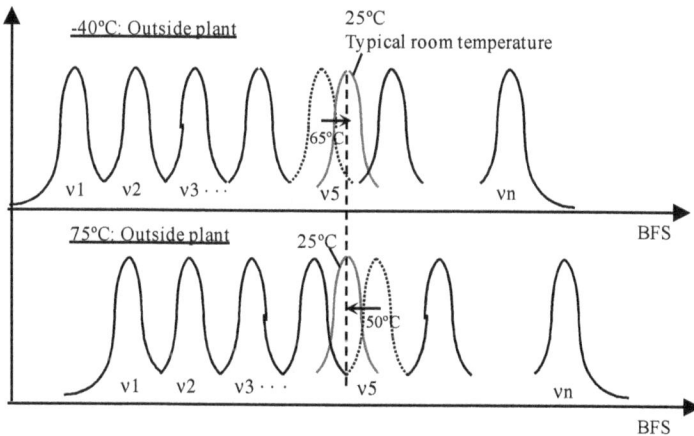

Fig. 4. Fluctuation in BFS spectra caused by temperature in outside plant environment

We can fabricate the large BFS fiber that we designed by adjusting the dopant concentration and taking account of the connection loss with single-mode fiber (SMF) by employing a germanium oxide (GeO_2) and fluorine (F) co-doping technique. We can assume that the saturation value of the aging loss increase of the fiber when F and GeO_2 are co-doped is rarely different from the value for SMF (Iino, et al., 1989). Figure 5 shows that the design of the BFS for optical fibers depends on the GeO_2 and F dopant concentrations in the fiber core. GeO_2 and F were used to fabricate the identification fibers because as dopants their effects on the refractive index are opposite, but their effects on the BFS are the same, and they are also the most widely used dopants (Iida, et al., 2007). Assuming the maximum BFS separation that we can design is 1.5 GHz and $\Delta \nu$ is 120 MHz, we can realize up to 13 kinds of BFS assigned fiber. This PON monitoring method employing BFS assigned fibers can be applied for a fiber network using 8-branched splitter in an outside plant.

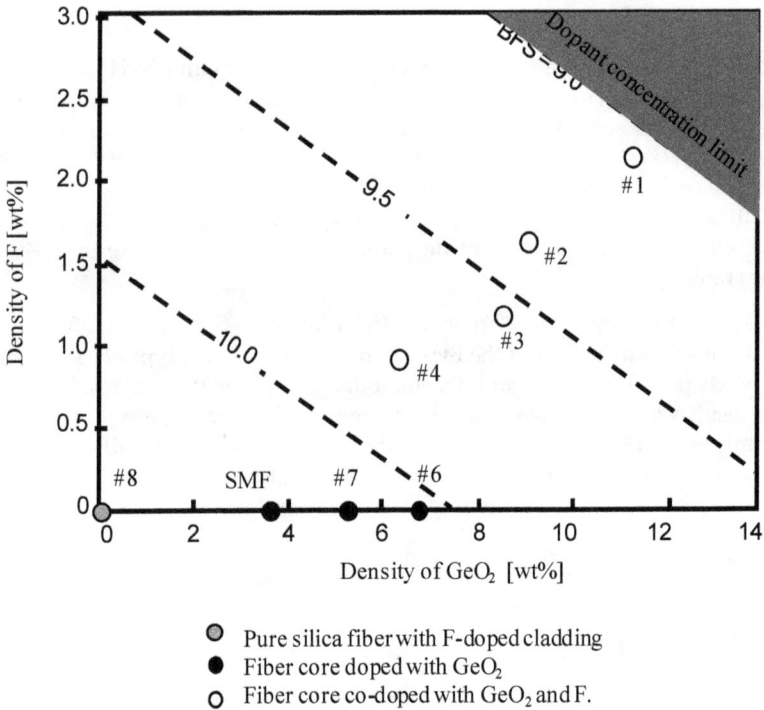

Fig. 5. Design of Brillouin frequency shift for optical fibers depends on GeO₂ and F concentrations in optical fiber core.

4. Performance of testing system employing 1650-nm B-OTDR

In this section, to clarify the operating area of the PON fiber monitoring system, we discuss the system performance derived from the 1650 nm B-OTDR dynamic range and the loss of the optical fiber network.

When an optical test module employs an optical amplifier, the amplifier can be used to compensate for the loss of the optical coupler that introduces the test light. Therefore, the single-way dynamic range (SWDR) of a B-OTDR does not include a term for optical coupler loss. The SWDR of a conventional B-OTDR for measuring strain and temperature is described in ref. (Horiguchi, et al., 1995) and defined by Eq. (2).

$$SWDR = \frac{1}{2}(Pt + BS + Ts - Lc - R_{min} + \frac{SNIR_{ave}}{2}) \qquad (2)$$

with

$$BS = 10log(0.5\ a_b\ S\ W\ vo) \qquad (3)$$

$$Ts = 10log\frac{2B}{\pi\Delta v} \qquad (4)$$

$$R_{min} = Bh\nu / \eta \tag{5}$$

where Pt is the incident peak power of the test light, Lc is the 1:1 coupler loss, and R_{min} is the sensitivity of a photo diode. BS is the Brillouin scattering coefficient. a_b is the Brillouin backscattering coefficient given by $7.1 \times 10^{-30}/\lambda^4$ m-1. Ts is the Brillouin scattering selection ratio, which indicates that the decrease in the received power is caused by a portion of the Brillouin scattering. To detect the frequency of a Brillouin scattering signal with a finite spectral linewidth, the signal is selected by an electrical band-pass filter with a bandwidth 2B, and detected. $SNIR_{ave}$ is the signal-to-noise (S/N) improvement achieved by averaging. The averaging number N improves the S/N by $\frac{1}{2} \cdot 10 \log \sqrt{N}$ dB.

W is the pulse width, v_o is the velocity of light in an optical fiber, B is the receiver bandwidth, Δv is the Brillouin linewidth, h is Planck's constant, v is the lightwave frequency, and η is the photodiode quantum efficiency. R_{min} is discussed as regards receiver sensitivity in coherent detection because it is suitable for measuring a small Brillouin scattering signal. When SNR=1, the minimum receiver sensitivity is as reported in (Koyamada, et al., 1992).

In (Horiguchi, et al., 1995), Horiguchi et al. consider the degradation caused by fluctuations in the strain and temperature in the single-way dynamic range (SWDR). However, when we use a B-OTDR to monitor the branched PON fibers, the BFSs of the fibers installed in the branched PON region are designed not to overlap each other. Therefore, the degradation caused by strain and temperature fluctuations can be negligible. The resulting large dynamic range is an advantage of this approach.

The required dynamic range for monitoring the branched PON fiber is derived from the maximum optical fiber loss between the OLT and ONT in a GE-PON, which is 24 dB (IEEE Std 802.3ah). With GE-PON, which provides the main FTTH service in Japan, an 8-branched optical splitter is deployed in an aerial closure near the customer's home, and an optical coupler, which introduces the test light, is installed after a 4-branched splitter. The insertion loss in a central office caused by a 4-branched optical splitter, a coupler and connectors is about 7 dB, and thus the required SWDR for a B-OTDR is more than 17 dB.

5. Experiments and discussion

To confirm the feasibility of the in-service line PON monitoring method, we measured the BFS spectra and B-OTDR traces and carried out an in-service line-monitoring test.

5.1 Branched PON fiber monitoring

The B-OTDR technique illustrated in Fig. 6 has a 1651.3 nm coherent light generated by a 10 kHz linewidth narrowed DFB laser diode that is divided into probe and reference lights. The probe light is modulated with a lithium niobate (LN) modulator. Then its signal power is amplified in a two-stage 1650 nm amplifier consisting of a thulium doped fiber amplifier (TDFA) and a Raman amplifier. The modulated probe light was amplified in Tm-Tb doped fiber (TDF) in the TDFA with a 1220 nm pump light of 49.7 mW. The core and cladding of the Tm-Tb doped fiber contained 2000 parts per million (ppm) of Tm and 4000 ppm of Tb, respectively, and the fiber length was 13 m. The refractive index difference between the core

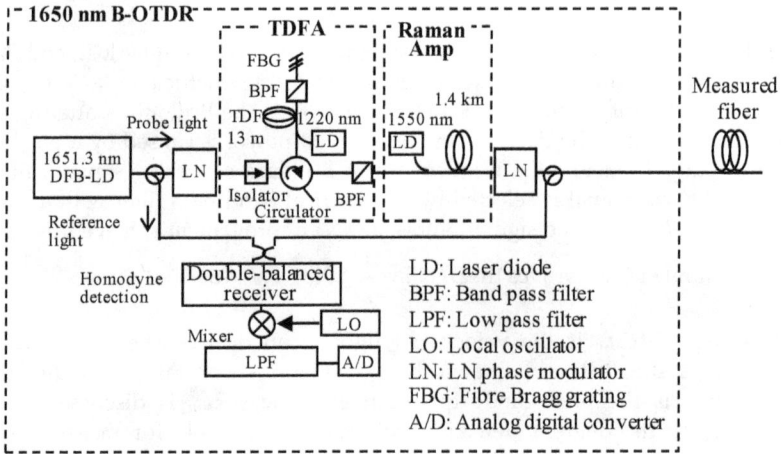

Fig. 6. Configuration for 1650 nm band B-OTDR employing TDFA and Raman amplifier

and the cladding was 3.7% and the core diameter was 1.8 μm. This configuration achieves a very high gain, and makes measurement with a 1650 nm B-OTDR possible. Figure 7 shows the measured B-OTDR trace at 10.17 GHz in an optical fiber consisting of 10 km of SMF, Sp-O(8), and 5 km of SMF with a test light pulse with a peak power of 26 dBm, a 100 ns pulse width, and a 2^{22} averaging time. We obtained an SWDR of 17.22 dB. For an input power of 26 dBm, the required averaging number using Eq. (2) ~ (5) is calculated to be 2^{12}. The difference between the experimental and calculated averaging numbers was because the two-stage high gain amplifier has a large noise figure, which meant that its amplified spontaneous emission power degraded the sensitivity of the receiver. A larger averaging number is needed to improve the SWDR.

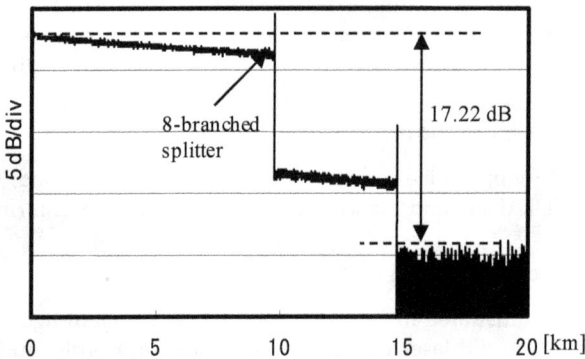

Fig. 7. Brillouin backscattered power measurement with 100 ns pulse width for incident pulse peak power of 26 dBm.

Figure 8 shows the experimental setup for measuring the Brillouin backscattering signal with a 1650 nm test light for an optical fiber line with an 8-branched optical fiber. The optical coupler loss for the test light was 1.06 dB and the incident peak power to the test fiber was controlled at 26 dBm. We used eight kinds of fibers for the lower region of the optical splitter as BFS assigned fibers and their dopant concentrations are plotted in Fig. 5. The cores of fibers 1-4 were co-doped with GeO_2 + F, fibers 5-7 were doped with GeO_2 and fiber 8 was pure silica fiber that had F doped in its cladding. We realized eight kinds of BFS fiber with a the maximum separation of 1.5 GHz. Figure 9 shows the BFS spectra measured 550 m from the B-OTDR. The eight BFS peaks at 9.055, 9.380, 9.535, 9.730, 9.825, 9.990, 10.095 and 10.435 GHz clearly indicate the existence of fibers 1-8. Figure 10 a) and b-i) show the B-OTDR traces obtained before and after the optical splitter, respectively. The traces in b)-i) clearly show the length of the BFS assigned fibers after the splitter. To simulate the fault identification process, the fiber of branch #6 fiber was intentionally bent with a loss of 4.1 dB

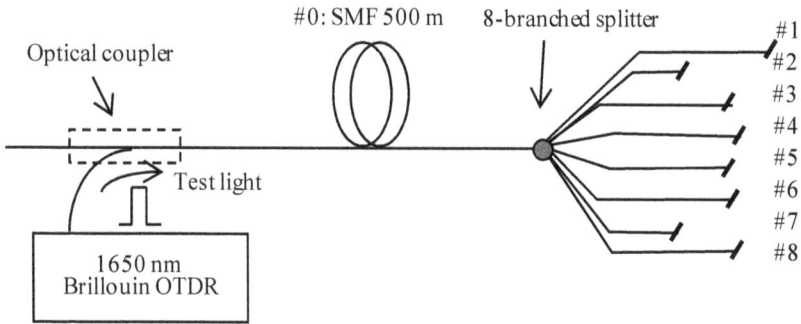

Fig. 8. Experimental setup with 1650 nm band B-OTDR for measuring PON branching region.

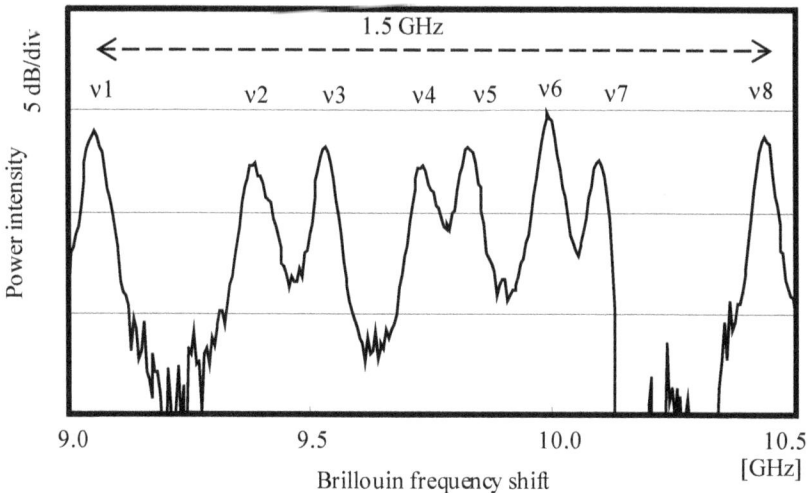

Fig. 9. Measured BFS spectra for 8-branched region 550 m from B-OTDR.

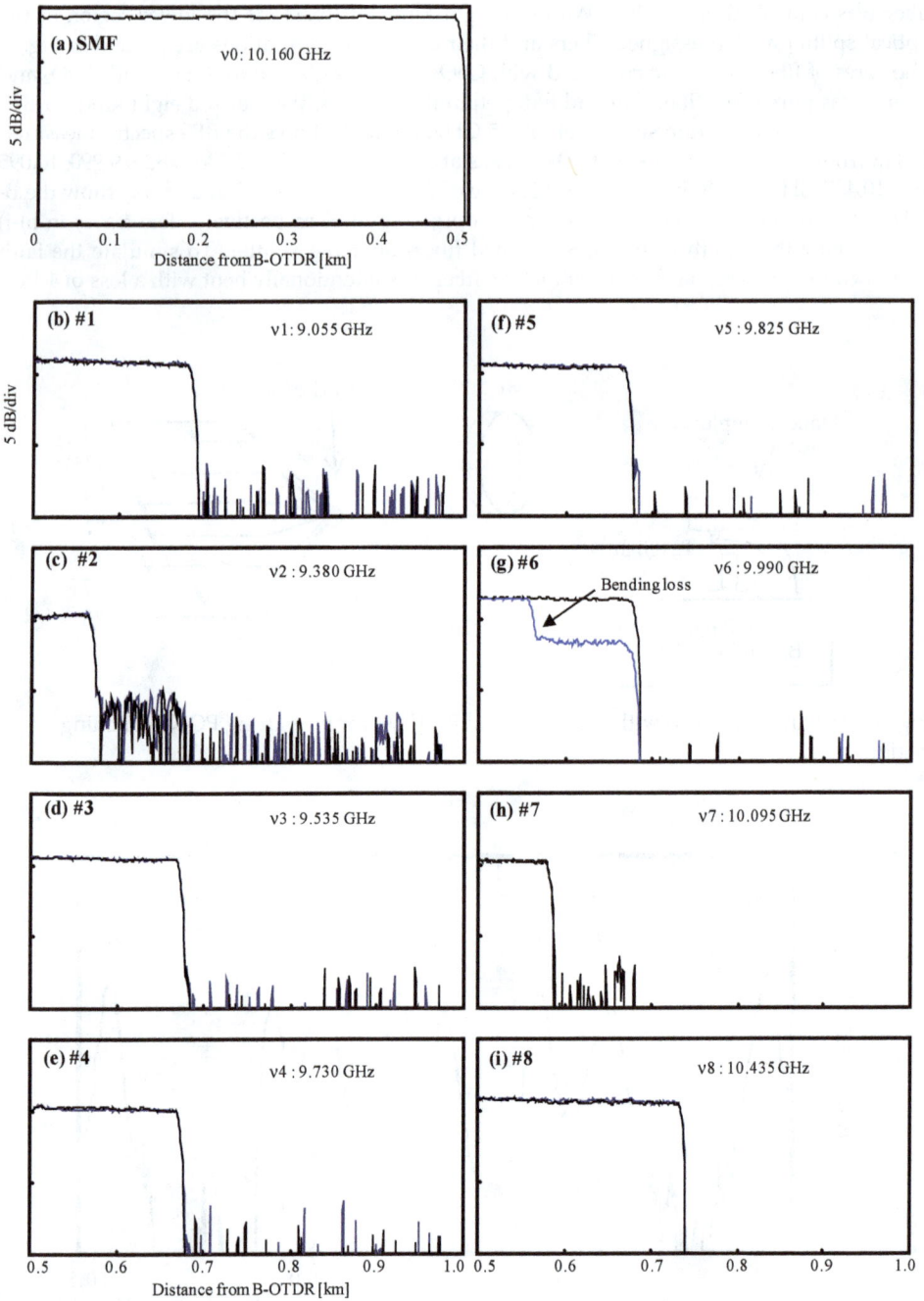

Fig. 10. B-OTDR trace a) between optical coupler and 8-branched splitter, b)-i) 8-branched PON fiber region

60 m from the optical splitter. Traces with the bending loss overlaid are shown in Fig. 9. There was hardly any difference between the BFS peak power traces of b)-f) and h)-i). However, trace g) shows a 4 dB bending loss. This loss agreed with that measured with an optical power meter. This reveals the great advantage of this PON fiber measurement method, namely that it can locate a fault and measure the loss value in branched PON fiber.

Figure 11 shows a measured trace of a branching fiber with two connections that had 0.32 and 0.18 dB losses caused by the MT connector. When we undertook measurements using a conventional OTDR, the calculated bending and connection losses were 0.078 and 0.044 dB, respectively, in the OTDR trace because of the accumulation of Rayleigh backscattered signals from all the branched optical fibers (Sankawa, et al., 1990). This reveals the great advantage of this measurement method using individually assigned BFS fibers in the PON branching region.

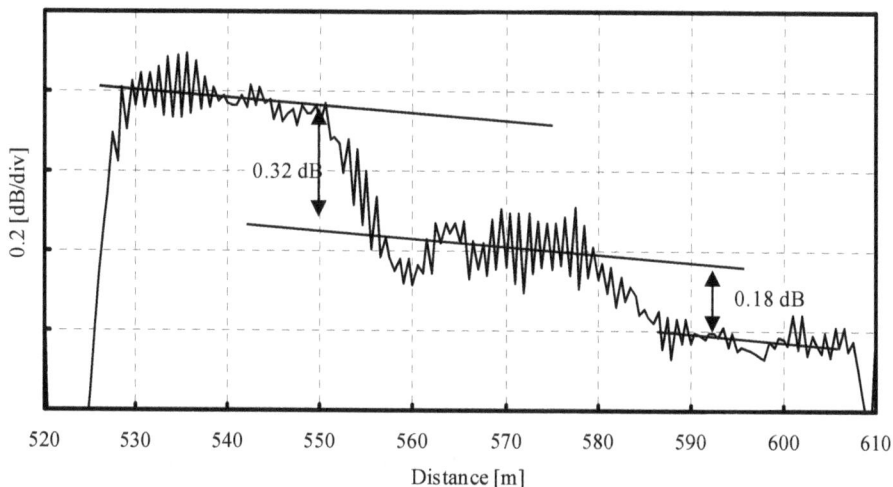

Fig. 11. Detail of B-OTDR trace in PON branching region with two connections

5.2 In-service line monitoring

The technique used for monitoring the PON branching fiber must allow in-service line monitoring because other branched fibers accommodated in the same optical splitter may still be active. To avoid any deterioration in the transmission, the difference between the optical powers of the communication light and the test light adjacent to the ONT should be sufficiently smaller than the signal to noise ratio (ITU-T Recommendation L.66). A termination filter that cuts off the test light is installed in front of the ONTs as mentioned in section II. The attenuation value of the termination filter LF for the test light is determined by considering the communication system margin (RS/X).

$$LF > Pt - Ps + RS/X, \tag{6}$$

These parameters were logarithmically transformed. Ps and Pt, respectively, are the optical powers of the OLT and the test light just after the optical coupler. As regards the communication and test lights, Hogari et al. reported that the optical losses are slightly different in optical access networks (Hogari, et al., 2003), namely, Pt - Ps is equal to the power difference between the communication and test lights at the optical coupler that coupled them. Each system has particular specifications and these can include the minimum output power of the transmitters, the minimum sensitivity of the receivers and the S/X ratio. In GE-PON, the minimum average power of the communication light from an OLT is defined as +2 dBm (peak power +5 dBm) (IEEE Std 802.3ah), and the insertion loss in a central office is about 7 dB as mentioned in Section III B. The maximum peak power of the test light was 26 dBm, and so LF> 28 + RS/X. When we use an ONT with an RS/X of 10 dB, LF must be more than 38 dB. The typical insertion loss of a termination cord with a fiber Bragg grating (FBG) filter is shown in Fig. 12.

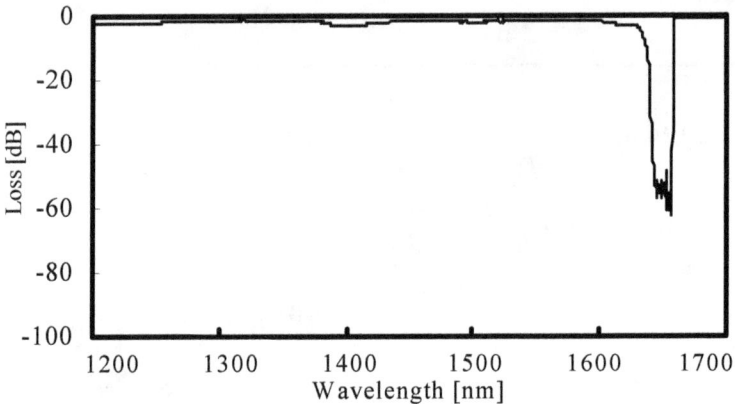

Fig. 12. Typical loss of FBG filter in front of ONT

Figure 13 shows the experimental setup for in-service line monitoring of GE-PON. A 1492.25 nm signal light with an average power of 4.9 dBm from an OLT was controlled at 2.0 dBm (minimum average launch power for an OLT) by an attenuator, and launched into the experimental fiber line. The fiber line consisted of 4-branched fiber, an optical coupler, a 10 km SMF, and a 200 m BFS assigned fiber (#6). We also installed an FBG filter in front of the receiver port of the ONT as shown in Fig. 13. The 1492.25 nm signal power in front of the ONT was changed by an attenuator to control the insertion loss of the fiber line. The 1650 nm test light was introduced by the optical coupler and coupled with the signal light. Characteristic features of a GE-PON receiver are that the maximum sensitivity and bit error rate (BER) are -24 dBm and 10^{-12} or better (IEEE Std 802.3ah). When we evaluate the transmission quality in a GE-PON system, Ethernet frame analysis using a LAN analyzer without retransmission processing is a simple and convenient method for measuring the frame error rate (FER). We used a LAN analyzer to create 64 byte packets of 100 Mb/s with the smallest possible packet interval of 0.96 μs, and simultaneously transmitted the packets at a maximum throughput of 148809 packets/sec. The FER characteristics with and without the B-OTDR test light are shown in Fig. 14. These results indicate that in-service fault location did not affect the transmission characteristics at either signal wavelength. The FER almost corresponds to the BER when certain conditions are satisfied: 1) each error packet has no error or only a 1-bit error, and 2) the packet layer protocol has only an error detection

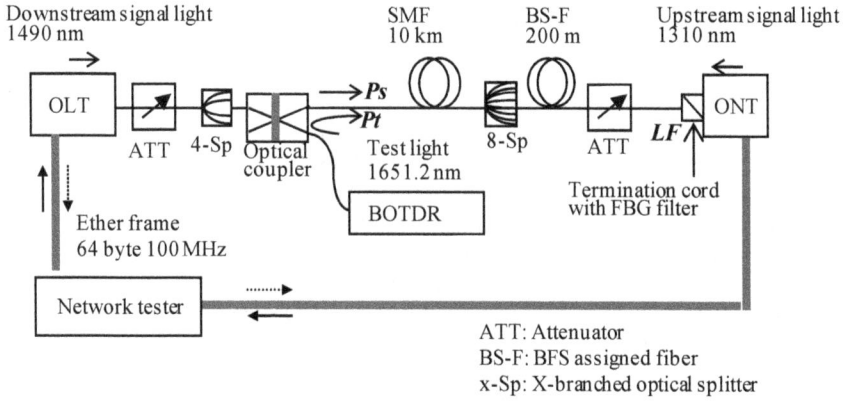

Fig. 13. Experimental setup for in-service line monitoring

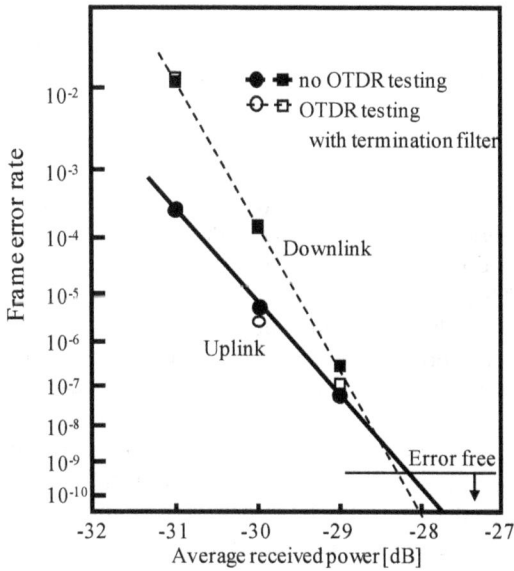

Fig. 14. FER measurement with and without in-service line monitoring using 1650 nm OTDR test light.

function and has no error recovery function such as forward error correction or retransmission (Ishikura, et al., 1999). The relationship between FER and BER is given by

$$FER = 1-(1-BER)^{Frame\ length[bits]} \tag{7}$$

A BER of 10^{-12} corresponds to an FER of 5.1×10^{-10}. In-service line monitoring was demonstrated under error free conditions in GE-PON. The PON monitoring method can perform the test with negligible degradation in transmission quality.

6. Conclusion

I introduced an in-service measurement technique for monitoring PON fibers with individually assigned Brillouin frequency shifts and an optical fiber line testing system including 1650 nm Brillouin-OTDR. BFS assigned fiber can be designed for 8-branched PON fibers that have a 1.5 GHz BFS separation. Moreover, the 1650 nm band optical fiber line testing system measured B-OTDR traces with a dynamic range of 17 dB. Fault location carried out in branching fiber demonstrated that we could determine the length of eight BFS assigned fibers after an optical splitter. The PON monitoring system can also detect and locate connection losses in branching regions with high sensitivity. In-service line monitoring of a GE-PON transmission was achieved using a 1650 nm test light, and the degradation in transmission quality was negligible. These techniques enable us to isolate a fiber fault, reduce maintenance costs and improve service reliability for access networks.

7. References

Azuma, Y.; Shibata, N.; Horiguchi, T. & Tateda, M. (1988) Wavelength dependence of Brillouin-gain spectra for single-mode optical fibers, Electron. Lett., 24, 5, pp. 251-252, Mar. 1988

Eur. Conf. Optical Communications (ECOC) work shop 8, Operation Expenditures Studies (2007)

Enomoto, Y.; Izumita, H. & Nakamura, M. (2005). "Highly developed fiber fault location technique for branched optical fibers of PONs using high spatial resolution OTDR and frequency domain analysis", The Review of Laser Engineering, Vol. 33, No. 9 September 2005, pp. 609-614

Hogari, K.; Tetsutani, S.; Zhou, J.; Yamamoto, F. & Sato, K. (2003). Opticaltransmission characteristics of optical-fiber cables and installed opticalfiber cable networks for WDM systems. J. Light. Technol., vol. 21, no. 2, pp. 540-545, Feb. 2003.

Horiguchi, T.; Shimizu, K.; Kurashima, T.; Tateda, M.; & Koyamada, Y. (1995). Development of a Distributed Sensing Technique using Brillouin Scattering, J. Lightwave Technol., vol. 13, No. 7, July 1995, pp 1296-1302

Honda, N.; Iida, D.; Izumita, H. & Ito, F. (2006). Optical fiber line testing system design considering outside environment for 8-branched PON fibers with individually assigned BFSs, Optical Fiber Sensors 2006, ThE8, 2006

Honda, N.; Iida, D.; Izumita, H. & Ito, F. Bending and connection loss measurement of PON branching fibers with individually assigned Brillouin frequency shifts, Optical Fiber Communication Conference 2006, OThP6, 2006

IDATE, (Jun., 2011), FTTx around the world, Available from http://www.idate.org/en /News/FTTx-around-the-world_679.html

Iida, D.; Honda, N.; Izumita, H. & Ito, F. (2007). Design of identification fibers with individually assigned Brillouin frequency shifts for monitoring passive optical networks, J. Lightwave Technol., vol. 25, No. 5, MAY 2007, pp1290-1297

Iino, A.; Matsubara, K.; Ogai, M.; Horiuchi, Y. & Namihira, Y. (1989). Diffusion of hydrogen molecules in fluorine-doped single-mode fibers., Electron. Lett. , 5th Jan. 1989 Vol. 25 No.1, pp78-79.

Ishikura, M.; Ito, Y.; Maeshima, O. & Asami, T. (1999). A traffic measurement tool for IP-based networks, IEICE Trans. Inf.&Syst., Vail. E82-D, No. 4, April 1999, pp 756-760.

ITU-T Recommendation (2003) G.984.1, Gigabit-capable Passive Optical Networks (GPON): General characteristics

ITU-T Recommendation L.25. (1996). Optical fibre cable network maintenance

ITU-T Recommendation L.37. (2007). Optical branching components (non-wavelength selective)

ITU-T Recommendation L.40. (2000) Optical fibre outside plant maintenance support, monitoring and testing system

ITU-T Recommendation L.41. (2000). Maintenance wavelength on fibres carrying signals

ITU-T Recommendation L.53. (2003). Optical fibre maintenance criteria for access networks

ITU-T Recommendation L.66. (2007). Maintenance criteria for in-service line monitoring

ITU-T Recommendation L.68. (2007). Fundamental requirement for optical fibre line testing system carrying high total optical power

ITU-T Recommendation L.85. (2010). Optical fibre identification for the maintenance of optical access networks

IEEE Std 802.3ah. (2004). Carrier sense multiple access with collision detection (CSMA/CD) access method and physical layer specifications

Izumita, H.; Horiguchi, T.; & Kurashima, T. (1997) "Distributed sensing technique using Brillouin scattering"., Optical Fiber Sensors 12th, OWD1-1.1997

Kawataka, J. ; Hogari, K.; Iwata, H.; Hakozaki, H.; Kanayama, M.; Yamamoto, H.; Aihara, T. & Sato, K. (2003) Novel optical fiber cable for feeder and distribution sections in access network", J. Lightwave Technol., vol. 21, no. 3, Mar. 2003, pp. 789–796, 2003.

Koyamada, Y.; Nakamoto, H. & Ohta, N. (1992). High performance coherent OTDR enhanced with erbium doped fiber amplifiers. J. Opt. Com, 13 (1992) 4, pp. 127-133

Nakao, N.; Izumita, H.; Inoue, T. ; Enomoto, Y. ; Araki, N. & Tomita, N. "Maintenance method using 1650-nm wavelength band for optical fiber cable networks", J. Lightwave. Technol. vol. 19, No. 10, Oct. 2001. pp 1513-1520

Sankawa, I.; Furukawa, S.; Koyamada, Y. & Izumita, H."Fault location technique for in-service branched optical fiber networks", IEEE Photon. Technol. Lett., Vol. 2, No. 10, October 1990, pp 766-768

Schmuck, H.; Hehmann, J.; Straub, M. & Th. Pfeiffer. (2006) Embedded OTDR techniques for cost-efficient fibre monitoring in optical access networks", Eur. Conf. Optical Communications (ECOC), Mo3.5.4, 2006

Shimizu, K.; Horiguchi, T.; & Koyamada, Y.(1995). Measurement of distributed strain and temperature in a branched optical fiber network by use of Brillouin Optical time-domain reflectometry, Optics Letters, vol. 20, No. 5, 1995, pp. 507–509.

Tanaka, K.; Tateda, M. & Inoue, Y., "Measuring individual attenuation distribution of passive branched optical networks", IEEE Photon, Technol. Lett., vol. 8, No. 7, pp. 915-917, July 1996,

12

A Comparative Study of Node Architectures with Add/Drop Constraints in WDM Networks

Konstantinos Manousakis and Emmanouel (Manos) Varvarigos

University of Patras, Department of Computer Engineering and Informatics / Research Academic Computer Technology Institute

Greece

1. Introduction

The advent of WDM technology has resulted in transmission capacities that have increased manifold in recent years. The most common architecture used for establishing communication in WDM optical networks is wavelength routing, where data are transmitted over all-optical WDM channels, called lightpaths, which may span multiple consecutive fibers. A lightpath is realized by determining a path between the source and the destination and allocating a free wavelength on all the links of the path. The selection of the path and the wavelength to be used by a lightpath is an important optimization problem, known as the routing and wavelength assignment (RWA) problem (Ramaswami & Sivarajan, 1995). In the absence of wavelength conversion, a lightpath must be assigned a common wavelength on each link it traverses; this restriction is referred to as the wavelength continuity constraint. However, two lightpaths may occupy the same wavelength, as long as they use disjoint sets of links; this property is known as wavelength reuse. The RWA problem is usually considered under two alternative traffic models. Static Lightpath Establishment (SLE) addresses the case where the set of connections is known in advance and Dynamic Lightpath Establishment (DLE) considers the case where connection requests arrive randomly, over an infinite time horizon, and are served one-by-one.

The key elements that make WDM technology feasible are optical line terminals, optical add/drop multiplexers, and optical cross-connectors (Ramaswami & Sivarajan, 2001). An optical line terminal (OLT) includes multiplexers/demultiplexers of wavelengths and transponders (TSPs). A TSP is responsible for adapting the signal to a form suitable for transmission over the optical network for originating traffic and for the reverse operation when traffic is terminated. An optical add/drop multiplexer (OADM) takes in signals at multiple wavelengths and selectively drops some of these wavelengths locally, while letting others to pass through. There are two types of OADMs: fixed (FOADM) and reconfigurable (ROADM). FOADMs are capable of adding or dropping fixed wavelengths, while ROADMs select the desired wavelengths to be dropped and added on the fly, a feature that is quite desirable. OADMS are useful network elements to handle simple network topologies, such as linear or ring topologies and utilize fibers with small number of wavelengths. Wavelength selective switch (WSS) technology has recently enabled the introduction of multi-degree ROADM and the deployment of cost-effective dynamic wavelength switched networks (Kaman et al., 2006).

In order to handle more complex topologies and utilize a large number of wavelengths optical cross-connects (OXC) are required. An OXC essentially performs functions similar to the ROADM but at much larger sizes. OXC node architectures can be distinguished based on whether colored or colorless and directed or directionless add/drop ports are utilized. Colored ports, unlike colorless add/drop ports, have a permanently assigned wavelength channel. Also, in a node equipped with non-directionless (directed) add/drop ports, a channel on a specific transmission fiber originating from or terminating at the node, can be added/dropped only by a particular multiplexing/demultiplexing element (port) connected to this transmission fiber. In order for such a port to switch to another wavelength or to another fiber respectively, manual intervention is required. In any case, the number and the type of transponders (TSPs) at each node are also important, since they determine not only the number of wavelengths that can be added or dropped but also the flexibility of the node.

In our work we evaluate how a routing and wavelength assignment (RWA) algorithm performs under optical cross-connect (OXC) node architectures with different levels of color- and direction-related flexibility. In particular, we concentrate on four node architectures that use add/drop ports with the following configurations: i) colored/non-directionless, ii) colored/directionless, iii) colorless/non-directionless, and iv) colorless/directionless. These node architectures come with a different cost; that is, the more flexible ones are also more expensive. As a result, an interesting tradeoff is introduced between the network performance achieved, in terms of network blocking and number of manual interventions (since the operators seldom refuse a connection, but undertake the necessary, manual work to serve it), and the cost of the node architecture used. In the process of comparing the node architectures of differing degrees of flexibility, we propose an adaptation of an online RWA algorithm that takes into account the lack of node flexibility, and aims at achieving using the more constrained node architectures, performance similar to that obtained with the more flexible and more expensive node architectures. Additionally, we evaluate different transponder (TSP) assignment policies and provision strategies and determine their effect on the network performance achieved. The term TSP assignment policy is referred to the correlation between a transponder and a network port. For example, the choice of a port in a directed architecture means that we have to decide in which fiber we want the TSP to be able to send/receive. Our simulation results show that the color constraint affects more negatively the network performance than the direction constraint. In addition, the performance of the architectures is highly affected by the transponder assignment policy used and the provision strategies.

The remainder of the paper is organized as follows. In Section 2 we report on previous work. In Section 3 we describe the network and node models used. In Section 4 we propose an RWA algorithm that accounts for the node limitations and also present various TSP assignment policies. Simulation results are presented in Section 5. Our conclusions are given in Section 6.

2. Previous work

ROADMs enable carriers to offer a flexible service and provide significant savings in Operational Expenditure (OpEx) and Capital Expenditure (CapEx) (Mezhoudi et al., 2006). (Mezhoudi et al., 2006) compare alternative ROADM network architectures and show that optimally deployed higher-degree ROADMs with optical bypass and grooming can significantly reduce the cost. In (Keyworth, 2005) the author describes the available technology options, and corresponding subsystem features, while highlighting the key advantages and implementation challenges associated with each of them. ROADM subsystems can be implemented using a variety of architectures and technologies, each with their own trade-offs in performance and functionality. (Roorda & Collings 2008) and (Homa & Bala, 2008), present different WSS-based ROADM architectures.

Currently, most of the RWA algorithms proposed either assuming a network with ideal physical layer (Zang et al., 2000) or a network with physical layer impairments (Saradhi & Subramaniam, 2009), assume that node architectures are fully flexible. Very few studies consider RWA algorithms assuming nodes that have architectural constraints. Authors in (Shen et al., 2003) study the performance of WDM networks with limited number of add/drop ports in OXCs. They consider the impact of the number of add/drop ports and conclude that only a limited number of add/drop ports are required at each node to achieve performance very close to that of a network where each node is equipped with the full number of widely tunable add/drop ports. The authors in (Zhu & Mukherjee, 2005) compare the design of metro optical WDM network architectures using two different ROADM architectures, namely, a switching-based architecture and a tuning-based architecture, and demonstrate that tuning-based architectures are more cost-effective for the metro networks under the current technologies. The same authors consider the tuning process of ROADMs with the constraint that it does not interfere with working wavelengths and provide heuristics to avoid such interference. Authors in (Turkcu & Subramaniam, 2009) investigate the blocking performance of all-optical reconfigurable networks with constraints on reconfigurable optical add/drop multiplexers (ROADMs) and transponders (TSP) that can be tuned to transmit and receive to a certain set of wavelengths (limited tunable). They, also, develop an analytical model to calculate the call blocking probability in a network of arbitrary topology for two different models of transponder sharing within a node: the share-per-link (SPL) and the share-per-node (SPN). The authors in (Staessens et al., 2010) assess the impact of node directionality on restoration in transparent networks.

3. Node models

In our study we consider the following types of add/drop ports and optical cross-connects (OXCs). We present four node architectures for the optical cross-connects (OXCs); their characteristics affect the final cost of the design strategy applied for serving a set of connection requests. All node architectures offer the same functionality for the transit traffic, namely, full remote re-configurability. They primarily differ in the way the traffic local to the node is treated. Therefore, categorizing OXCs mainly refers to the features/flexibility of add/drop ports.

3.1 Add/Drop ports

Colorless add/drop ports do not have a permanently assigned wavelength channel but rather are provisioned as to which wavelength channel will be added/dropped. In contrast, colored add/drop ports have a permanently assigned wavelength channel.

In directed add/drop ports, a channel on a specific transmission fiber originating from or terminating at the node can be added/dropped only by a multiplexing/demultiplexing element connected to this transmission fiber. In OXCs with directionless add/drop ports, traffic can be added/dropped to/from arbitrary transmission fibers.

3.2 Optical cross-connects

Node architectures (OXCs) are categorized based on the type of add/drop ports, which are used to implement them. We evaluate four different node architectures: colored/directed, colored/directionless, colorless/directed, and colorless/directionless. In general, architectures using add/drop ports without color and direction limitations are more flexible, while adding more constraints on the add/drop ports makes the architectures less flexible. All the OXC architectures can remotely configure all transit traffic with a broadcast and select architecture. The incoming channels (for example from box A in Fig.1-4) are broadcasted (with a splitter) to all other network interfaces (for example boxes B and C in Fig. 1-4). A WSS is connected to the outgoing fiber and can select which wavelength from which other network interface or add/drop terminal it wants to add. Each network interface (NI) consists of a WSS and a splitter.

3.2.1 Colored/Directed node architecture

As previous mentioned, OXC architectures are primarily differ in the way the add/drop terminals are implemented. In colored/directed architecture (Fig. 1) each add/drop

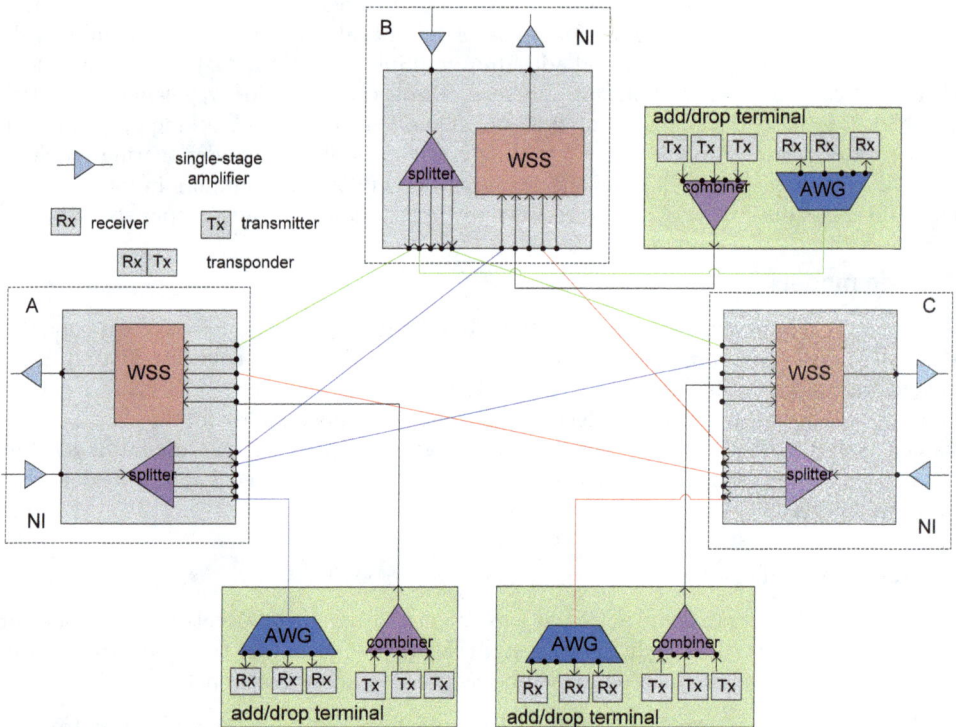

Fig. 1. Colored/directed node architecture

terminal (green box) is connected to a specific network interface and as a consequence to a specific optical fiber (directed feature). Moreover, each transponder located in the add/drop terminal is connected through fixed wavelength demultiplexing element (for example an arrayed waveguide grating (AWG)). The AWG is used to separate the drop channels to the receivers. Using an AWG imposes fixed wavelengths on the drop (colored feature). Also in this add/drop terminal, a combiner is used to collect all the added traffic. The advantage of this architecture is that there is no need for WSS equipment in the add/drop terminals and only an AWG and a combiner are needed to implement each add/drop terminal.

3.2.2 Colored/Directionless node architecture

To avoid having separate add/drop terminals for each network interface and to be able to share the add/drop ports between the network interfaces, one common add/drop terminal can be used. In the colored/directionless architecture (Fig. 2) one add/drop terminal (green box) per node is utilized and each port of the add/drop terminal is physically connected to all the fibers of NIs of the node. In this architecture the drop part of the add/drop terminal contains of a WSS and an AWG. The WSS is used to collect the drop traffic from the different directions and the AWG to separate the drop channels to the receivers. Channels

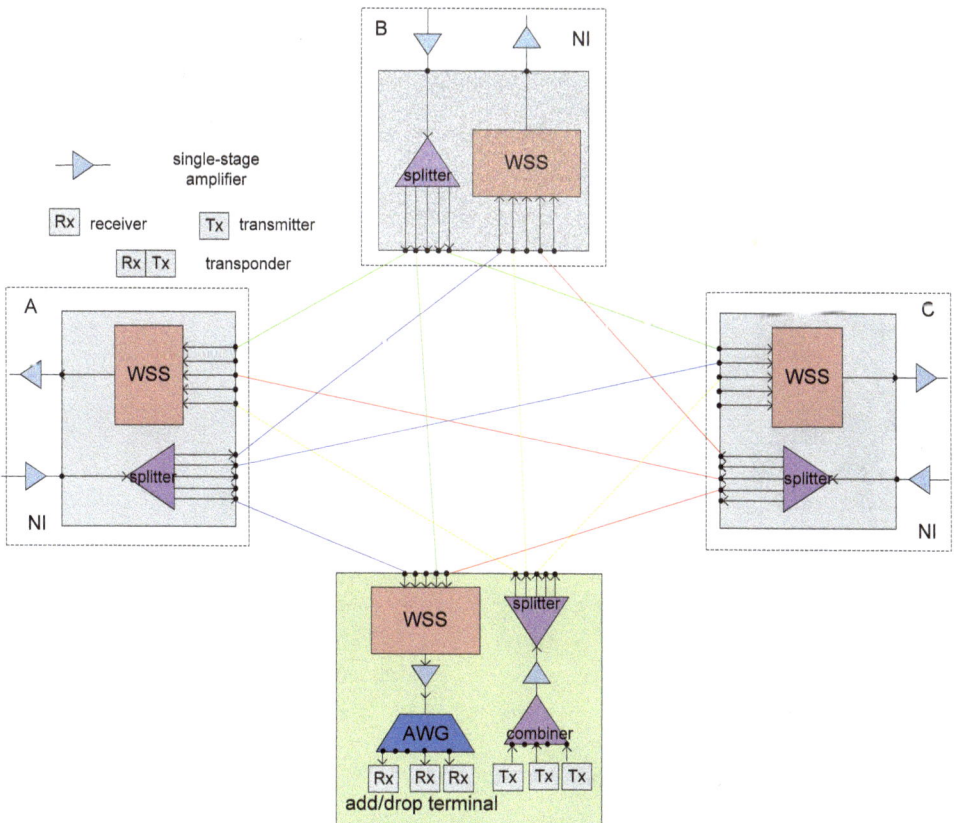

Fig. 2. Colored/directionless node architecture

from a bank of (different) wavelengths can be switched towards any network port connected to the OXC, as shown in Fig. 2. The drawback of this architecture is that only those wavelengths equipped in the channel bank can be used for traffic connections, and every wavelength can only be used to terminate a single network interface, only one wavelength per color can be used.

3.2.3 Colorless/Directed node architecture

To avoid having to spend a lot of CapEx upfront to allow a more flexible routing, the architecture according to Fig. 3 can be used. Colorless ports are generally created by replacing a fixed wavelength demultiplexing element (AWG) with a wavelength selective switch (WSS). A WSS can steer each optical channel present on its input port toward one of its output ports. The dropped channels from a network port are distributed to the corresponding wavelength transponders via a 1xN WSS. In the add direction, all wavelengths are combined and sent to the corresponding network interface. This allows the reduction of the number of equipped transponders while still maintaining routing flexibility.

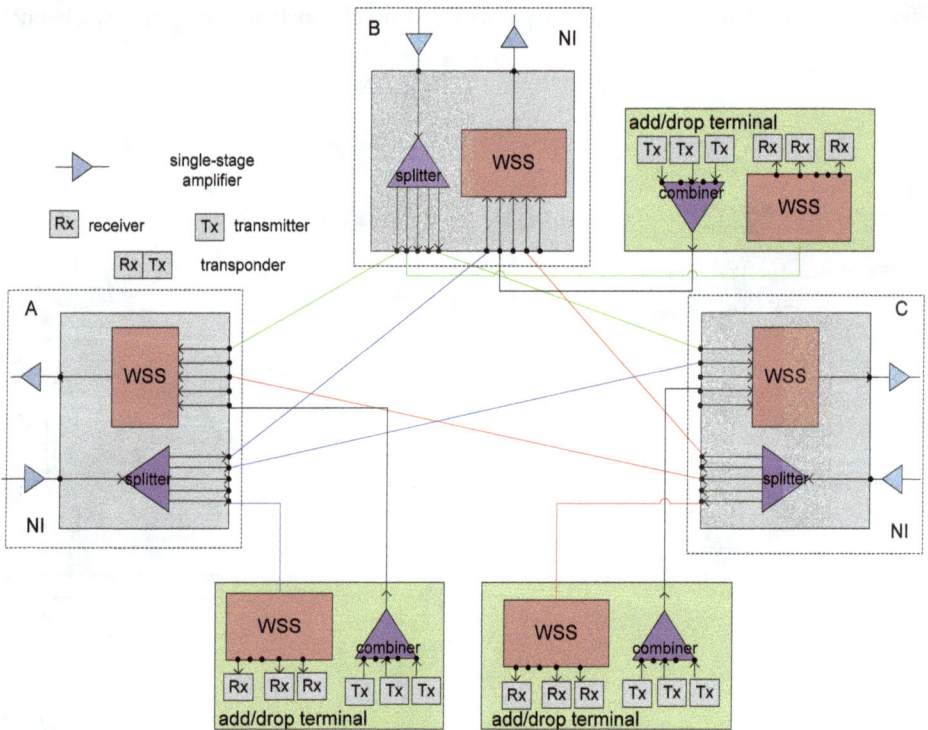

Fig. 3. Colorless/directed node architecture

3.2.4 Colorless/Directionless node architecture

To add more flexibility at the add/drop side of the architecture of Fig. 2, it is possible to use an extra network interface instead of an AWG component to select which wavelengths have to be dropped from the other network interfaces. This extra network interface adds the directionless feature of the node (see Fig. 4), and is the first stage of the add/drop terminal. The colorless feature is implemented by the second stage of the add/drop terminal. In that box a combiner is used to add all the wavelengths and a WSS is used to select which wavelength you want to drop at which port.

If one add/drop terminal is used in this node configuration, only one unique wavelength can be dropped at an add/drop terminal, because a WSS can only drop the same wavelength channel once to its output port (contention feature). If more than one add/drop terminals (Fig. 4) are used then wavelength blocking at the add/drop terminal is prevented.

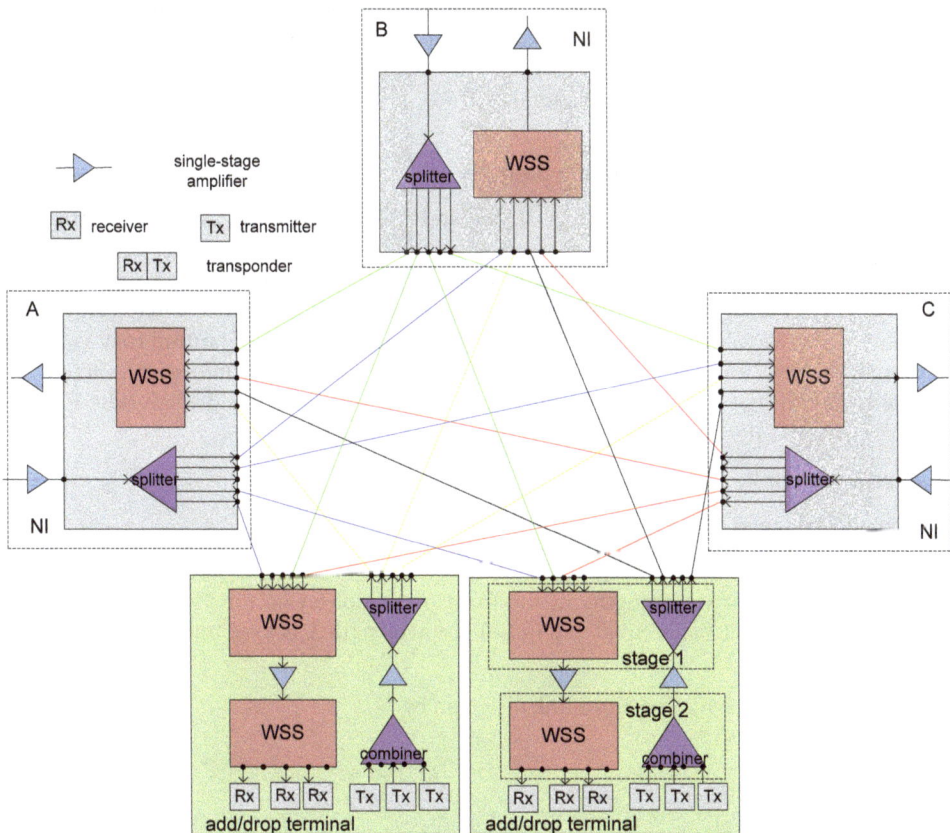

Fig. 4. Coloreless/directionless node architecture

4. RWA problem

The Routing and Wavelength Assignment (RWA) problem is usually considered under two alternative traffic models. Offline (or static) lightpath establishment addresses the case where the set of connections is known in advance, usually given in the form of a traffic matrix that describes the number of lightpaths that have to be established between each pair of nodes. Dynamic (or online) lightpath establishment considers the case where connection requests arrive at random time instants, over a prolonged period of time, and are served upon their arrival, on a one-by-one basis. We focus our study on the online RWA problem.

4.1 RWA algorithm with full flexibility

The proposed multi-cost RWA algorithm consists of two phases. In contrast to traditional single-cost approach, where each link is characterized by a scalar, in the multi-cost approach a vector of cost parameters is assigned to each link, from which the parameter vectors of candidate lightpaths are calculated. In our work, we assume that nodes are equipped with TSPs that can be tuned to transmit and receive at any wavelength (widely tunable TSPs). In particular, the number of TSPs each node n is equipped with, depends on its degree D_n. The number of TSPs of node n, that are assigned to each link l is assumed to be constant and equal to T and as a result node n has a total of $T_n = D_n \cdot T$ TSPs.

4.1.1 Computing the cost vector of a path

We consider a WDM network with N nodes and L fiber-links, each of which carries m wavelengths. Each fiber is able to support a common set $C=\{\lambda_1, \lambda_2,..., \lambda_m\}$ of W distinct wavelengths. The WDM network employs no wavelength conversion. We also assume that the node where the algorithm is executed (in a decentralized or centralized architecture) has a picture of the wavelengths' utilization of all links. Although the algorithm may run in a decentralized way, and thus due to propagation delays utilization information might be outdated, we will not focus on such problems. We assume that all nodes are fully flexible (colorless/directionless nodes) without add/drop constraints.

Cost vector of a link

Each link l is assigned a cost vector that contains $m+1$ cost parameters:

i.   the length L_l of the link(scalar);
ii. the availability of wavelengths in the form of a Boolean vector $\overline{W_l} =(w_{l1}, w_{l2},...,w_{lm})$, whose ith element w_{lm} is equal to 0 (false) if wavelength λ_i is used and equal to 1 (true) when λ_i is free.

Thus, the cost vector characterizing a link l is given by

$$V_l = (L_l, \overline{W_l})$$

Cost vector of a path

Similarly to a link, a path has a cost vector with $m+1$ parameters, in addition to the list of labels of the links that comprise the path. Assume a path p with cost vector

$$V_p = (L_p, \overline{W_p}, {}^*p),$$

where L_p, and $\overline{W_p}$ are as previously described, and $^*p$ is the list of identifiers of the links that comprise path p. The cost vector of p can be calculated by the cost vectors of the links $l=1,2,..,k$, that comprise it as:

$$V_p = \left(\sum_{l=1}^{k} L_l, \underset{l=1}{\overset{k}{\&}} \overline{W_l}, (1,2,...,k) \right),$$

where the operator & denotes the bitwise AND operation. Note that all operations between vectors have to be interpreted component-wise.

Checking if the path is further extendable

We check if path p has at least one available wavelength.

If $\overline{W_p} = 0$ (all zero vector), then path p is rejected.

Domination relationship

We also define a *domination* relationship between two paths that can be used to reduce the number of paths considered by the RWA algorithm. In particular, we will say that

$$p_1 \text{ dominates } p_2 \text{ (notation: } p_1 > p_2 \text{) iff } \quad L_{p_1} \le L_{p_2} \text{ and } \overline{W_{p_1}} \ge \overline{W_{p_2}}$$

The " $\ge$ " relationship for vectors $\overline{W}$, should be interpreted component-wise. A path that is dominated by another path has larger length and worse wavelength availability than the other path and there is no reason to consider it or extend it further.

4.1.2 Multi-cost RWA algorithm

The proposed multi-cost RWA algorithm consists of two phases:

Phase 1: Computing the set of non-dominated paths P_{n-d}

The algorithm that computes the non-dominated paths from a given source to all network nodes (including the destination) can be viewed as a generalization of Dijkstra's algorithm that only considers scalar link costs. The basic difference is that instead of a single path, a set of non-dominated paths between the origin and each node is obtained. Thus a node for which one path has already been found is not finalized (as in the Dijkstra case), since we can find more "non-dominated" paths to that node later. An algorithm for obtaining the set P_{n-d} of non-dominated paths from a given source to all nodes is given in (Varvarigos et al., 2008). By definition, for the given source and destination, the non-dominated paths that the algorithm returns have at least one available wavelength.

Phase 2: Choosing the optimal lightpath from P_{n-d}

In the second phase of the proposed algorithm we apply an optimization function or policy $g(V_p)$ to the cost vector, V_p, of each path $p \in P_{n-d}$. The function g yields a scalar cost per path and wavelength (per lightpath) in order to select the optimal one. Given the connections already established, we order the wavelengths in decreasing utilization order and choose the lightpath whose wavelength is most used. This approach is the well known "most used wavelength" algorithm (Zang et al., 2000), proven to exhibit good network–layer blocking assuming ideal physical layer. In the end, the algorithm establishes the decided lightpath if there are available transponders (TSPs) in the source/destination nodes of the connection, assuming colorless/directionless node architectures.

4.2 RWA algorithm with limited flexibility

A network topology is represented by a connected graph $G=(V,E)$. V denotes the set of OXCs-nodes.

4.2.1 Colored vs. colorless architecture

Colored add/drop ports in network nodes limit the flexibility of the RWA algorithm, mainly regarding which channels/wavelengths it can use for serving a connection request. This is because the node ports are permanently assigned to specific wavelengths. In this case, the links' wavelength availability vectors $\overline{W_l}$, used by the RWA algorithm, are updated according to these wavelengths. If the algorithm cannot find a lightpath for serving a connection request, then manual intervention can be performed. In particular, manual intervention corresponds to the assignment of an available TSP to a different port than the one already provisioned. If no TSPs are available, then the demand is finally blocked.

Figure 5a shows how the definition of the wavelength availability vector $\overline{W_l}$ of link l has to be modified to account for the color related constraints. If node d is the destination of a connection request, then the availability vectors of the node's incoming links are modified according to its available receivers - drop ports (that are tuned to specific wavelengths). For example in Figure 5a, the original vector of link l is $\overline{W_l} = [0\,1\,1\,1\,1]$, implying that the available

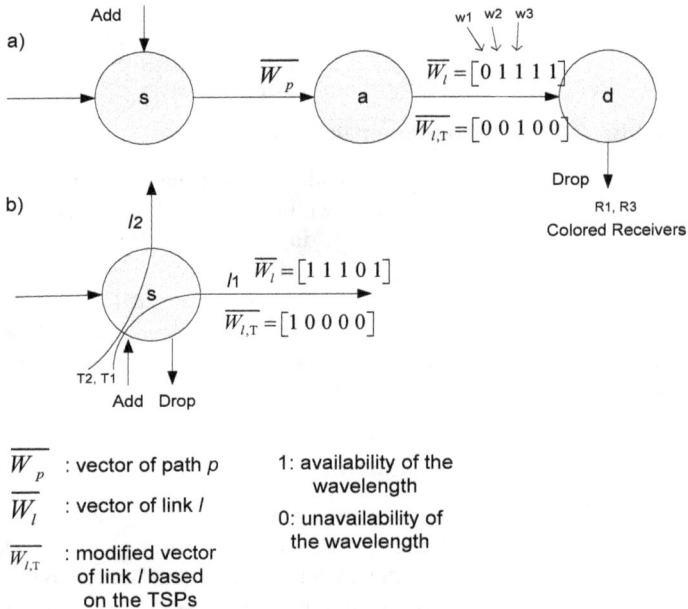

$\overline{W_p}$: vector of path p

$\overline{W_l}$: vector of link l

$\overline{W_{l,T}}$: modified vector of link l based on the TSPs

1: availability of the wavelength

0: unavailability of the wavelength

Fig. 5. a) Availability vectors of the RWA algorithm when considering colored ports. Receivers / drop ports R_1, R_3 can only receive wavelengths w_1, w_3, respectively.
b) Availability vectors of the RWA algorithm when considering directed ports, where transmitter / input port T_1 can only send traffic to link l_1 and transmitter / input port T_2 can only send traffic to link l_2.

wavelengths of link l are the w_2, w_3, w_4, w_5 (the example assumes five wavelengths per fiber). In case the RWA algorithm attempts to find a lightpath that terminates at node d, then all the availability vectors of the links incoming to d are modified based on the way node's d drop ports are colored. In our example, node d can only receive on wavelengths w_1, w_3 because only receivers / drop ports R_1 and R_3 are available and therefore, the original availability vector is updated to $W_{l,T} = [0\ 0\ 1\ 0\ 0]$. This means that only wavelength w_3 of link l is actually available for use by the RWA algorithm in order to end the lightpath in node d.

If the RWA algorithm cannot find a lightpath, either due to the unavailability of a path and/or wavelength from source to destination or due to the color constraint, manual intervention is necessary. In this case the RWA algorithm is re-executed for deciding the lightpath that will serve the request, assuming that there are not color constraints. Next, based on the RWA algorithm's decisions manual intervention is performed so as to plug a TSP at the decided (input or output) port. As mentioned, the RWA algorithm (that does not consider color constraints) is executed only if there are free TSPs at the source and destination nodes of the connection request; otherwise the connection is blocked.

4.2.2 Directed vs. directionless architecture

Colored Non-Directionless ports limit the routing choices available to the RWA algorithm, mainly regarding the first and the last link of the path to be used for serving a connection. For example, assume there is only one free input port (with a plugged TSP) connected to a specific fiber in a node s. This free input port can only be used by a connection request, which originates from s and uses this fiber as its first hop. This constraint must be accounted for by the corresponding RWA algorithm. If a lightpath cannot be found, the connection is either blocked, or manual intervention is performed to connect an available TSP to another fiber. In this case, an RWA algorithm that does not consider direction-related constraints will point out which fiber-link is most efficient to use. In the case where there are no available TSPs then the connection will be blocked.

In Figure 5b, if node s is the source of a connection request, then we can only set up a connection from transmitter / input port T_1 to link l_1 and from T_2 to l_2. Also, the wavelength availability vectors of the links are again modified, in a way similar to that used for colored ports. In case we also have color constraints (that is, the ports are not colorless), the RWA algorithm will have to find a solution under both constraints.

5. TSP assignment policy

An important factor affecting network efficiency in case colored node architectures are used, is the way the transponders (TSPs) of a link are provisioned to specific wavelengths. Next, we present a number of such TSP assignment policies.

In Figure 6, we illustrate an abstraction of node architectures based on the configuration of add/drop ports. Also in this figure we depict the way the TSPs are connected to the optical fibers (in which wavelength and direction). For example, Fig. 6a presents four add/drop ports connected statically to Fibers 1 and 2 and wavelengths 1 and 2 respectively, while Fig. 6d presents four add/drop ports that can switch on the fly to any of the two fibers, serving any wavelength.

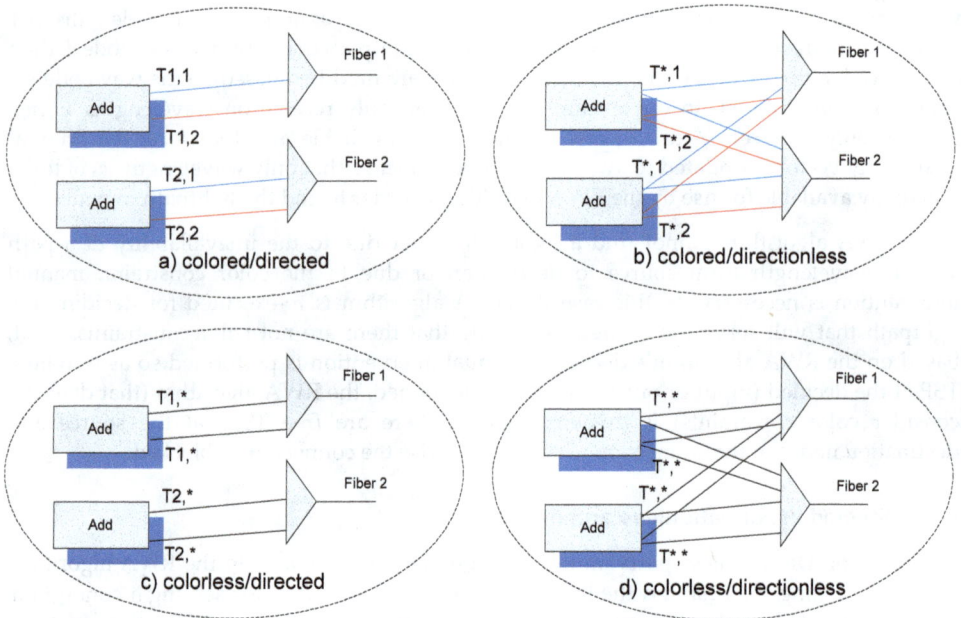

Fig. 6. a) Different node architectures: a) colored/non-directionless, b) colored/non-directionless, c) colored/directionless, d) colorless/directionless. Tx,y express the ability of add/drop port: x is the fiber and y is the wavelength that the transponder (TSP) is plugged in. The symbol '*' denotes that there is no limitation.

5.1 Colored architectures - policy 1: Lowest wavelength count first

The provision of wavelengths in the TSPs of a link can be performed according to the "lowest available wavelength count first" rule. That is, assuming there are T available TSPs per link and no connections are already established, the TSPs can be provisioned to the first T wavelengths of the link (Figure 6a and 6b). This is the simplest TSP assignment policy that can be used in colored architectures.

5.2 Colored/directed architecture - policy 2: Cyclic wavelength rotation

In this policy, the T available TSPs of each link are provisioned based on a cyclic rotation process. That is, the TSPs of the first link of a node are provisioned to wavelengths 1 to T, the TSPs of the second link are provisioned to wavelengths $T+1$ to $2T$, and the provisioning procedure continues similarly to the remaining links, until all the TSPs are provisioned (Figure 7a). The sense behind this policy is that the available TSPs of a node have to be provisioned in as many wavelengths as possible, so as each connection originating/terminating from/to that node to be able to use all the available wavelengths.

5.3 Colored/directionless architecture - policy 2: Full wavelength cover

Under this policy (Figure 7b), all the available TSPs of a node are provisioned to wavelengths 1 to $T_n = D_n \cdot T$, assuming $T_n \leq W$. In case $T_n > W$, then $\lfloor T_n / W \rfloor$ TSPs are provisioned to all the wavelengths and the remaining $T_n \bmod W$ TSPs are provisioned to wavelengths 1 to $T_n \bmod W$. This policy has the same logic as the previous and also taking into advantage the directionless feature of the node.

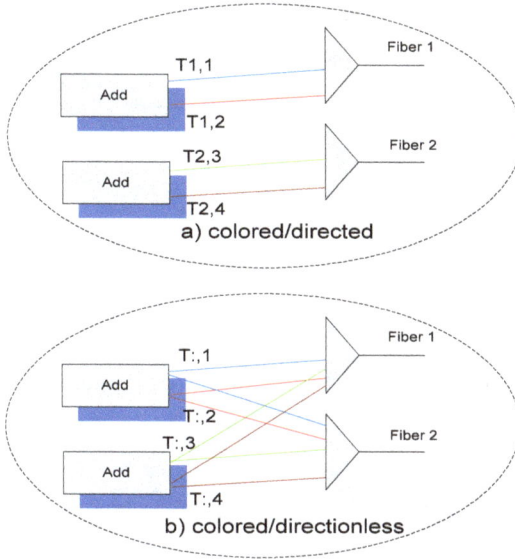

Fig. 7. TSP assignment policy 2 for: (a) the colored/directed architecture, (b) the colored/directionless architecture (as opposed to policy 1 in Fig. 6a and 6b)

6. Simulation results

The network topology used in our simulations was the generic Deutsche Telekom network (DTnet) that has 14 nodes and 23 links (Fig. 8). The capacity of a wavelength was assumed equal to 10Gbps. We performed two different sets of simulations: In the first set, we have limited resources and we report on blocking performance, while in the second set we have enough resources to establish all the requested connections and we report on required manual interventions.

6.1 Impact of node flexibilities in blocking probability

In this set of simulations, connection requests (each requiring bandwidth equal to 10Gbps) are generated according to a Poisson process with rate λ (requests/time unit). The source and destination of a connection are uniformly chosen among the nodes of the network. The duration of a connection is given by an exponential random variable with average $1/\mu$ (time units). Thus, λ/μ gives the total network load in Erlangs. In this set we also assumed that widely tunable TSPs are plugged into specific ports, while the number of TSPs is constant during the network operation. That is, we cannot add extra TSPs and if a connection cannot be served due to limited resources then it is blocked.

In Fig. 9 we examine the performance of the various TSP assignment policies proposed in conjunction with the node's architectures considered, assuming network load equal to 100

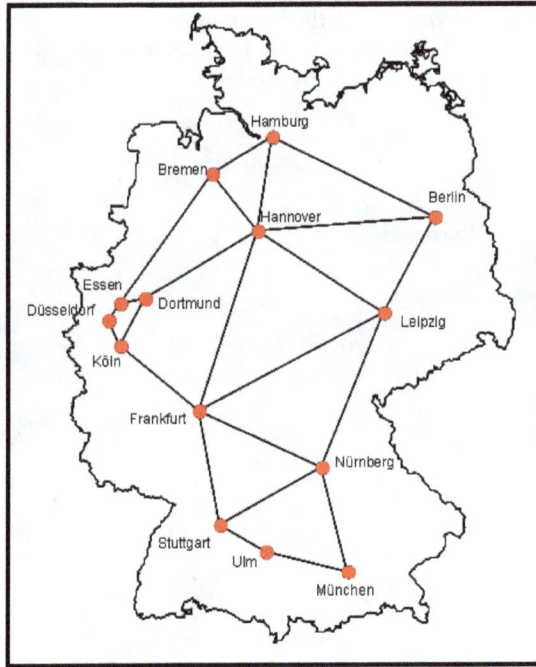

Fig. 8. DT network: 14 nodes, 23 links

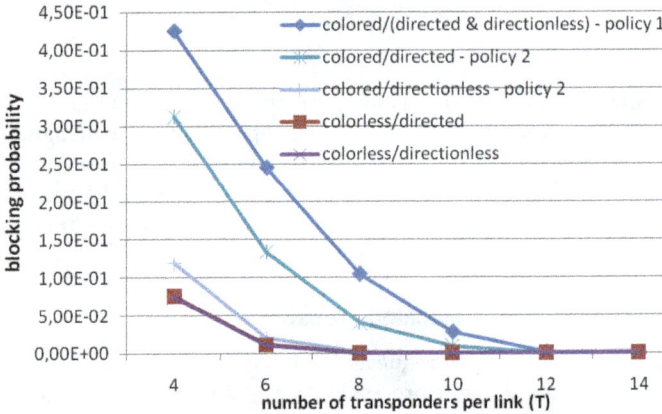

Fig. 9. Blocking probability vs. number of transponders when no manual interventions are allowed, assuming 14 available wavelengths per link and network load equal to 100, for various node architectures and TSP assignment policies

Erlangs and 14 available wavelengths. We assumed that no MIs are allowed and as a result if the wavelength of the transmitter (source) does not fit with the wavelength at the receiver (destination), then the connection is blocked. We observe that the colored/directed and colored/directionless architectures exhibit the same, bad performance when the TSP assignment policy 1 is used. This is due to the fact that under this policy not all the available wavelengths are actually utilized. On the other hand the performance of these architectures, and especially that of the colored/directionless architecture, is improved when TSP assignment policy 2 is used. Colorless/directed architecture exhibits similar performance with the most flexible architecture (colorless/directionless) and this can be explained by the characteristics of the DT network. In particular, the average node degree of DT network is small and as result the direction related constraint is not as restrictive as the color related one.

Fig. 10 illustrates the blocking probability versus the number of TSPs per link for different number of available wavelengths. We assume that each fiber has the same number of wavelengths and TSPs. In the cases where we do not have fully flexible architecture and an available TSP has to be assigned to a different port than the one originally assigned, so as to serve a new connection, then a manual intervention is performed, for changing the direction and the color of a port. For this reason the results of blocking probability presented in Fig. 10 hold for all the node architectures under consideration. Small variations in blocking probability is possible, because in different architectures the differences in ports flexibilities lead to different wavelength assignment by the RWA algorithm, which assigns the wavelengths based on the already provisioned TSPs.

In general, the performance of the RWA algorithm is constrained by the number of transponders; however, as this number increases, then the number of wavelengths becomes the performance bottleneck. In particular, we note that in order to achieve zero blocking probability 8 TSPs and 14 wavelengths per link/fiber are required. When having only 10 available wavelengths per fiber, we cannot achieve zero blocking for load equal to 100 Erlangs, irrespectively of the number of TSPs.

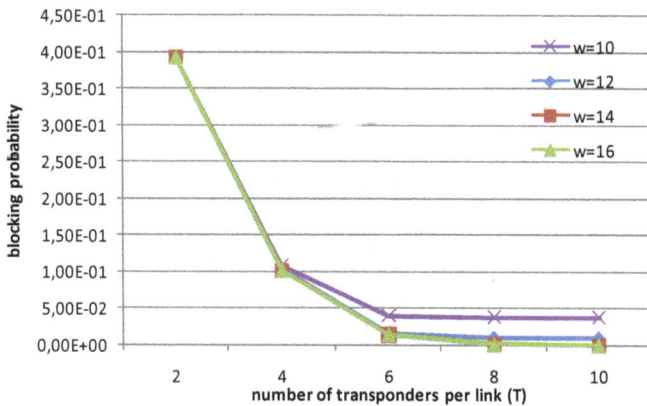

Fig. 10. Blocking probability vs. number of transponders for different number of available wavelengths per link, assuming network load equal to 100. Blocking probability is the same irrespective of the node architecture used.

6.2 Impact of node flexibilities in operational cost

In this study we evaluate a realistic operational scenario of the DT core network. Initially, we assumed that in year 2008, 270 demands were present. For the year 2008, the network was provisioned with 460 transponders. We made the assumption that new demands arrive during the next years leading to an increase of 50% in the requested connections per year. In this set of simulations we allow manual interventions in order to change the port of an already installed TSP or to install new TSPs.

We define two different types of manual interventions. The type 1 of manual intervention is the switching of an available transponder from one to another port of the same node. Manual intervention of type 2 is referred to the installation of extra transponders. We consider different pre-provisioning strategies (manual interventions of type 2). All strategies start with 10 TSPs per link, which results in 460 TSPs in total. The first strategy is when there are no more TSPs available at a particular node to establish a connection, and only a new TSP is installed. We call this approach one TSP. In the other approaches a certain amount of TSPs are installed per link (bank of transponders). For example in case of one TSP per link, we will install 3 extra transponders if the node degree is 3. We have also similar approaches with 5, 7 and 9 TSPs per link.

In Fig. 11, we show the sum of the manual interventions of type 1 (MI1) and those of type 2 (MI2) cumulated over three years. The results for different node flexibilities are depicted to point the differences between them. In Fig. 11a) we can observe that provisioning of more transponders has only a little impact on the amount of manual interventions. In Fig. 11b), the architecture with the directionless feature is depicted with TSP assignment policy 2. This results in a lower number of manual interventions as compared to the previous architecture. In Fig. 11c), it is clear that provisioning of more transponders has huge impact on the manual interventions. The difference between three and nine transponders per link is really small. So there is no reason to provision more than 3 transponders per time because the cost will be increased. In Fig. 11d), we consider the colorless/directionless architecture, which has the best performance in terms of MIs because all transponders provisioned in the node can be used for every new demand. There are no constraints in terms of color or fiber anymore. When provisioning only one TSP per link instead of one TSP, the MIs are decreased from 270 to 100.

Based on these remarks we are interested in the operational processes that involve several actions/activities that need to be performed by the operator's staff. The duration of the activity determines, to an important extent, the cost of the action. The costs for transport (going to the location of the node where an intervention is needed) are calculated from the topology characteristics. We assume that technical teams are present on average 2 links away from one another, this is every 340 km. The average distance to the failure location is therefore 85 km. One way and return adds to 170 km, with an average speed of 50 km/h, this means 3.4 hours for transport.

With the number of MI1 and MI2 (Fig. 11) we can calculate the total transport time and the real intervention time that is the time to switch a transponder in case of MI1 and the time to install new transponders in case of MI2. The duration of transport is 3.4 hours and the duration of switching/installing a transponder is 1 hour.

In Fig. 12 we depict the working hours over the years from 2009 to 2011 needed for manual intervention purposes. In this figure we present two blocks for the node architectures, where in the first block we assume that one TSP is installed, while in the second three TSPs per link are installed. We can see that the colored/directed node and the colored/directionless node

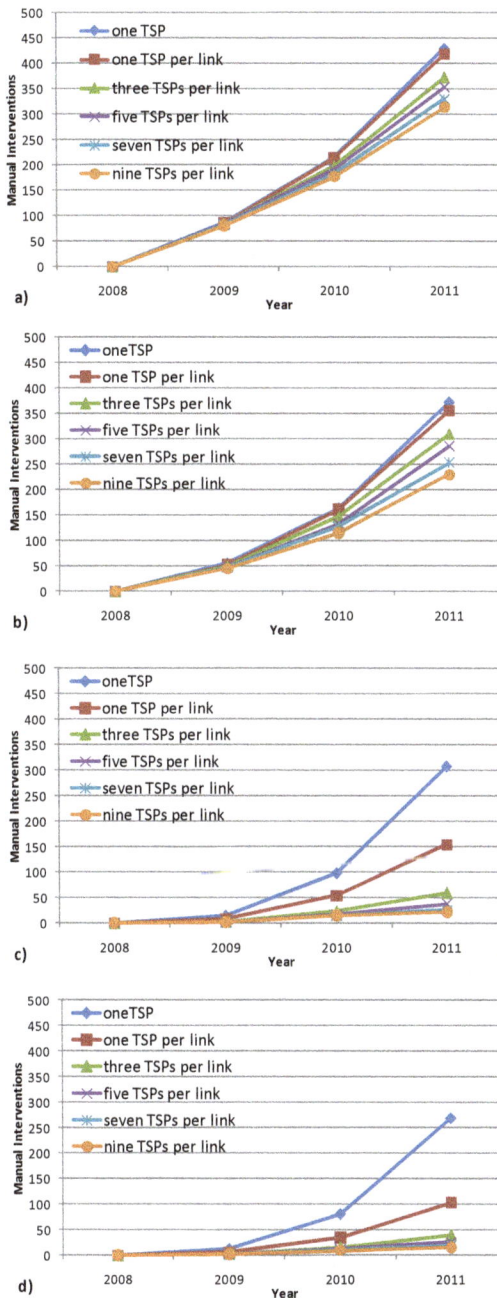

Fig. 11. Cumulative sum of number of manual interventions for a) colored/directed (TSP policy 2), b) colored/directionless (TSP policy 2), c) colorless/directed and d) colorless/directionless.

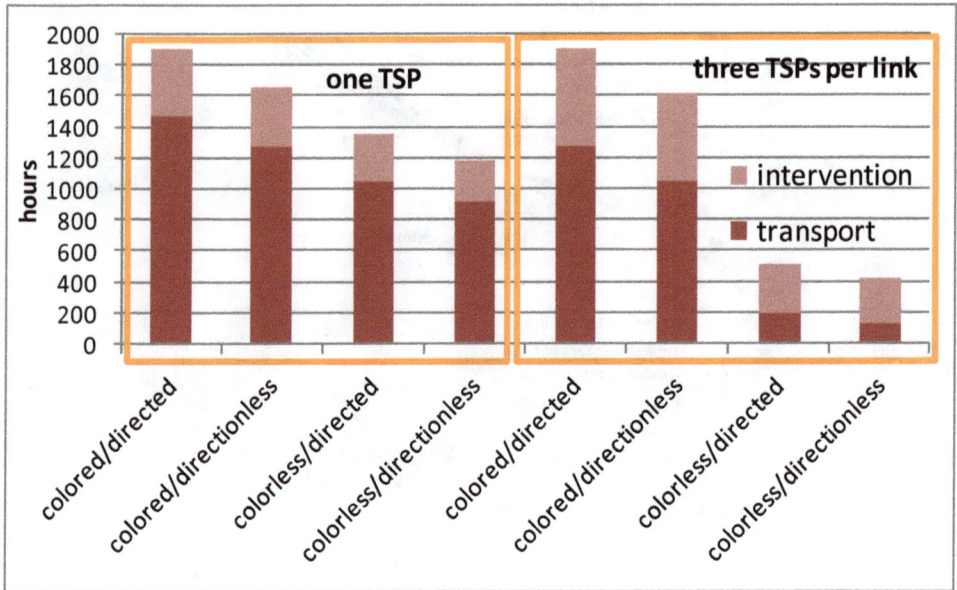

Fig. 12. Working hours of manual interventions

have little improvement in transport times but the intervention times are worst. This happens because when installing more TSPs at once maybe additional MIs of type 1 will be necessary to switch a TSP in different port. The improvement of working hours in case of colorless/directed and colorless/directionless is obvious when installing three TSPs per link instead of one TSP at once, and this happens because of the saving in transport time.

The colorless/directed architecture has almost the same performance as the colorless directionless with the provisioning of 3 TSPs per fiber. The benefits of the directionless architectures are almost negligible due to higher cost and because similar performance can be achieved with directed architectures when appropriate provision strategies and TSP assignment policies are used.

7. Conclusion

We evaluated and compared the performance of several node architectures with color and direction related constraints used in a WDM network. In comparing the node architectures, we also proposed an adaptation of an RWA algorithm that accounts for the lack of node flexibility, and aims at achieving performance similar to that obtained with fully flexible node architectures. Our results demonstrated that in topologies where the node degree is small, the colored constraint is a more dominant performance limiting factor than the direction related one. In addition, we observed that even if a sufficient number of transponders exist in each node, a small number of wavelengths can also be a bottleneck of the network's performance. Finally, we illustrated that the way and the number of transponders are assigned to wavelengths are important and assignment policies utilizing all the available wavelengths should be used.

8. References

Christodoulopoulos, K., et al. (2009). A Multicost Approach to Online Impairment-Aware RWA, *Proceedings of* ICC, 2009

Homa, J & Bala K., (2008). ROADM architectures and their enabling WSS technology, *IEEE Communications Magazine*, vol. 46, no. 7, pp. 150 –154, 2008

Kaman, V.; Helkey, R. & Bowers, J. (2007) "Multi-degree ROADM's with agile add-drop access", *Proceedings of Photonics in Switching conference*, San Francisco (US), Aug. 2007

Keyworth, B., (2005) "ROADM subsystems and technologies", *Proceeding of OFC/NFOEC 2005 Optical Fiber communication/National Fiber Optic Engineers Conference,*, Vol. 3, pp. 1–4, 2005

Mezhoudi, M., et al. (2006), "The value of multiple degree ROADMs on metropolitan network economics", *Proceeding of OFC/NFOEC 2008 Optical Fiber communication/National Fiber Optic Engineers Conference*, pp. 1–8, 2006

Ramaswami, R. & Sivarajan K., (1995). Routing and Wavelength Assignment in All-Optical Networks, IEEE/ACM Transactions on Networking, vol. 3, no. 5, pp. 489-500, Oct. 1995

Ramaswami, R. & Sivarajan K., (2001). Optical Networks: A Practical Perspective, 2nd ed., Morgan Kaufmann, 2001

Roorda, P. & Collings B. (2008). Evolution to colorless and directionless roadm architectures, *Proceeding of OFC/NFOEC 2008 Optical Fiber communication/National Fiber Optic Engineers Conference*, 2008

Saradhi, C. & Subramaniam, S. (2009). Physical layer impairment aware routing (PLIAR) in WDM optical networks: issues and challenges, *IEEE Communications Surveys & Tutorials*, Vol. 11, No. 4, pp.109-130

Shen, G.; Bose, S.; Cheng, T.; Lu, C. & Chai T. (2003). The impact of the number of add/drop ports in wavelength routing all-optical networks, *Optical Networks Magazine*, pp. 112–122, 2003

Staessens, D.; Colle, D.; Pickavet, M. & Demeester, P. (2010). Impact of node directionality on restoration in translucent optical networks, *Proceedings of Optical Communication (ECOC), 2010 36th European Conference and Exhibition*, 2010.

Turkcu O. & Subramaniam S. (2009). Performance of optical networks with limited reconfigurability, *IEEE/ACM Transactions on Networking*, Vol. 17, No. 6, pp. 2002 - 2013, 2009

Varvarigos E.; Sourlas V. & Christodoulopoulos K. (2008). Routing and Scheduling Connections in Networks that Support Advance Reservations, *Computer Networks* (58), 2008.

Zang H.; Jue, J. & Mukherjee, B. (2000). A Review of Routing and Wavelength Assignment Approaches for Wavelength-Routed Optical WDM Networks, *Optical Networks Magazine*, Vol. 1, 2000

Zhu, H. & Mukherjee B. (2005). Online connection provisioning in metro optical WDM networks using reconfigurable OADMs (ROADMs), *IEEE/OSA Journal of Lightwave Technology*, Vol 23, No. 10, pp. 2893–2901, 2005.

13

Accurate Receiver Model for Optical Fiber Systems with Polarization Induced Performance Degradation

Aurenice Oliveira
Michigan Technological University
USA

1. Introduction

Polarization-mode dispersion (PMD) and polarization-dependent loss (PDL) are the main polarization effects that degrade intermetropolitan and transoceanic high-speed optical fiber communication systems [Huttner et al., 2000]. As a result of the stochastic nature of PMD [Khosravani et al., 2001], it is very difficult to compensate the performance degradation due to PMD, which leads to waveform distortions and signal depolarization. Because PMD causes random fluctuations of the polarization state of the light, the performance degradation due to PDL also becomes stochastic; leading to power fluctuation in wavelength-division multiplexed (WDM) systems, and producing additional waveform distortions.

In this chapter, we demonstrate that one can use a semi-analytical receiver model to accurately estimate the performance of on-off-keyed (OOK) optical fiber communication systems, taking into account the impact of the choice of the modulation format, arbitrarily polarized noise, and the receiver characteristics [Lima Jr. et al., 2005]. We initially validate our semi-analytical model by comparing the results obtained with this model against experiments and extensive Monte Carlo simulations for cases in which the signal does not suffer significant waveform distortions, as in the case of negligible intra-channel PMD [Wang & Menyuk, 2001], [Lima Jr. et al. , 2003a]. For that case, we extend the work by [Marcuse, 1990], [Humblet & Azizoglu, 1991], and [Winzer et al., 2001] through the derivation of an expression that shows how the Q factor depends on both the electrical signal-to-noise ratio (SNR) and the optical signal-to-noise ratio (OSNR) for arbitrary modulation format and receiver characteristics. Marcuse's results [Marcuse, 1990], which have been widely used in the calculation of the Q-factor, only consider two extreme cases that the noise is unpolarized or copolarized with the signal. How the partially polarized noise, which happens in many optical systems with significant PDL [Wang & Menyuk, 2001], [Sun et al., 2003a], affects the system performance remains unclear. Therefore, in our next step we extend the Q-factor derived expression for the case in which the optical noise is partially depolarized due to PDL in long-haul optical fiber systems [Wang & Menyuk, 2001],[Lima Jr. et al., 2003a],[Sun et al., 2003a], [Sun et al., 2003b]. We systematically investigate effects of partially polarized noise in a receiver and compute the Q-factor using a general and accurate receiver model that takes into account the effect of partially polarized noise as well as the optical pulse format immediately prior to the receiver and the shapes of the optical and electrical filters. Our results show that the system performance depends on both the degree of polarization of the noise (DOP) and the random angle between the polarization states of the signal and of the polarized part of the noise, i.e., the Stoke's vectors of the signal and the noise [Lima Jr. et al., 2005]. We also demonstrate that the relationship between the OSNR and the Q factor is not unique when the noise is partially polarized.

Finally, we show how to use our developed semi-analytical model to calculate the performance degradation in the presence of PMD-induced waveform distortions and the performance dependence on the receiver characteristics for different modulation formats [Lima Jr. & Oliveira, 2005]. In this study we focus on OOK optical fiber communication systems, which are the ones most widely used today because of their cost-effectiveness.

2. Modelling systems with negligible amount of intra-channel PMD

Undersea WDM systems that operate with speeds of up to 40 Gbit/s using ultra-low PMD fiber are not subject to waveform distortions due to PMD, but can suffer power fluctuations. In this case, PMD is not large enough to drift the spectral components within a single channel, but is sufficient to drift apart the polarization states of the WDM channels as the optical signal propagates down the transmission fiber [Wang & Menyuk, 2001]. The inter-channel polarization drift combines with PDL in the isolators and couplers of the erbium-doped optical amplifier subsystems, which leads to fluctuation in the power level of the channels. This power fluctuations cause performance degradations that can lead to outages [Lima Jr. et al., 2003a].

In the absence of waveform distortions due to PMD, and operation in the quasi-linear regime (that prevents inter-channel cross talk), the marks have a pulse shape that does not change overtime. We generalize a procedure introduced earlier by Winzer, et al. [Winzer et al., 2001] to show how one can derive an expression that determines the variance of the electric current due to arbitrarily polarized noise at the receiver. In this study, we neglect electrical noise at the receiver because optical transmission systems operate in the optimum regime with the use of optically preamplified receivers, which boost both the signal and the optical noise well above the electrical noise floor. The variance of the electric current σ_i^2 in the receiver has two components: one due to the noise-noise beating, and another due to the signal-noise beating. Therefore, the variance of the current at any time t has the form:

$$\sigma_i^2(t) = \langle i^2 \rangle (t) - \langle i \rangle^2 (t) = \sigma_{ASE-ASE}^2(t) + \sigma_{S-ASE}^2(t) \tag{1}$$

The first component on the right isde of Eq. (1) is the variance of the electric current due to the noise-noise beating in the receiver, and is given by

$$\sigma_{ASE\text{-}ASE}^2 = \frac{1}{2} R^2 N_{ASE}^2 \frac{I_{ASE\text{-}ASE}}{\Gamma_{ASE\text{-}ASE}} \tag{2}$$

Where

$$\Gamma_{ASE\text{-}ASE} = \frac{1}{1 + DOP_n^2} \tag{3}$$

and

$$I_{\text{ASE-ASE}} = \int_{-\infty}^{+\infty} |r_o(\tau)|^2 r_e(\tau) d\tau \tag{4}$$

and the expressions

$$r_o(\tau) = \int_{-\infty}^{+\infty} h_o(\tau') h_o^*(\tau + \tau') d\tau' \tag{5}$$

and

$$r_e(\tau) = \int_{-\infty}^{+\infty} h_e(\tau') h_e(\tau + \tau') d\tau' \tag{6}$$

are, respectively, the autocorrelation function of the optical and of the electrical filter at the receiver. In Eq. (3), DOP_n is the degree of polarization of the optical noise after the optical filter, and the noise-noise beating factor $\Gamma_{\text{ASE-ASE}}$ is the ratio between the variance of the current due to noise-noise beating (in the case that the noise is unpolarized) to the actual variance of the current due to noise-noise beating.

The second component of the variance of the electric current is due to the signal-noise beating, and is given by

$$\sigma_{\text{S-ASE}}^2(t) = R^2 N_{\text{ASE}} \Gamma_{\text{S-ASE}} I_{\text{S-ASE}}(t) \tag{7}$$

where

$$I_{\text{S-ASE}}(t) = 2\int_{-\infty}^{+\infty} e_{s_o}(\tau) h_e(t - \tau)$$
$$\times \int_{-\infty}^{+\infty} e_{s_o}^*(\tau') h_e(t - \tau') r_o(\tau - \tau') d\tau' d\tau \tag{8}$$

The coefficient ,

$$\Gamma_{\text{S-ASE}} = \frac{1}{2}\left[1 + \text{DOP}_n\left(\mathbf{s}_s \cdot \mathbf{s}_n^{(p)}\right)\right] \tag{9}$$

is the signal-noise beating factor, which is the fraction of the noise that beats with the signal.

The performance of optical fiber systems is typically quantified by the bit-error-ratio (BER) or by the Q factor [Marcuse, 1990]. The Q factor, which is defined as a function of the mean and of the variance of the electric current at the receiver for the marks and for the spaces, is given by

$$Q = \frac{\langle i_1 \rangle - \langle i_0 \rangle}{\sigma_1 + \sigma_0} \tag{10}$$

Using the Gaussian approximation, which was validated in [Winzer et al., 2001], we can use the Q factor to calculate the BER by BER $= erfc(Q/\sqrt{2})/2 \cong \exp(-Q^2/2)/(\sqrt{2\pi}Q)$. The current mean is given by

$$\langle i \rangle(t) = i_s(t) + \langle i_n \rangle(t) \tag{11}$$

where $\langle \cdot \rangle(t)$ is the average over the statistical realizations of the noise at time t. Substituting Eq. (11) and Eq. (1) into Eq. (10), we now obtain Eq. (12), where t_1 and t_0 are the sampling times of the lowest mark and the highest space, respectively [Lima Jr. et al., 2005].

$$Q = \frac{[i_s(t_1) + \langle i_n \rangle] - [i_s(t_0) + \langle i_n \rangle]}{\left(\sigma^2_{S\text{-}ASE}(t_1) + \sigma^2_{ASE\text{-}ASE}\right)^{1/2} + \left(\sigma^2_{S\text{-}ASE}(t_0) + \sigma^2_{ASE\text{-}ASE}\right)^{1/2}} \tag{12}$$

Applying the expressions that we derived for the variance of the electric current at the receiver, which accounts for arbitrary modulation format, noise polarization state, extinction ratio α_e, and receiver characteristics, the Q factor can be expressed as

$$Q = \frac{(1 - \alpha_e)\xi OSNR(\Gamma_{ASE\text{-}ASE}\mu)^{1/2}}{\left(2\Gamma_{S\text{-}ASE}\Gamma_{ASE\text{-}ASE}\kappa_1\xi OSNR + 1\right)^{1/2} + \left(2\Gamma_{S\text{-}ASE}\Gamma_{ASE\text{-}ASE}\kappa_0\alpha_e\xi OSNR + 1\right)^{1/2}} \tag{13}$$

In Eq. (13),

$$\kappa_j = \frac{RB_o I_{S-ASE}(t_j)}{i_s(t_j) I_{ASE-ASE}} \tag{14}$$

κ_j (Eq. 14) is the signal-noise beating parameter for the marks ($j = 1$) and for the spaces ($j = 0$), and

$$\mu = \frac{2B_o^2}{I_{ASE\text{-}ASE}} \tag{15}$$

μ (Eq. 15) is the effective number of noise modes for the equivalent case in which the noise is unpolarized. The expression in Eq. (15) converges to the one in [Marcuse, 1990] for the simplified integrate and dump receiver with unpolarized noise that has been widely used in the literature.

The OSNR in Eq. (13) is defined by

$$OSNR = \frac{\langle |e_s(t)|^2 \rangle_t}{N_{ASE} B_{OSA}} \tag{16}$$

Where $\langle |e_s(t)|^2 \rangle_t$ is the time-averaged noiseless optical power per channel prior to the optical filter, and B_{OSA} is the noise equivalent bandwidth of an optical spectrum analyzer (OSA) that is used to measure the optical power of the noise. The parameter ξ in Eq. (13) and Eq. (17) is the enhancement factor [Lima Jr. et al., 2003b], which is used to express the Q-factor as a function of the OSNR, and is defined the as the ratio between the signal-to-noise ratio of the electric current of the marks SNR_1 and the OSNR at the receiver. The parameter ξ' in Eq. (17) is the normalized enhancement factor, which is equal to ξ when $B_{OSA} = B_o$.

$$\xi = \frac{SNR_1}{OSNR} = \frac{i_s(t_1)}{<i_n>} \frac{N_{ASE} B_{OSA}}{<|e_s(t)|^2>_t} = \xi' \frac{B_{OSA}}{B_o} \tag{17}$$

$$\xi' = i_s(t_1) / \left[R\langle |e_{in}(t)|^2 \rangle_t \right] \tag{18}$$

For a fixed SNR, the Q-factor is a function of the DOP of the noise and of the angle between polarization states of the signal and the polarized part of the noise. If the polarization state of the signal is fixed and the polarization states of the polarized part of the noise uniformly cover the Poincaré sphere, $\hat{s} \cdot \hat{p}$ is uniformly distributed between -1 and $+1$. In this situation, the probability density function (pdf) of the Q-factor is given by [Sun et al., 2003b]

$$f_Q(q) = \frac{1}{\kappa DOP_n} \sqrt{\frac{\mu}{\Gamma_{ASE-ASE}}} \left(\frac{SNR\sqrt{\Gamma_{ASE-ASE}}\,\mu}{q^3} - \frac{1}{q^2} \right), \quad q \in [Q_{min}, Q_{max}] \tag{19}$$

where Q_{max} and Q_{min} are given by substituting $\hat{s} \cdot \hat{p} = -1$ and $\hat{s} \cdot \hat{p} = +1$ in Eq. (9) and Eq. (13).

2.1 Modelling validation with simulations

In Figs. 1 and 2, we show the validation of Eq. (13) by comparison to Monte Carlo simulations with a large number of realizations in which the Q factor is computed using the standard time-domain formula $Q = (<i_1> - <i_0>)/(\sigma_1 + \sigma_0)$. For the results in Fig. 1, we used a back-to-back 10 Gbit/s optical system with unpolarized optical noise that was added prior to the receiver using a Gaussian noise source that has a constant spectral density within the spectrum of the optical filter. Since our study is focused on the combined effect that the pulse shape and the receiver have on the system performance, we did not include transmission effects here, such as those due to nonlinearity and dispersion.

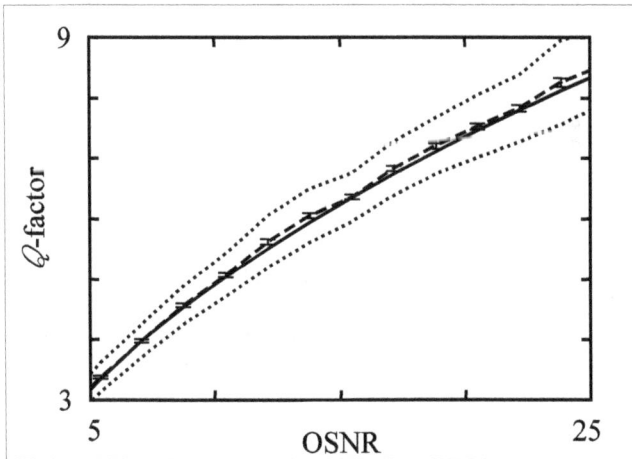

Fig. 1. Q factor as a function of the OSNR, in which the optical spectrum analyzer has a noise-equivalent bandwidth of 25 GHz. Validation of Eq. (13) (**solid line**) for the RZ raised-cosine format against Monte Carlo simulations with 100 Q samples each with 128 bits (**dashed line**). The **dotted line** shows the confidence interval in a single Monte Carlo simulation. The confidence interval is defined by the mean Q-factor plus and minus one standard deviation of the Q-factor, which gives an estimate of the error in the computation of the Q-factor using the time domain Monte Carlo method with a single string of bits.

In Fig. 1, we show the results using Eq. (13) with a solid line, which were obtained using only a single mark and a single space of the transmitted bit string. The results for the time-domain Monte Carlo method are shown with a dashed line. We obtained these results by averaging over 100 samples of the Q-factor, where for each sample the means and standard deviations of the marks and spaces were estimated using 128 bits. The agreement between the two methods is excellent.

For the results in Fig. 2, we used another back-to-back 10 Gbit/s system with partially polarized optical noise with $DOP_n = 0.5$ prior to the receiver. The partially polarized optical noise was obtained by transmitting unpolarized noise through a PDL element. We plot the Q-factor versus the OSNR for a linearly-polarized RZ raised-cosine signal with an optical extinction ratio of 18 dB. The curves show the results obtained using Eq. (13) and the symbols show the results obtained using Monte Carlo simulations. The solid curve and circles show the results when the polarized part of the noise is co-polarized with the signal.

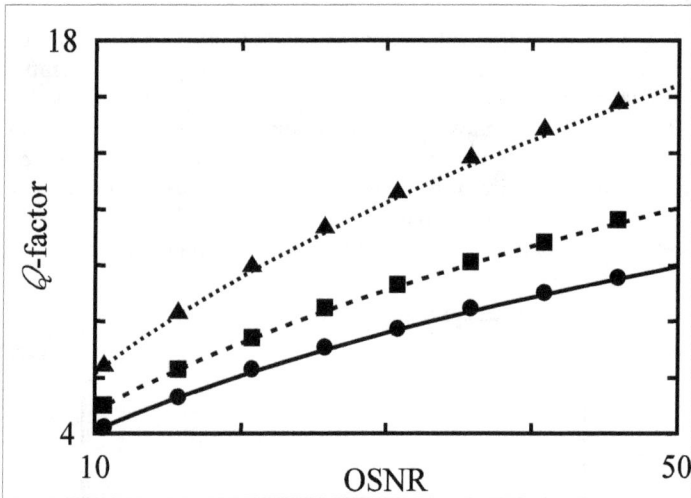

Fig. 2. Q factor as a function of the OSNR, in which the optical spectrum analyzer has a noise-equivalent bandwidth of 25 GHz. Validation of Eq. (13) (lines) with for the RZ raised-cosine format for different noise polarization states with $DOP_n = 0.5$. The solid line and the circles show results when the polarized part of the noise is co-polarized with the signal. The dashed lines and the squares and the dotted lines and triangels show results when the polarized part of the noise is in the left-circular and orthogonally polarized states to the signal, respectively.

The dashed curve and the squares, and the dotted curve and the triangles show the results when the polarized part of the noise is in the left circular and orthogonal linearly polarized states, respectively. Similarly to the results in Fig. 1, the agreement between Eq. (13) and Monte Carlo simulations in Fig. 2 is also excellent. When $DOP_n = 0.5$, the Q-factor varies by about 60% as we vary the polarization state of the noise. This variation occurs because the signal-noise beating factor $\Gamma_{S\text{-ASE}}$ in Eq. (9) depends on the angle between the Stokes vectors of the signal and the polarized part of the noise. The parameters in for this system are the same ones in Fig.1 except that $\Gamma_{ASE-ASE}$ = 0.8 and Γ_{S-ASE} = 1 for the solid line, Γ_{S-ASE} = 0.5 for the dashed line, and Γ_{S-ASE} = 0.25 for the dotted line. These results illustrate the significant impact that partially polarized noise can have on the performance of an optical fiber transmission system. Typical values for the PDL per optical amplifier in optical fiber systems range from 0.1 dB to 0.2 dB, which can partially polarize the optical noise in the transmission line.

2.2 Modelling validation with experimental results

In Fig. 3 we present a validation of Eq. (13) by comparison with back-to-back 10 Gbit/s experiments. The Q-factor versus the OSNR is obtained using both simulations and experiments for RZ and NRZ signals with unpolarized optical noise (DOP_n < 0.05) that is generated by an erbium-doped fiber amplifier without input power [Lima Jr. et al., 2005],[Sun et al., 2003b]. In Fig.3, the curves show results obtained using Eq. (13) and the symbols show the experimental results. The dot-dashed curve and the diamonds show the results for an RZ format with the electrical filter. The solid curve and circles show the results for the RZ format without the electrical filter. The dashed curve and squares show the results for the NRZ format with the electrical filter, and the dotted curve and triangles show the results for the NRZ format without the electrical filter. The parameters in Eq. (13) for the modulation formats shown in Fig.3 are described in Table 1.

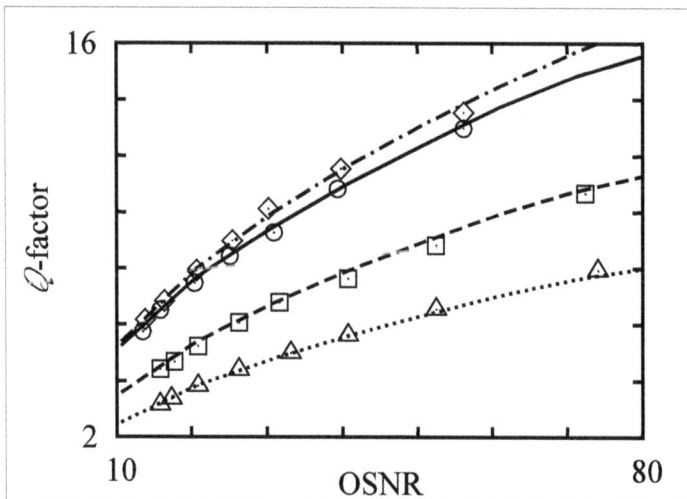

Fig. 3. Validation of Eq. (13) (lines) with experimental results (symbols). The dotted–dashed curve and the diamonds show the results for the RZ format with an electrical filter with a 3-dB bandwidth of 7 GHz. The solid curve and circles show the results for the RZ format without the electrical filter. The dashed curve and the squares show the results for the NRZ format with an electrical filter with a 3-dB bandwidth of 7 GHz. The dotted curve and the triangles show the results for the NRZ format without the electrical filter.

In Fig.3, we show that the performance of the RZ format is less sensitive than is the performance of the NRZ format to variations in the characteristics of the receiver. Since the

Format	α_e (dB)	ξ'	ξ	K_1	K_0	M
RZ with EF	−18.0	3.49	0.44	3.51	3.51	38.8
RZ w/o EF	−18.0	5.91	0.74	3.17	3.17	17.7
NRZ with EF	−11.3	1.89	0.24	2.88	2.68	38.8
NRZ w/o EF	−11.9	1.95	0.25	2.81	2.79	17.7

Table 1. Parameters of the modulation formats used in Fig. 3 with and without electrical filter (EF).

noise is unpolarized, $\Gamma_{ASE\text{-}ASE} = 1$, and $\Gamma_{S\text{-}ASE} = 0.5$. The results that we obtain using the formula Eq. (13) are in good agreement with the experimental results shown in this figure. An increase of the bandwidth of the electrical filter increases the amount of noise in the decision circuit which degrades the system performance. On the other hand, for systems with a 10 Gbit/s RZ format, increasing the electrical bandwidth from 7 to 15 GHz also reduces the broadening of the RZ pulses, and thereby increases the electric current due to the signal in the marks. However, this same effect does not occur in systems that use the NRZ format, since the NRZ pulses have a much narrower bandwidth.

In Fig. 4, we plot the Q-factor versus $\hat{s} \cdot \hat{p}$ when the noise is highly polarized and when it is partially polarized. The details of the experimental setup and schematic diagram are given in [Sun et al., 2003b].

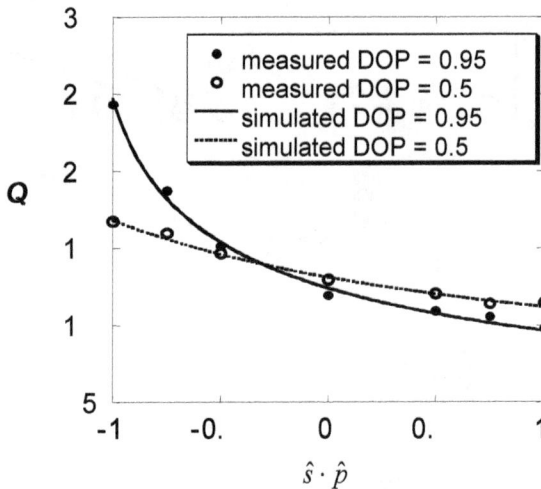

Fig. 4. The Q-factor plotted as a function of $\hat{s} \cdot \hat{p}$ [Sun et al., 2003b].

The experimental and analytical results we obtained when the DOP of the noise was set to 0.95 are shown with filled circles and a solid curve respectively. The corresponding results when the DOP of the noise is 0.5 are shown with open circles and a dotted curve. The agreement between theory and experiment is excellent. In both cases, the largest Q value occurs when the signal is antipodal on the Poincaré sphere to the polarized part of the noise and the signal-noise beating is weakest. Similarly, the smallest Q value occurs when the signal is co-polarized with the polarized part of the noise and the signal-noise beating is strongest. Furthermore, as $\hat{s} \cdot \hat{p}$ is varied from −1 to +1 the variation in Q is less when the noise is partially polarized than when it is highly polarized.

In Fig.5, we measured the distribution of the Q-factor where the samples were collected using 200 random settings of the polarization controller (PC), chosen so that the polarization state of the polarized part of the noise uniformly covered the Poincaré sphere. The details of the experimental setup and schematic diagram are given in [Sun et al., 2003b]. We measured the Q-distribution when the DOP of the noise was $DOP_n = 0.05, 0.25, 0.5, 0.75$ and 0.95 when SNR = 12.3. In Fig. 5, we show the histogram of the measured Q-factor distribution with bars when $DOP_n = 0.5$, the corresponding result obtained using Eq. (19) with a solid curve, and the results obtained using a Monte Carlo simulation with 10,000 samples with a dotted curve. In the simulation, we chose the polarization states of the signal and of the polarized noise prior to the PC to be (1, 0, 0) in Stokes space and we used a random rotation after the polarized noise to simulate the PC. The 10,000 random rotations were chosen so that the polarization state of the polarized noise uniformly covered the Poincaré sphere. The theoretical and simulation results both agree very well with the experimental result. The sharp cut-offs in the Q-distribution at $Q = 11.4$ and $Q = 17$ correspond to the cases that the signal is respectively parallel and antipodal on the Poincaré sphere to the polarized part of the noise. The width $Q_{max} − Q_{min}$ of the Q-distribution depends on the DOP of the noise.

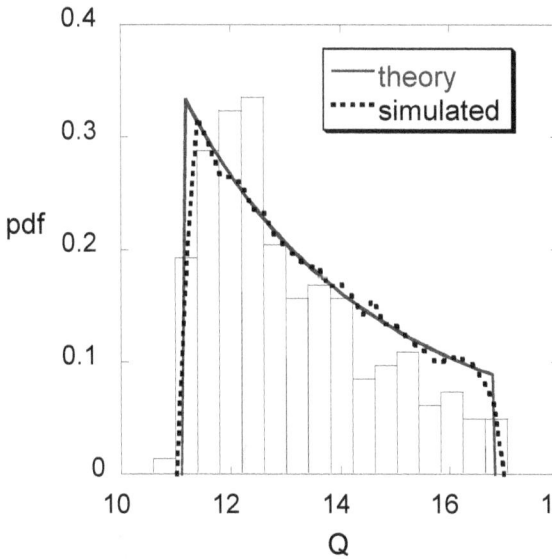

Fig. 5. The Q-factor distribution when $DOP_n = 0.5$ [Sun et al., 2003b].

In Fig. 6, we show the Q_{max}, Q_{min} and average Q factors as a function of the DOP of the noise, obtained both from measurements and analytically Eq. (19) [Sun et al., 2003b]. Although the average Q is not sensitive to a change in the DOP of the noise, the maximum and minimum Q values change dramatically with the DOP of the noise, especially the maximum Q values. The results shows that highly polarized noise will cause larger system variation than unpolarized noise.

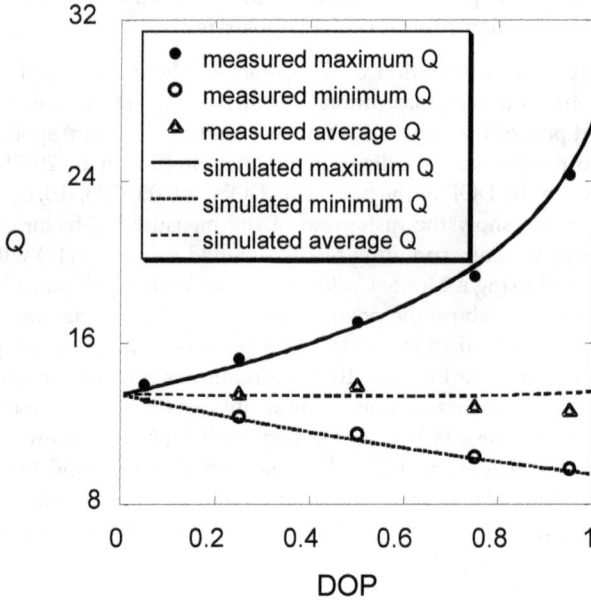

Fig. 6. The variation of the Q-factor as a function of the DOP of the noise. [Sun et al., 2003b].

The application of Eq. (13) for a particular system can enable the calculation of the power margin that can be allocated to different impairments and the calculation of the outage probability. This semi-analytical model can be combined with the reduced Stokes parameters model in [Wang & Menyuk, 2001], [Lima Jr. et al., 2003a] to determine the performance degradation that results from the combination of PDL and inter-channel PMD in transoceanic optical fiber transmission systems.

3. Modelling systems with significant intra-channel PMD

PMD is a polarization impairment that limits the data rate increase to 40 Gbit/s in a significant number of the optical fiber links built with high PMD coefficient fibers. PMD causes random waveform distortions that can produce outages in the communication channel. Because PMD distorts the waveform and leads to pattern dependences and even to inter-symbol interference, the BER cannot be calculated through the direct application of Eq.(10) and Eq.(13). Using the Gaussian approximation for each bit of a sufficiently long bit string enables the BER to be accurately calculated by [Lima Jr. & Oliveira, 2009]

$$
\begin{aligned}
\mathrm{BER}(t_s, i_{\mathrm{th}}) = \\
\frac{1}{N_0} \sum_{j=1}^{N_0+N_1} I_0(t_s + jT) \mathrm{erfc}\left[\frac{i_{\mathrm{th}} - i_s(t_s + jT) - \langle i_n \rangle}{\sqrt{2}\sigma_i(t_s + jT)}\right] \\
+ \frac{1}{N_0} \sum_{j=1}^{N_0+N_1} I_0(t_s + jT) \mathrm{erfc}\left[\frac{i_s(t_s + jT) + \langle i_n \rangle - i_{\mathrm{th}}}{\sqrt{2}\sigma_i(t_s + jT)}\right]
\end{aligned}
\tag{20}
$$

The instantaneous variance of the electric current in the receiver is given by,

$$\sigma_i^2(t) = \sigma_{\text{s-ASE}}^2 + \sigma_{\text{ASE-ASE}}^2 + \sigma_{\text{elec}}^2 \tag{21}$$

The first two terms in the right-hand-side of Eq. (21) are the signal-noise beating, and the noise-noise beating, respectively, the third term is due to the electrical noise in the receiver. Both the mean current due to noise in Eq. (20) and the noise-noise beating in Eq. (21) were computed as in Section 2. Because intra-channel PMD depolarizes the signal, the signal-noise beating must be computed using any two orthogonal decomposition of the Jones vector of the signal, which for unpolarized signal is given by [Lima Jr. & Oliveira, 2009]

$$\sigma_{\text{s-ASE}}^2(t) = R^2 N_{\text{ASE}} \times \left\{ \begin{array}{l} \iint e_x(\tau) h_e(t-\tau) e_x^*(\tau') h_e(t-\tau') r_o(t-\tau') d\tau' d\tau \\ + \iint e_y(\tau) h_e(t-\tau) e_y^*(\tau') h_e(t-\tau') r_o(t-\tau') d\tau' d\tau \end{array} \right\} \tag{22}$$

In Eq. (22), $e_x(t)$ and $e_y(t)$ are the horizontally and the vertically polarized components of the optically filtered noise-free signal, respectively, N_{ASE} is the noise spectral density prior to the optical filter, and R is the responsivity of the photodetector. The function $r_o(t)$ is the autocorrelation function of the impulse response of the optical filter and $h_e(t)$ is the impulse response of the electrical filter.

3.1 Simulation results

The power penalty was used as the performance measure. Once the BER in Eq. (20) is computed, the power penalty is calculated. The power penalty is defined as the input power increase in the system that produces the same performance observed in a PMD-free system that has optimized receiver filter bandwidths. The electrical filter bandwidth is defined as the 3-dB bandwidth and the optical filter bandwidth is specified as the full-width at half maximum (FWHM). The outage probability is the probability that the power penalty will exceed a specified penalty margin.

Using Eq. (21) into the value of σ_i^2 in Eq. (20), and considering unpolarized optical noise, we calculate the BER for 10 Gbit/s NRZ and raised-cosine RZ systems with optimized receiver filters. We consider -8 dBm of input optical signal, an optical noise spectral density of 0.60μW/GHz, and assuming a receiver with an equivalent electrical noise density of 31.5pW/Hz$^{1/2}$. The inclusion of the electrical noise is necessary in this study because its contribution increases with the electrical bandwidth. Since PMD is a linear effect, these results can be rescaled to 40 Gbit/s or to any other data rate. In Fig. 7, we show results of the power penalty with respect to the optimized receiver as a function of the receiver filter bandwidths. The optimized performances without PMD were obtained with optical filters with FWHM of 10 GHz for the NRZ format and 12 GHz for the RZ format, which are so narrow that they could result in additional penalty to the system due to detuning of the laser source wavelength, and the 3-dB electrical filter bandwidth was 12 GHz for both modulation formats. These results agree with earlier studies indicating the greater robustness of RZ systems when compared with NRZ systems with respect to the receiver characteristics [Winzer et al., 2001]. The performance advantage of RZ over the NRZ format is due to the larger enhancement factor that is characteristic of modulation formats with short duty cycle.

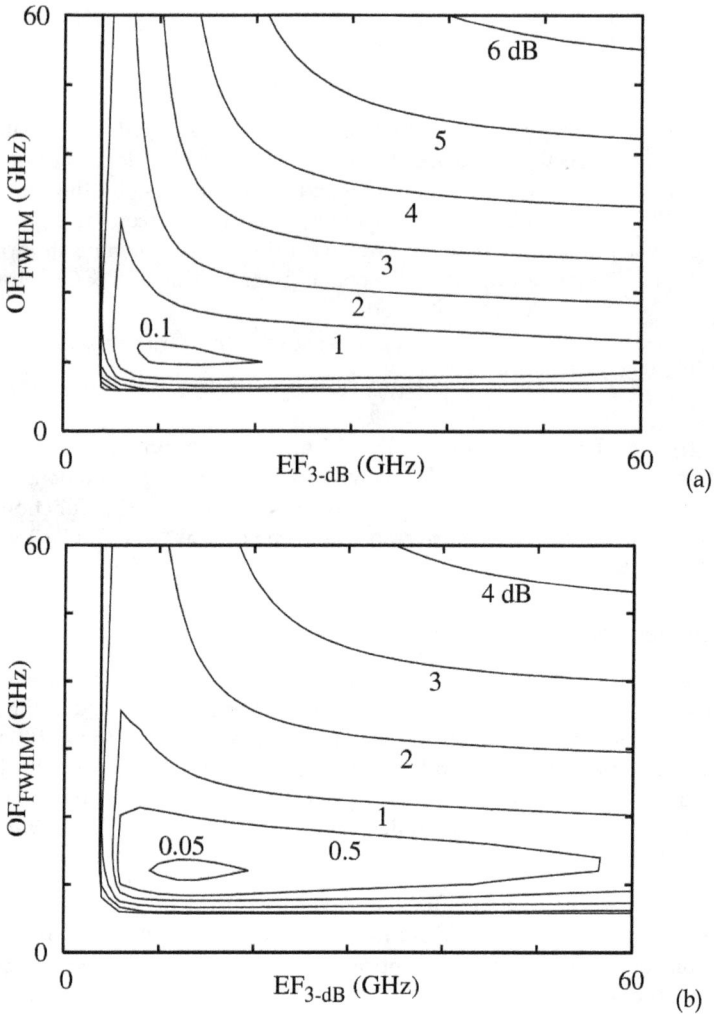

Fig. 7. Power penalty for (a) an NRZ system and (b) and RZ system with 10 Gbit/s without PMD as a function of the receiver filter bandwidths. The horizontal axis is the 3-dB bandwidth of the electrical filter and the vertical axis is the FWHM of the optical filter.

In Fig. 8, we use importance sampling in the Monte Carlo simulations of PMD [Biondini et al., 2002], [Oliveira et al., 2003] combined with the semi-analytical model in Eq. (13) to calculate the power penalty with respect to the optimized receiver at 10^{-5} outage probability level for the NRZ and raised-cosine RZ systems operating in a transmission fiber system with 10 ps of mean DGD (10% of the bit period). We observed that there is little difference between the optimum receiver filter bandwidths in the system with PMD and with PMD-free operation. In Fig. 8, we also observed a decrease of the robustness of the RZ system with respect to the receiver filter bandwidths. This effect results from the PMD-induced pulse broadening, which makes the RZ pulses to become similar to NRZ pulses.

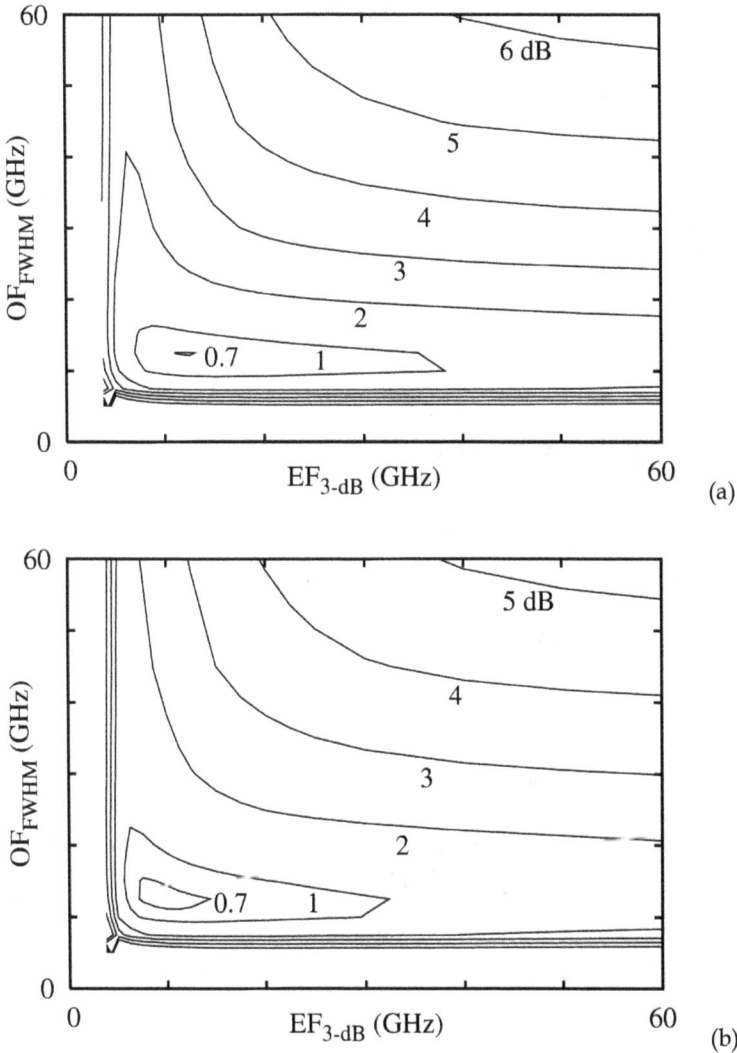

Fig. 8. Power penalty for (a) an NRZ system and (b) and RZ system with 10 Gbit/s with a mean DGD of 10 ps as a function of the receiver filter bandwidths. The horizontal axis is the 3-dB bandwidth of the electrical filter and the vertical axis is the FWHM of the optical filter.

In Fig. 9, we show how an NRZ system with the receiver filters optimized for PMD-free operation and the receiver filters optimized for operation with mean DGD of 10 ps perform under different mean DGD values. Therefore, this system optimized for operation in the presence of PMD is operating in the sub-optimum regime in the cases in which the actual mean DGD is different from 10 ps. We observed only a small difference in the performance with the two optimized sets of filters, which reflects the small difference of the optimized receiver filter bandwidths for these two cases.

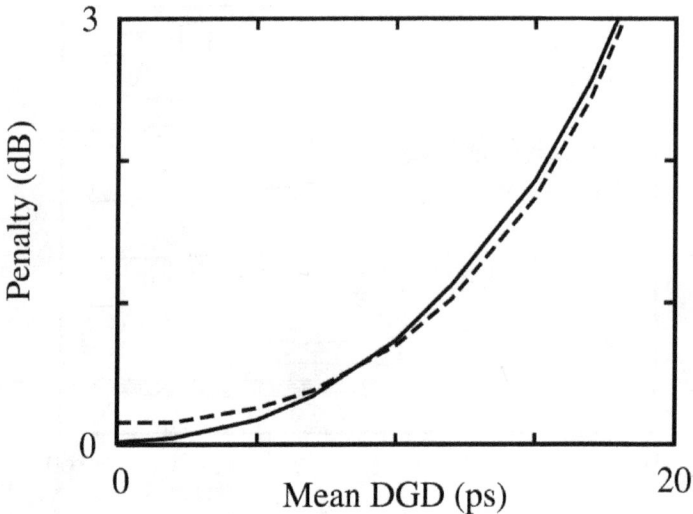

Fig. 9. Power penalty as a function of the mean DGD for an NRZ system. The solid line shows results for this system with optimized filter bandwidths in the absence of PMD. The dashed line shows results for the optimized filters for 10^{-5} outage probability in a system with mean DGD of 10 ps (10% of the bit period).

4. Conclusions

We used laboratory experiments and Monte Carlo simulations to show how one can use a semi-analytical receiver model to accurately calculate the Q factor for systems with arbitrary optical pulse shapes, arbitrary receiver characteristics, and arbitrary polarized noise. Our results showed that the system variation caused by partially polarized noise depends not only on the angle between the signal and polarized part of the noise but also on the DOP of the noise. Highly polarized noise will cause larger variation in the system performance. Our results suggest that in order to reduce the variation of the system performance, one needs to keep the noise unpolarized. The receiver model that we developed is also used to determine the performance degradation due to intra-channel PMD in optical fiber communication systems, and to show that the receiver filter bandwidths optimized for optical fiber systems at 10^{-5} outage probability due to PMD are very close to the ones optimized for the same systems in the absence of PMD. We observed that the PMD-induced waveform distortions significantly reduce the robustness of the RZ formats to the receiver characteristics. The receiver model that we developed can also be used to efficiently determine the performance degradation of optical fiber communication systems due to the combination of inter-channel PMD and PDL using the simplified reduced Stokes model.

5. Acknowledgment

The author thanks Dr. Yu Sun and Dr. Ivan Lima Jr. for helpful discussions, and for allowing the results of joint publications with the author (Dr. Aurenice Oliveira) to be used in this book chapter.

6. References

G. Biondini, W. L. Kath, and C. R. Menyuk, "Importance sampling for polarization-mode dispersion," *IEEE Photon. Technol. Lett.*, vol. 14, pp. 310-312, 2002.

B. Huttner, C. Geiser, and N. Gisin, "Polarization-induced distortions in optical fiber networks with polarization-mode dispersion and polarization-dependent losses," *IEEE J. Selec. Topics Quantum Electron.*, vol. 6, no. 2, pp. 317-329, Mar.-Apr. 2000.

P. A. Humblet and M. Azizoglu, "On the bit error rate of lightwave systems with optical amplifiers," *IEEE/OSA J. Lightwave Technol.*, vol. 9, no. 11, pp. 1576-1582, Nov. 1991.

R. Khosravani, I. T. Lima Jr. P. Ebrahimi, E. Ibragimov, A. E. Willner, and C. R. Menyuk, "Time and frequency domain characteristics of polarization-mode dispersion emulators," *IEEE Photon. Technol. Lett.*, vol 13, no. 2, Feb. 2001.

I. T. Lima Jr., A. M. Oliveira, Y. Sun, H. Jiao, J. Zweck, C. R. Menyuk, and G. M Carter, "A receiver model for optical fiber communication systems with arbitrarily polarized noise," IEEE/OSA J. Lightwave Technol., vol. 23., no. 3, pp. 1478-1490, Mar. 2005.

I. T. Lima Jr., A. M. Oliveira, J. Zweck, and C. R. Menyuk, "Efficient computation of outage probabilities due to polarization effects in a WDM system using a reduced Stokes model and importance sampling," *IEEE Photon. Technol. Lett.*, vol. 15, no. 1, pp. 45-47, Jan. 2003.

I. T. Lima Jr. and A. M. Oliveira, "Optimum receiver filters for optical fiber systems with polarization mode dispersion," IEEE/OSA J. Lightwave Technol., vol. 27, no. 14, pp. 2886-2891, Jul. 2009.

I. T. Lima Jr., A. M. Oliveira., J. Zweck, and C. R. Menyuk, "Performance characterization of chirped return-to-zero modulation format using an accurate receiver model," *IEEE Photon. Technol. Lett.*, vol. 15, no. 4, pp. 608-610, Apr. 2003.

D. Marcuse, "Derivation of analytical expression for the bit-error-probability in lightwave systems with optical amplifiers," *IEEE/OSA J. Lightwave Technol.*, vol. 8, no. 12, pp. 1816-1823, Dec. 1990.

A. M. Oliveira, I.T. Lima Jr., C. R. Menyuk, G. Biondini, B. S. Marks, and W. L. Kath, "Statistical analysis of the performance of PMD compensators using multiple importance sampling," *IEEE Photon. Technol. Lett.*, vol. 15, no. 12, pp. 1716-1718, Dec. 2003.

Y. Sun, A. M. Oliveira, I. T . Lima Jr., J. Zweck, L. Yan, C. R. Menyuk, and G. carter, "Statistics of the system performance in scrambled recirculating loop with PDL and PDG," *IEEE Photon. Technol. Lett.*, vol. 15, no. 8, pp. 1067-1069, Aug. 2003.

Y. Sun, I. T. Lima Jr., A. M. Oliveira, H. Jiao, J. Zweck, L. Yan, C. R. Menyuk, and G. Carter, "System performance variations due to partially polarized noise in a receiver," *IEEE Photon. Technol. Lett.*, vol. 15, no. 11, pp. 1648-1560, Nov. 2003.

D. Wang and C. R. Menyuk. "Calculation of penalties due to polarization effects in a long-haul WDM system using a stokes parameter model," *IEEE/OSA J. Lightwave Technol.*, vol 19, no. 4, pp. 487-494, Apr 2011.

P. Winzer, M. Pfnnigbauer, M. M. Strasser, and W. R. Leeb, "Optimum filter bandwidth for optically preamplified NRZ receivers," IEEE/OSA J. Lightwave Technol., vopl. 19, no. 9, pp. 1263-1273, Sep. 2001.

14

Designing WAN Topologies Under Redundancy Constraints

Pablo Sartor Del Giudice and Franco Robledo Amoza
Engineering School - Universidad de la República
Uruguay

1. Introduction

The huge amount of data that can be transported by fiber lines when compared to former existing networks of telephone lines introduced many new challenges when it comes to the design of network topologies. Given the important costs incurred when deploying and then operating such lines and their unprecedented bandwith capacities, "tree-like" topologies are usually sufficient to provide the required information flow while having minimal costs. But such topologies are extremely vulnerable; the loss of one single fiber link (or even worst, the failure of a switching site) might split the entire network into two or more disconnected components. Therefore the problem of designing or expanding an existing fiber Wide Area Network (WAN) involves two antagonistic objectives. A certain level of redundancy is to be achieved to keep certain sites connected in case of eventual failures in components; while at the same time, it is desirable to lower as much as possible the costs associated with fiber deployment and operation, thus leading to the problem of choosing which of subset of the feasible links to deploy. Depending on the particular application, redundancy requirements can consider that switch sites could fail, or assume that these are fault-tolerant and that only the failure of fiber lines is possible. Graph Theory is a field of mathematics useful for designing networks and analyzing their properties. In particular, the problem known as "Generalized Steiner Problem" (GSP) is very suitable for modelling the mentioned antagonistic objectives. It has been shown to be a quite complex NP combinatorial complexity class problem, for which the use of heuristic algorithms is mandatory to solve real general cases with reasonable usage of computer resources. In this chapter it is shown how the GSP can be solved by applying combinatorial optimization metaheuristics both for the node-connected and the edge-connected versions. The underlying context is that a number of existing sites that we will call "fixed sites" are to be connected among themselves (making optional use of existing intermediate switch entities if convenient) through fiber lines whose deployment and operation involve specific costs that are to be minimized; while at the same time the amount of component failures to tolerate is a specific requirement for every pair of fixed sites. Suitable algorithms are proposed for generating low cost designs with reasonable use of computer resources and some of their properties are analyzed. Results of test involving real network topologies are presented showing that this approach generates optimal or near-optimal topologies. Finally limitations, conclusions and current research lines on these topics are presented.

2. Context and problem definition

In general, a typical WAN backbone network has a meshed topology, and its purpose is to allow efficient and reliable communication between the switch sites of the network that act as connection points for the local access networks (eventually incorporating other switch sites for efficiency purposes). The topological design of a WAN basically consists of finding a minimum cost topology which satisfies some additional requirements, generally chosen to improve the survivability of the network (that is, its capacity to resist the failures of some of its components). One way to do this is to specify a connectivity level, and to search for topologies which have at least this number of disjoint paths (either edge disjoint or node disjoint) between pairs of switch sites. In the most general case, the connectivity level can be fixed independently for each pair of switch sites (heterogeneous connectivity requirements). This problem can be modelled as a *Generalized Steiner Problem* (denoted by GSP) and it is an NP-Complete problem (Steiglitz et al., 1969; Winter, 1986; 1987). We present the formal definition of this problem later in this section. Some references in this area are (Agrawal et al., 1995; Baïou, 1996; Balakrishnan et al., 2004; Chopra, 1992; Goemans & Bertsimas, 1993; Grötschel et al., 1995; Ko & Monma, 1989; Robledo & Canale, 2009). Most of these works are either focused on the edge-disjoint flavor of the problem, or on the exploration of particular cases, for example, when it is required to have two disjoint paths between all pairs of distinguished switch sites, which is called the 2-survivability problem (Baïou, 1996). In (Kerivin & Mahjoub, 2005; Stoer, 1992), extensive surveys over high survivability models are introduced. We will denote by GSP-NC and GSP-EC the GSP versions with node-connectivity constraints and edge-connectivity constraints respectively. Topologies verifying edge-disjoint path connectivity constraints assure that the network can survive to failures in the connection lines; whereas node-disjoint path constraints assure that the network can survive to failures both in switch sites as well as in the connection lines.

Winter (Winter, 1985; 1986; 1987) demonstrated that the GSP can be solved in linear time if the network is series-parallel, outerplanar or a Halin graph. Here follows a summary of the survivability problems related to the GSP. Gröstchel, Monma and Stoer (Grötschel & Monma, 1990) consider a particular case of the GSP working on a slightly different context where different types of node exist, representing a hierarchy of fault-tolerance requirements; they called it the NCON problem. In (Stoer, 1992), Stoer gives an extensive survey for the NCON and the ECON (the version with edge-connectivity constraints), and some particular cases. In the NCON (resp. ECON) each node i has an associated nonnegative integer r_i, the **type** of i (the survivability requirement or "importance" of a node is modeled by node types). The GSP model generalizes the NCON(ECON) model since in the GSP there exist general survivability requirements r_{ij} that are specified for each pair i, j of fixed nodes independently. Nevertheless, Grötschel, Monma and Stoer (Grötschel et al., 1991; Grötschel et al., 1992a;b; 1995) introduce the use of node types to define survivability requirements based on the premise that these adequately express the relative importance placed on maintaining connectivity between offices and they classify the different problem types according to the largest occurring node type and according to whether the node types represent node or edge connectivity requirements. Let us note that there exist many specializations of the survivability problems which can be formulated by varying its parameters (the required amount of disjoint paths to connect pairs of sites, general, euclidean, uniform or other hipothesis about costs, etc). There exist polynomially solvable cases of the NCON and ECON problems. They result from relaxing the original problem with restrictions like uniform costs, 0/1 costs, restricted node types, and special underlying graphs such as outerplanar, series-parallel, and Halin graphs. All these particular cases are referenced and briefly exposed in (Stoer, 1992). On the other hand, lower bounds and heuristics with worst-case guarantees for kECON[1] problems were found for restricted costs, e.g., uniform costs or costs satisfying the triangle inequality, as well as very important results on the structure of optimal survivable networks for this cost structure. Details of these works can be seen in (Bienstock et al., 1990; Cheriyan et al., 2001; Chou & Frank, 1970; Frank & Chou, 1970; Frederickson & Jàjà, 1982; Goemans & Bertsimas, 1993; Goemans & Williamson, 1992; Monma et al., 1990) and in a summarized form in (Stoer, 1992). Unfortunately, there exist few exact algorithms for the NCON and ECON for general costs. Christofides and Whitlock (Christofides & Whitlock, 1981) introduce a cutting plane algorithm together with branch-and-bound for ECON problems where the connection levels are specified for each pair of nodes. Chopra and Gorres (Chopra, 1992) give a cutting plane algorithm mixed with branch-and-bound for solving 2ECON problems.

In the literature there are several works related to approximation algorithms for the GSP and different particular cases. Next, we will introduce a survey of the main existing algorithms based on this approach. In (Ravi & Klein, 1993) the authors show how to obtain approximately optimal solutions to 2-edge-connected versions of the problems addressed in (Goemans & Williamson, 1992). Subsequent papers (Gabow et al., 1993; Goemans et al., 1994; Williamson et al., 1995) extended these methods to give approximation algorithms for the GSP-EC without link duplication. Agrawal, Klein and Ravi (Agrawal et al., 1995) developed an algorithm for the GSP-EC with performance guarantee of $2\lceil \log_2(r_{max} + 1)\rceil$, where r_{max} is the highest requirement value. More recently Jain (Jain, 2001) presented a factor 2 approximation algorithm for the GSP-EC. Kortsarz, Krauthgamer and Lee (Kortsarz et al., 2004) introduced the first strong lower bound on the approximability of the GSP when there are no Steiner nodes (i.e. all sites are fixed). An important special case of the GSP occurs when we are searching the minimum-cost k-node-connected subgraph spanning all the nodes. In first place, let us see the general case. In (Cheriyan et al., 2001; 2002; Czumaj & Lingas, 1999; Kortsarz et al., 2004; Kortsarz & Nutov, 2003; Ravi & Williamson, 1997; 2002) the authors propose several approximation algorithms for the problem of finding a minimum-cost k-node-connected spanning subgraph, besides they give their respective approximation ratios. For $k \leq 7$ an approximation ratio of $\lceil (k + 1)/2 \rceil$ is known; see (Khuller & Raghavachari, 1996) for $k = 2$, (Auletta et al., 1999) for $k = 2, 3$, (Jain, 1999) for $k = 4, 5$, and (Kortsarz & Nutov, 2003) for $k = 6, 7$. Other approximations for $k = 2$ can be seen in (Böckenhauser et al., 2002; Csaba et al., 2002). Furthermore, in (Czumaj & Lingas, 1999), (Cheriyan & Thurimella, 2000) and (Kortsarz & Nutov, 2003) the authors respectively supply approximation algorithms for the following special cases: the graph has complete Euclidean topology, uniform costs, and metric costs (i.e. when the costs satisfy the triangle inequality).

Finally, let us see works related to the particular case named "Steiner two-node-survivable network problem", (denoted by STNSNP). In (Baïou, 1996) the author mentions different problems related directly to the STNSNP. In particular, the problems known as the Steiner 2-edge-connected subgraph problem (STECSP), the Steiner 2-node-connected subgraph problem (STNCSP) and the Steiner 2-edge-survivable network problem (STESNP). The STNSNP (resp. STESNP) also corresponds to the problem kNCON (resp. kECON) in the case where all nodes have a connectivity level requirement belonging to $\{0, 2\}$. Given a graph $N = (X, U)$, a subset $T \subseteq X$ and a matrix C of connection costs associated to U;

---

[1] ECON problems where there are at least two nodes with connectivity requirement k.

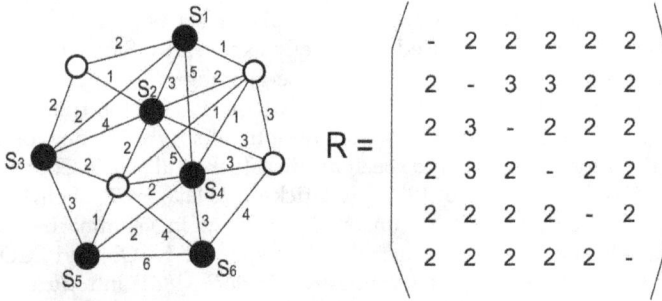

Fig. 1. Example instance for the GSP

the objective in the STNCSP (resp. STECSP) is to find a minimum-cost 2-node-connected (resp. 2-edge-connected) subgraph spanning the set of nodes T. If the matrix C is positive, the sets of optimal solutions associated to the STNSNP and STNCSP are equal. Idem the sets of optimal solutions associated to the STESNP and STECSP. If all the nodes are fixed (there are no Steiner nodes) the problems STESNP and STECSP coincide, and also the STNSNP with the STNCSP. Moreover, it is easy to see that all feasible solution of the STNCSP (resp. STECSP) is also feasible for the STNSNP (resp. STESNP). In (Coullard et al., 1991) the authors developed a linear algorithm to solve the STNCSP in the case of graphs without W_4 (a wheel graph with four nodes) and Halin graphs. The authors of this chapter have previously developed a parallel method (of worst case exponential complexity) for the general case (Cancela et al., 2005). Other works related to particular cases of the STNCSP, e.g. when $T = X$ or uniform costs, already have been mentioned above.

2.1 Problem formalization and definitions

We will formalize our optimal network design problem by using the following notation:

- $G = (V, E, C)$: Simple undirected graph with weighted edges, modelling feasible links;
- V : Nodes of G, representing fixed sites and intermediate optional sites to connect;
- E : Edges of G, representing feasible links between nodes;
- $C : E \rightarrow \mathbb{R}^+$: Edge weights, representing the cost of deploying and operating each link;
- $T \subseteq V$: Terminal nodes (representing the set of fixed sites, i.e. the ones that have non-zero connectivity requirements with at least one other node);
- $R : R \in \mathbb{Z}^{|T| \times |T|}$: Symmetrical integer matrix of connectivity requirements; $r_{ij} = r_{ji} \geq 0, \forall i, j \in T; r_{ii} = 0, \forall i \in T$.

We will model our design problem as a Generalized Steiner Problem (GSP) whose definition is as follows.

Definition 2.1. *GSP. Given the graph G with edge weights C, the teminals set T and the connectivity requirements matrix R, the objective is to find a minimum cost subgraph $G_T = (V_T, E_T, C_T)$ of G where C_T is the restriction of C to the subset T and every pair of terminals i, j is connected by r_{ij} disjoint paths.*

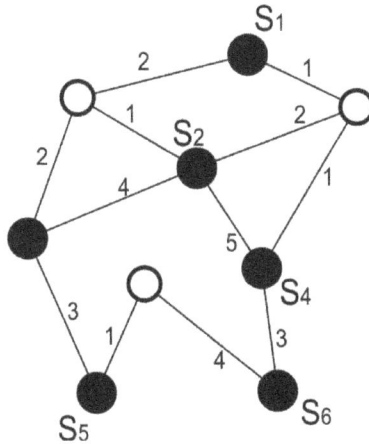

Fig. 2. Solution for the GSP instance

Two different versions of the problem arise depending on the way "disjoint" is interpreted above. If it refers to node-disjoint paths we will denote it as GSP-NC (node-connected); if it refers strictly to edge-disjoint paths (allowing to share nodes) then we will denote the problem as GSP-EC (edge-connected). This versions will allow us to model situations in which only link-failure tolerance is required (GSP-EC) or situations in which site-failure tolerance is required (GSP-NC). An example instance of the GSP is shown in Figure 1. There are six fixed switch sites, colored black and labeled S1, S2, S3, S4, S5 and S6, and four non-fixed switch sites, colored white. The connections that can be potentially deployed are shown in the figure, annotated with their costs. The matrix R shows the connectivity requirements among the fixed sites, ranging in this case from 2 to 3. Figure 2 shows a solution of this instance having cost 29; note that only three of the four non-fixed sites were used. Due to the enormous intrinsic complexity of the GSP, exact algorithms to solve it (i.e. that guarantee that optimal solutions are built) can only be applied under specific circumstances and/or on small instances (a few sites); it is known to be an NP complexity class combinatory problem. Therefore, to deal with real general problems, the use of heuristic algorithms conceived to generate good quality solutions within reasonable time and use of computing power resources turns to be mandatory.

2.2 The GRASP metaheuristic

GRASP (Greedy Randomized Adaptive Search Procedure) is a metaheuristic which proved to perform very well for a variety of combinatorial optimization problems; we will make use of it to solve the GSP. A GRASP is a "multistart local optimization" procedure which performs two consecutive phases in each iteration:

- Construction Phase: it builds a feasible solution that chooses (following some randomized criterion) which elements to add from a list of candidates defined with some greedy approach;

Procedure GRASP(*MetaParams, MaxIter, RndSeed*)

1: *bestSol* ← *NIL*
2: **for** $k = 1$ to *MaxIter* **do**
3:     *greedySol* ← *ConstPhase*(*MetaParams, RndSeed*)
4:     *localSearchSol* ← *LocalSearchPhase*(*greedySol*)
5:     **if** *cost*(*localSearchSol*) < *cost*(*bestSol*) **then**
6:         *bestSol* ← *localSearchSol*
7:     **end if**
8: **end for**
9: **return** *bestSol*

Fig. 3. GRASP pseudo-code

- Local Search Phase: it explores the neigborhood[2] of the feasible solution delivered by the Construction Phase to reach a local optimum.

Figure 3 presents a generic GRASP pseudo-code.   The procedure inputs include metaparameters *MetaParams* which set the size of the list of candidates and other behaviour of the *ConstPhase* procedure; the amount of iterations to run *MaxIter*; and a seed for random number generation.   After having run *MaxIter* iterations the procedure returns the best solution found. Details of this metaheuristic can be found in (Resende & Ribeiro, 2003). In the next sections we introduce algorithms for implementing the Construction and Local Search Phases suitable to solve the GSP-EC (edge-connected version) as well as comment any changes necessary for adapting them also to the GSP-NC problem.

2.3 Construction phase algorithm

Our construction phase algorithm proceeds by building a graph which satisfies the requirements of the matrix R; it starts with an edgeless graph and in each iteration one new path is added to the solution under construction. The algorithm is shown in Figure 4. It takes as inputs the graph G of feasible edges, the edge costs C, the set of terminal nodes T and the matrix of requeriments R. In line 1 we initialize the solution graph under construction G_{sol} with the nodes of T and no edges; the matrix $M = (m_{ij})_{i,j \in T}$ which records the amount of connection requirements not yet satisfied in G_{sol} between the terminal nodes i and j; the sets P_{ij} that will be used to record the r_{ij} disjoint paths found for connecting the nodes i, j; and an auxiliary matrix $A = \{A_{ij}\}$ used to record how many times it was impossible to find one more path between two terminal nodes i, j whose requirements r_{ij} were not yet covered. In line 2 we alter the costs of the matrix C in order to make the algorithm satisfy a property that we describe below (together with the altering function used) and introduce random.

Loop 3-15 is repeated until all terminal nodes have their connectivity requirements satisfied, or until for a certain pair of terminals i, j, the algorithm fails to find a path a certain number of times MAX_ATTEMPT. Each iteration works the following way. Line 4 selects two terminal nodes i, j at random for which there are pending connectivity requirements. Line 5 computes the graph obtained by removing from G the edges of all paths already computed to connect i and j; thus, any path computed in G' will be edge-disjoint from the former $(i...j)$ paths in P_{ij}.   In the case of the GSP-NC, not only the edges should be supressed but also the

---
[2] Set of solutions that can be obtained by well-defined replacement of parts of the current solution

Procedure ConstPhase(G, C, T, R)

1:    $G_{sol} \leftarrow (T, \varnothing); m_{ij} \leftarrow r_{ij} \forall i, j \in T; P_{ij} \leftarrow \varnothing \forall i, j \in T; A_{ij} \leftarrow 0 \forall i, j \in T$
2:    $C \leftarrow$ alter-costs(C)
3:    **while** $\exists m_{ij} > 0 : A_{ij} <$ MAX_ATTEMPT **do**
4:        let i, j be any two terminals with $m_{ij} > 0$
5:        $G' \leftarrow G \setminus P_{ij}$
6:        let $C' = (c'_{uv}) : c'_{uv} \leftarrow [0$ if $(u, v) \in G_{sol}; c_{uv}$ otherwise $]$
7:        $p \leftarrow$ shortest-path(G', C', i, j)
8:        **if** $\not\exists p$ **then**
9:           $A_{ij} \leftarrow A_{ij} + 1; P_{ij} \leftarrow \varnothing; m_{ij} \leftarrow r_{ij}$
10:      **else**
11:          $G_{sol} \leftarrow G_{sol} \cup \{p\}$
12:          $P_{ij} \leftarrow P_{ij} \cup \{p\}; m_{ij} \leftarrow m_{ij} - 1$
13:          $[P, M] \leftarrow$ general-update-matrix(G_{sol}, P, M, p, i, j)
14:      **end if**
15:    **end while**
16:    **return** G_{sol}, P

Fig. 4. ConstPhase pseudo-code

nodes of the former $(i...j)$ paths in order to generate node-disjoint paths. In line 6, the edges already present in the solution under construction are given cost 0; by doing this, they will be taken as costless when considering the cost of any new path, enabling edge-reusing among different pairs of terminals. Line 7 computes the shortest path (regarding costs) connecting i and j, considering as feasible the edges from G' and with costs given by C'. In case this turns to be impossible, this is acknowledged in line 9 by incrementing the counter A_{ij} and resetting the path set P_{ij}, hoping that computing a different sucession of paths for i, j allow to satisfy the r_{ij} requirements. In case a path p was found, it becomes part of the solution under construction (lines 11-12), and the general-update-matrix procedure on line 13 updates the pending connection requirements of the matrix M, by applying the Ford-Fulkerson's algorithm with all capacities equal to 1, to detect if the adoption of the new path turned to satisfy other requirements besides the one for the pair i, j. Finally, the algorithm ends by returning the feasible solution G_{sol} together with the path set P which "certifies" that all requirements R were satisfied.

2.3.1 Altering costs

The algorithm here proposed satisfies the property given below, provided an appropriate function alter-costs is used in line 2 (unlike similar construction phases previously proposed for the GSP-NC in (Robledo & Canale, 2009) as certain trivial instances can attest):

$$\lim_{iterations \to \infty} probability(get\ an\ optimal\ solution) = 1$$

In other words, we can guarantee that any desired level of certainty of getting an optimal solution can be reached provided as many iterations as needed are run.

We proved that this property is verified if the alter-costs function is such that all edges have their costs altered independtly from the others and the altered costs take values in $(0, +\infty)$

with any probability distribution that assigns non-zero probabilities to any open subinterval of $(0, +\infty)$. In our tests we used an exponential distribution with parameter $1/real\_cost$.

Moreover, by altering costs the proposed algorithm proceeds by just computing the shortest path in its main loop, instead of computing a set of "simultaneous disjoint shortest paths" and then randomly choosing one (as in previous algorithms), thus involving less computing.

2.4 Local search phase algorithms

The local search phase starts with a feasible solution obtained from the construction phase and proceeds by consecutively moving to neighbour solutions which reduce the cost of the solution graph until it reaches a local optimum. Any local search algorithm needs a precise definition of the neighbourhood concept; we propose two different ones, which we chain inside our suggested LocalSearchPhase algorithm. They are defined in terms of a certain structural decomposition of graphs that we define below together with some other auxiliary definitions.

Definition 2.2. key-node: *Given a GSP-EC instance and a feasible solution G_{sol}, we define a **key-node** as a non-terminal node with degree at least three in G_{sol}.*

Definition 2.3. key-path: *Given a GSP-EC instance and a feasible solution G_{sol}, we define a **key-path** as a path in G_{sol} such that all intermediate nodes are non-terminal with degree two in G_{sol} and whose endpoints are either terminal nodes or key-nodes.*

Definition 2.4. key-tree: *Given a GSP-NC instance, a feasible solution G_{sol} and a key-node v of G_{sol}, we define as the **key-tree** associated to v the subgraph of G_{sol} obtained through the union of all key-paths with v as an endpoint.*

Definition 2.5. key-star: *Given a GSP-EC instance, a feasible solution G_{sol} and any node v of G_{sol}, we define as the **key-star** associated to v the subgraph of G_{sol} obtained through the union of all key-paths with v as an endpoint.*

2.4.1 Path-based local search neighbourhood

Our first neighbourhood is based on the replacement of any key-path k by another key-path with the same endpoints, built with any edge from the feasible connections graph G (even some of G_{sol}), provided no connectivity levels are lost when reusing edges. Let k be a key-path of a certain solution G_{sol} and P a set of paths which "certificates" its feasibility (as the one returned by ConstPhase). We will denote by $J_k(G_{sol})$ the set of paths $\{p \in G_{sol} : k \subseteq p\}$. These are the paths which contain the key-path k. We will also denote by $\chi_k(G_{sol})$ the edge set

$$\chi_k(G_{sol}) = \bigcup_{q=i...j \in J_k(G_{sol})} E(P_{ij} \setminus q)$$

These are the edges that, if used to replace the key-path k in P (obtaining a path set P'), would turn to be shared by some paths from G_{sol} with the same endpoints, thus invalidating the resulting set P' as a feasibility certificate. We can now define our first neighbourhood.

Definition 2.6. Neighbourhood1: *Given a GSP-EC instance and a feasible solution G_{sol}, it is the set of all graphs obtained by replacing any key-path k of G_{sol} by another path p such that $cost(p) < cost(k)$ and the edges of p are chosen from the set $E \setminus \chi_k(G_{sol})$ and/or k. (Recall that E represents the feasible edges between nodes).*

Procedure LocalSearchPhase1(G, C, T, S)

```
 1: improve ← TRUE
 2: κ ← k-decompose(S)
 3: while improve do
 4:     improve ← FALSE
 5:     for all kpath k ∈ κ with endpoints u, v do
 6:         G' ← the subgraph induced from G by E(k) ∪ (E \ χₖ(S))
 7:         C' ← (c'ᵢⱼ)/c'ᵢⱼ = 0 if (i, j) ∈ S \ k; c'ᵢⱼ = cᵢⱼ otherwise
 8:         k' ← shortest-path(G', C', u, v)
 9:         if cost(k', C') < cost(k, C') then
10:             improve ← TRUE
11:             update S : ∀p ∈ Jₖ(S)(p ← (p \ k) ∪ k')
12:             if ∃z ∈ V(k'), z ∉ {u, v}, degree(z) ≥ 3 in S then
13:                 remove-cycles(Jₖ(S))
14:                 κ ← k-decompose(S)
15:             else
16:                 κ ← κ \ {k} ∪ {k'}
17:             end if
18:         end if
19:     end for
20: end while
21: return S
```

Fig. 5. LocalSearchPhase1 pseudo-code

Based on these definitions we built the path-based local search algorithm LocalSearchPhase1 shown in Figure 5. The algorithm receives as inputs the graph G of feasible connections, the edge cost matrix C, the terminals set T and a path set S which build up a feasible solution. Line 1 initializes the flag *improve* which indicates wheter an improved solution has been found or not. Line 2 computes the decomposition in key-nodes and key-paths of the set S. Loop 3-20 looks for successive cost improvements until no more can be done. Each iteration proceeds as follows. The loop 5-19 analyzes each key-path k trying to find a suitable replacement with lower cost. Line 6 computes the edge set $E(k) \cup (E \setminus \chi_k(S))$, where edges are to be chosen from to build the replacing key-path. This set is such that, as seen above, ensures no loss of connectivity levels in the new solution obtained, while allowing the reuse of edges already present in the current solution S. Line 7 computes a new cost matrix C', zeroing the cost of all edges of S that are not included in the key-path k, to reflect the fact that using any of those edges to build the replacing key-path adds no extra cost to the modified solution. Line 8 computes the path with lower cost according to the matrix C' over the subgraph computed in line 6. Line 9 verifies if the adoption of the new key-path implies a cost reduction. If so, it is acknowledged by the flag *improve* and k is replaced in all paths of S which included k. Care is taken to remove cycles and recompute the k-decomposition if a certain node happens to have a degree greater than two after the replacement (lines 12-14); if the latter does not happen, line 16 simply updates the k-decomposition by replacing the key-path (thus avoiding computing a new k-decomposition from scratch). After exiting the main loop, line 21 returns a feasible solution whose cost can no more be reduced by moving to neighbour solutions. As we commented in the Construction Phase, for the GSP-NC problem a small change in the

Procedure LocalSearchPhase2(G, C, T, S)

```
1: improve ← TRUE
2: κ ← k-decompose(S)
3: while improve do
4:     improve ← FALSE
5:     for all kstar k ∈ κ do
6:         [k', newCost] ← BestKeyStar(G, C, T, S, k)
7:         if newCost < cost(k, C) then
8:             improve ← TRUE
9:             replace k by k' in all paths from S
10:            κ ← k-decompose(S)
11:            abort for all
12:        end if
13:    end for
14: end while
15: return S
```

Fig. 6. LocalSearchPhase2 pseudo-code

definition of J_k and χ_k must be made, supressing also the nodes involved rather than only the edges.

2.4.2 Key-star-based local search neighbourhood

Our second neighbourhood is based on the replacement of key-stars, which frequently allow to improve feasible solutions that are locally optimal when only considering Neighbourhood1. In the case of the GSP-NC, as no node sharing is allowed among disjoint paths, all key-stars are trees (named *key-trees*); a key-tree replacement neighbourhood for the GSP-NC can be found in (Robledo & Canale, 2009). Due to the possibilty of sharing nodes among edge-disjoint paths, when working with GSP-EC problems, we have to work with key-stars, and unlike (Robledo & Canale, 2009) we will allow the root node to be a terminal node in order to get a broader neighbourhood. In the GSP-NC any key-tree can be replaced by any tree with the same leaves with no loss of connectivity levels. In the GSP-EC, if the replacing structure is also a key-star the same holds true; but it does not for other general structures (non-star trees included). We propose an algorithm that given a key-star k, deterministically seeks for the lowest cost replacing key-star k' able to "repair" the paths from P broken when removing the edges of k.

Let k be a key-star in a certain feasible solution G_{sol} and P a set of paths which "certificates" its feasibility (as the one returned by ConstPhase). For allowing as much reusing of edges as possible, we can extend our previous definition of $J_k(G_{sol})$ and $\chi_k(G_{sol})$ to consider key-stars k instead of key-paths; and thus we can define the key-star based neighbourhood as follows.

Definition 2.7. Neighbourhood2: *Given a GSP-EC instance and a feasible solution G_{sol}, it is the set of all graphs obtained by replacing any key-star k of G_{sol} by the lowest possible cost key-star k' such that k' preserves the same connectivity among the leaves of k and its terminal nodes, and the edges of k' are chosen from the set $E \setminus \chi_k(G_{sol})$ and/or k.*

We present the star-based local search algorithm LocalSearchPhase2 in Figure 6. The algorithm receives as inputs the graph G of feasible connections, the edge cost matrix C, the

Procedure BestKeyStar(G, C, T, S, k)

1: $G' \leftarrow$ the subgraph induced from G by $E(k) \cup (E \setminus \chi_k(S))$
2: $C' \leftarrow (c'_{ij})/c'_{ij} = 0$ if $(i, j) \in S \setminus k; c'_{ij} = c_{ij}$ otherwise
3: add a "virtual node" w to G'
4: $\Omega \leftarrow \psi_k$
5: **if** $\theta_k \in T$ **then**
6:      $\Omega \leftarrow \Omega \cup \{\theta_k\}$
7: **end if**
8: **for all** $m \in \Omega$ **do**
9:      add $\hat{\delta}_{k,m}$ parallel edges (w, m) to G' with cost 0
10: **end for**
11: $c_{min} \leftarrow 0; k_{min} \leftarrow k$
12: **for all** $z \in V(G)$ **do**
13:     $k' \leftarrow$ simult-shortest-paths$(G', \delta_{G,w}, z, w)$
14:     **if** k' has $\delta_{G,w}$ paths $\wedge \; cost(k', C') < c_{min}$ **then**
15:         $c_{min} \leftarrow cost(k', C'); k_{min} \leftarrow k'$
16:     **end if**
17: **end for**
18: **return** $[k_{min}, c_{min}]$

Fig. 7. BestKeyStar pseudo-code

terminals set T and a path set S which build up a feasible solution. Line 1 initializes the flag *improve* which indicates wheter an improved solution has been found or not. Line 2 computes the decomposition in key-nodes and key-paths of the set S. Loop 3-14 looks for succesive cost improvements until no more can be done. Each iteration proceeds as follows. The loop 5-13 analyzes each key-star k trying to find a suitable replacement with lower cost. Line 6 determines the lowest cost key-star k' that could replace k and its cost (computed assuming that edges from the current solution not in k have no cost to promote edge reusing). To do so it uses the procedure BestKeyStar described later. Line 7 verifies if the replacing key-star has lower cost than k; if it does, lines 8-11 acknowledge the fact, the replacement is done over the set S, the k-decomposition is recomputed and the "for all" loop is aborted to restart looking for improvements. After exiting the main loop, line 15 returns a feasible solution whose cost can no more be reduced by moving to neighbour solutions.

Figure 7 presents the algorithm BestKeyStar. Given a key-star k we denote by θ_k its root node; by ψ_k the set of its leaf nodes; and by $\hat{\delta}_{k,m}$ (being m the root node of k or one of its leaves) the highest amount of key-paths that join m in k with any other node that is root or leaf in k. The algorithm is based on the idea of building key-stars by employing the simult-shortest-paths algorithm (that can be found in (Bhandari, 1997) as *k-shortest-path*). Lines 1-2 compute the subgraph of G obtained by removing the edges that could cause loss of connectivity level if reused; the altered cost matrix C' with cost zero for reused edges; and adds a virtual node w whose purpose is explained below. Lines 4-7 determine the set of leaf nodes that the key-star to build must have. Lines 8-10 connect each of the latter to w with an appropriate number of parallel zero-cost edges totalling $\delta_{G,w}$ (degree of w in G) edges. The loop 12-17 considers nodes of G that could be potencial roots z of the key-star to be found, and then builds the lowest-cost one with root node z through the application of the simult-shortest-path algorithm in line 13; $\delta_{G,w}$ edge-disjoint paths connecting z and w are requested. If found (lines 14-16) and with

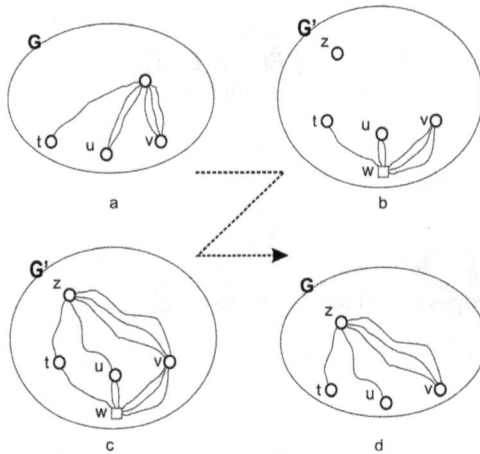

Fig. 8. Computing the best key-star

lower cost than k then the new key-star and its associated cost are recorded as the best ones so far found. After having considered all possible root nodes, line 18 returns both the best key-star and its cost according to C'.

Figure 8 depicts the process of determining which the best key-star to replace a given one is. It illustrates (a) the feasible graph G with a key-star that keeps connected the leaf nodes t, u, v; (b) the graph G' obtained after adding the virtual nodes w linked with no cost to t, u, v by the appropriate amount of edges and choosing a "candidate" root node z; (c) the shortest paths found to connect z and w; and (d) the new key-star obtained after removing the virtual node w.

2.5 GRASP algorithm description

Now we are able to put the pieces together and build a GRASP algorithm for solving the GSP-EC. Figure 9 shows the resulting pseudo-code. Basically the local search phase of this algorithm applies key-path replacement based movements until no further improvements are possible; then it tries to apply the best key-star replacement movement (once); if the latter is done with a cost reduction, then key-path replacements are tried again, and so on, until no further improvements are possible for both kinds of movements.

The algorithm receives as inputs the graph G of feasible connections, the cost matrix C, the terminals T, the redundancy requirements matrix R and a the number of iterations $iters$ to perform. In line 1 the minimum cost found c_{min} is initialized to ∞ and an empty path set S_{opt} is initialized. The main loop (2-18) is executed $iters$ times and then the best solution found is returned. Line 3 builds a feasible solution employing our ConstPhase greedy randomized adaptive algorithm; being S the path set that certifies feasibility. If the set S has less paths than the so far found best solution S_{opt} (line 4) this iteration is discarded. This could happen if the last call to ConstPhase was not able to satisfy all requirements of R and a previous call was able to do it (or at least to satisfy a greater number); our first objective is to satisfy as many requirements of R as possible. Line 6 applies the key-path based movements by calling LocalSearchPhase1. If this was the first local search or if a cost reduction was achieved then

Procedure GRASP_GSP$(G, C, T, R, iters)$

1: $c_{min} \leftarrow \infty; S_{opt} \leftarrow \emptyset$
2: **for** $i = 1$ to $iters$ **do**
3:     $[G_{sol}, S] \leftarrow$ ConstPhase(G, C, T, R)
4:     **if** $|S| \geq |S_{opt}|$ **then**
5:        $flag \leftarrow$ TRUE
6:        OptLoop: $[G_{sol}, S'] \leftarrow$ LocalSearchPhase1(G, C, S)
7:        **if** $flag \lor \text{cost}(S') < \text{cost}(S)$ **then**
8:           $flag \leftarrow$ FALSE
9:           $[G_{sol}, S''] \leftarrow$ LocalSearchPhase2(G, C, S')
10:           **if** $\text{cost}(S'', C) < \text{cost}(S', C)$ **then**
11:              $S \leftarrow S''$; go to OptLoop
12:           **end if**
13:        **end if**
14:        **if** $\text{cost}(S', C') < c_{min}$ **then**
15:           $c_{min} \leftarrow \text{cost}(S'', C); S_{opt} \leftarrow S$
16:        **end if**
17:     **end if**
18: **end for**
19: **return** S_{opt}

Fig. 9. GRASP_GSP pseudo-code

a best key-star movement is tried in line 9. In case the latter succeeds in reducing the cost (verified in line 10), the execution flow resumes at line 6, for trying a new cycle of chained improvements. When no further local improvements are possible, lines 14-16 update the best known solution in case an improvement was achieved.

3. Performance tests

This section presents the results obtained after testing our algorithms with twenty-one test cases. The algorithms were implemented in C/C++ and tested on a 2 GB RAM, Intel Core 2 Duo, 2.0 GHz machine running Microsoft Windows Vista. All instances were run with the parameter *iters* set to 100.

3.1 Test set description

To our best knowledge, no library containing benchmark instances related to the GSP-NC nor GSP-EC exists; we have built a set of twenty-one test cases that are based in cases found in the following public libraries:

- steinlib (Koch et al., 2000): instances of the Steiner problem; in many cases the optimal solution is known, in others the best solution known is available;

- tsplib (Reinelt, 2004): instances of diverse graph theory related problems, including a "Traveling Salesman Problem" section.

The main characteristics of the twenty-one test cases are shown in Table 1.. For each case we show the amount of nodes (V), feasible edges (E), terminal nodes (T) and Steiner (non terminal) nodes (St). We also show the level of edge-connectivity requirements (one, two,

Case	V	E	T	St	Redund.	Opt
b01-r1	50	63	9	41	1-EC	82
b01-r2	50	63	9	41	2-EC	NA
b03-r1	50	63	25	25	1-EC	138
b03-r2	50	63	25	25	2-EC	NA
b05-r1	50	100	13	37	1-EC	61
b05-r2	50	100	13	37	2-EC	NA
b11-r1	75	150	19	56	1-EC	88
b11-r2	75	150	19	56	2-EC	NA
b17-r1	100	200	25	75	1-EC	131
b17-r2	100	200	25	75	2-EC	NA
cc3-4p-r1	64	288	8	56	1-EC	2338
cc3-4p-r3	64	288	8	56	3-EC	NA
cc6-2p-r1	64	192	12	52	1-EC	3271
cc6-2p-r2	64	192	12	52	2-EC	NA
cc6-2p-r123	64	192	12	52	1,2,3-EC	NA
hc-6p-r1	64	192	32	32	1-EC	4003
hc-6p-r2	64	192	32	32	2-EC	NA
hc-6p-r123	64	192	32	32	1,2,3-EC	NA
bayg29-r2	29	406	11	18	2-EC	NA
bayg29-r3	29	406	11	18	3-EC	NA
att48-r2	48	300	10	38	2-EC	NA

Table 1. Characteristics of the Test Cases

three or mixed) and the optimal costs when available. GSP poblems solved with connectivity level one are Steiner problems and in those cases we got the optimal solution cost from steinlib. Problems b01, b03, b05, b11 and b17 were taken from steinlib's problem instances set "B" and are cases randomically generated with integer uniform costs ranging from 1 to 10. The case cc3-4p belongs to steinlib's instance set "PUC"; eigth terminal nodes are terminal and we solved two instances with uniform connectivity requirements one and three. The cases cc6-2p and hc-6p belong also to steinlib's instance set "PUC"; twelve and thirty-two nodes are terminal and we solved three instances for each one with connectivity requirements one, two, and a mix of one to three. Finally the cases bayg29 and att48 were taken from the library tsplib; both correspond to real cases (twenty-nine cities from Bavaria, Germany; and 48 cities from USA).

3.2 Numerical results

Computational results of the tests can be found in Table 2. Here follows the meaning of each column:

- Reqs.: total amount of requirements satisfied by the best solution found
- t(ms): the average running time (in ms) per iteration
- Cost: the cost of the best solution found

Case	Reqs.	t(ms)	Cost	%LSI
b01-r1	36	77	82	3.0
b01-r2	42	80	98	3.4
b03-r1	300	2611	138	10.6
b03-r2	378	3108	188	4.1
b05-r1	78	298	61	9.2
b05-r2	144	1389	120	5.2
b11-r1	171	1477	88	13.8
b11-r2	324	4901	180	3.4
b17-r1	300	6214	131	10.2
b17-r2	531	15143	244	3.0
cc3-4p-r1	28	388	2338	10.0
cc3-4p-r3	84	2221	5991	4.6
cc6-2p-r1	66	2971	3271	2.4
cc6-2p-r2	132	4801	5962	10.2
cc6-2p-r123	140	6317	8422	9.8
hc-6p-r1	496	25314	4033	6.8
hc-6p-r2	992	28442	6652	3.5
hc-6p-r123	957	26551	7930	5.2
bayg29-r2	110	975	6856.88	4.6
bayg29-r3	165	2413	11722	4.2
att48-r2	90	1313	23214	13.0
Averages	265	6524	-	6.7

Table 2. Numerical Test Results

- LSI: "local search improvement" - the percentage of cost improvement achieved by the local search phase when compared to the cost of the solution delivered by the construction phase, for the best solution found

In all cases with connectivity requirements equal to one (1-EC) for all pairs of terminals (for which the optimal costs are known) every best solution found is optimal, with the exception of the case hc-6p-r1 (found cost 4033 being the optimal cost 4003). Note also that the average cost improvement over the solution delivered by ConstPhase (LSI) amounts to 6.7% (when computed only for the best solutions found). All solutions found are edge-minimal regarding feasibility (no edge can be supressed without losing required connectivity levels); and in all cases the maximum possible number of requirements are satisfied (i.e. for all pairs of terminals i, j, whether their requirement r_{ij} was satisfied or f_{ij} disjoint paths were found being f_{ij} the maximum achievable amount of disjoint paths joining i and j given by the topology of the feasible connections graph G). Figure 10 show the best 3-connected network found to connect the sites of bayg29, as well as the best 2-connected network found to connect the sites of att48.

4. Current research

4.1 Relationship among the GSP-NC and GSP-EC problems

There is a strong relationship among the GSP-NC and the GSP-EC problems. We have demonstrated that any GSP-EC instance can be transformed in polinomial time into a GSP-NC

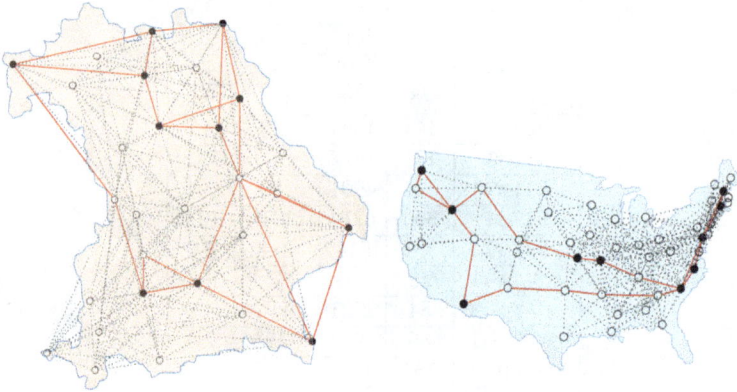

Fig. 10. Best Solutions for cases bayg29 and att48

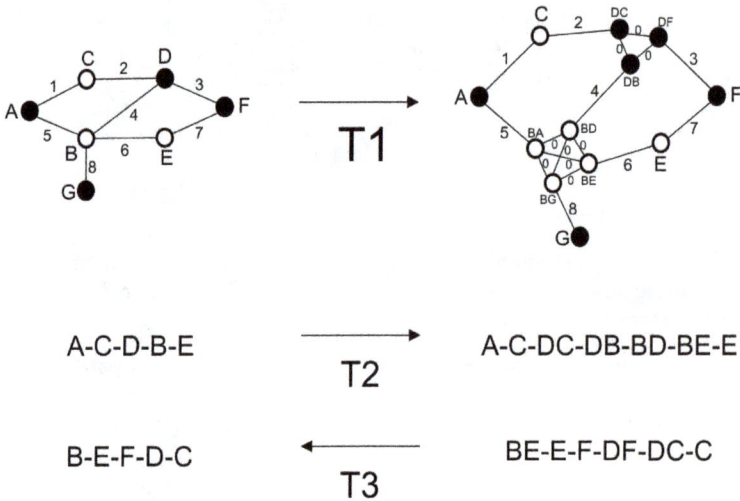

A-C-D-B-E $\xrightarrow{\text{T2}}$ A-C-DC-DB-BD-BE-E

B-E-F-D-C $\xleftarrow{\text{T3}}$ BE-E-F-DF-DC-C

Fig. 11. Transforming GSP-EC into GSP-NC instances

instance, whose optimal solutions can in turn be transformed back (in polinomial time) into optimal solutions of the original GSP-EC instance. To do so a well-known node-splitting technique is applied as shown in Figure 11.. There, a GSP-EC instance involving four terminal nodes and three Steiner nodes is transformed (through $T1$) into a GSP-NC instance with six terminal nodes and six Steiner nodes. Every node with degree $d > 2$ is transformed into a d-clique with all edge costs equal to zero. The transformations $T2$ and $T3$ show how to translate a path from the GSP-EC instance to its corresponding path in the GSP-NC and vice-versa. After solving the GSP-NC instance, finding a solution S with cost c and a set P of certificate paths, these can be translated through $T3$ into a set of paths that happens to be a valid solution of the GSP-EC instance and with the same cost c. We proved also that if S is optimal for the GSP-NC instance, then $T3(S)$ will be an optimal solution of the GSP-EC instance. This opened the question of wether it is best to focus on solving efficiently GSP-NC instances and using these transformations to solve GSP-EC instances, or focus also on developing GSP-EC specific algorithms which make use of the particulars introduced by the chance of node reusing among paths.

4.2 Altering costs

As we have seen above, altering the costs under certains hipothesis guarantees that the construction phase algorithm can have as high a chance of building an optimal solution as wanted provided the needed amount of iterations are run. In our tests we have used the exponencial distribution to alter the costs; it satisfies those hipothesis. Nonetheless it is pending to investigate which other distributions can be applied, eventually with better results. The exponential distribution just takes into account the expected value of the edge cost as parameter, not considering other factors that characterize the problem like costs dispersion. During testing we found that for those cases with little cost differences among edges the exponential distribution tends to generate too randomized values (i.e. with high dispersion); for many of those cases better results were attained using alternative ways of randomization that took under consideration also the standard deviation of the edge costs.

4.3 A recursive neigbourhood approach

The neigbourhoods that we used to define local search movements are based on the k-decomposition of graphs into key-paths, key-trees and key-stars. When studying the performance of the algorithms we have found several other types of structures that could be optimized yet using more complex movements i.e. not decomposable into sequences of key-paths, key-trees or key-stars replacements. We have also seen that the deterministic computation of the optimal replacing structure rapidly gains complexity when considering more complex structures; in the case of key-stars we have presented an algorithm for computing the best replacing key-star that made processing time soar because of the need of computing k-shortest disjoint paths. Therefore we consider relevant the creation of neighbourhood mechanisms that encompass heuristic criteria with complex structures replacement. We are working on an extension of the GSP that we named EGSP and can be solved by recursively invoking smaller instances of itself, determined by the supression of any desired edge set; this should allow to choose at will the structures to replace no matter their size or complexity. Current questions regarding this topic include: which kind of structures to replace given a certain (sub)instance of the problem for the next recursive calls; and at which point the use of more direct algorithms like the ones in this chapter turns to be more convenient.

5. Conclusion

In this chapter we introduced a framework and algorithms suitable to address the design of minimal cost networks under connectivity constraints, modelled as Generalized Steiner Problems both node-connected (GSP-NC) and edge-connected (GSP-EC). Our algorithm GRASP_GSP was shown to find good quality solutions to the GSP-EC when applied to a series of heterogeneous test cases with up to 100 nodes and up to 406 edges. For all cases with known optimal cost the algorithm was able to find solutions with costs no more than

0,74% higher than the optimal cost. Significant cost reductions were achieved after applying the local search phase over the greedy solutions built by the construction phase. Execution times of the tests run for the GSP-EC were comparable to the ones of previous similar works (Robledo & Canale, 2009) for the node-connected version of the GSP-NC and similar sizes. We also mentioned current research lines and extensions on the ideas treated within this chapter.

6. References

Agrawal, A., Klein, P. & Ravi, R. (1995). When trees collide: an approximation algorithm for the generalized Steiner problem on networks, *SIAM Journal on Computing* 24(3): 440–456.

Auletta, V., Dinitz, Y., Nutov, Z. & Parente, D. (1999). A 2-approximation algorithm for finding an optimum 3-vertex-connected spanning subgraph, *Journal of Algorithms* 32(1): 21–30.

Baïou, M. (1996). *Le problème du suos-graphe Steiner 2-arête connexe: approache polyédrale*, PhD thesis, Université de Rennes I.

Balakrishnan, A., Magnanti, T. & Mirchandi, P. (2004). Connectivity-splitting models for survivable network design, *Networks* 43(1): 10–27.

Bhandari, R. (1997). Optimal physical diversity algorithms and survivable networks, *Computers and Communications, 1997. Proceedings., Second IEEE Symposium on*, pp. 433–441.

Bienstock, D., Brickell, E. & Monma, C. (1990). On the structure of minimum weight k-connected spanning networks, *SIAM Journal on Discrete Mathematics* 3(3): 320–329.

Böckenhauser, H., Bongartz, D., Hromkovič, J., Klasing, R., Proietti, G., Seibert, S. & Unger, W. (2002). On the hardness of constructing minimal 2-connected spanning subgraphs in complete graphs with sharpened triangle inequality, *Proceedings of the 22nd Conference Kanpur on Foundations of Software Technology and Theoretical Computer Science*, Springer-Verlag, pp. 59–70.

Cancela, H., Robledo, F. & Viera, O. (2005). A parallel algorithm for the Steiner 2-edge-survivable network problem, *Journal of ICHIO (Chilean Institute of Operations Research)* 7(1): 15–27.

Cheriyan, J., Jordan, T. & Nutov, Z. (2001). On rooted node-connectivity problems, *Algorithmica* 30(3): 353–375.

Cheriyan, J. & Thurimella, R. (2000). Approximating minimum-size k-connected spanning subgraphs via matching, *SIAM Journal on Computing* 30: 528–560.

Cheriyan, J., Vempala, S. & Vetta, A. (2002). Approximation algorithms for minimum-cost k-vertex-connected subgraphs, *Proceedings of the 34th Annual ACM Symposium on the Theory of Computing*, pp. 306–312.

Chopra, S. (1992). Polyhedra of the equivalent subgraph problem and some edge connectivity problems, *SIAM Journal on Discrete Mathematics* 5(3): 321–337.

Chou, W. & Frank, H. (1970). Survivable communication networks and the terminal capacity matrix, *IEEE Transactions on Circuit Theory* 17: 192–197.

Christofides, N. & Whitlock, C. (1981). An algorithm for the design of optimal invulnerable networks, *Technical Report IC-OR-81-6*, Imperial College, London.

Coullard, R., Rais, A., Rardin, R. & Wagner, D. (1991). Linear-time algorithm for the 2-connected Steiner subgraph problem on special classes of graphs, *Technical Report No. 91-25*, School of Industrial Engineering, Purdue University.

Csaba, B., Karpinski, M. & Krysta, P. (2002). Approximability of dense and sparse instances of minimum 2-connectivity, TSP and path problems, *Proceedings of the 13th Annual ACM-SIAM Symposium on Discrete Algorithms*, pp. 74–83.

Czumaj, A. & Lingas, A. (1999). On approximability of the minimum-cost k-connected spanning subgraph problem, *Proceedings of the 10th Annual ACM-SIAM Symposium on Discrete Algorithms*, pp. 281–290.

Frank, H. & Chou, W. (1970). Connectivity considerations in the design of survivable networks, *IEEE Transactions on Circuit Theory* 17: 486–490.

Frederickson, G. & Jàjà, J. (1982). On the relationship between biconnectivity augmentation and traveling salesman problem, *Theoretical Computer Science* 19: 189–201.

Gabow, H., Goemans, M. & Williamson, D. (1993). An efficient approximation algorithm for the survivable network design problem, *Proceedings of the 3rd MPS Conference on Integer Programming and Combinatorial Optimization*, pp. 57–74.

Goemans, M. & Bertsimas, D. (1993). Survivable networks, linear programming relaxations and the parsimonious property, *Mathematical Programming* 60: 143–166.

Goemans, M., Goldberg, A., Plotkin, S., Tardos, E. & Williamson, D. (1994). Improved approximation algorithms for network design problems, *Proceedings of the 5th ACM-SIAM Symposium on Discrete Algorithms*, pp. 223–232.

Goemans, M. & Williamson, D. (1992). A general approximation technique for constrained forest problems, *SIAM Journal on Computing* 24(2): 296–317.

Grötschel, M. & Monma, C. (1990). Integer polyhedra associated with certain network design problems with connectivity constraints, *SIAM Journal on Discrete Mathematics* 3: 502–523.

Grötschel, M., Monma, C. & Stoer, M. (1991). · Polyhedral Approaches to Network Survivability, *in* F. Roberts, F. Hwang & C. Monma (eds), *Reliability of Computer and Communication Networks, Proc. Workshop 1989, New Brunswick, NJ/USA*, Vol. 5 of *Series in Discrete Mathematics and Theoretical Computer Science*, American Mathematical Society, pp. 121–141.

Grötschel, M., Monma, C. & Stoer, M. (1992a). Computational results with a cutting plane algorithm for designing communication networks with low-connectivity constraints, *Operations Research* 40(2): 309–330.

Grötschel, M., Monma, C. & Stoer, M. (1992b). Facets for polyhedra arising in the design of communication networks with low-connectivity constraints, *SIAM Journal on Optimization* 2(3): 474–504.

Grötschel, M., Monma, C. & Stoer, M. (1995). Polyhedral and computational investigations for designing communications networks with high survivability requirements, *Operations Research* 43(6): 1012–1024.

Jain, K. (1999). A 3-approximation algorithm for finding optimum 4,5-vertex-connected spanning subgraphs, *Journal of Algorithms* 32(1): 31–40.

Jain, K. (2001). A factor 2 approximation algorithm for the generalized Steiner network problem, *Combinatorica* 21: 39–60.

Kerivin, H. & Mahjoub, R. (2005). Design of survivable networks: A survey, *Networks* 46: 1–21.

Khuller, S. & Raghavachari, B. (1996). Improved approximation algorithms for uniform connectivity problems, *Journal of Algorithms* 21(2): 434–450.

Ko, C. & Monma, C. (1989). Heuristics methods for designing highly survivable communication networks, *Technical report*, Bellcore.

Koch, T., Martin, A. & Voß, S. (2000). SteinLib: An updated library on Steiner tree problems in graphs, *Technical Report ZIB-Report 00-37*, Konrad-Zuse-Zentrum für Informationstechnik Berlin, Takustr. 7, Berlin.
URL: *http://elib.zib.de/steinlib*

Kortsarz, G., Krauthgamer, R. & Lee, J. R. (2004). Hardness of approximation for vertex-connectivity network-design problems, *SIAM Journal on Computing* 33(3): 704–720.

Kortsarz, G. & Nutov, Z. (2003). Approximating node connectivity problems via set covers, *Algorithmica* 37(2): 75–92.

Monma, C., Munson, B. & Pulleyblank, W. (1990). Minimum-weight two connected spanning networks, *Mathematical Programming* 46: 153–171.

Ravi, R. & Klein, P. (1993). When cycles collapse: a general approximation technique for constrained 2-connectivity problems, *Proceedings of the 3rd Symposium on Integer Programming and Combinatorial Optimization*, pp. 39–55.

Ravi, R. & Williamson, D. (1997). An approximation for minimum-cost vertex-connectivity problems, *Algorithmica* 18(1): 21–43.

Ravi, R. & Williamson, D. (2002). Erratum: An approximation for minimum-cost vertex-connectivity problems, *Proceedings of the 13th Annual ACM-SIAM Symposium on Discrete Algorithms*, pp. 1000–1001.

Reinelt, G. (2004). TSPLIB library, http://elib.zib.de/pub/Packages/mp-testdata/tsp/tsplib/tsplib.html.

Resende, M. & Ribeiro, C. (2003). Greedy randomized adaptive search procedures, *in* F. Glover & G. Kochenberger (eds), *Handbook of Metaheuristics*, Kluwer Academic Publishers, pp. 219–249.

Robledo, F. & Canale, E. (2009). Designing backbone networks using the Generalized Steiner Problem, *Proceedings of the International Workshop of Design of Reliable Communication Networks (DRCN'09)*, Washington DC, pp. 327–334.

Steiglitz, K., Weiner, P. & Kleitman, D. (1969). The design of minimum-cost survivable networks, *IEEE Transactions on Circuit Theory* 16: 455–460.

Stoer, M. (1992). *Design of survivable networks*, Vol. 1531 of *Lecture Notes in Mathematics*, Springer-Verlag.

Williamson, D., Goemans, M., Mihail, M. & Vazirani, V. (1995). A primal-dual approximation algorithm for the generalized Steiner network problem, *Combinatorica* 15: 435–454.

Winter, P. (1985). Generalized Steiner problem in outerplanar graphs, *BIT* 25(3): 485–496.

Winter, P. (1986). Generalized Steiner problem in series-parallel networks, *Journal of Algorithms* 7: 549–566.

Winter, P. (1987). Steiner problem in networks: A survey, *Networks* 17(2): 129–167.

15

Design of Advanced Digital Systems Based on High-Speed Optical Links

J. Torres, R. García, J. Soret, J. Martos, G. Martínez, C. Reig and X. Román
Department of Electronic Engineering, University of Valencia, Valencia
Spain

1. Introduction

Optical fiber links offer very important benefits as EMI immunity, low losses, high bandwidth, etc, so an increasing number of communication applications are being developed and deployed. At both sides of these optical links, the optical data signal has to be converted to (or from) the electronic domain. The processing of such a high speed optical signals is not straightforward in most cases, and special considerations need to be taken into account for a proper electronic design.

In this chapter the main considerations for the design of digital electronic systems based on optical links are going to be presented. In the first section, the optical fiber links components are described, emphasizing the main advantages of optical links for digital data transmission and discussing how high speed optical links are handled in the electronic domain.

In the second section, the fundamentals of Printed Circuit Board (PCB) design will be reviewed, including trace design and routing, multilayer PCBs, electromagnetic interferences and clock signal management.

The pre and post-layout studies required for a proper design will be described in the third section, illustrating the explanation with some considerations about real designs for electronic experiments using high speed optical links.

2. High speed optical link components

In this section the main components of high-speed optical links are identified and described. These components are, from optical to electronic domain, fiber optic as transmission medium, high-speed optical transceivers, electronic serializers/deserializers and digital signal processors, typically Field Programmable Gate Arrays (FPGAs). The main characteristics of each component will be identified, and their impact on the total system will be discussed.

2.1 Optical fiber link

The main advantages of optical fiber data links are those inherited from the optical nature of the transmission medium. These advantages and drawbacks of optical fibers for data communication are summarized in Table 1.

Advantages	Drawbacks
Inmunity to EMI	Unsuitable for electrical power transmission
Free from electrical short-circuits and ground loops.	Fragile when handling
Do not produce sparks. Suitable for explosive environment	Not easy to reconnect when broken
Secure from external monitoring	
Low loss, can reach large distances without signal regeneration.	
Large bandwidth, multiplexing capability	
Small size and light weight	
Inexpensive	

Table 1. Advantages and drawbacks of optical fiber for data communication.

From the perspective of high speed digital data transmission systems, the immunity to electromagnetic interferences (EMI) is greatly appreciated. In modern electronic systems it is often necessary to run bundles of wires over considerable distances. These wires can act as antennas, so the electromagnetic fields surrounding the wires can generate by induction undesired electical signal that degrades the transmitted data information. These electromagnetic fields may be, for instance, stray fields from adjacent wires, radio waves present in the environment, or even gamma radiation released during high energy nuclear experiments. Optical fibers have inherent inmunity to most forms of EMI, since no metallic wires are present. So, the optical fiber links ability of operating under severe EMI conditions is extremly important for a great number of applications, especially in defense, health and telecommunication sectors.

The second most important advantage of optical fiber links for high speed data transmission is their extremly low-losses (0.2 dB/Km @ 1550 nm) and their very high bandwidth. Moreover, these low losses are relatively independent of frequency, while those of competitive high speed data links increase rapidly with frequency. New generation of Wavelength Division Multiplexing (WDM) systems operating at 40 Gbps per channel can reach more than 2 Tbps along distances longer than 1000 Km without signal regeneration. However, WDM technology and optical equipment in C-band (1550 nm) is relatively expensive, and optical transceivers typically used in digital data transmission electronic systems offer much lower data rates, up to 10 GHz, and a reach of a few hundreds of meters.

2.2 Optical transceivers

Optical transceivers are the interfaces between optic and electronic worlds. They perform the optical data transmission and reception, so they integrate a semiconductor laser, an optical photo-detector, an optical modulator, and all the required electronic circuitry for proper signal conditioning, as the laser driver, a limiting amplifier for reception, etc. They can work at different wavelengths (typically 850 nm or 1330 nm) depending on the kind of optic fiber used in the high speed optical data link. In Figure 1 it is shown a comercial Small Form Factor (SFF) LC Optical Transceiver from CS Electronics. This optical transceiver works up to 1.25 Gbps at a wavelength of 850 nm over multimode fiber, and can reach 550 meters without optical regeneration.

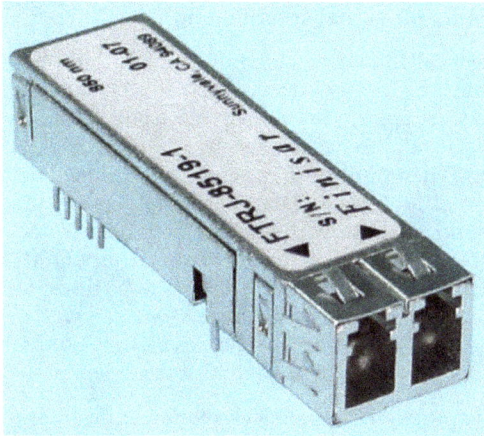

Fig. 1. Image of a comercial SFF optical transceiver.

In order to achieve faster switching and to increase immunity to EMI, crosstalk and noise, high-speed data links work over differential signals. In the case of high speed optical data links, both the electrical input and output signals are typically LVPECL signals (Low-voltage positive emitter-coupled logic). LVPECL is a power optimized version of PECL (uses 3.3 V instead of 5V supply), and both are differential signalling systems mainly used in high speed and clock distribution systems. This technology achieve high speed data rates by using an overdriven BJT differential amplifier with single-ended input, whose emitter current is limited to avoid the slow saturation region of the transistor operation. In Figure 2, a block diagram of a generic optical transceiver is shown.

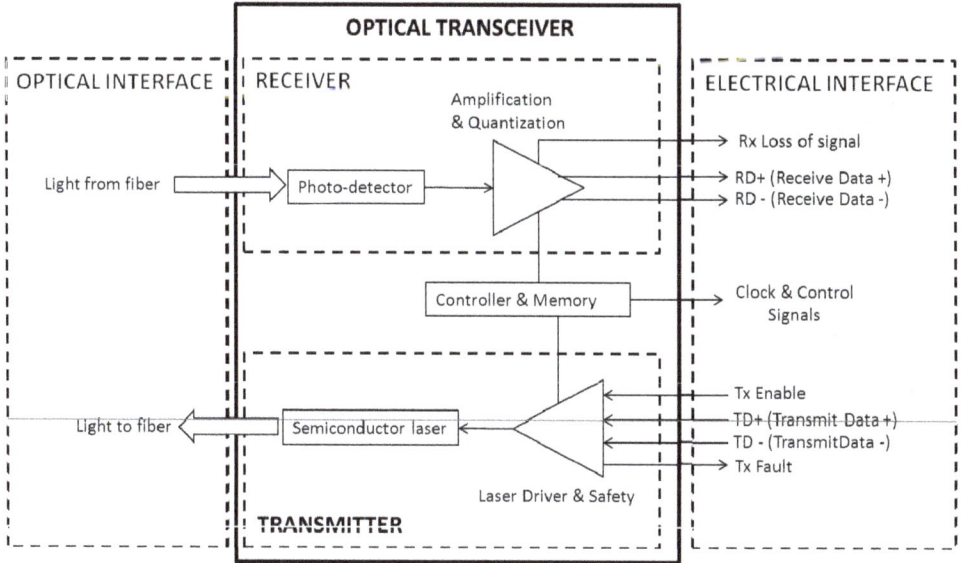

Fig. 2. Transceiver functional diagram.

2.3 Electronic components for signal conditioning

As mentioned before, optical links can reach quite high data rates, up to 10 Gbps, that can not be easily handled by the electronic circuitry. Usually, the serial data received by the optical data link is split into several electronic data channels, each of them working at a lower data rate, so they can be properly processed by the electronic components and devices. When transmitting, several electronic data channels are combined onto a single data channel at a high data rate and then is optically transmitted. This aggregation/disaggregation process is performed by electronic serializer/deserializer devices. A serializer receives data information from N inputs at a given data rate and combine them into a single data channel at a data rate N times faster. When working as deserializer the process is the opposite. The deserializer receives a single data chunk and breaks it into N data channels at a data rate N times slower. So the PCB and the electronic circuitry do not have to operate at the high data rates provided by the optical link.

To serialize data at high speeds, the serial clock rate must be an exact multiple of the clock for the parallel data, so most of electronic designs for high speed optical links use a PLL to multiply a reference clock running at the desired parallel rate to the required serial rate. Moreover, when serial data are received, the optical transceiver must use the same serial clock that serialized the data to deserialize it. At high line rates, providing the serial clock with a separate wire is very impractical because even the slightest difference in length between the data line and the clock line can cause significant clock skew. Instead, optical transceivers recover the clock signal from the data directly, using transitions in the data to adjust the rate of their local serial clock so it is locked to the rate used by the other optical transceiver. Systems using Clock Data Recovery (CDR) can operate over much longer distances at higher speeds than their non-CDR counterparts. However, if transmitted data has too few transitions, the receiving optical transceiver can be unable to apply CDR techniques, so the electronic implementation of encoding schemes is required, as 8B/10B, in which each octet of data is mapped to a 10 bit sequence, or 64B/66B, in which data are grouped into sets of 64 bits, scrambled, then prepended with a 2 bit header.

Additionally, most systems require some form of error detection and correction, as encoding-based error detection or Cyclic Redundancy Checks (CRCs). All this electronic signal conditioning hardware requires quite complex design tasks in order to properly connect the data received/transmitted by the high speed optical link and the processing unit. A simplified block diagram of an optical data link is shown in the figure below.

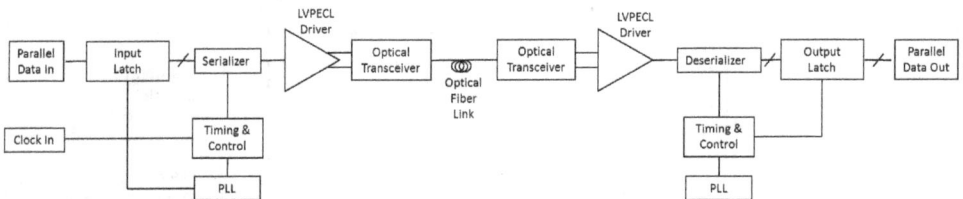

Fig. 3. Optical data link electronic blocks.

2.4 High speed digital data processing

The information carried by the data signals has to be generated and processed. To handle the huge amount of information transmitted by the high speed optical links parallel processing is required, so FPGAs are used. A FPGA is an integrated circuit designed to be configured by the designer after manufacturing. FPGAs contain programmable logic components called "logic blocks", and a hierarchy of reconfigurable interconnects that allow the blocks to be "wired together", so many logic gates that can be inter-wired in many different configurations. Logic blocks can be configured to perform a huge amount of complex combinational functions. In most FPGAs, the logic blocks also include memory elements, which may be simple flip-flops or more complete blocks of memory.

A FPGA can also include multipliers, so it can be used for digital signal processing functions. Once the FPGA internal circuits have been hardware interconnected to perform a given functionality, the FPGA can perform parallel data processing at a very high speed, in the order of hundreds of MHz. So that is the reason for its wide use together with high speed data links.

Moreover, very impressive advances have been done in recent years related to FPGAs configuration capabilities. Nowadays it is possible to implement inside the FPGA a great variety of electronic building blocks and peripherals, for instance, 32-bits hardware microprocessors as MicroBlaze® from Xillinx. These advances include the implementation of MultiGigabit Transceivers (MGT) inside the FPGA. These transceivers performs all the electronic signal conditioning described in section 2.3, so the optical transceiver can be connected almost directly to the FPGA input/output ports. These MGTs manage all the aspects of communication with optical transceivers, as signal integrity, serialization/deserialization, terminations and coupling, line rates, encoding, clock correction, latencies, etc. So the FPGA embedded MGTs are a great improvement for high speed optical links because they simplify the electronic hardware design and reduce the project development costs. Figure 4 shows a building block of the typical MGT functionality.

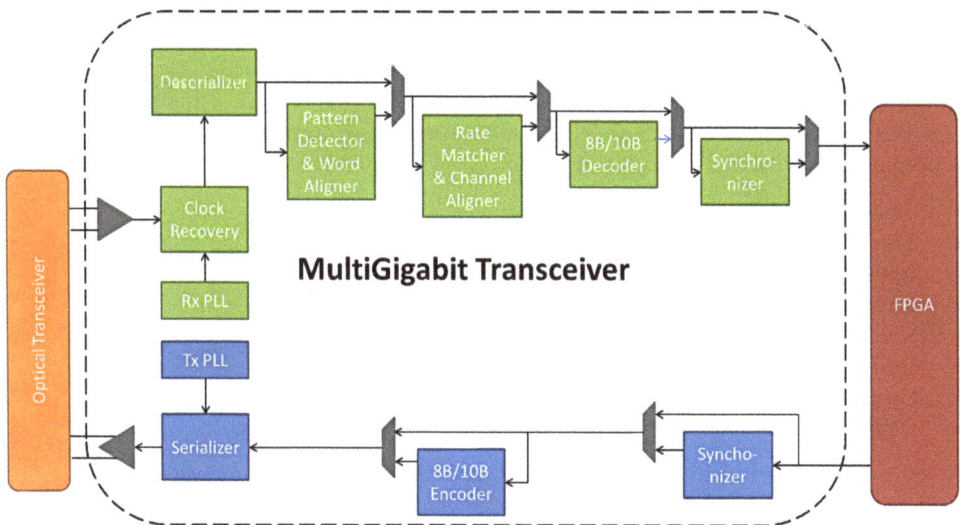

Fig. 4. MGT building block configuration.

3. Optical links design considerations

Optical fiber links are typically used for high-speed data transmissions. In such scenarios, several specific issues need to be taken into account. In this section, the most common problems in high-speed PCB design will be enumerated and briefly described. Some suggestions will be also given in order to minimize them. The key points to be considered are the transmission lines, crosstalk, differential traces, decoupling and power system, EMI, clock signals and other specific considerations.

3.1 Transmission lines

Lossy transmission lines are common on printed circuit boards. Signals travelling at high frequencies through narrow strips are affected by the skin effect and the dielectric losses, producing signal distortion. In a basic model, transmission lines can be described as formed by a network of inductors, capacitors and resistances, as it is shown in Figure 5. At high frequency, these effects appear, causing reflections and attenuations of the signal.

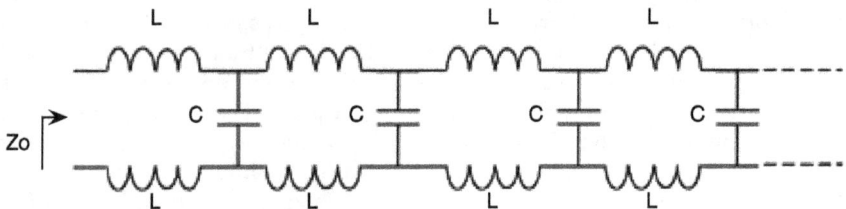

Fig. 5. Distributed parameters transmission line model.

In this sense, the characteristic impedance of a transmission line needs to be introduced as its fundamental parameter. It is defined by the following expression:

$$Z_0 = \sqrt{L/C}$$

In order to reduce the reflections, the characteristic impedance (Z_0) of the line should be matched to the source impedance (Z_s) as well as to the load impedance (Z_L). This matching procedure can be carried out by using several types of matching networks:

- Single Parallel Termination.
- Thevenin Parallel Termination.
- Active Parallel Termination.
- Series-RC Parallel Termination.
- Series Termination.
- Differential Pair Termination.
- On-Chip Termination.
- Diode Termination.

A common problem in high-speed PCB design is the formation of undesired stubs. These stubs are non-terminated transmission line segments that generate impedance mismatching, and then, undesired reflexions. Stubs can appear from single non-terminated lines, pins, unfinished IC's or non-terminated segments acting as antennas, as illustrated in Figure 6. In order to avoid unexpected stubs, the length or the strips must be reduced at maximum and all the unused pins should be connected to ground or power.

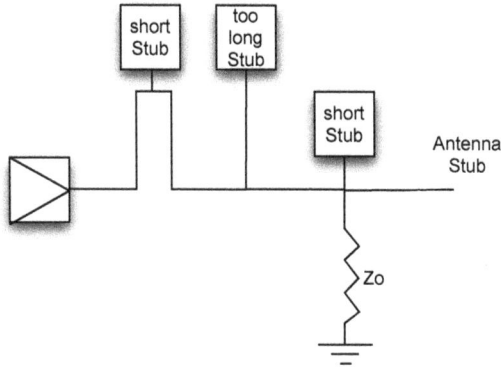

Fig. 6. Stubs examples.

3.2 Crosstalk and differential traces

Coupling between signals appears due to induced voltage from one line to another one. The magnetic coupling is produced by the mutual inductance, while the electric coupling is represented by mutual capacitances. This is represented in Figure 7. The undesirable energy coupled between lines is called crosstalk. Switching signals travelling in the same direction and driving the same current are in even mode. Otherwise they are in odd mode.

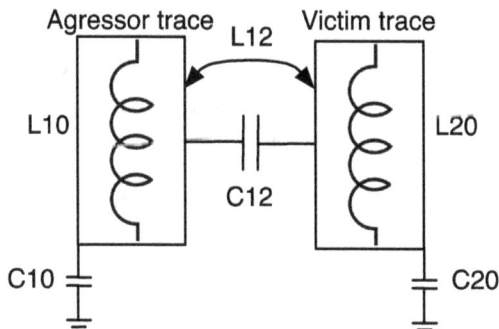

Fig. 7. Crosstalk between traces.

For reducing this effect a number of considerations are listed:

- Use, when possible, strip-lines. They are strips placed between planes, acting as shielding.
- Use, when possible, proper stack-up, by placing the traces as close as possible from their reference planes. This will help to uncoupling nearby signals and will couple it to the reference plane.
- Separate tracks as far as possible. Use the rule: the distance between the middle of traces must be four times the trace width.

- Use terminations to reduce the crosstalk.
- Minimize the signal return loops. If it exists a significant coupling between signals of contiguous layers, both should be orthogonal to each.

Another very important consideration in order to reduce crosstalk is using differential routing techniques (see Figure 8) since, in this way, ground noise related problems are avoided by providing high noise margins. In addition, the inductance influences are cancelled. Due the differential signals have the same length and they are opposite, there is not signal return through ground. Switching times can be more accurate if these kinds of signals are used instead of single-ended signal.

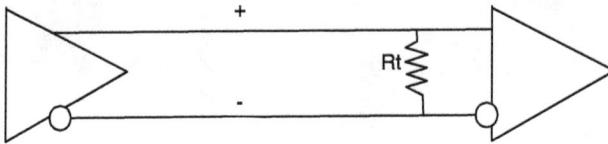

Fig. 8. Differential signal.

The key point when dealing with this kind of signals is setting the lines with the same distance in order to keep the signals in phase. Otherwise, the power integrity should be affected. It is possible to give a number of recommendations regarding the design of this kind of lines:

- The traces must have the same length. This is because the delay must be minimum. Otherwise, it could generate serious EMI problems due to the appearance of common mode currents. Another problem is caused by the induced current on the plane, acting as crosstalk.
- Keep the loop area minimum. The traces must be routed as close as possible, even eliminating the planes that are below of differential traces and removing induced loops.
- When dealing with differential traces very close to each other, terminations for reducing coupling should be used. For selecting the appropriate termination, impedance calculations can be required.
- The separation between lines must be constant along them. Try to avoid layer changes, so routing in the same layer, and try to avoid using traces between two lines forming a differential pair.
- Try to avoid the use of vias. They introduce losses and an impedance steps. If we need to use them, in transitions, place them next to each other for maintaining the differential impedance ratio.

3.3 Decoupling and power systems

It is indispensable to know which is the current return of a high-speed signal. An effect, called ground bounce, will produce to cause a reference level increase. The effect is caused by the short switching times. They cause high transients current and discharge load capacitances. Load capacitance, inductance of the connectors and the number of switching are the predominant factors to increase the effect. For this reason, capacitors should be placed near devices and parasitic inductances that contribute to the ground bounce. This effect is shown in Figure 9.

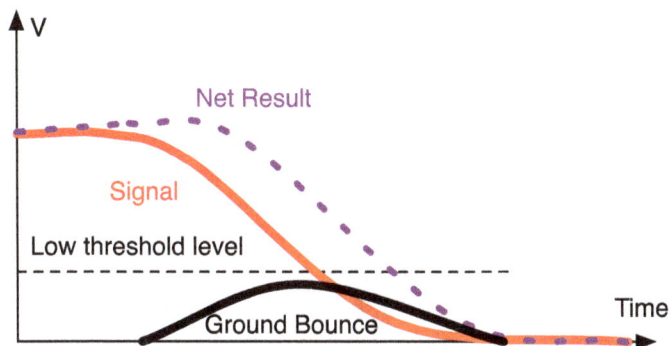

Fig. 9. Ground bounce effect over signal.

The uncoupling of power supplies provides benefits on power integrity, signal integrity and highly reduces EMI.

Small capacitors usually display better performance at high frequency regimes, but they also usually display higher inductances than bigger ones. Each capacitor has a recommended frequency band usage, described by its equivalent series resistance (ESR) and the quality factor (Q). To reduce its inductance, the capacitor should be placed as close as possible to the power source. It is recommended to connect it directly to the power and reference plane, avoiding any surrounding traces around it. The distance should not be more than quarter of wavelength.

In the frequency response of a capacitor there is a point called resonance point where the value of the impedance of the equivalent LC circuit is zero, as shown in Figure 10. From this point, the capacitor behaves like an inductor rather than a capacitor. The use of multiple capacitors in parallel does not change the resonance frequency but increases the capacitance effect, so reducing the individual inductance and the ESR.

The impedance response in power systems can be improved by the increasing of the number of capacitors and by considering capacitors with moderate ESR.

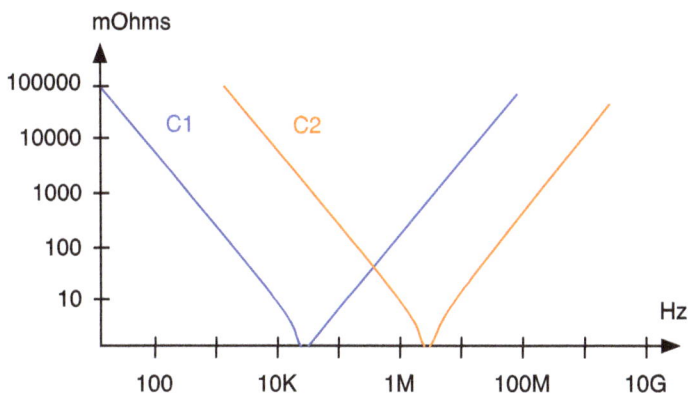

Fig. 10. Frequency response of the inductance of capacitors in parallel.

Other considerations that we should follow are:

- Eliminate connectors when possible.
- Use multilayer PCBs.
- Connect the capacitor pad in the plane through big vias to minimize the inductance and so helping the current flow.
- The traces travelling from power pins to planes must be wide and short in order to reduce the serial inductance, so decreasing the ground bounce.
- Connect each ground pin or via in the plane individually.
- The signal returns that go through connectors must have ground connections with the same potential. For this, the use of several returns (ground) for each one or two signals is necessary.
- Using antipads reduces coupling between the connector and the ground or power plane.

For isolating the high frequency noise, local filtering is recommended. Ferrites, requiring a big size capacitor to keep the output impedance in a reasonable level, are usually used.

Another point is which considerations take into account when using analog and digital sections in power systems. Variations in voltage gradients can be produced, due to the high frequency of returning currents or to the current that flows through the planes from regulated sources. These variations produce the charge and discharge of the bypass and planar capacitors of the circuit, therefore generating noise. To avoid the noise generated for another circuit sections different power supplies distributed in different regions of the circuit can be used. It can be used different voltage power distributed in the same plane. If each power section requires its own distribution plane, then they should have their own reference plane. All reasons to use planes go in the same direction: noise control.

Some of the rules regarding the use of planes are the following:

- Planes must be routed separately, in star. When multiples islands of power supply are routed in the board, they must be connected to a single point through 0-ohm resistors or ferrites. Often, the analog ground is joined to the digital ground, in this way.
- Do not allow sections of analog power to be placed above or below a region of digital plane. The components must be efficiently placed and grouped with their planes without overlap with other circuits (Figure 11).
- Be careful when uncoupling. Do not bypass erroneous references that may cause noise coupling between planes.
- Do not track traces if its current return has a discontinuity or gap. They will have a big loop and EMI problems can appear.
- If the power plane shares analog and digital supply, both sections must be separated. Then, the components should be placed in their respective planes.
- Each high-speed line must be referenced with its contiguous plane, for reducing loop, controlling the impedance and the crosstalk.

The layer stack is very important for reducing loops and having the control of the capacitance between planes as well as having EMI control. A good design is characterized by having each trace referenced to nearby planes and each power supply, providing a capacitance between planes. A good layer stack example is shown in Figure 12.

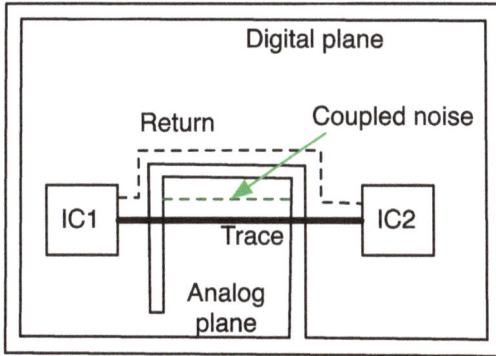

Fig. 11. Trace overlapping a not related plane.

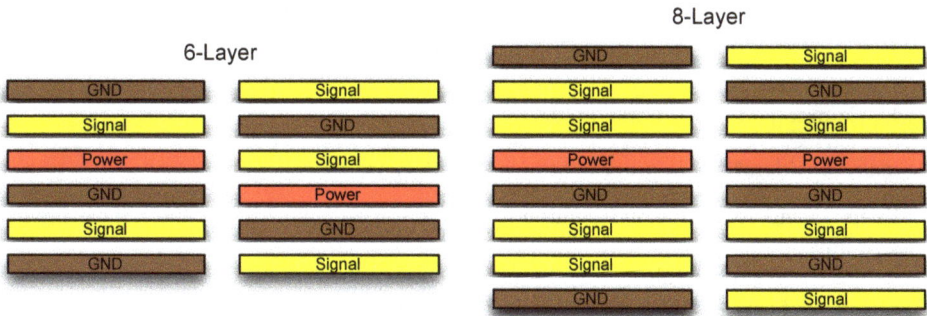

Fig. 12. Good stack layer for 6 and 8 layers.

As it can be seen, it is always preferred a layer stack where the signal layers are placed between two planes. If using many signal layers is necessary, two signal layers can be contiguously placed, although they should be orthogonally routed, in order to avoid couplings.

3.4 Electromagnetic Interferences (EMI)

Electromagnetic interferences are directly proportional to the change in current or voltage as a function of the time and the serial inductance of the circuit. PCBs always generate EMI, so a number of considerations for minimizing them should be taken into account.

- Place each signal layer between ground plane and power plane. Inductance is directly proportional to the distance. The shorter distance, the lower inductance.
- Select low inductance components, like surface mount devices (SMD).
- Reduce return paths by using solid ground planes. Keep the signal and the return as close as possible each to other. Remember that current return travels through the minimum impedance path.
- Place capacitors near connectors or devices.
- The use of strip lines adds an extra control on EMI.
- Avoid the use of stubs. They can behave like antennas.

One of the main sources of EMI is the current loop. The other is common mode problems. The differential mode is the mode where the signals travel forming a path and a return in opposite direction. When signals travel in the same direction, both signal and return, is called common mode. This occurs because the ground is not a perfect driver and there is an undesired associated inductance. This effect is illustrated in Figure 13.

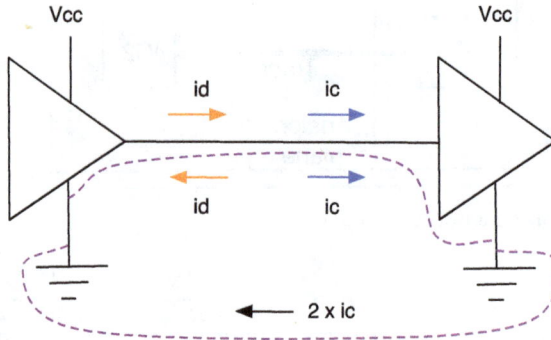

Fig. 13. Common and differential mode.

The main considerations in order to reduce this effect are the following:

- Keep a solid reference and continuous plane for each line. Trace the critical lines as striplines.
- Reduce presence of stubs.
- Ensure that exists a good capacitive coupling between planes.

3.5 Clock transmission line

We have mentioned differential lines but we have not introduced single-ended lines connecting a source with load or receiver. They are used in point-to-point routing, signal clock routing, low-speed lines and non-critical I/O lines. Signal clock routing is the most remarkable point in single-ended lines. The following considerations are given to improve signal integrity in clock signals:

- Keep lines as straight as possible. Use rounded shapes instead of sharp angled ones.
- Do not use multiple signal layers for clock signals.
- Do not use vias in clock lines. They change the impedance and they cause reflexions.
- Place a ground plane next to the outer layer to minimize the noise. If you use inner layers for routing a clock trace, form a sandwich with both layers.
- Use terminations to minimize reflexions.
- Use point-to-point traces.

The clock signals can be routed in several ways. If a daisy chain routing is used, undesired stubs or short traces can appear, so degrading the signal quality and producing reflexions. When considering a star routing, the clock signal arrives to all devices at the same time, so lines must have the same length. Each load must be identical for minimizing signal integrity problems. To design traces with the same length, serpentine techniques for time adjusting are used. Several types of clock signal routing are illustrated in Figure 14.

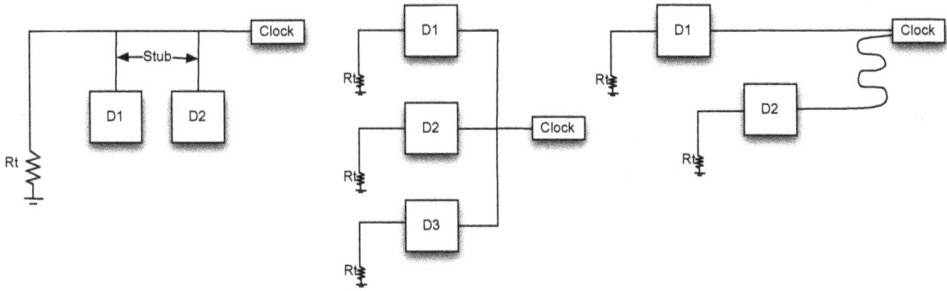

Fig. 14. Types of clock signal routing.

3.6 Other considerations

In high-speed designs other considerations are taken into account, depending on the amount of information needed to be sent.

For example, at data rates of 622 Mbps and higher, the skin effect is extremely important. This could cause signal attenuations over long distances. The result is a low pass filter with attenuation, which increases with frequency. For this reason, the traces should be wider.

On the other hand, connectors and vias cause impedance discontinuities. In order to reduce them, specific connectors with shielding features with many ground connections must be used.

High-performance cables are also very important to have a good bandwidth and for minimizing attenuation losses.

When working at 2.5 Gbps and beyond, the problems become substantially more difficult to eliminate. There are many copper and dielectric losses. It is needed to pay attention to every consideration for PCB and layout design. A backplane thickness of less than 0.200 inches gives the best result. Vias used for interconnection between layers create line discontinuities. These routed layers should be as few as possible and thus limiting the number of vias, so the vias are shorter, minimizing line discontinuities. Buried vias can be used to reduce this problem in thicker boards.

The material used in boards is very important. FR4 dielectric commonly used has significant losses above 2 Gbps. For this reason, low-loss dielectric PCB material can be used such as Rogers 4350, GETEK or ARLON.

4. Signal integrity studies

Digital system design has therefore moved deep into the Multi-GHz range, with signal rise and fall-times of the order of 100ps or less. It is a well-known fact that Signal Integrity (SI) simulations become necessary when system designs break the 50-100MHz barrier, and are virtually mandatory in the GHz range. SI simulations are used to ensure the quality and accurate timing of electrical signals. The benefits of SI analysis early in the design cycle are well established, as it allows the identification and resolution of SI problems like overshoot, ringing, crosstalk, delay mismatches, etc. before the first prototype is built.

4.1 Pre-layout studies

The pre-layout simulations are required at the earliest stages of the PCB design. In this stage the designer evaluates several topologies and selects the one that fulfils all the specifications such as size, component number and performance. In addition, the results of these simulations help to set crucial parameters for the transmission structure, such as trace width, trace spacing, maximum trace length, and critical component placement. It is important to understand these simulations are intended for selecting the components and topologies as well as for fine tuning of the signal path. The results analysis is used to set the rules that will be incorporated into the layout.

As an example, and taking into account the design considerations described in the previous section, a real design is going to be studied. The main elements in this design are:

- SFP Optical Connector with two receiver channels (1)
- Transmitter/Receiver Chip Set (2)
- FPGA (3)

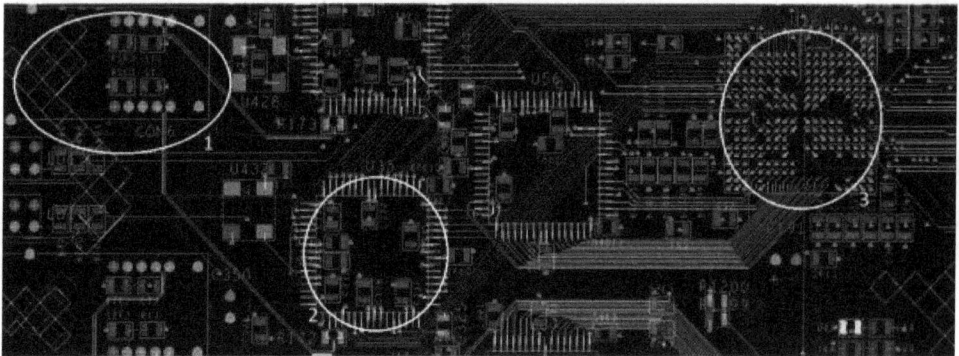

Fig. 15. Elements of Advanced Digital Systems based on High-Speed Optical Links in PCB.

The optical connector has a differential output (LVPECL), routed to the entrance of Transmitter/Receiver Chip Set. These transmission lines are the most critical of the study because is a point-to-point serial data transmission with a very high bit rate.

Another point to be taken into account is the connection between data acquired in the Transmitter/Receiver Chip Set and the FPGA. In this case, the importance of the study is not based in data frequency. We must avoid the data bus crosstalk.

4.1.1 Differential lines

This signal has a critical jitter and the topology, geometry, length, and termination impedance must be carefully studied. In this case, it is observed the differential line without being routed.

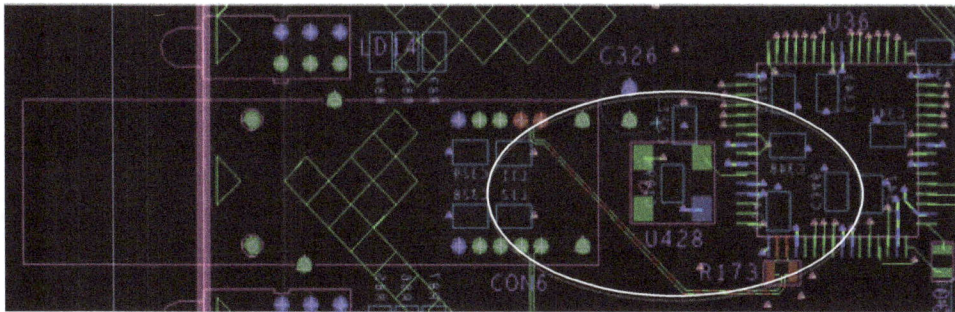

Fig. 16. Unrouted differential lines.

The topology is extracted according to the manufacturer's datasheet optical connector.

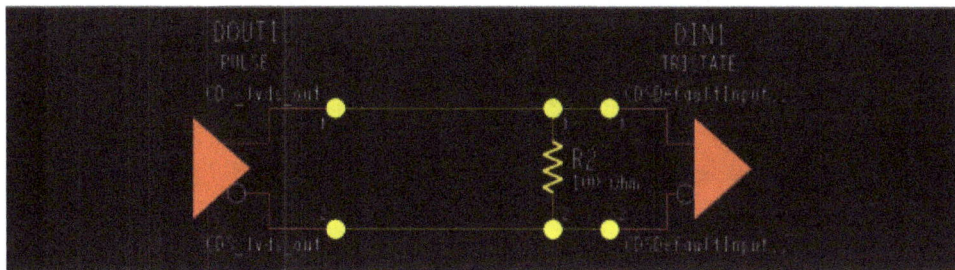

Fig. 17. Extracted topology.

An ideal study without transmission line is performed. The simulation for this case is:

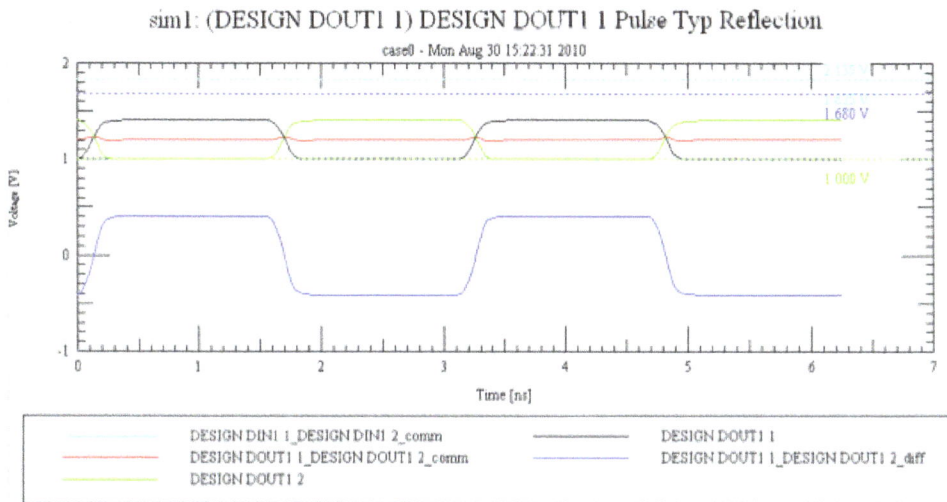

Fig. 18. Waveforms for differential lines.

In black and green color RX+ and RX- signals are observed and in blue color the resulting differential signal is presented. The voltage values obtained are the same that the manufacturer recommendations.

The next step, seeing the topology is correct, it would add a transmission line and observe the maximum length that could have the routing. In the next section another types of line will be studied.

4.1.2 Parallel data bus

In this case, the connection between the transmitter/receiver chip set and the FPGA is studied. Its speed of transmission is not comparable with that of the differential line. Even so, the maximum length must be checked and a termination may be needed.

First, a study of a single line to detect maximun length and optimal termination has to be performed. The study was performed for a bus of 32 lines at 80 MHz.

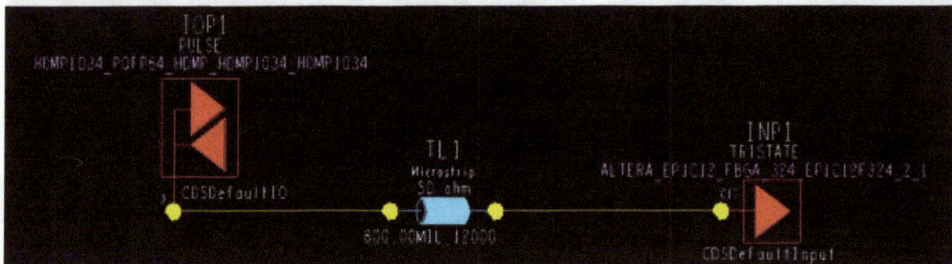

Fig. 19. Topology of single-ended line.

The waveforms for each length (800, 5000, 9000 y 12000 mils) are as follows:

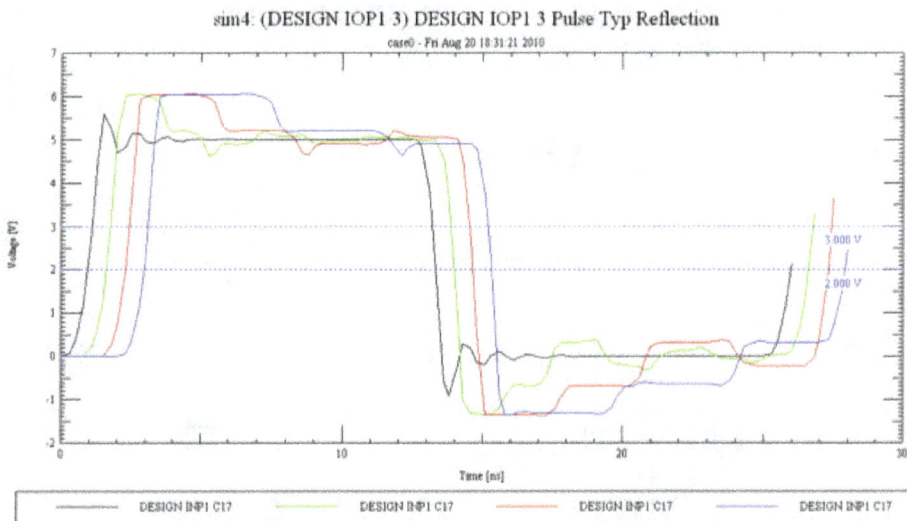

Fig. 20. Waveform for single-ended line.

The waveforms 800, 5000, 9000 y 12000 mils in black, green, red and blue provide a delay to the length, which increases to approximately 2.14 ns in the worst case. This shows a mismatched line. Also the waveforms will degrade depending on the length of the track due to the increasing influence of the reflections, since an overshoot and ringing growing.

The worst case above (12000 mils) but with a termination is now studied. The topology is:

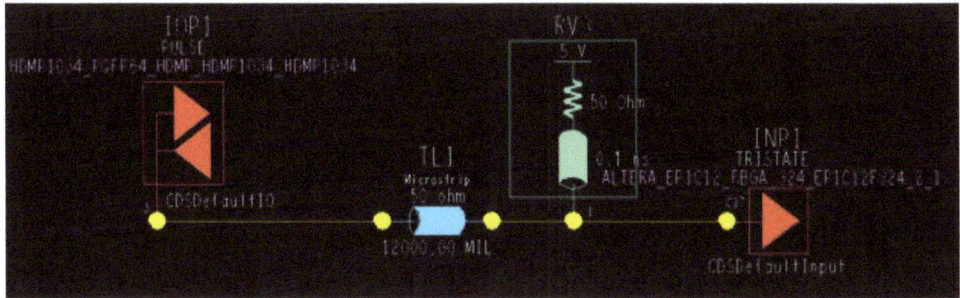

Fig. 21. Topology of single-ended line with active termination.

The results are:

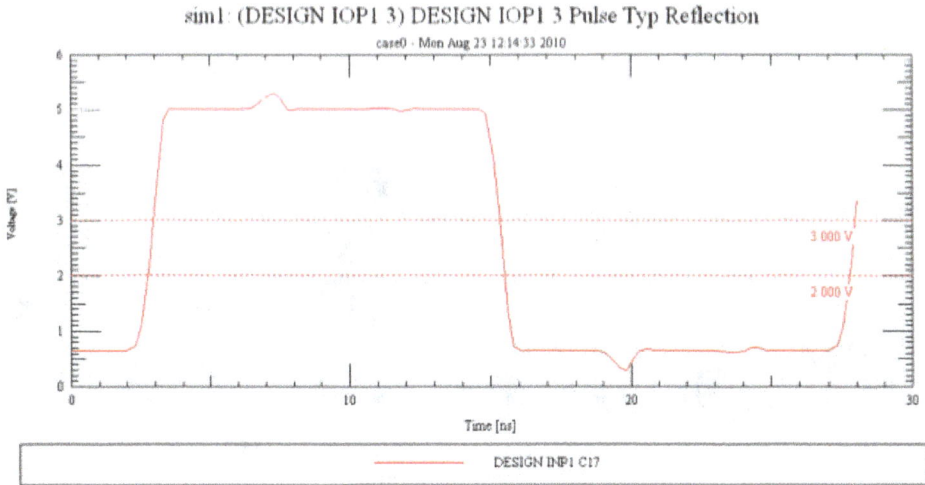

Fig. 22. Waveform of single-ended line with active termination.

Which greatly improves the previous case:

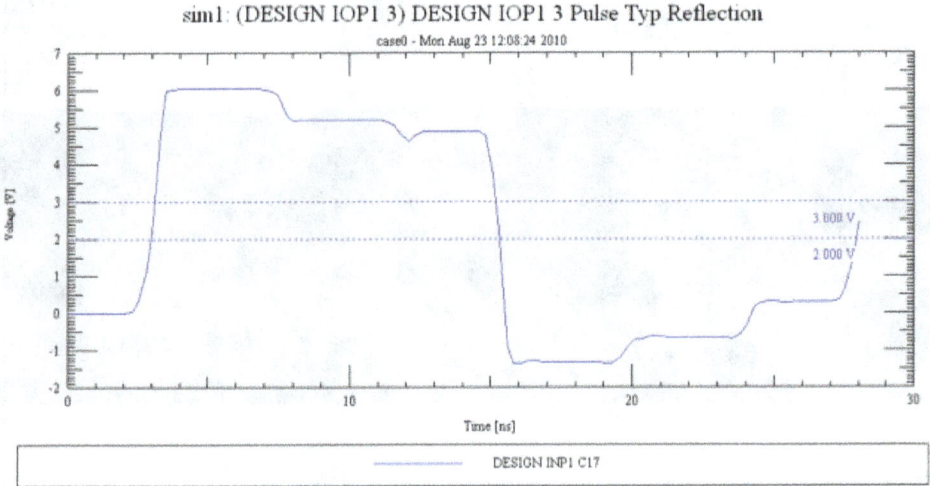

Fig. 23. Waveform of single-ended line without termination.

Working with a data bus is essential to examine the crosstalk between the lines. The topology studied is as follows:

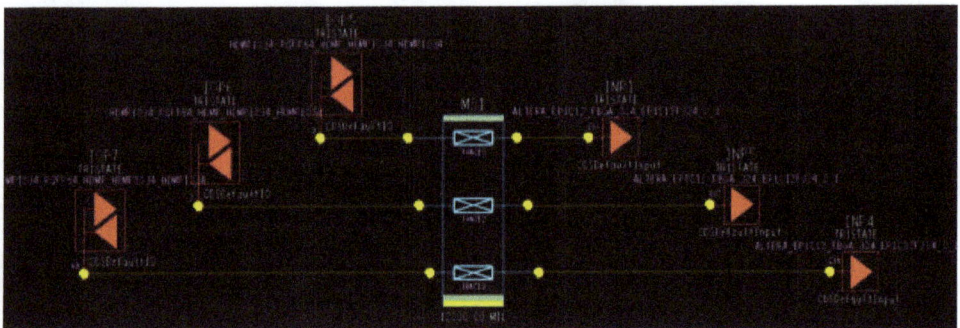

Fig. 24. Topology of three parallel lines.

Where there are three bus lines, taking the top and bottom as aggressors and the middle line as victim.

The results for high steady state are:

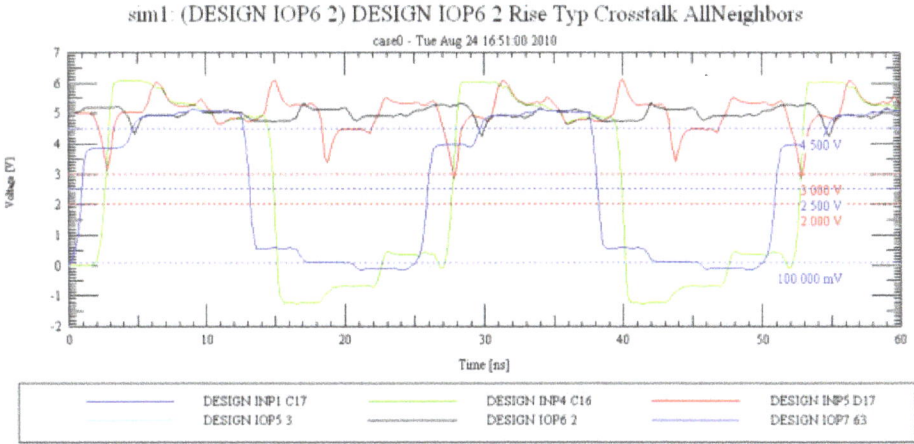

Fig. 25. Waveforms of three parallel lines.

In blue and green color, we have waveforms of the FPGA inputs, that correspond to the agressor signal. The red one is the receiver input that flowing by the victim trace, and the black one is the output driver of the victim trace. In this case a crosstalk about 2149mV is obtained and the simulation does not pass the test (the victim signal is in an undefined logic state, below 3V).

In the previous section is commented that the solution is adapting the transmission lines using terminations. If an active termination is added to the lines and then the simulation is made in each line, the result is as follows.

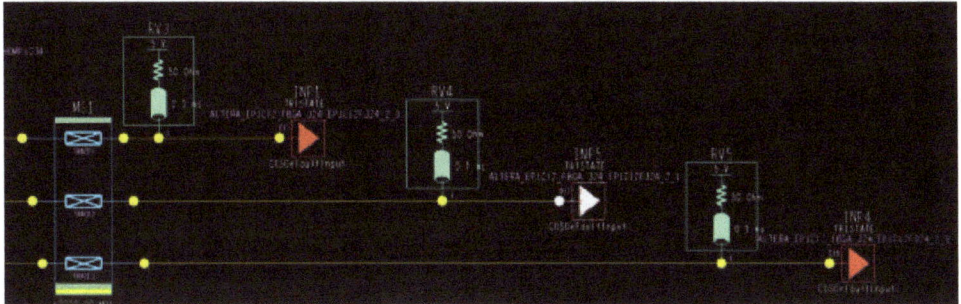

Fig. 26. Topology of three parallel lines with active termination.

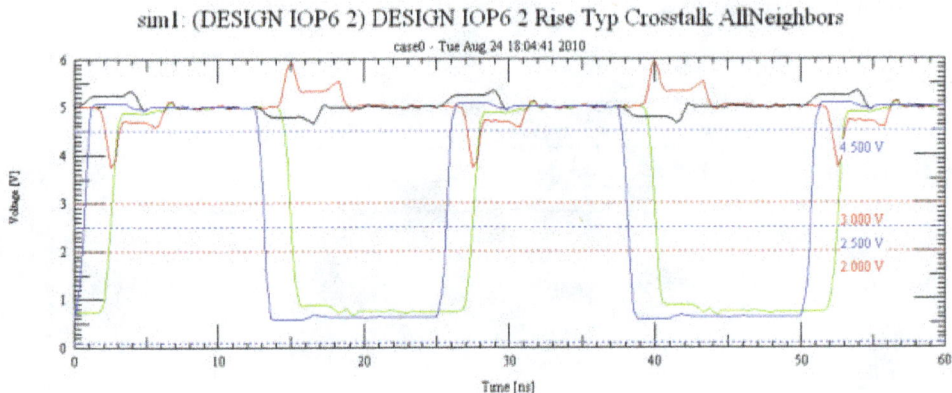

Fig. 27. Waveforms of three parallel lines with active termination (high level).

It can be observed a clear improvement in the crosstalk level, keeping the victim signal inside the permitted high level range. However, at low level it is shown a crosstalk of 1421 that causes the victim signal introduces in the forbidden region for low level, and this can cause errors.

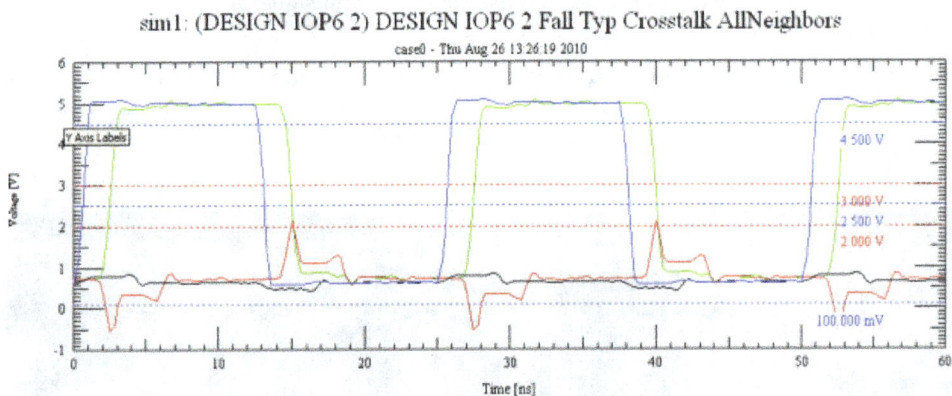

Fig. 28. Waveforms of three parallel lines with active termination (low level).

Then, the initial solution of placing an active termination is not suitable. As theory says, we should test a serial termination and check what happen.

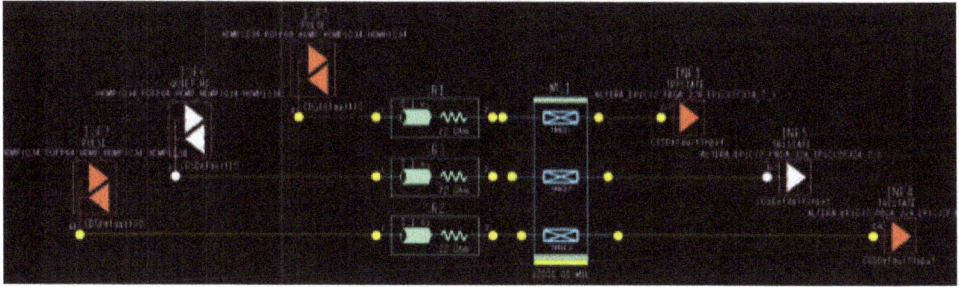

Fig. 29. Topology of three parallel lines with serial termination.

In this case, crosstalk level keeps the victim signal within the permissible limits either in/for high level (figure 30) or low level (figure 31). The values are 1365 mV for high level and 1514 mV for low level.

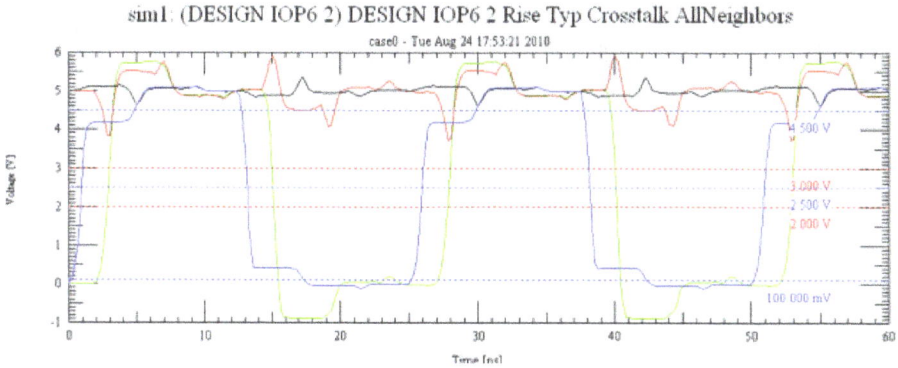

Fig. 30. Waveforms of three parallel lines with serial termination (high level).

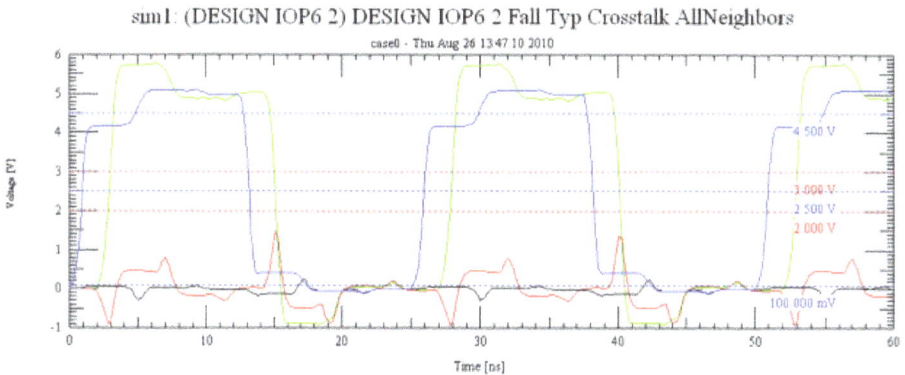

Fig. 31. Waveforms of three parallel lines with serial termination (low level).

4.2 Post-layout studies

When the layout is complete, a post-layout simulation is performed on the critical sections of the board to ensure there are no major signal integrity problems. Based on the results of the post-layout simulation, any changes required are incorporated in the layout and the layout is released to the fabrication house for board manufacturing. The post-layout simulation process requires the layout of all active layers on the board as well as the physical properties of the dielectric and metal layers. The post-layout simulations use the physical properties supplied by the fabrication house that may differ from those published.

4.2.1 Differential lines

Taking the pre-layout studies conducted on the differential lines and after to route them with the defined conditions, we extract the real topology that will be sent to the manufacturer for verification.

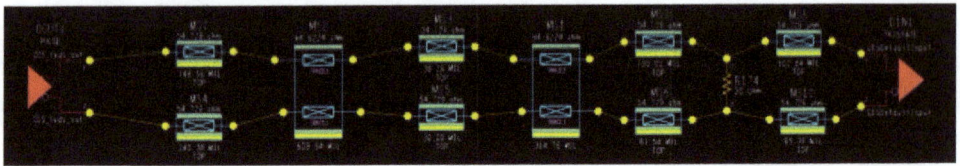

Fig. 32. Real topology of differential lines.

We can see that actual results match those obtained in the ideal simulation (Figure 18):

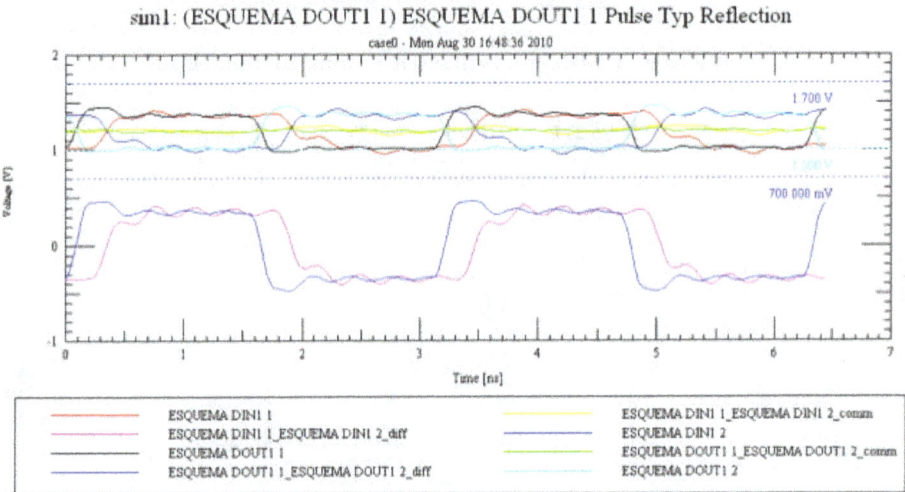

Fig. 33. Real waveforms of differential lines.

4.2.2 Parallel data bus

In this case, as in the above, we extract the real topology. We studied a serial termination in bus line.

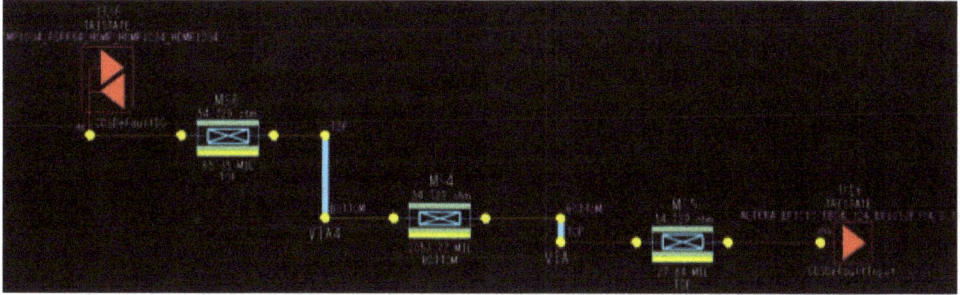

Fig. 34. Real topology of parallel data bus.

The results are as expected, we see that the data sent between the transmitter/receiver chip set and the FPGA function properly at the selected frequency and the distance routed.

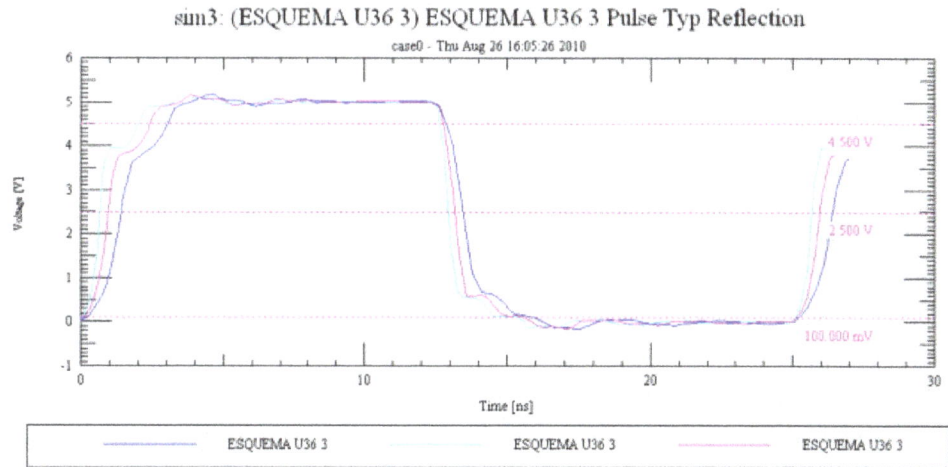

Fig. 35. Real waveforms of parallel data bus.

5. Conclusions

The wide range of applications of optical fibers has been continuously supported by their friendly integration with classic electronics. Being the optical transceiver the key element in such hybrid systems, their design is not straightforward, and need to take into account a good number of particularities.

In this chapter the main handicaps when developing electronics specifically for high-speed fiber optic communications have been highlighted. Some of these issues fully lay in the field of electronic engineering. Other aspects, and due to the high frequency involved in such systems, need some additional analysis. In this sense, the electromagnetic theory, as used for wave propagation in dielectric materials, has to be applied to the electric transmission lines present in some cases.

Being experiments the common analysis tools in hybrid opto-electronic systems, some simulation tools can be sometimes used for fixing specific problems like cross-talking. These numerical tools are of particular interest for proper routing design in PCBs.

After all, these systems have demonstrated their applicability in a huge number of scenarios including communications, medicine, nuclear research, etc.

6. References

[1] Douglas Brooks, "Signal Integrity Issues and Printed Circuit Board Design", Prentice Hall PTR, 2003. ISBN: 0-131-41884-X

[2] Stephen C. Thierauf, "High-Speed Circuit Board Signal Integrity", Artech House Inc., 2004, ISBN: 1-58053-131-8

[3] Mark. I. Montrose, "Printed Circuit Board Design Techniques for EMC", Wiley-Interscience-IEEE, 2000, ISBN: 0-7803-5376-5

[4] Lattice Semi, "High-Speed PCB Design Considerations", Technical Note TB1033, 2011.

[5] Altera, "High-Speed Board Layout Guidelines", Application Note 224, 2009.

[6] Sackinger, E., "Broadband circuits for optical fiber communication", Wiley, 2005. ISBN: 9780471712336

[7] Cox, C.H., "Analog optical links: Theory and practice", Cambridge University Press, 2004. ISBN: 0-521-62163-1

[8] Muller, P., Leblebici, Y., "CMOS Multichannel Single-Chip Receivers for Multi-Gigabit Optical Data Communications (Analog circuits and signal processing)", Springer, 2010. ISBN: 978-90-481-7473-7

[9] ATLAS Trigger and DAQ steering group, "Trigger and Daq Interfaces with FE systems: Requirement document. Version 2.0", DAQ-NO-103, 1998.

[10] Dowell, M. Pearce, "ATLAS front-end read-out link requirements", ATLAS internal note, ATLAS-ELEC-1, July 1998

[11] J. Torres, "Estudio, diseño e implementación del módulo de preprocesado de datos del Sistema Read Out Driver para el calorímetro TileCal del experimento ATLAS/LHC del CERN", Tesis Doctoral, Servicio de Publicaciones de la Universidad de Valencia, Junio 2005.

[12] J. Torres et al, "Signal integrity studies at optical multiplexer board for tilecal system", 2007 IOP Publishing Ltd and SISSA.

[13] Lynne Green, "Signal Integrity", IEEE Circuits and Devices, November 1999.

[14] Jim Lipman, "Models make the difference in high-speed pc-board design", Electronic Design News, 15th April 1999.

16

Physical Layer Impairments in the Optimization of the Next-Generation of All-Optical Networks

Javier E. Sierra

GIDATI Research Group, Universidad Pontificia Bolivariana, Medellin, Colombia

1. Introduction

The Internet traffic is constantly growth and the applications of type unicast/multicast with different Quality of Service (QoS) requirements. The increase of internet traffic over the last couple of years is well known (the rate of increase is reported to be 60% to 100% per year (Malik, 2011). Applications like multicast are used more frequently than ever (for example HDTV, videoconferencing, IPTV, interactive games among others)(Kamat, 2006).

For this reason, the Optical Transport Networks (OTN) must continue the evolution towards All-optical networks (without optical-to-electrical-to-optical conversion). OTN employ Wavelength Division Multiplexing (WDM) in order to transmit great deals of information. WDM allows the multiplexing of different wavelengths along the same fiber, each one can transmit at speeds of around 40 Gbps and can achieve speeds in the range of Tbps along a single fiber. Currently, the equipment needed to carry out the transmission (including routing) of information functions in an optical environment between two nodes and at each one of these an optical-electronic-optical (OEO) conversion is carried out when it is needed in order to add or drop traffic. Optical Cross Connects (OXC) are systems that allow for the commutation of traffic at each of these nodes.

New applications (both unicast and multicast) do not yet have the capacity provided by a wavelength, therefore, by allotting a wavelength in the range of Gbps to an application of a couple of Mbps one is underutilizing the full bandwidth available in one wavelength. To solve the underutilization problem researchers have proposed the concept of Traffic Grooming (TG). TG came about in order to improve the utilization of bandwidth and optimize OTN systems (Solano et al., 2007). TG is the ability given to a WDM network to combine several slow speed traffics (in the range of Mbps or a few Gbps, example: OC-1, OC-3) into one of greater speed (OC-192 or greater). To accomplish TG all of the nodes must have some special characteristics, more so if it is needed for multicast traffic. The network design problem that support TG efficiently is not an insignificant one and the solution may have a great impact on the cost of the network. TG is ability to support unicast traffic has been widely researched (Bermond et al., 2006).

The routing of unicast traffic is accomplished using the concept of the lightpath, which is a virtual channel in a completely optical environment between two nodes (Zhang et al., 2008). The intermediate nodes do not carry out OEO conversions for routing. The concept of the light-tree is employed in order to support Multicasting Traffic Grooming (MTG). The transport of traffic point-multipoint is achieved in an entirely optical medium (without OEO conversions). This kind of transmission is called transparent and it is possible to carry it out using optical cross-connect (OXC). The architecture for the support of light-trees is presented by Khalil et al. (2006). When light trees perform grooming of unicast and multicast traffic they can use a lot of bandwidth in routing unicast sessions toward unwanted destinations. This is done in order to avoid OEO conversions in information transmission which, from a transparency point of view, are very expensive (Sreenath et al., 2006). With the purpose of improving on the resources available (wavelength and available capacity) in an optical transport network and to accomplish this in a completely optical medium, Sierra et al. (2008) have proposed the Stop and Go (S/G) Light-tree architecture. S/G light-tree allow optimal routing and grooming of unicast and multicast sessions.

Currently, there are different architectures for optical transport nodes that allow the optimal routing and/or traffic management unicast/multicast using the concept of Traffic Grooming in optical networks. However, grooming techniques and the assignment and routing algorithms proposed do not account for phenomena that can be provided in the optical fiber (Bastos-Filho et al., 2011), which mitigate or added interference between the different wavelengths in WDM Networks.

The chapter of the book describe various optical transport architectures that performs unicast/multicast traffic grooming. Routing and wavelength assignment are analyzed taking into account the effects of linear and nonlinear optical fiber. The model presented optimizes network resources taking into account the blocking probability in all-optical transport networks. Traffic has different levels of service quality. The work presented shows different optimization models and algorithms.

1.1 Background multicast traffic grooming

There is a tendency in telecommunications toward an increase in multicast traffic, for this reason many researchers have been interested in examining and providing better solutions. In the design of optical networks WDM mesh that are used by Billah et al. (2003) a heuristic algorithm for the efficient use of bandwidth and improvement of the throughput of the network is proposed. It is divided into two steps: *i*) find the light-tree and *ii*) assign the wavelength. They apply an algorithm for a WDM mesh network with sparse splitter capacity. They show that the heuristic algorithm accomplishes a significant reduction in the number of wavelengths needed in a connection and in the total wavelengths required. The nodal architecture used is Multicast Grooming Capable Wavelength Router. The node has wavelength conversion (efficient in the optical domain), splitting, grooming, and amplifiers. The heuristic algorithm takes into account the amount of hops, used by Dijkstra to determine the Shortest Path and the assigning of wavelengths are accomplished through the First Fit strategy.

The term light-tree is often used to refer to the design on multicast networks with grooming capabilities. It was introduced in *wavelength-routed optical networks* by Sahasrabuddhe & Mukherjee (1999). In their article they focus on unicast and broadcast traffic. They present the light-tree as an optimization problem, given a virtual topology, how to find a traffic matrix with the following functions: limit packet delay, average in the jump distance for a wide area network and, limit the number of total transceivers in the network. They explain that a light-tree supports as much unicast traffic as it does multicast, although it has better performance for multicast by using splitting light.

Vishwanath & Liang (2005) examine the problem of online multicast routing in mesh transport networks without the capability for conversion of wavelengths, by dividing wavelengths in multiple time slots and multiplexing the traffic. The goal is to route the multicast traffic efficiently by using grooming while balancing the connection loads. Likewise in Sahasrabuddhe & Mukherjee (1999), they point out that multicast applications can be efficiently routed using light-tree (this improves throughput and network performance).

Sreenath et al. (2006) address the problem of routing and the assigning of wavelengths in multicast sessions with low capacity demands in WDM networks with sparse splitting capacity. For this reason only a few nodes on the network are able to split traffic. Nevertheless those nodes not able to split can do so with OEO conversions. They point out that the splitting of traffic is more expensive at the electronic level than at the optic level because of the delays caused by OEO conversion.

Liao et al. (2006) explore the dynamic problem of WDM mesh networks with MTG to analyze and improve the blocking probability, by proposing an algorithm based on light-tree integrated with grooming. The results after using it show its usefulness. The blocking probability is reduced while taking advantage of the resources of the network under low restrictions of non conversion of wavelength and a limited number of wavelengths and transceivers. They divide the problem into three sub-sections: *i*) defining the virtual topology using light tree, *ii*) routing the connection applications across the physical topology and optimally assigning the wavelengths for the multicast tree and, *iii*) grooming low speed traffic in the virtual topology.

Khalil et al. (2006) explore the problem of providing dynamic low speed connections unicast and multicast in mesh WDM networks. They focus on the dynamic construction of the logic topology, where the lightpath and the light-tree are configured according to the traffic demands. They also propose using all resources efficiently in order to decrease the blocking probability. This is how they propose several heuristic sequential techniques, by breaking down the problem into four parts:

1. Routing problem
2. Logic topology design
3. Problem of providing wavelengths
4. TG problem

Huang et al. (2005) also analyze the blocking probability. Nevertheless, they also analyze when there are sparse splitting capacities. The algorithm that they proposed is based on light-tree dynamics that support multihop. The algorithm can be dropped and branched and can establish a new path when an application is received or alter itself when there are existing path free of traffic.

The components mentioned carry out the process of grooming by using OEO conversions when multicast and unicast traffics are jointly multiplexed.

1.2 Routing unicast and multicast traffic together

In WDM networks, there are two typical all-optical communication channels, lightpaths and light-trees (Kamat (2006)). A lightpath is an all-optical communication channel that passes through all intermediate nodes between a source and a single destination without

OEO conversion. A light-tree is an all-optical channel between a single source and multiple destinations. Like the lightpath, there is no OEO conversion at any intermediate node on a light-tree.

Using a light-tree to carry multicast traffic is a natural choice in WDM mesh networks. Many researches have addressed the very fundamental multicast routing and wavelength-assignment problem, such as in (Liao et al., 2006; Singhal et al., 2006; Sreenath et al., 2006; Ul-Mustafa & Kamal, 2006). In these studies, proposals for handling static and dynamic traffic has been made. Proposals have focused on mathematical models based on ILP (Integer Linear Programming) and heuristic techniques based on minimum-cost steiner tree. All these studies used a node architecture similar to that employed in Singhal et al. (2006), which employs Optical Splitters for the duplication of traffic. However, these proposals do not take into account the optimal routing of unicast and multicast traffic together.

Huang et al. (2005) tackled the problem of routing traffic unicast/multicast together. They address the online multicast traffic grooming problem in wavelength-routed WDM mesh networks with sparse grooming capability. The architecture node that employ them provide: optical multicasting and electronic grooming. The basic component of the architecture is a SaD Switch, which has configurable Splitters.

The routing, allocation and grooming problem has been initially resolved with off-line techniques. Sahasrabuddhe & Mukherjee (1999) presents a mathematical model (MILP) with opaque nodes (OEO conversions) and wavelength continuity constraint for the type broadcast traffic. Billah et al., 2003; Zsigri et al., 2003 employs heuristics that use Shortest path and First Fit for the routing and allocation of wavelengths. Additionally, it must be taken into account that not all nodes have multicast capabilities (sparse splitting).

Recently the work has been focused on the analysis of dynamic traffic. Vishwanath & Liang (2005) proposes an Adaptive Shortest Path Tree (ASPT) using Dijkstra's algorithm that takes into account a function of cost to minimize implementation costs. Khalil et al. (2006) divides the problem into: *i*) routing, *ii*) logical topology, *iii*) provisioning and *iv*) traffic grooming. This makes it possible to minimize the blocking probabilities in transparent networks.

In previous works, different algorithms have been used to handle the traffic unicast and multicast together but taking into account electronical grooming and OEO conversions. Below, we describe the problems of using the architectures mentioned.

1.2.1 Problem definitions

In this section, an example is used to explain the disadvantages of the classical methods used for routing unicast and multicast traffic. Let us consider a subset of the *NSFNet* network of 14 nodes interconnected through optical links (Figure 1). Three sessions are considered: *i*) S_1 being a unicast session $\{N_3\} \rightarrow \{N_6\}$, where the node N_3 is the source node and the node N_6 is the destination; *ii*) S_2 being a multicast session $\{N_3\} \rightarrow \{N_6, N_7\}$, where N_6 and N_7 are the destinations nodes, and *iii*) S_3 being a unicast session $\{N_5\} \rightarrow \{N_7\}$, where the node N_5 is the source node and the node N_7 is the destination. Routing these two sessions can be performed in the following ways:

Light-trees (Singhal et al. (2006), Figure 2): sessions S_1 and S_2 are both routed through the same wavelength. In this case, no OEO conversions are used but traffic cannot be differentiated. As a consequence, all groomed traffic in a light-tree is routed to all

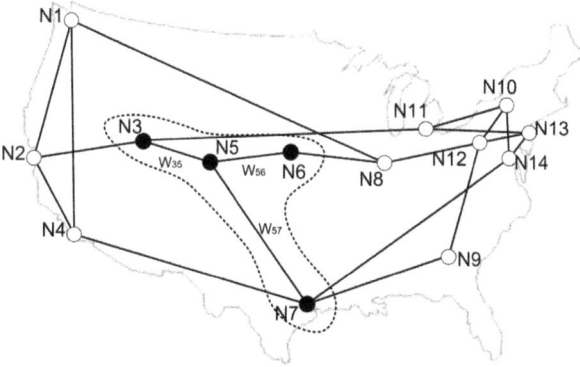

Fig. 1. NSFNet network. Sessions S_1 and S_2 in nodes N_3, N_5, N_6 and N_7

destinations. In this example, since the S_1 traffic should not be sunk at node N_7, there is bandwidth wastage. When a new request arrives (S_3) a new lightpath ($N_5 \rightarrow N_7$) is set up.

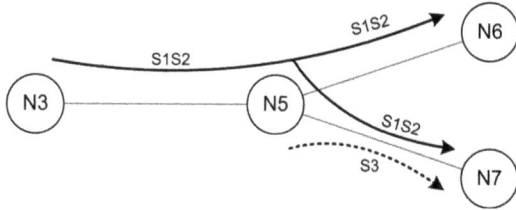

Fig. 2. Example Light-tree, Unicast $S_1 : \{N_3\} \rightarrow \{N_6\}$, Multicast $S_2 : \{N_3\} \rightarrow \{N_6, N_7\}$, and Unicast $S_3 : \{N_5\} \rightarrow \{N_7\}$

Lightpaths (Solano et al. (2007); Zhu & Mukherjee (2002), Figure 3): two lightpaths are needed for routing both sessions S_1 and S_2. The first lightpath follows the path $N_3 \rightarrow N_5 \rightarrow N_6$ routing the sessions S_1 and S_2. The second lightpath routes session S_2 using the path $N_6 \rightarrow N_5 \rightarrow N_7$. It requires an additional wavelength, even though both demands could fit within one wavelength. In this case, there is also a waste of bandwidth, since spare bandwidth cannot be used. As in Light-tree, this scheme requires an additional lightpath to route S_3.

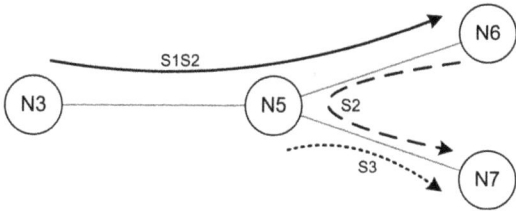

Fig. 3. Example Lightpath, Unicast $S_1 : \{N_3\} \rightarrow \{N_6\}$, Multicast $S_2 : \{N_3\} \rightarrow \{N_6, N_7\}$, and Unicast $S_3 : \{N_5\} \rightarrow \{N_7\}$

Light-trails (Wu & Yeung (2006), Figure 4): one light-trail is required for routing sessions (S_1, S_2, S_3). A light-trail is an unidirectional optical bus. In the example, we can setup one between nodes N_3 and N_7 as $N_3 \rightarrow N_5 \rightarrow N_6 \rightarrow N_5 \rightarrow N_7$. The disadvantage of light-trails is that the path may contain repeated nodes and the length of a light-trail is limited. Note that in our example, a wavelength is used in $N_5 \rightarrow N_6$ and another one in $N_6 \rightarrow N_5$.

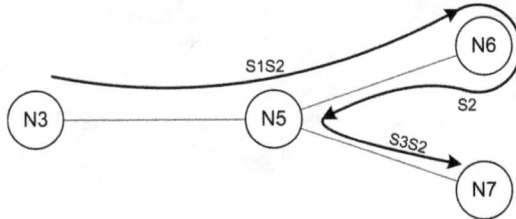

Fig. 4. Example Light-trail, Unicast $S_1 : \{N_3\} \rightarrow \{N_6\}$, Multicast $S_2 : \{N_3\} \rightarrow \{N_6, N_7\}$, and Unicast $S_3 : \{N_5\} \rightarrow \{N_7\}$

Link-by-Link (Huang et al. (2005), Figure 5): this scheme routes traffic allowing OEO conversions on all nodes. Three lightpaths are used: $N_3 \rightarrow N_5$, $N_5 \rightarrow N_6$ and $N_5 \rightarrow N_7$. A lightpath routes sessions S_1 and S_2 together from node N_3 to node N_5. Node N_5 processes electronically the traffic and forwards sessions S_1 and S_2 together through the lightpath $N_5 \rightarrow N_6$ and, S_2 and S_3 through the lightpaths $N_5 \rightarrow N_7$. The wavelength bandwidth is efficiently used, however it requires more electronic processing and OEO conversions.

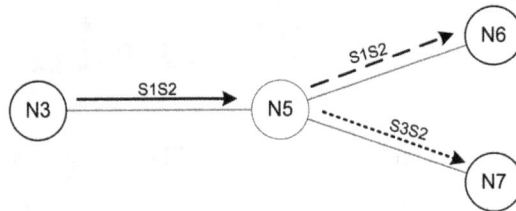

Fig. 5. Example Link-by-Link routing, Unicast $S_1 : \{N_3\} \rightarrow \{N_6\}$, Multicast $S_2 : \{N_3\} \rightarrow \{N_6, N_7\}$, and Unicast $S_3 : \{N_5\} \rightarrow \{N_7\}$

In particular, the problem arises when there are two (or more) sessions such as in: *a*) both are originated in the same root node, *b*) the wavelength capacity is enough for both sessions but, *c*) destination nodes of one session is a subset of the other. As we could see by our example, there is no optical architecture that can efficiently route such traffic: either residual bandwidth is wasted, or more OEO conversions are needed. While bandwidth plays an important role in the revenues of any service provider, the cost incurred by OEO conversion is the dominant cost in setting up the OTN. In general, the tendency is to setup a light-tree spanning to all possible destinations of a set of sessions, as shown in Figures 2-5.

Several studies tackle this problem. Huang et al. (2005) proposes an on-line technique called MulTicast Dynamic light-tree Grooming Algorithm (MTDGA). MTDGA is an algorithm that performs multicast traffic grooming with the objective of reducing the blocking probability by multiplexing unicast and multicast together. Khalil et al. (2006) also sets out to reduce the blocking probability, however it uses separate schemes for routing and grooming multicast and unicast traffic.

1.3 Stop-and-Go Light-tree (S/G Light-tree) architecture

We use *Stop-and-Go Light-tree (S/G Light-tree)* (Sierra et al., 2008). S/G Light-tree allows grooming unicast and multicast traffic together in a light-tree, hence reducing bandwidth wastage. An S/G Light-tree allows a node to optically drop part of the multiplexed traffic in a wavelength without incurring on OEO conversions. Hence, once the traffic is replicated, it prevents or *stops* the replicas from reaching undesirable destinations. Moreover, it enables a node to aggregate traffic in a passing wavelength without incurring on OEO conversions. More detailed information can be found in Sierra et al. (2008).

Figure 6 shows the solution to the previous problem using an S/G Light-tree. Session S_1 is dropped at node N_5 without the need of OEO conversions of the routed traffic in the wavelength. Session S_3 is added on the same wavelength of the S/G Light-tree at node N_5. While Link-by-link (Figure 5) and S/G Light-tree (Figure 6) efficiently use the bandwidth, the first needs OEO conversions.

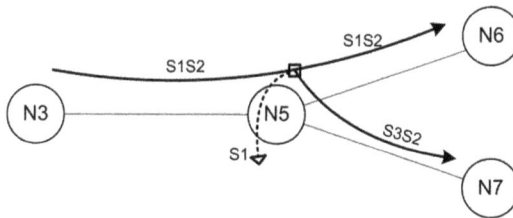

Fig. 6. *S/G Light-tree* scheme

The Stop-and-Go functionality is supported by optical labels or "Traffic Tags" (TT). Each packet in a wavelength contains a header carrying a TT field. Both unicast and multicast traffic can be marked with a TT. A TT can be inserted orthogonally to the packet data. The label information is FSK modulated on the carrier phase, and the data is modulated on the carrier amplitude. Figure 7 shows this procedure. The architecture has been designed for easy detection and processing of the TT. We assume that the bit pattern interpreter in the architecture has low configuration times. Moreover, the bit pattern has to be configured for the traffic of each multicast tree.

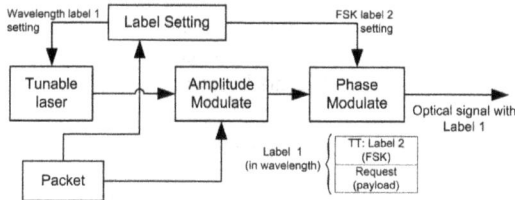

Fig. 7. *S/G Light-tree* Labels

Figure 8 shows the used node architecture. Initially, the optical fiber traffic flows are demultiplexed in the wavelength channel (Demux). λ_2 carries the request S_1 and S_2 multiplexed electronically. S_1 is marked with a TT indicating that it should be stopped from going to N_5. λ_2 is switched (OSW1) in the Splitter and Amplifier Bank. The splitter replicates the incoming traffic to all the node's neighbors, regardless of the TT field. Then, for each packet replica, the TT field is extracted in order to decide whether the packet should be stopped from being forwarded to an undesired destination.

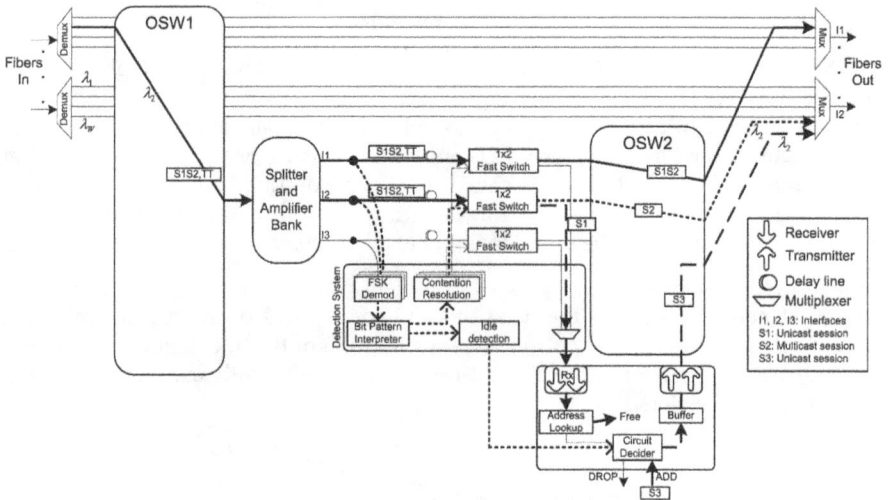

Fig. 8. *Stop-and-Go Light-tree (S/G Light-tree), node N$_5$*

A detection system consists of FSK Demod, 1x2 Fast Switch, Bit pattern Interpreter, Contention Resolution, Idle detection and fiber delay lines (A similar detection system was proposed in Van Breusegern et al. (2006); Vlachos et al. (2003)). A small amount of power is tapped from the wavelength and redirected to the FSK Demod, where the label gets demodulated and ready for interpretation. FSK Demod sends the TT field to the Bit pattern Interpreter. The TT-field is analyzed by an all-optical correlator in the Bit pattern Interpreter block.

If the interpreter-block identifies that the TT field has stopped, it communicates to its corresponding 1x2 Fast Switch in order to either drop or switch the packet towards the receiver (Rx). A multiplexer is used to reduce the number of receivers. These packets are later analyzed to decide whether they must be dropped (FREE), groomed in another S/G Light-tree or, dropped to the local network.

A S/G Light-tree node allows to add traffic to the wavelength as well, only when free capacity is detected (Idle Detection). In our example, session S_3 can employ wavelength 2 with tunable lasers. S/G Light-tree also allows to add sessions using the traditional way.

2. Physical phenomena in optical fibers and the importance in WDM networks

Grooming algorithms, routing and wavelength assignment (GRWA) work with the assumption that all wavelengths in the optical media have the same characteristics of transmission of bits - no bit error (Azodolmolky et al., 2011). However, the optical fiber presents some phenomena that impair the transmission quality of the light-trees. Physical phenomena that may occur in the fiber is divided into two:

1. Linear optical effects: spontaneous amplification, spontaneous emission (ASE), polarization mode dispersion (PMD), chromatic dispersion.
2. Non-linear optical effects: Four-wave mixing (FWM), Selfphase modulation (SPM), Cross-phase modulation (XPM), Stimulated Raman scattering (SRS).

Current work studying PMD, ASE, FWM algorithms applied to routing and wavelength assignment (without grooming), taking into account the effect of power, frequency, wavelength and length of the connection (Ali Ezzahdi et al., 2006).

In this chapter, we propose a predictive model of allocation of wavelengths based on Markov chains. The model takes into account the residual dispersion in WDM networks with traffic grooming and support the applications unicast/multicast with QoS requirements.

2.1 Allocation model wavelengths, taking into account chromatic dispersion

Some definitions and/or parameters used:

- We define 3 classes of service (CoS) for different traffic or sessions that will use the transport network. The CoS are: High Priority (CoS_A), Medium Priority (CoS_M) and Low Priority (CoS_B). The CoS of each session to be sent by the network depends on the type of protocol or traffic, for example, if a video session will require a better deal on the network, so their priority is high (CoS_A). In case, for example, a data session will be low priority (CoS_B).

- Λ is the set of wavelengths available to allocate. Where $\Lambda = \lambda_\alpha, \lambda_\beta, \lambda_\gamma$. λ_α is the subset of wavelengths with low dispersion, λ_β the subset of wavelengths with a mean dispersion, λ_γ the subset of wavelengths with high dispersion.

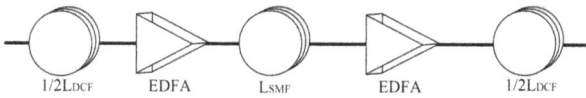

Fig. 9. Standard section

The model is based on the Residual Dispersion (RD), which is defined as the total dispersion in optical fiber transmission in a given fiber compensation. The model takes into account a standard section (Figure 9) and contains the following elements:

- Single Mode Fiber (SMF): optical fiber designed to carry a single ray of light. The fiber may contain different wavelengths. It is used in DWDM.

- Dispersion Compensating Fiber (DCF): Fibers responsible for controlling/improving the chromatic dispersion. It works by preventing excessive temporary widening of the light pulses and signal distortion. The DCF compensates the distortion accumulated in the SMF.

- Length of SMF (L_{SMF})

- DCF length (L_{DCF})

- EDFA Amplifiers

The model is intended to find the percentage of wavelengths with low (λ_α), medium (λ_β) and high dispersion (λ_γ), comparing the value of RD with a threshold. The model is defined as follows:

Inputs:

- B: Compensation Factor (Dispersion Slope) [ps/nm^2km].

- Λ: set of wavelengths available to allocate. $\Lambda = \lambda_1, \lambda_2, ..., \lambda_w$. Where w is the number of wavelengths.

- λ_{ref}: reference wavelength [nm]. It depends on the bandwidth of the channels. The parameters are available in the Rec G.694.1.
- *Threshold*: threshold of acceptance [ps/nm]. *Threshold* = 1000 ps/nm for speeds of 10 Gbps.
- D_{smf}: Coefficient of dispersion in the SMF for the reference wavelength [$ps/nm.Km$].
- D_{dcf}: Coefficient of dispersion in the DCF for the reference wavelength [$ps/nm.Km$].
- L_{SMF}: SMF length [km].
- L_{DCF}: DCF length [km].

Outputs:

Equations 1,2,3 help to obtain the parameters of RD, as shown in Equation 4.

$$\Delta\lambda_w = \lambda_w - \lambda_{ref} \; ; \forall w \tag{1}$$
$$\Delta D_w = \Delta\lambda_w \times B \; ; \forall w \tag{2}$$
$$D_w = D_{\lambda_{ref}} + \Delta D_w \; ; \forall w \tag{3}$$

$$RD_w = D_w(SMF) \times L_{SMF} + D_w(DCF) \times L_{DCF} \tag{4}$$

The RD parameter will be used for the allocation of wavelengths. The proposal seeks to allocate the wavelengths less DR sessions with higher priority (CoS_A). We used the cost function proposed in Ali Ezzahdi et al. (2006) (Threshold = 1000, other parameters were taken from Zulkifli et al. (2006)) to determine the value of RD (Equation 5).

$$d_{ij} \times RD_w \leq threshold \tag{5}$$

Given the analysis performed, we conclude that the first 15% of the wavelengths have less residual dispersion, the dispersion medium below 60%, while the remaining 25% has high dispersion. These parameters will then be used for the assignment.

2.1.1 Proposed allocation model

The WDM network is modeled by a connected directed graph $G(V, E)$ where V is the set of nodes in the network with $N = |V|$ nodes. E is the set of network links. Each physical link between nodes m and n is associated with a L_{mn} weight, which can represent the cost of fiber length, the number of transceivers, the number of detection systems or other. The total cost of routing sessions unicast/multicast in the physical topology is given by equation 6:

$$TotalCost = \sum_{i \epsilon k} \sum_{w \epsilon W} \sum_{(m,n) \epsilon N} L_{mn} \cdot f_i \cdot \chi_{mn}^{iw} \tag{6}$$

Where:

- N: Number of nodes in the network.
- W: Maximum number of wavelengths per fiber.
- bw_i: Bandwidth required per session unicast/multicast i.
- C_w: Capacity of each channel or wavelength. For example, $C_w = $ OC-192 or OC-48.

- f_i: Fraction of the capacity of a wavelength used for the session i. $f_i = bw_i/C_w$.

- k: a group of unicast or multicast sessions.

- χ_{mn}^{iw}: Boolean variable, which equals one if the link between nodes m and n is occupied by the session i on wavelength w. Otherwise $\chi_{mn}^{iw} = 0$.

K sessions are considered unicast/multicast denoted by $R_i(S_i, D_i, \Delta_i)|i = 1, 2, ..., k$. Each session R_i is composed of a source node S_i, node or set of destination nodes and a parameter D_i class of service associated $\Delta_i = CoS_A, CoS_M, CoS_B$. Δ_i be determined by a model presented in the next subsection.

Let $T_i(S_i, D_i, \Delta_i, \lambda_i)$ tree routing for the session R_i in λ_i wavelength. When R_i is multicast, the message source S_i to D_i a tree along the t_i is divided (split) on different nodes to route through the various branches of the tree to wound all nodes D_i. The architecture of S/G Light-tree allows this operation. Regarding the degree of the node is supposed to be unlimited (bank splitter architecture S/G unlimited). In addition, the wavelength conversion are not considered. The wavelength conversion in all-optical half are expensive and are still under development.

The objective of grooming, routing and allocation algorithm is to minimize the cost of the tree taking into account the dispersions present in the wavelengths. That is, the network has a set $\Lambda = \lambda_1, \lambda_2 ... = \lambda_\alpha, \lambda_\beta, \lambda_\gamma$ of wavelengths, which: λ_α is the set of wavelengths of low dispersion, λ_β is the set of half wavelength dispersion and λ_γ all wavelengths of high dispersion. As obtained in the previous section: λ_α is the first 15%, λ_β 15% to 75% and λ_γ the last 25% of wavelengths. The wavelength is assigned to a particular R_i depend on the type of service required for that session Δ_i. The main objective is given by the equation 7.

$$Minimize \sum_{i \in k} \sum_{w \in W} \sum_{(m,n) \in N} L_{mn} \cdot f_i \cdot \chi_{mn}^{iw} \tag{7}$$

The problem of routing unicast/multicast is basically a minimum Steiner Tree problem, which is NP-hard. We propose a heuristic to find the tree predictive routing taking into account QoS (through CoS) and dispersions in all wavelengths. Another feature of the heuristic is trying to keep more spare capacity in the low wavelength dispersion for the sessions r_i with $\Delta_i = CoS_A$ are most likely to access this resource.

2.1.2 Prediction using Markov chains

Markov chains are a tool to analyze the behavior of some stochastic processes, which evolve in a non-deterministic over time to around a set of states. Using Markov chains to predict in different systems has been tested and validated for their efficiency in different systems of telecommunications. We use Markov chains to predict the possible CoS that come with the next session (in $t + D_t$). The states are defined as class of service (CoS) of a given session. The model applies for n types of CoS as shown in Figure 10. For the case study (3 CoS), we obtained the transition probabilities (P_{xy}, where x and y are states that define the CoS) taking into account the available data traces of ACM SIGCOMM (Acm, 2000). From this data was obtained the following transition matrix:

$$P_{xy} = \begin{bmatrix} 0.1009 & 0.3082 & 0.5910 \\ 0.1007 & 0.3089 & 0.5905 \\ 0.1009 & 0.3083 & 0.5908 \end{bmatrix} \tag{8}$$

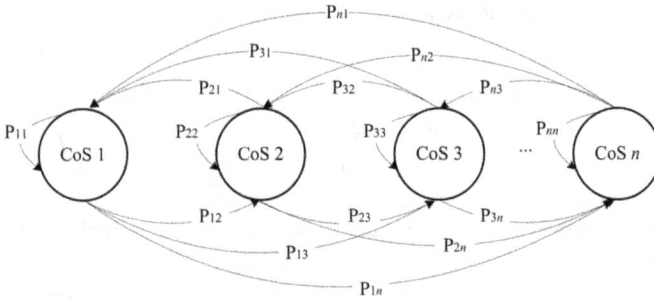

Fig. 10. Markov chain diagram for n CoS

Markov chain with transition probabilities will be used to determine the type of packet (CoS) that come in the following application (session).

2.1.3 Heuristic proposed

We propose a heuristic on-line that deals with the optimal routing, wavelength assignment and grooming, taking into account quality of service for the various sessions and the effects of dispersion in the wavelengths available for allocation. The heuristic aims to probabilistically assign the wavelengths with lower dispersion sessions that have higher priority or CoS. The algorithm is called PredictionTG-QoS and is shown in Figure 11. The algorithm uses Assignmentgrooming function which is shown in Figure 12. The input parameters of the algorithm are:

- N: is the number of nodes in the network.

- X: set of sessions, $k = |X|$ is the number of sessions. $k = 1.2, ...i$.

- Set $\Lambda = \lambda_1, \lambda_2 ... = \lambda_\alpha, \lambda_\beta, \lambda_\gamma$ of wavelengths. $W = |\Lambda|$ is the number of lengths.

- $T_i(S_i, D_i, \Delta_i, \lambda_i)$ is the routing tree for the session R_i in wavelength λ_i .

- Class of Service (CoS) associated $\Delta_i = \{Cos_A, CoS_M, CoS_B\}$

- P_{mn}: physical topology, where $P_{mn} = P_{mn} = 1$ indicates an optical fiber direct link between nodes m and n. If no fiber link between nodes m and n, then $P_{mn} = 0$.

- Each link between nodes m and n is an associated weight L_{mn}.

- C: capacity of each wavelength. Assume $C = OC - 48$.

- S_i: source node for session i.

- D_i: set of destination nodes for each session. D_i includes unicast and multicast traffic.

- bw_i: bandwidth required for each session.

PredictionTG-QoS algorithm initially with session information R_i determines the class of service (Δ) and the set of lengths ($\lambda \epsilon \Delta$) in which the session can be routed (including grooming) taking into account the prediction through the Markov chain. With this information we proceed to apply the routing, allocation and grooming algorithm shown in Figure 11. The assignment and grooming algorithm is based on the known minimun steiner tree to determine the routing tree. Once it is determined the tree routing (in this case the time) it is found that the wavelength being tested have the capacity available for the session can

Function PredictionTG-QoS(*n,s,D,   *)
1 | *Lambda*= Determine the set of lengths that can be assigned (   ) given Markov chain
2 | If
3 | T=Assignmentgrooming(*n,s,D,bw,lambda,*);
4 | If Could not allocate
5 | | Blocking
6 | end
7 |
8 | Elseif
9 | T=Assignmentgrooming(*n,s,D,bw,lambda,*);
10 | If Could not allocate
11 | | Blocking
12 | end
13 |
14 | If
15 | T=Assignmentgrooming(*n,s,D,bw,lambda,*);
16 | If Could not allocate
17 | | Blocking
18 | end
19 | end

Fig. 11. PredictionTG-QoS algorithm

Función Assignmentgrooming (*n,s,D,   *));
1 |
2 | While Assigned ==false &&
3 | Search tree set of available lambdas (steiner minimun tree)
4 | If capacity is available for the whole tree
5 | | Generated routing tree in the lambda
6 | | Available capacity decreases lambda:
    | | Assigned =true
7 | Else
8 | | Assigned =false
9 | End
10 | t--;
11 | End

Fig. 12. Assignmentgrooming Function

access that resource. In case of available capacity is allocated to that wavelength the session and is included in *T*. If it is not possible to assign that wavelength is tested in the next, until you find available capacity or until the wavelengths are exhausted. If it is not possible to assign any wavelength, we proceed to eliminate this session is marked as blocked traffic. The advantage of the algorithm is to use the CoS cycles are reduced search when looking for that wavelength can be assigned.

2.2 Analysis and results of the proposed model

The simulations are performed using NSFnet transport network, in which the physical topology consists of 14 nodes with 21 bidirectional links. In order to obtain results as close to reality, we decided to get a model coming session to the optical transport network as well as their duration. We used traces of data available in ACM SIGCOMM Acm (2000), which contain traffic carried on the transport network with duration of 30 days between the Lawrence Berkeley Laboratory, California and the world. The data used have information about the timing, duration, protocol, bytes transferred, and others.

The proposed allocation model (PredictionTG-QoS) is compared with the case when given the same treatment to the different sessions (regardless of QoS, called in this case standard assignment) and when it does not take into account the QoS (TG -QOS). The article compares the blocking probability (blocking) and the ability to average available bandwidth of each wavelength. The analysis is done taking into account the following simulation parameters:

- Number of wavelengths: 10
- Wavelengths Capacity: $OC - 48$
- Possible bandwidth: $bw = \{OC - 1, OC - 3, OC - 12, OC - 48\}$, generated with a uniform distribution $OC - 1 : OC - 3, OC - 12, OC - 48 = 1 : 11 : 1$.
- Maximum number of sessions: 10000
- Group of wavelengths with low dispersion $\lambda_\alpha = [1:2]$.
- Group half-wavelength dispersion $\lambda_\beta = [3:7]$.
- Group of wavelengths with high dispersion $\lambda_\gamma = [8:10]$.
- The arrival rate of session (λ) and the duration (μ) of these were modeled as $\mu = 1$ and λ to vary the load in Erlangs. The load in Erlangs is defined as Load (Erlang) = $bw \cdot \lambda / \mu$.

In Figure 13 shows the blocking probability of link sessions with CoS_A. The proposed heuristic improves by 16% approx. to TG-QoS heuristics and 11% approx. when performing standard assignment for different traffic loads. As noted the allocation taking into account only the QoS does not improve the standard setting, but all traffic is treated the same way leading to the sessions with CoS_A not routed by half with less dispersion.

In the case when you have sessions with CoS_M (Figure 14), shows a better performance when using TG-QoS, but PredictionTG-QoS enhancement to the standard assignment. The reason for TG-QoS provides better performance is due to 60% of available wavelengths are to be assigned only to all traffic with CoS_M. Moreover, the heuristic-QoS PredictionTG you are looking to improve the QoS sessions mainly CoS_A giving any kind of traffic can access a wavelength less chromatic dispersion. It is noteworthy that the blocking probability for CoS_M remains at approximately 32% as for CoS_A sessions.

As expected, the traffic CoS_B is penalized by both TG-QoS-QoS as PredictionTG (Figure 15). Importantly, however PredictionTG-QoS blocking probability remains in about 40% for this type of traffic, close to CoS_A and CoS_M supplied.

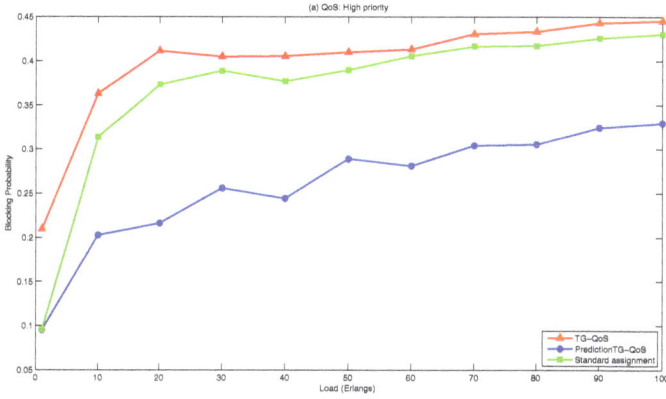

Fig. 13. Blocking Probability for CoS_A, QoS: High priority

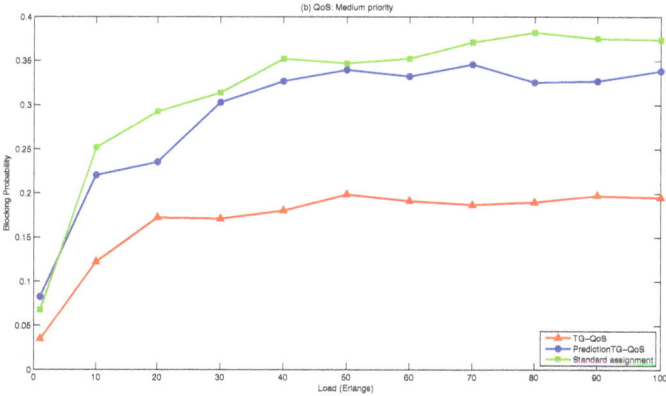

Fig. 14. Blocking Probability for CoS_M, QoS: Medium priority

Regarding the capacity of available bandwidth in each wavelength, as shown in Figure 16, PredictionTG-QoS on average available capacity remains higher when compared with the other two allocation algorithms. In addition, the algorithm meets its primary objective: to keep the wavelengths with less dispersion available for traffic with CoS_A. The wavelengths of 3 to 7 are those who remain less available capacity due to more traffic coming into a system is CoS_M.

2.3 Nonlinear model: Four Wave Mixing

Four Wave Mixing (FWM) is one of the main phenomena induced nonlinear crosstalk in WDM networks (Agrawal, 2001). In WDM networks, FWM phenomenon generates a new wave frequency $w_f = w_i + w_j - w_k$, where w_i, w_j, w_k channels are used in the network. For a system with M-channel i, j, k range from 1 to M, which produces up to $M^2(M-1)/2$ new frequencies.

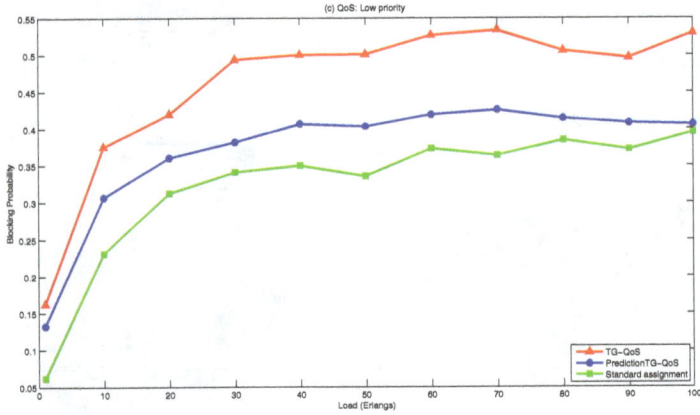

Fig. 15. Blocking Probability for CoS_B, QoS: Low priority

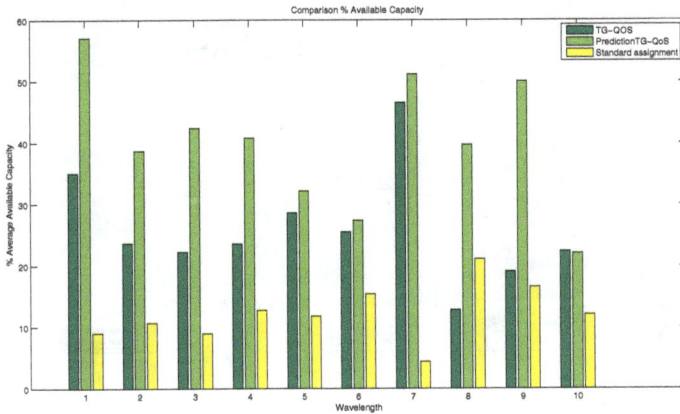

Fig. 16. Average available capacity for each wavelength

In All-Optical Networks (AONS) is important to consider this phenomenon because it does not use OEO conversion at intermediate nodes. This leads to the lightpath and the lighttree signal receives interference by not regenerating (Fonseca et al., 2003; Xin & Rouskas, 2004). When the separation of the channels in the network is the same, it generates new frequencies coincide with frequencies enabled in the system. This leads to the occurrence of interference depends on the bit patterns and the receivers receive different signal fluctuations.

To explain the concept consider a WDM network with 3 channels, with initial wavelength $\lambda_0 = 1.45\mu$ and channel separation 0.1μ. Figure 17 shows an example, where in (A) observed the 3 channels used in the system. The phenomenon generates 9 components, however, some matches several times in the channels being used. Figure 17(B) shows the new components.

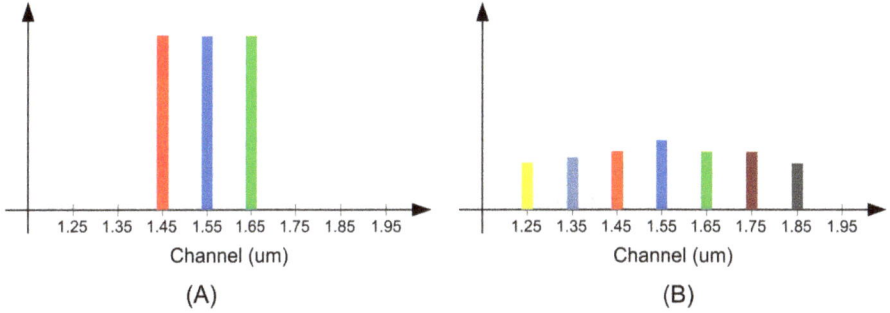

Fig. 17. Example FWM. (A) channels used, (B) signals generated by FWM effect

2.3.1 Physical parameters

The system is characterized by the interaction of multiple channels w_i, w_j, w_k with $k \neq i, j$. The new components are generated by w_φ given by equation 9.

$$w_\varphi = w_{ijk} = w_i + w_j - w_k; \forall k \neq i, j \tag{9}$$

The power of the frequency component in the w_φ is calculated using the expression used by Fonseca et al. (2003) and Agrawal (2001), shown in equation 10.

$$P_{w_\varphi}(L) = \frac{\eta}{9} D^2 \gamma^2 P_i P_j P_k e^{-\alpha L} L_{eff}^2 \tag{10}$$

Where:

- L: is the length of optical fiber.
- P_i, P_j, P_k: transmission power of each channel.
- D: deterioration factor. $D=3$ for $i = j$, $D= 6$ for $i \neq j$.
- α: fiber attenuation.
- γ: Nonlinear coefficient. γ can be determined as $\gamma = 2\pi n_2/\lambda A_{eff}$, where n_2 is the nonlinear refractive index of the fiber, A_{eff} is the effective area of the core of the fiber and λ the wavelength in vacuum.
- L_{eff}: effective length of the fiber. $L_{eff} = 1 - e^{-\alpha L}/\alpha$.
- η: FWM efficiency.

Considering that in the OTN link has several hops before reaching the destination should be considered that the power is the sum of the components in each hop, so the total power for each component is given by equation 11 (h is the number of hops). P_{TOTAL} represents the FWM noise accumulated over the link.

$$P_{TOTAL} = \sum_h \sum_{i,j,k} P_{w_\varphi} \tag{11}$$

The efficiency η depends on the separation of channels, chromatic dispersion D_c (dispersion slope $dD_c/d\lambda$) and the fiber length and can be determined as shown in equation 12.

$$\eta = \frac{\alpha^2}{\alpha^2 + \Delta\beta^2} \left[1 + \frac{4e^{-\alpha L} \sin^2(\Delta\beta \cdot L/2)}{(1 - e^{-\alpha L}/2)} \right] \tag{12}$$

Where:

$$\Delta\beta = \left(\frac{2\pi\lambda_0^2}{c}\right)(w_i - w_k)(w_j - w_k)$$
$$\times \left(\left[D_c + \frac{\lambda_0^2}{2c}\frac{dD_c}{d\lambda}\right]\left[(w_i - w_0) + (w_j - w_0)\right]\right) \tag{13}$$

c is the speed of light in vacuum and λ_0 is the wavelength on zero dispersion. The term used to determine which wavelength is assigned to certain traffic is Q-factor (Fonseca et al., 2003). To determine taking into account Gaussian noise using On-Off Keying (OOK) and calculating the BER as shown in equation 14.

$$BER = \frac{1}{\sqrt{2\pi}}\int_Q^\infty exp(-t^2)dt \tag{14}$$

Assuming thermal noise and shot noise can be ruled out in the presence of FWM distortion, Q-factor can be represented as shown in equation 15.

$$Q = \frac{bP_s}{\sqrt{N_{FWM}}} \tag{15}$$

$$N_{FWM} = 2b^2\frac{P_{FWM}}{8} \tag{16}$$

Where, b is the responsibility of the receiver, $P_S = P_i e^{-\alpha L}$ is the received power y P_i the transmission power of the channel i.

2.3.2 Proposed allocation model

The proposed allocation model is shown in Figure 18. The model is divided into two modules: 1) network layer and 2) physical layer. The network layer is responsible for determining the routing tree (applies to both lighttree to SG). The physical layer is responsible for determining if the routing tree found in certain wavelength can satisfy the QoS requirements of traffic.

The proposed model is called QoSImproved-FWM. QoSImproved-FWM takes into account that a percentage of links to destinations not meet the QoS parameters. In this case if the percentage of links that are acceptable to route the session is over 70, it proceeds to search for those who do not meet again another way. If you do not find the session is blocked. If you are under 70 do not assign that wavelength to the session unicast/multicast (the value 70 is used as an example, this value can be changed).

A variation of QoSImproved-FWM does not take into account the percentage and is called GroomingQoS-FWM. When all branches of lighttree meet the threshold for QoS immediately locks independent of the number of destinations that have good reception.

2.4 Simulation and analysis

The analysis was performed for the network NSFnet considering dynamic unicast/multicast traffic with QoS requirements. Was analyzed for 3 classes of service: CoS_A, CoS_M, CoS_B.

The physical and network parameters used for the analysis are shown in Tables 1 and 2 respectively. The model is analyzed in terms of blocking probability and average capacity available in the network for each CoS. Grooming, GroomingQoS-FWM y QoSImproved-FWM are analyzed.

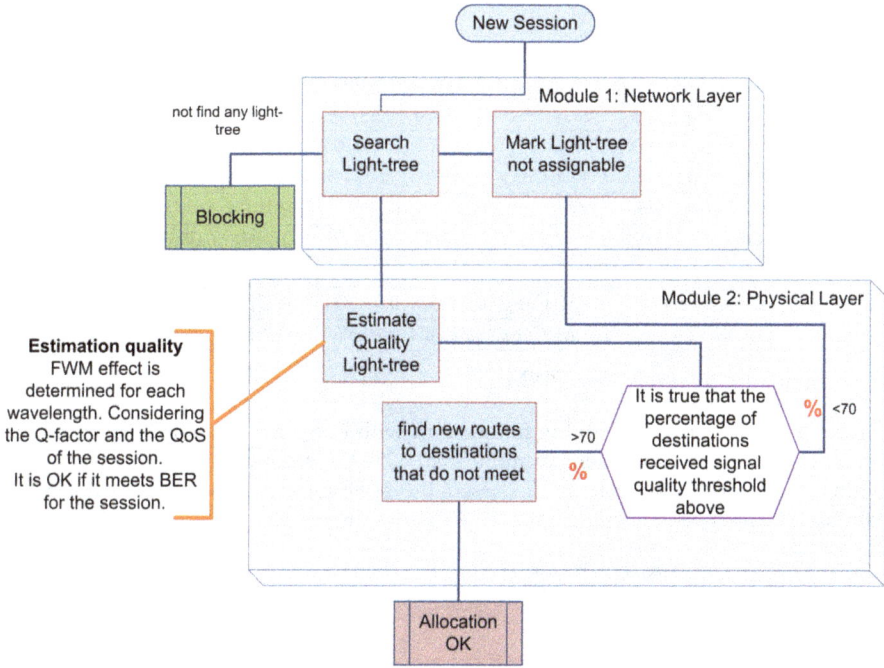

Fig. 18. Flowchart allocation model considering FWM

Parameter	Value
Fiber type	Dispersion Shift Fiber
Zero-dispersion wavelength λ_0	$1549nm$
Chromatic dispersion slope	$0.67ps/[nm^2km]$
Nonlinear coefficient γ	$2.3(Wkm)^{-1}$
Fiber attenuation α	$0.23dB/Km$
Transmit power P_s	$0dBm$
Channel Separation	$100GHz$
BER or threshold for CoS_A	10^{-9}
Receptor responsivity	1

Table 1. Physical parameters of simulatio model FWM

parameter	Value
Number of nodes	14
Number of sessions	1000
Number of wavelengths	8
Traffic generation model	Poisson
Duration model	Exponential

Table 2. FWM model simulation parameters

Algorithm y	μ		σ	
CoS	Min	Max	Min	Max
$Grooming, CoS_A$	0,360844	0,489065	0,0666778	0,167471
$Grooming, CoS_M$	0,37086	0,49784	0,0610475	0,162028
$Grooming, CoS_B$	0,43367	0,484205	0,0199832	0,0615137
$GroomingQoS - FWM, CoS_A$	0,540755	0,642445	0,0488886	0,129757
$GroomingQoS - FWM, CoS_M$	0,352686	0,506732	0,0801074	0,201202
$GroomingQoS - FWM, CoS_B$	0,278883	0,323472	0,019591	0,0555651
$QoSImproved - FWM, CoS_A$	0,263923	0,41164	0,0768166	0,192936
$QoSImproved - FWM, CoS_M$	0,426429	0,514891	0,0425289	0,112878
$QoSImproved - FWM, CoS_B$	0,323474	0,397966	0,0358129	0,0950524

Table 3. Confidence Intervals 95%. FWM

In analyzing the blocking probability for sessions with CoS_A the proposed allocation model QoSImproved-FWM improvement in more than 12% the algorithm Grooming and 20% to GroomingQoS-FWM. Note that as discussed the analysis seeks to improve the blocking probability for this type of traffic. Figure 19(A) shows the results.

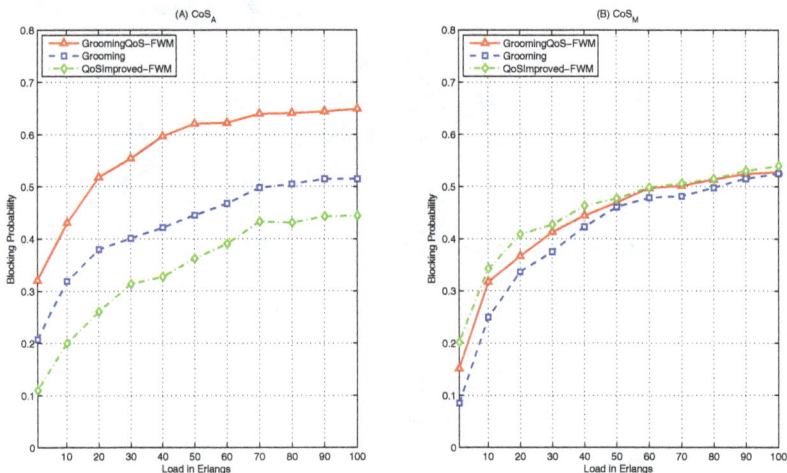

Fig. 19. Blocking probability, (A) CoS_A and (B) CoS_M. Parameters: NSFnet, k=1000, c_w =OC-48, w =8, BW =OC-[1 3 12 48]

The algorithms showed a similar result for trades with CoS_M. Approximately have a blocking probability of 50% as shown in Figure 19(B).

By using QoSImproved-FWM blocking probability for traffic with CoS_B was not good compared to GroomingQoS-FWM. It should be noted that the analysis found that the algorithm is enhanced by Grooming proposals for this project.

When analyzing the average available capacity per wavelength, we found that the wavelengths 1 and 8 have more available capacity when using QoSImproved-FWM. These

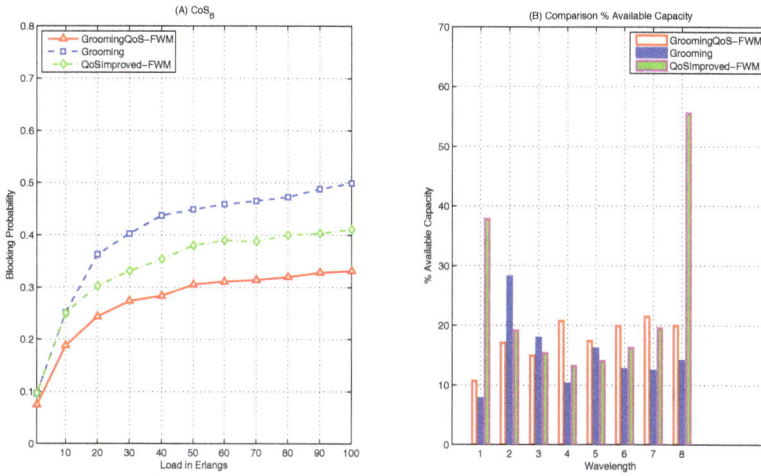

Fig. 20. Blocking probability, (A) CoS_A and (B) Available capacity. Parameters: NSFnet, k=1000, c_w =OC-48, w =8, BW =OC-[1 3 12 48]

wavelengths are reserved for traffic requiring QoS improvement. For other wavelengths the average available capacity is similar for all algorithms.

Table 3 summarizes the confidence intervals for the results.

3. Conclusions

In this chapter, we propose a predictive model based on Markov chains. The allocation, routing and grooming algorithm takes into account the phenomena occurring in the optical fiber as well as parameters of quality of service (QoS) in traffic of unicast and multicast type.

The proposed allocation model significantly improves the blocking probability for high priority traffic, while maintaining a similar range to other types of traffic. The model also keeps most available capacity in the low wavelength dispersion, which will allow traffic with high quality requirements may be more likely to have access to good resources.

This chapter analyzes dynamic traffic networks using OTN architecture SG. Heuristics are proposed that seek to minimize the blocking probability for these networks. Furthermore it is noted that the traffic have different characteristics related to QoS. Given this, it is proposed to note that the physical environment in AONs has limitations that the systems are evident and alter the signal propagating in different lighttree. Models are proposed that take into account linear and nonlinear distortions. Results show that it is important to analyze the physical effects.

4. References

Acm (2000). Traces In The Internet Traffic Archive.
    URL: *http://ita.ee.lbl.gov/html/traces.html*
Agrawal, G. P. (2001). *Applications of Nonlinear Fiber Optics*, Academic Press.

Ali Ezzahdi, M., Al Zahr, S., Koubaa, M., Puech, N. & Gagnaire, M. (2006). LERP: a Quality of Transmission Dependent Heuristic for Routing and Wavelength Assignment in Hybrid WDM Networks, *Computer Communications and Networks, 2006. ICCCN 2006. Proceedings.15th International Conference on*, pp. 125–136.

Azodolmolky, S., Perello, J., Angelou, M., Agraz, F., Velasco, L., Spadaro, S., Pointurier, Y., Francescon, A., Saradhi, C. V., Kokkinos, P., Varvarigos, E., Zahr, S. A., Gagnaire, M., Gunkel, M., Klonidis, D. & Tomkos, I. (2011). Experimental Demonstration of an Impairment Aware Network Planning and Operation Tool for Transparent/Translucent Optical Networks, *Journal of Lightwave Technology* 29(4): 439–448.
URL: *http://dx.doi.org/10.1109/JLT.2010.2091622*

Bastos-Filho, C. J. A., Chaves, D. A. R., Silva, Pereira, H. A. & Martins-Filho, J. F. (2011). Wavelength Assignment for Physical-Layer-Impaired Optical Networks Using Evolutionary Computation, *Journal of Optical Communications and Networking* 3(3): 178+.
URL: *http://dx.doi.org/10.1364/JOCN.3.000178*

Bermond, J. C., Coudert, D., Munoz, X. & Sau, I. (2006). Traffic Grooming in Bidirectional WDM Ring Networks, *Transparent Optical Networks, 2006 International Conference on*, Vol. 3, pp. 19–22.
URL: *http://dx.doi.org/10.1109/ICTON.2006.248390*

Billah, A. R. B., Wang, B. & Awwal, A. A. S. (2003). Multicast traffic grooming in WDM optical mesh networks, *Global Telecommunications Conference, 2003. GLOBECOM '03. IEEE*, Vol. 5.

Fonseca, I. E., Almeida, R. C., Ribeiro, M. R. N. & Waldman, H. (2003). Algorithms for FWM-aware routing and wavelength assignment, *Proc. SBMO/IEEE MTT-S International Microwave and Optoelectronics Conference IMOC 2003*, Vol. 2, pp. 707–712.

Huang, X., Farahmand, F. & Jue, J. P. (2005). Multicast traffic grooming in wavelength-routed WDM mesh networks using dynamically changing light-trees, *Lightwave Technology, Journal of* 23(10): 3178–3187.
URL: *http://dx.doi.org/10.1109/JLT.2005.856244*

Kamat, A. E. (2006). Algorithms for multicast traffic grooming in WDM mesh networks, *Communications Magazine, IEEE* 44(11): 96–105.
URL: *http://dx.doi.org/10.1109/MCOM.2006.248171*

Khalil, A., Hadjiantonis, A., Assi, C. M., Shami, A., Ellinas, G. & Ali, M. A. (2006). Dynamic provisioning of low-speed unicast/multicast traffic demands in mesh-based WDM optical networks, *Lightwave Technology, Journal of* 24(2): 681–693.
URL: *http://dx.doi.org/10.1109/JLT.2005.861922*

Liao, L., Li, L. & Wang, S. (2006). Dynamic multicast traffic grooming in WDM mesh networks, *Next Generation Internet Design and Engineering, 2006. NGI '06. 2006 2nd Conference on*.
URL: *http://dx.doi.org/10.1109/NGI.2006.1678264*

Malik, O. (2011). Internet Keeps Growing! Traffic up 62% in 2010.
URL: *http://gigaom.com/2010/10/06/internet-keeps-growing-traffic-up-62-in-2010/*

Sahasrabuddhe, L. H. & Mukherjee, B. (1999). Light trees: optical multicasting for improved performance in wavelength routed networks, *Communications Magazine, IEEE* 37(2): 67–73.
URL: *http://dx.doi.org/10.1109/35.747251*

Sierra, J. E., Caro, L. F., Solano, F., Marzo, J. L., Fabregat, R. & Donoso, Y. (2008). All-Optical Unicast/Multicast Routing in WDM Networks, *Proc. IEEE Global Telecommunications Conference IEEE GLOBECOM 2008*, pp. 1–5.
URL: *http://dx.doi.org/10.1109/GLOCOM.2008.ECP.493*

Singhal, N. K., Sahasrabuddhe, L. H. & Mukherjee, B. (2006). Optimal Multicasting of Multiple Light-Trees of Different Bandwidth Granularities in a WDM Mesh Network With Sparse Splitting Capabilities, *Networking, IEEE/ACM Transactions on* 14(5): 1104–1117.
URL: *http://dx.doi.org/10.1109/TNET.2006.882840*

Solano, F., Caro, L. F., de Oliveira, J. C., Fabregat, R. & Marzo, J. L. (2007). G+: Enhanced Traffic Grooming in WDM Mesh Networks using Lighttours, *Selected Areas in Communications, IEEE Journal on* 25(5): 1034–1047.

Sreenath, N., Palanisamy, B. & Nadarajan, S. R. (2006). Grooming of Multicast Sessions in Sparse Splitting WDM Mesh Networks using Virtual Source Based Trees, *Systems and Networks Communication, 2006. ICSNC '06. International Conference on*, p. 11.
URL: *http://dx.doi.org/10.1109/ICSNC.2006.42*

Ul-Mustafa, R. & Kamal, A. E. (2006). Design and provisioning of WDM networks with multicast traffic grooming, *Selected Areas in Communications, IEEE Journal on* 24(4).

Van Breusegern, E., Cheyns, J., De Winter, D., Colle, D., Pickavet, M., De Turck, F. & Demeester, P. (2006). Overspill routing in optical networks: a true hybrid optical network design, *Selected Areas in Communications, IEEE Journal on* 24(4): 13–25.

Vishwanath, A. & Liang, W. (2005). On-line multicast routing in WDM grooming networks, *Computer Communications and Networks, 2005. ICCCN 2005. Proceedings. 14th International Conference on*, pp. 255–260.
URL: *http://dx.doi.org/10.1109/ICCCN.2005.1523861*

Vlachos, K., Zhang, J., Cheyns, J., Sulur, Chi, N., Van Breusegem, E., Monroy, I. T., Jennen, J. G. L., Holm-Nielsen, P. V., Peucheret, C., O'Dowd, R., Demeester, P. & Koonen, A. M. J. (2003). An optical IM/FSK coding technique for the implementation of a label-controlled arrayed waveguide packet router, *Lightwave Technology, Journal of* 21(11): 2617–2628.

Wu, B. & Yeung, K. L. (2006). Light-Trail Assignment in WDM Optical Networks, *Global Telecommunications Conference, 2006. GLOBECOM '06. IEEE*, pp. 1–5.
URL: *http://dx.doi.org/10.1109/GLOCOM.2006.374*

Xin, Y. & Rouskas, G. N. (2004). Multicast routing under optical layer constraints, *Proc. INFOCOM 2004. Twenty-third AnnualJoint Conference of the IEEE Computer and Communications Societies*, Vol. 4, pp. 2731–2742.
URL: *http://dx.doi.org/10.1109/INFCOM.2004.1354691*

Zhang, B., Zheng, J. & Mouftah, H. T. (2008). Fast Routing Algorithms for Lightpath Establishment in Wavelength-Routed Optical Networks, *IEEE/OSA Journal of Lightwave Technology* 26(13): 1744–1751.
URL: *http://dx.doi.org/10.1109/JLT.2007.912530*

Zhu, K. & Mukherjee, B. (2002). Traffic grooming in an optical WDM mesh network, *Selected Areas in Communications, IEEE Journal on* 20(1): 122–133.
URL: *http://dx.doi.org/10.1109/49.974667*

Zsigri, A., Guitton, A. & Molnar, M. (2003). Construction of light-trees for WDM multicasting under splitting capability constraints, *Telecommunications, 2003. ICT 2003. 10th International Conference on*, Vol. 1.
URL: *http://dx.doi.org/10.1109/ICTEL.2003.1191206*

Zulkifli, N., Okonkwo, C. & Guild, K. (2006). Dispersion Optimised Impairment Constraint Based Routing and Wavelength Assignment Algorithms for All-Optical Networks, *Proc. International Conference on Transparent Optical Networks*, Vol. 3, pp. 177–180. URL: *http://dx.doi.org/10.1109/ICTON.2006.248430*

17

Fiber Optic Temperature Sensors

S. W. Harun[1,2], M. Yasin[1,3], H. A. Rahman[1,2,4], H. Arof[2] and H. Ahmad[1]
[1]Photonic Research Center, University of Malaya, Kuala Lumpur
[2]Department of Electrical Engineering,
University of Malaya, Kuala Lumpur
[3]Department of Physics, Faculty of Science and
Technology, Airlangga University, Surabaya
[4]Faculty of Electrical Engineering, Universiti
Teknologi MARA (UiTM), Shah Alam
[1,2,4]Malaysia
[3]Indonesia

1. Introduction

The need for temperature measurement exists in many applications such as in automated consumer products, automated production plants and high performance processors. Recent works have mainly focused on temperature sensors that satisfy user requirements for specific applications, and the main considerations are performance, dimension and reliability. In fact, traditional low-cost solutions, such as thermocouples and resistance temperature detectors (RTDs), do not always yield satisfactory performance, e.g., when the fluid temperature has to be measured in hostile environments, in the presence of electromagnetic, chemical, and mechanical disturbances. Since signals from the thermoelectric sensors are normally mixed with intrinsic noise and extrinsic interferences, they may contain intolerable errors if not properly filtered. Therefore, this type of sensors is inept for gauging temperature in microfluidic or nano-sized devices, in extreme marine environments, and underground geological sites where long distance measurement with precision is required. For such applications, fiber optical sensors offer a better alternative since the optical signal does not suffer from interference by electromagnetic fields and can be transmitted over extremely long distances without any significant loss [Yasin et al., 2010; Li et al., 2010; Ahmad et al., 2009; Lim et al., 2009; Ahmad et al., 2009; You et al., 2005; Xu et al., 2005]. Furthermore, they are relatively small in size, and compatible with other optical fiber devices.

To date, various types of fiber optic temperature sensors have been reported in the literatures and they are mostly based on fiber interferometric [Choi et al., 2008] and fiber Bragg grating (FBG) [Han et al., 2004]. However, the first type of sensors are rather expensive to produce and complicated to implement on-site [Golnabi, 2000]. Fiber Bragg gratings are very efficient at temperature sensing and are easy to implement; however, they always need additional techniques to discriminate the Bragg shifts by temperature and by strain/compression and they also require expensive phase-masks. In this chapter, a temperature sensor is demonstrated based on four different techniques; intensity modulated fiber optic displacement sensor (FODS), lifetime measurements, microfiber loop resonator (MLR) and stimulated brillouin scattering. The first sensor is based on a rugged, low cost and very efficient FODS utilizing a plastic optical fiber (POF)-based coupler as a probe and a linear thermal expansion of aluminum. The second temperature sensor, which is based on fluorescence decay time in Erbium-doped silica fiber has the advantage of incorporating a time based encoding system, which is less sensitive to system losses such as those associated with optical cables and connectors. The MLR is formed by coiling a microfiber, which was obtained by heating and stretching a piece of standard silica single-mode fiber (SMF). The MLR is embedded in a low refractive index material for use in temperature measurement. The MLR-based temperature sensor has a low loss splicing with a standard SMF. Lastly, a temperature sensor is demonstrated using an SBS effect, which requires measurement of frequency shift. In the proposed sensor, a Brillouin pump is injected into one end of a ring cavity resonator, in which a sensing fiber is located, and then the frequency shift between the BP and the Brillouin fiber laser (BFL) output is measured using a heterodyne method.

2. FODS based temperature sensing

POFs have widespread uses in the transmission and processing of optical signals for optical fiber communication system compatible with the Internet. POFs also have potential applications in WDM systems, power splitters and couplers, amplifiers, sensors, scramblers, integrated optical devices, frequency up-conversion, and etc. [Yasin et al., 2009; Yang et al., 2011]. Recently, an intensity modulated FODSs have been demonstrated to be efficient for many applications including sensor. They are relatively inexpensive, easy to fabricate and suitable for deployment in harsh environments. In this section, a low cost temperature sensor is demonstrated using POF-based coupler as a probe based on a linear thermal expansion of aluminum. The temperature sensor is schematically shown in Fig. 1. The sensor is essentially a FODS with a 3 dB multimode fiber coupler as a probe. A 594 nm He-Ne beam is launched into port 1 of the coupler. Light travels to port 3 and is scattered when it exits the fiber end. It is then reflected by the top surface of an aluminium rod with dimensions of 0.5 cm diameter and 7 cm length. The port 3 probe is held in position about 1 mm perpendicular to the top surface of the aluminium rod so that the reflected light can be easily launched back into the same port. The collected light is sent to port 2 by the 3dB coupler and measured by a silicon photo-detector. The detector converts the light into electrical signal, which is then processed by the lock-in amplifier and finally displayed and stored onto the computer.

In the calibration stage, the static operating range of the probe is identified and this process requires the probe to be mounted on a translational stage, which is rigidly attached to a vibration free table. Firstly, the output from port 2 at zero point is measured, where the aluminum rod and the probe are in close contact. Then the aluminum rod is moved away from the probe in 50 μm steps and at each position, under vibrationless condition, the output voltage is recorded. A graph of displacement (gap) against output voltage is drawn and a linear range on the graph is identified. A position at the center of the linear range is chosen and the gap between the probe and the top surface of the aluminum rod is fixed at the chosen displacement point. Then an experiment is carried out where the aluminum is fixed onto a hotplate for heating purpose. A thermocouple placed at the upper region of the aluminum rod is used to display and monitor the temperature of the aluminum rod. The

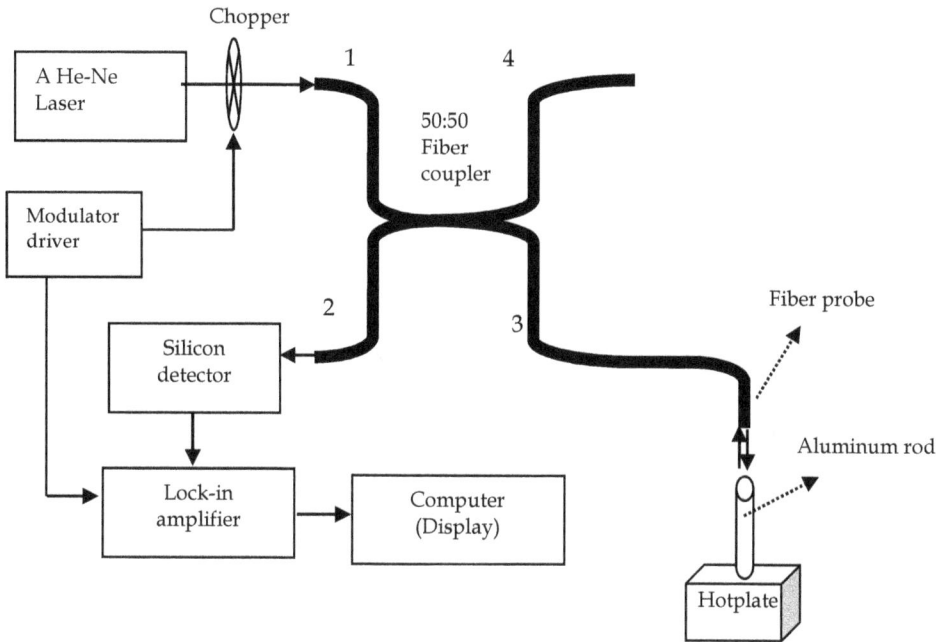

Fig. 1. Experimental setup for the proposed temperature sensor using a POF-based coupler

thermocouple has a resolution of 10C and a temperature range of -500C to 13000C. The heat to the aluminum rod is controlled by varying the heat intensity produced by the hotplate ranging from room temperature (250C) to 900C.

Fig. 2 show the efficiency of the FODS as a function of displacement obtained both experimentally and theoretically without the temperature effect. The characteristic of the proposed sensor can be compared with the case of coupling two similar collinear fibers which are separated at the end-faces and with both axis aligned as discussed in [Yang et al., 2011]. Given that the the distance between the parallel end-faces is called z, a and NA is the radius and numerical aperture of the fiber respectively, the efficiency η for small values of z/a is [Van Etten, 1991]

$$\eta = 1 - \frac{z}{a}\frac{2}{\pi(NA)^2}\left[arcsin(NA) - NA\sqrt{1 - NA^2}\right]$$ (1)

for z/a <<1

The fiber receives the maximum light when the gap between the tip of fiber probe and the reflected surface is zero, and thus the measured intensity of the reflected light is maxima as shown in the figure. However, the measured intensity of the reflected light decreases almost linearly as the distance or gap increases especially for close distance target. Theoretically, the distance and the reflected power vary according to the inverse square law and the ratio between the reflected power and the transmitted power is given by [Kulkarini et al., 2006]

$$\frac{P_r}{P_t} = \frac{d^2}{(2x \tan \theta)^2} \tag{2}$$

where Pr, Pt, d, x, and θ are the reflected power, transmitted power, core diameter, axial displacement and fiber's acceptance angle, respectively. The characteristic of the displacement curve is summarized in Table 1 where the sensitivity is obtained at 0.0005 mV/μm and the slope shows a good linearity of more than 99% within the displacement range of 1400 μm. The displacement sensor is observed to be very stable with the measurement error of less than 0.8%.

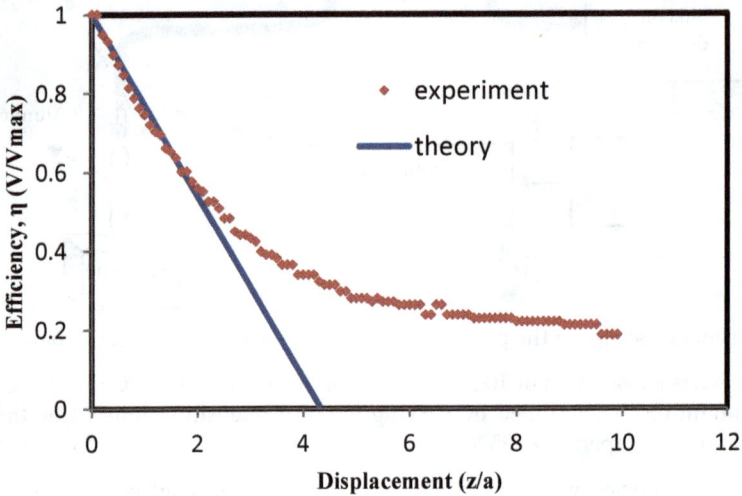

Fig. 2. Efficiency as a function of displacement obtained both experimentally and theoretically.

Parameter	Value
Sensitivity	0.0005 mV/μm
Linear Range	0 – 1400 μm
Linearity	More than 99%
Standard Deviation	0.01 mV (0.8%)
Resolution	19 μm

Table 1. The Performance of fiber optic displacement sensor using the aluminum rod.

Fig. 3 show the linear function of the output signal against the aluminum rod temperature for two different runs. The two different runs were taken because repeatability of results is a crucial factor in the operation of any sensor system. In the experiment, the gap between the aluminum rod and fiber tip of port 3 is fixed at 1 mm, which is within the linear range of the displacement response without the temperature effect. The temperature of the aluminum rod is then varied from 250C to 900C, resulting in an output signal ranging from 0.95 mV to

1.20 mV. The output signal starts to significantly increase with increasing rod temperature at 420C, which henceforward forms a linear function until the rod temperature reaches 900C. It is observed that the maximum difference between the two runs is about 0.04 mV, which is small compared to the full range of 1.20 mV. The temperature sensor is observed to be very stable with the measurement error of less than 0.8% which is obtained at 900C (see inset of Fig. 3). The output voltage are recorded for 200 seconds and the standard deviation obtained at 250C, 600C and 900C are 0.5%, 0.5% and 0.8% respectively. The performance of the temperature sensor is summarized in Table 2 where the sensitivity of the linear function for the first run is 0.0044 mV/0C with 98% linearity whereas the sensitivity of the linear function for the second run is 0.0041 mV/0C with 96% linearity.

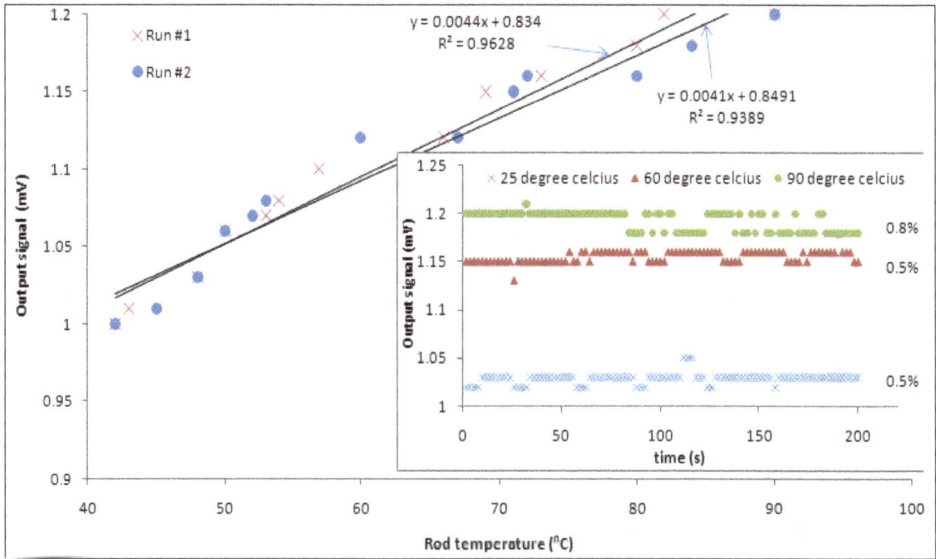

Fig. 3. Linear relationship between the output voltage and aluminum rod temperature for 25⁰C to 90⁰C. Inset shows the standard deviation values for output voltage versus rod temperature at 25⁰C, 60⁰C and 90⁰C.

Parameter	Value (Run No. 1)	Value (Run No. 2)
Sensitivity	0.0044 mV/ºC	0.0041 mV/ºC
Linear Range	42 - 90 ºC	42 - 90 ºC
Linearity	More than 98%	More than 96%
Standard Deviation	± 0.01 mV	± 0.01 mV
Resolution	2.4 ºC	2.4 ºC

Table 2. The performance of the temperature sensor using FODS.

The temperature sensor has a higher sensitivity (0.0044 mV/0C) compared to the displacement sensor (0.0005 mV/μm) even though the initial gap between the aluminum rod and fiber tip of port 3 is fixed at 1 mm, which is within the linear range of the displacement response without temperature effect. This is because as temperature increase, the reflectivity of aluminum will also increase [Guo et al., 2003], hence sensitivity increases when increasing temperature is applied to the aluminum rod. The possible sources of error in the sensor operation can be due to light source fluctuation, stray light and possible mechanical vibrations. To reduce these effects a well-regulated power supply is used for the 594 nm He-Ne laser and this minimizes the fluctuation of source intensity. The sensor fixture is also designed so that the stray light cannot interfere with the source light and room light does not have any effect on the output voltage. To reduce the mechanical vibrations, the experimental set-up is arranged on a vibration free table

3. Fiber optic temperature sensor based on lifetime measurement

Fluorescence-based sensors are widely used for measuring various parameters due to its relatively independent of ambient conditions. This approach is widely used for temperature sensor, which is normally based on the detection of fluorescence lifetimes in various rare-earth-doped silica fibers [Zhang et al., 2009; Lopez et al., 2004; Seat et al., 2002; Baek et al., 2006]. In principle, fluorescence is induced by pump power at a certain wavelength that is suitable for the doped ions, and only a straightforward detection procedure is needed. An additional advantage of such rare-earth-doped fiber sensors is that they are compatible with a wide range of existing fiber-optic multiplexing schemes that can simultaneously detect multiple physical parameters. The underlying principle behind the ability of the rare-earth doped materials to be used as temperature sensors [McSherry et al., 2005] is their properties of emission and absorption that are dependent on the temperature. This behavior is due to the homogeneous broadening of the line width and the changing population of the energy levels with temperature. In the earlier work, a remote temperature sensor has been proposed using fluorescence intensity-ratio technique [Castrellon-Uribe, 2005]. In this section, a temperature sensor is proposed based on fluorescence decay time in Erbium-doped fiber (EDF).

Fig. 4 shows the schematic diagram of the proposed sensor set-up. The 980nm laser pump beam is launched into a piece of 90 cm long Erbium-doped fiber (EDF) via a wavelength division multiplexing (WDM) coupler. The EDF is placed in a vacuum oven which allows us to vary the fiber temperature within 25 to 200oC interval. The fluorescence signal from the forward pumped EDF is detected by a Ge photo-detector and processed with a digital oscilloscope. The 980 nm pump beam is chopped so as to generate a square-wave modulated signal with pulse width of 2.2ms, peak power of approximately 124mW and frequency of 45Hz. When the erbium-doped fiber is pumped with the photon energy of 980 nm, the $^4I_{11/2}$ erbium level is excited and the $^4I_{13/2}$ metastable level is quasi-instantaneously populated due to the non-radiative transition. The population inversion between $^4I_{13/2}$ and $^4I_{11/2}$ level is responsible for the emission of fluorescence at around 1550 nm. When the EDF is pumped at a fixed rate, the fluorescence variation including a lifetime change can reflect corresponding temperature. The temperature dependent fluorescence lifetimes of the spontaneous emission of the EDF is investigated and studied in this work.Fig. 5 shows the output spectrum of the amplified spontaneous emission (ASE) of the forward pumped EDF when the continuous wave pump power is fixed at 31 mW. The ASE spectrum peaks at 1529 nm with the average power of around -58 dBm. The temperature sensing mechanism in this work is based on the temperature dependence of the Erbium fluorescent lifetime decay. The Erbium fluorescence lifetime is measured using a modulated pump laser with power of

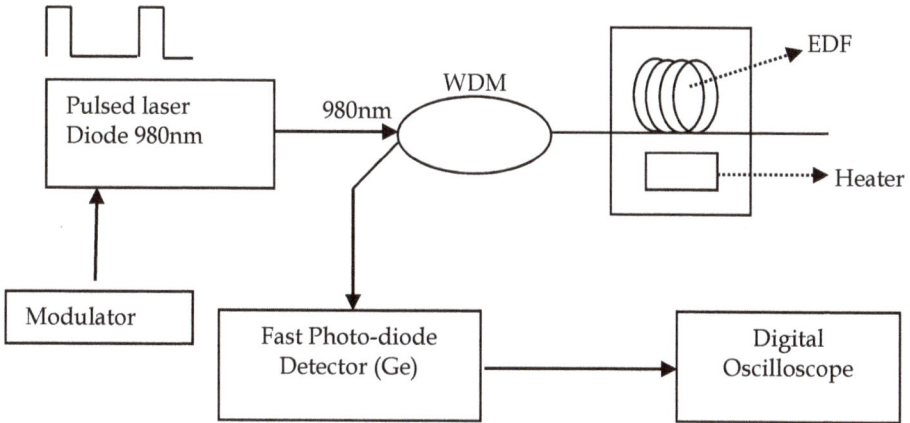

Fig. 4. Experiment setup of fiber optic high temperature sensor.

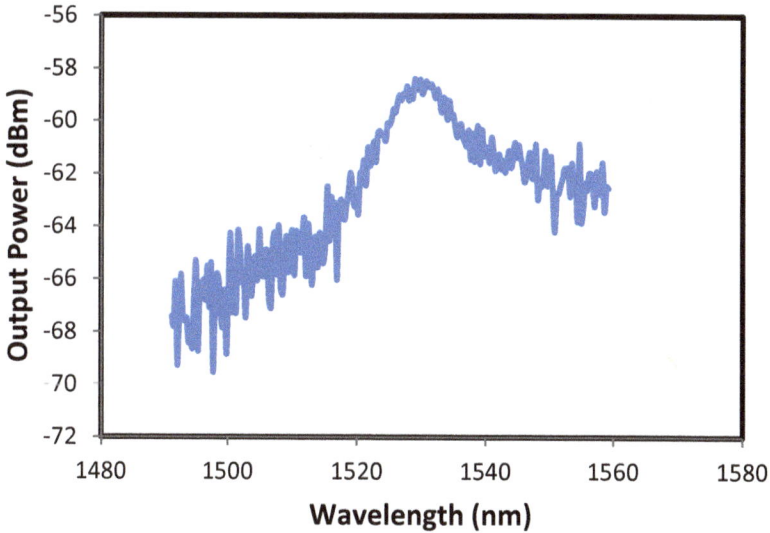

Fig. 5. Output spectrum of the generated ASE with 980nm pump power of 31mW.

approximately 2 mW and the results for temperature measurements of 85°C and 130°C are shown in Figs. 6(a) and (b), respectively. Fitting the exponential curve produces lifetimes of 4.85 and 4.75 ms for temperatures of 85°C and 130°C accordingly. The fluorescence lifetime data taken over the range of 25°C to 160°C are presented in Fig. 7. The graph shows the existence of an inverse linear relationship between the Erbium lifetime and the temperature. The fluorescence lifetime data shown in Fig. 7 are mean values of consecutive measurements at corresponding stabilized temperature. The sensitivity of the sensor is obtained at 0.009 ms/°C with a linearity more than 94%. The lifetime reduction is attributed to the quenching of the Erbium luminescence, which results in less efficient excitation. The quenching is mainly due to a decrease in the absorption coefficient of Erbium ion as the temperature is increased.

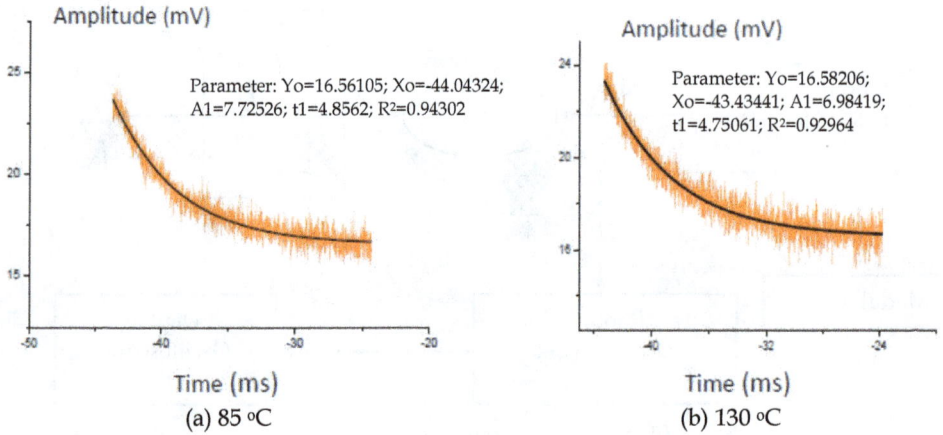

Fig. 6. Fluorescence lifetime decay of EDF at (a) 85 °C and (b) 130 °C from a digital oscilloscope.

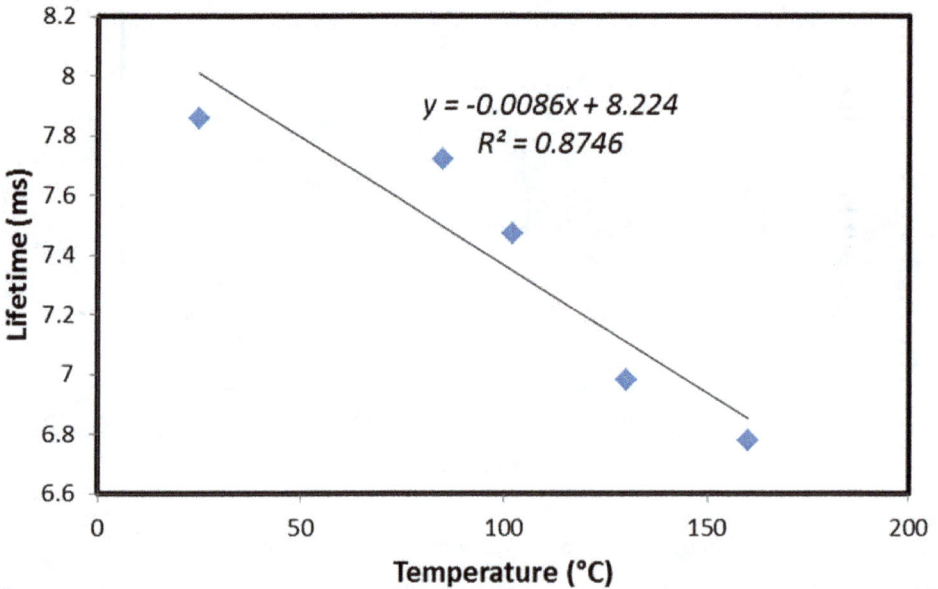

Fig. 7. Lifetime of EDF as function of Temperature.

4. Microfiber loop resonator based temperature sensor

Optical ring resonators have attracted considerable attention owing to their simple structure, compact size, and many applications in communication, optical signal processing and optical devices [Amarnath et al., 2005; Harun et al., 2010]. Planar waveguide micro-ring resonators have been well developed, but suffer from larger connection-losses with fibers and higher cost. Recently, research on low-loss microfibers has opened up new opportunities for developing micro-photonic devices such as resonators [Jiang et al., 2006], couplers [Tong et al., 2006], and sensors [Li & Tong, 2006]. As one of the basic functional elements, microfiber resonators, in forms of loops, knots, or coils, have shown high intrinsic optical quality and have exhibited promising applications such as filters and lasers [Jiang et al., 2006]. Microfiber-based sensors are amongst the simplest of such devices, as they do not require expensive and complex fabrication procedures. Recently, strain, high temperature and refractive index sensing using tapered micro-structured optical fiber has been reported [Nguyen et al., 2005; Minkovich et al., 2005; Villatoro et al., 2006]. However, the authors used tapered micro-structured fiber which is expensive and difficult to work with in term of getting low loss splicing with standard optical fiber.

A microfiber fabrication set-up is shown in Fig. 8. A coating removed standard SMF was held by two fiber holders which were fixed on two translation stages. One stage is fixed and another stage is a motorized stage that can be moved in one dimensional direction and the speed of the motor can be controlled. The fiber ends were connected to an amplified spontaneous emission (ASE) source and an optical spectrum analyzer (OSA). The microfiber was fabricated by heating the fiber to its softening temperature, and then pulling the ends apart to reduce the fiber's diameter down to a around 1-3 mm. A high temperature and stable micro-burner fueled by clean butane gas is used in the tapering. Mounted on the movable stand, the micro-burner can be swung vertically to flame brush the tapered fiber. The flame should be clean and the burning gas flow should be controlled carefully so that the air convection does not break the fiber during the drawing process. During the tapering process, we monitored both the inter-modal interference and the insertion loss of the fiber using the ASE source in conjunction with the OSA. The flame is applied to the fiber in an optimized angle to make sure the heat distribution is homogeneous.

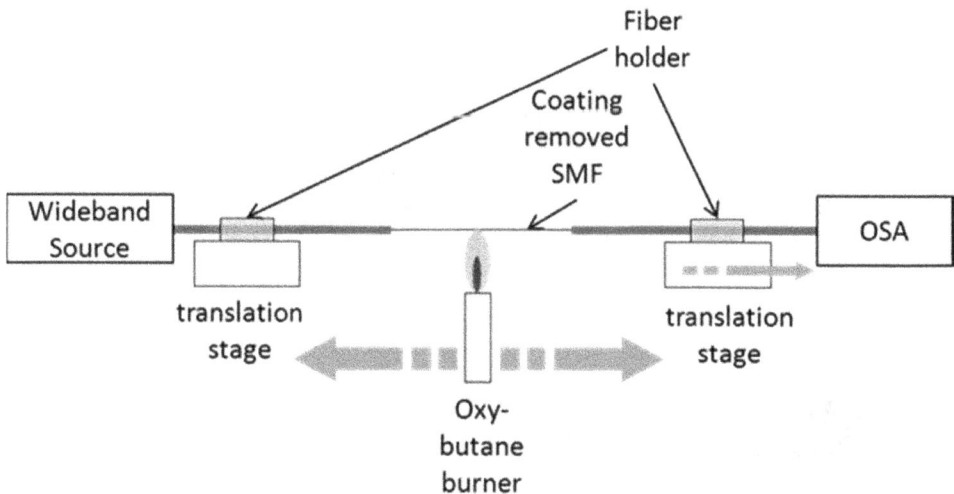

Fig. 8. Schematic diagram of the microfiber fabrication set-up.

The MLR is fabricated by coiling the microfiber onto itself using two surface attractions, van der Waals force and electrostatic force, which keep the loop stable since the forces overcome the elastic force to make the fiber straight. The fabricated MLR is laid on an earlier prepared glass plate with a thin and flat layer of low refractive index material to address the temporal stability of the device as shown in Fig. 9. The thickness of the low refractive index material is approximately 0.5mm which is thick enough to prevent leakage of optical power from the microfiber to the glass plate. Some uncured resin is also applied on surrounding the MLR before it is sandwiched by another glass plate with the same low refractive index resin layer from the top. It is essentially important to ensure that minimum air bubbles and impurity are trapped around the fiber area between the two plates. This is to prevent refractive index non-uniformity in the surrounding of microfiber that may introduce loss to the system. The uncured resin is solidified by the UV light exposure for 3 ~7 minutes and the optical properties of the MLR are stabilized. The packaged MLR is located in an oven with temperature control to investigate how the comb spectrum changes with the temperature.

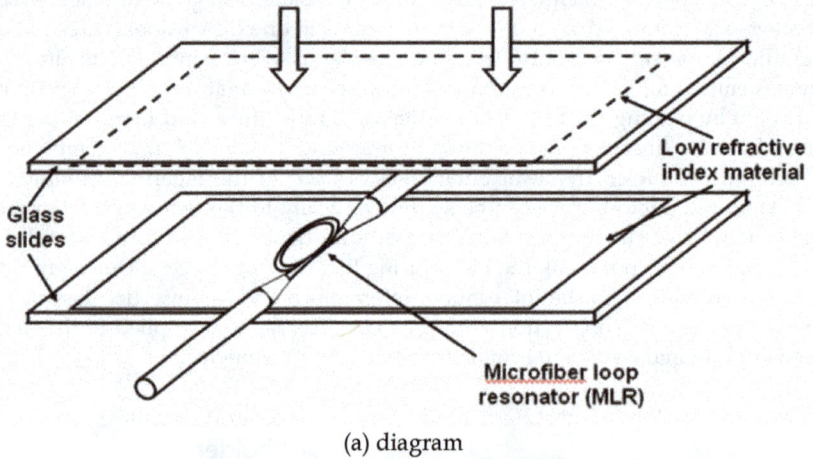

(a) diagram

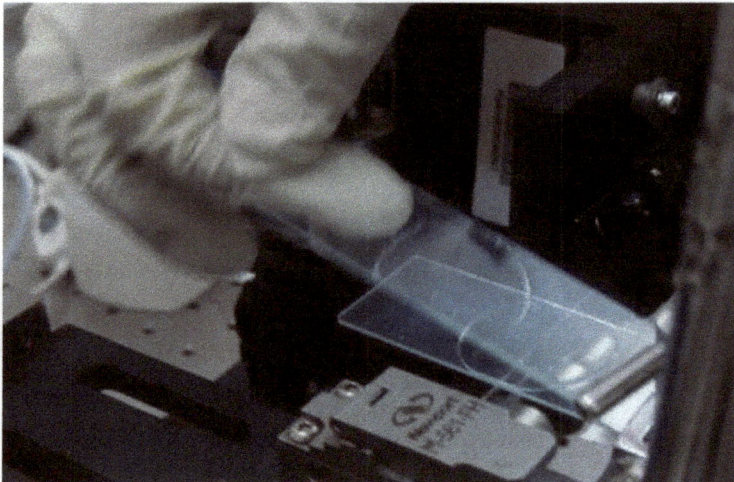

(b) Snapshot

Fig. 9. Packaging of the MLR (a) schematic diagram (b) snapshot

A MLR has a comb transmission spectrum similar to that of a Fabry–Perot filter. The resonant wavelength must satisfies $\lambda_m = 2\pi R n_{eff}/m$, where R is the radius of the ring, n_{eff} is the effective index of the ring, m is the resonant mode number. The resonant frequency spacing, i.e. Free Spectral Range (FSR), is given by [Sumetsky et al., 2006];

$$\Delta v_{FSR} = \frac{c}{2\pi R n_{eff}} \tag{3}$$

where c is the velocity of light in vacuum. The quality factor of a ring resonator can be expressed as

$$Q = \frac{\lambda_m}{\Delta\lambda_{FWHM}} \tag{4}$$

where $\Delta\lambda_{FWHM}$ is the 3 dB bandwidth of the resonant peak. The lower the coupling coefficient between the ring and straight waveguide, the higher the Q is. If the radius and the coupling coefficient of the ring are chose properly, the desired resonant frequency spacing and bandwidth of the resonant peak can be obtained. The measured comb transmission spectrum of the MLR is shown in Fig. 10, which was obtained by using the ASE source in conjunction with OSA. The resonant response of the MLR is obvious. The extinction ratio is about 4.0 dB, and the FSR is 0.08 nm.

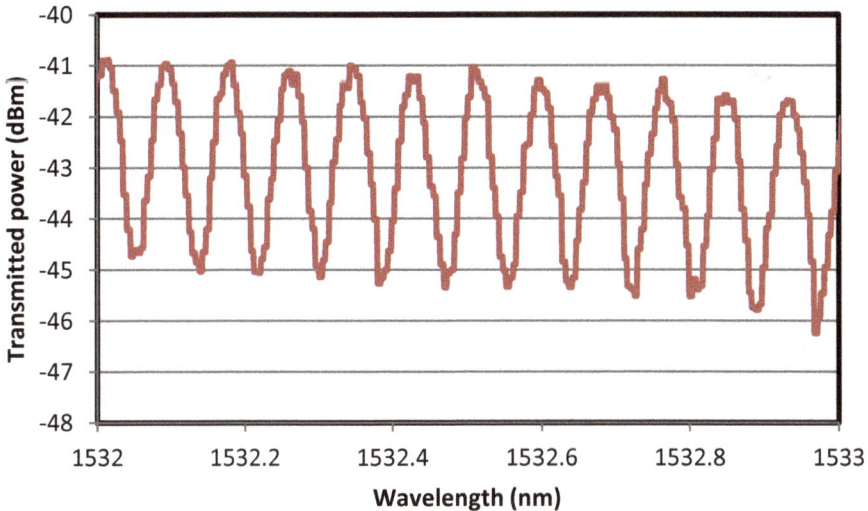

Fig. 10. Transmission spectrum of the fabricated MLR

The temperature response of the packaged MLR is then investigated. Fig. 11 shows the transmission spectrum of another packaged MLR at various temperatures. As shown in this figure, the spacing of the transmission comb spectrum is unchanged with the temperature. However, we observed a linear dependence of the extinction ratio of the MLR on temperature; higher temperature MLRs had a smaller extinction ratio. Fig. 12 shows an

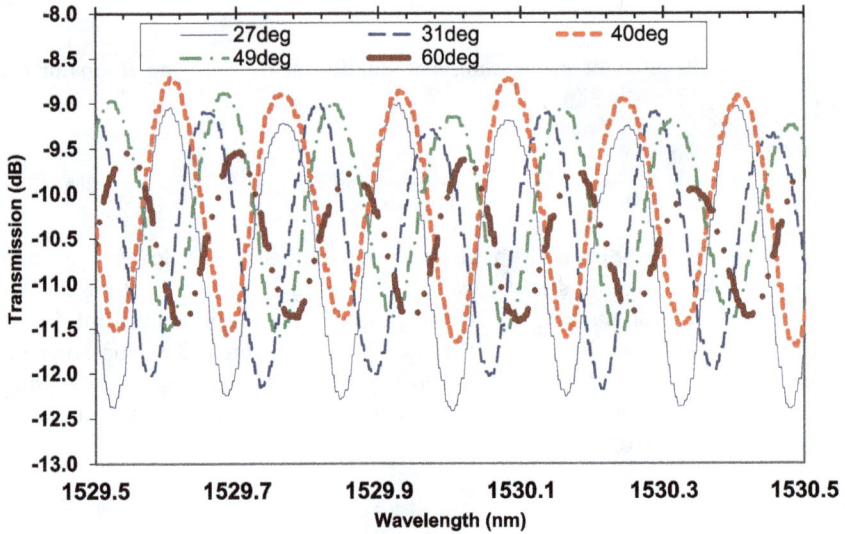

Fig. 11. Transmission spectra of the fabricated MLR obtained at various temperatures

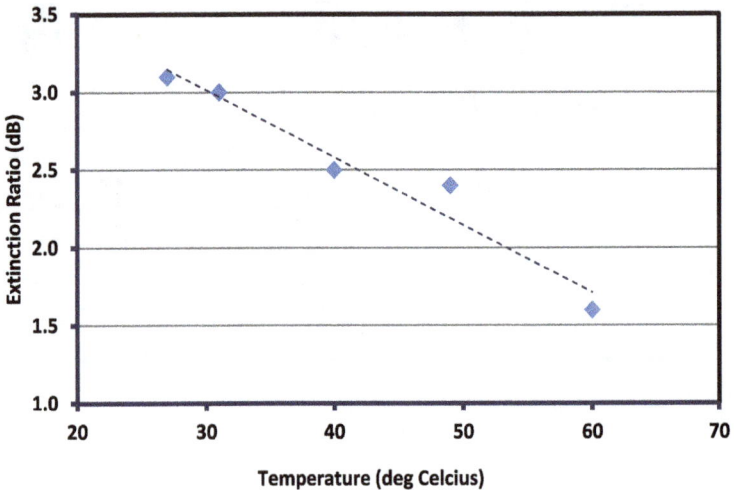

Fig. 12. Extinction ratio of the MLR as a function of the temperature

extinction ratio against the temperature. As seen, the slope of the extinction ratio reduction against temperature was about 0.043dB/°C. The dependence of the extinction ratio on temperature is due to the change in the material's refractive index, which increase the loss at higher temperature and thus reduces the extinction ratio. In the proposed sensor, the optical path length is not much changed with the temperature and therefore, the comb spacing is unchanged as shown in Fig. 11.

5. Stimulated Brillouin scattering based temperature sensor

Brillouin scattering is an effect caused by the nonlinearity of a medium, whereby an incident pump photon can be converted into a scattered photon of slightly lower energy, usually propagating in the backward direction, and a phonon. At high optical pump powers, the Stimulated Brillouin scattering (SBS) is one of the most important nonlinear effects in fibers and has a significant influence on the operation of optical transmission systems as well as fiber sensors. Brillouin-based optical fiber sensors have been widely reported in recent years owing to their possibilities to perform distributed strain and temperature sensing along an optical fiber [Bao et al., 1993; Nikles et al., 1996; Culverhouse et al., 2008; Kee et al., 2002; Rathod et al., 1994]. Several techniques have been proposed for effective sensing using either spontaneous Brillouin scattering (SpBS) or stimulated Brillouin scattering (SBS) effects. Brillouin reflectometers are the most widely exploited in Brillouin-based optical sensing due to their simplicity. They are implemented by injecting a light beam into one of the sensing fibers' ends and are not affected by nonlocal effect, thus appearing more suitable for long range sensing applications. However, these type of sensors suffer from a small amplitude of the signal due to the use of spontaneous Brillouin scattering.

The configuration of the proposed sensor is shown in Fig. 13. It consists of a Brillouin pump (BP), an Erbium-doped fiber amplifier (EDFA), a ring cavity Brillouin gain block and a frequency shift measurement setup. The gain block consists of a piece of sensing fiber and a 3 dB coupler, which the Brillouin gain oscillates in an anti-clockwise direction to generate a BFL. The sensing fiber is located in an oven with temperature control and is shown in the dashed square in Fig. 13. An external cavity tunable laser source (TLS) with a line-width of approximately 20MHz is used as the BP. The BP light operating at 1555 nm is split into reference and probe signals, by using a 10dB coupler. The probe signal from 10% part is optically amplified by an EDFA to provide sufficient power for this study. This amplified signal is launched into a piece of sensing fiber through an optical circulator, which is also used to couple into the photodiode the backward propagating light from the ring cavity, including a BFL operating at wavelength downshifted by 0.08 nm from the BP wavelength. The output Brillouin signal is combined with the reference or BP signal by a 3 dB coupler for Brillouin shift measurement using a heterodyne technique. The combined optical signal is converted into an electrical signal by a phodiode which is connected to an RF spectrum analyzer.

To compare the performance achieved by various sensing fibers, a standard single-mode fiber (SMF), non-zero dispersion shifted fiber (NZ-DSF) and photonic crystal fiber (PCF) are alternatively employed as a gain medium in the BFL. The lengths of SMF, NZ-DSF and PCF are fixed at 25 km, 10 km and 100 m, respectively. The NZ-DSF has a core diameter of 4.2 μm and cut-off wavelength of 1150nm. The PCF has a triangular core with average diameter of 2.1 ± 0.3 μm and cladding diameter of 128 ± 5 μm. The average air hole diameter of the

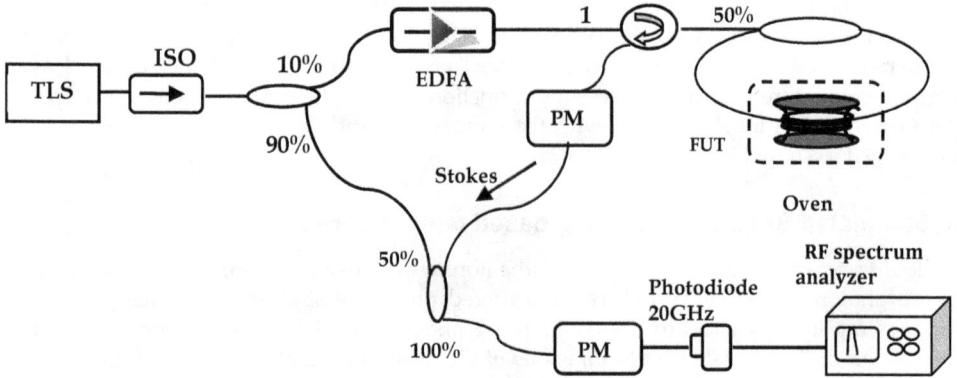

Fig. 13. Experimental setup for the spontaneous Brillouin scattering measurements.

fiber is 0.8 μm with 1.5 μm pitch. The PCF is made from pure silica with 17.4 wt% of Ge-doped core region. The SMF and NZ-DSF have core diameters of 9 and 4.2 μm, respectively. The sensing fiber is incorporated in a ring BFL configuration to reduce the required BP power for the operation. The Brillouin threshold is relatively lower for the SBS.

Fig. 14 shows the backward propagating Brillouin spectrum obtained from the ring cavity BFL, which is measured after the optical circulator by an OSA with resolution of 0.015 nm. In this experiment, the sensing fiber is NZ-DSF and the temperature is varied from 25oC to 80oC. The Brillouin Stokes laser is generated from interactions between the optical mode and acoustic modes in the fiber core. Thermally excited acoustic waves (acoustic phonons) produce a periodic modulation of the refractive index. Brillouin scattering occurs when light is diffracted backward on this moving grating, giving rise to frequency shifted Stokes and anti-Stokes components. This process can be stimulated when the interferences of the laser light and the Stokes wave reinforce the acoustic wave through electrostriction. Since the scattered light undergoes a Doppler frequency shift, the Brillouin shift depends on the acoustic velocity and is given by [Alahbabi et al., 2005]

$$v_B = \frac{2nV_a}{\lambda} \tag{5}$$

where V_a is the acoustic velocity within the fiber, n is the refractive index and λ is the vacuum wavelength of the incident lightwave. As shown in Fig. 14, the Brillluin Stokes is slightly shifted to a longer wavelength as the temperature is increased from 25oC to 80oC. This is attributed to the refractive index of the sensing medium, which increases with the temperature. Fig. 14 shows the Brillouin shift for the BFL with NZ-DSF, measured by the RF spectrum analyzer at two different temperatures; 25oC and 80oC. As shown in the figure, the Brillouin shift peak power shifted toward a higher frequency with increasing temperature. For a sensing fiber of NZ-DSF, at room temperature of 25oC, a value for v_B is experimentally obtained at 10.7 GHz at 1550 nm wavelength region as shown in Fig. 15.

The strong attenuation of sound waves in silica determines the shape of the Brillouin gain spectrum. Actually, the exponential decay of the acoustic waves results in a gain presenting a Lorenzian spectral profile [Xiao-qiang et al., 2009].

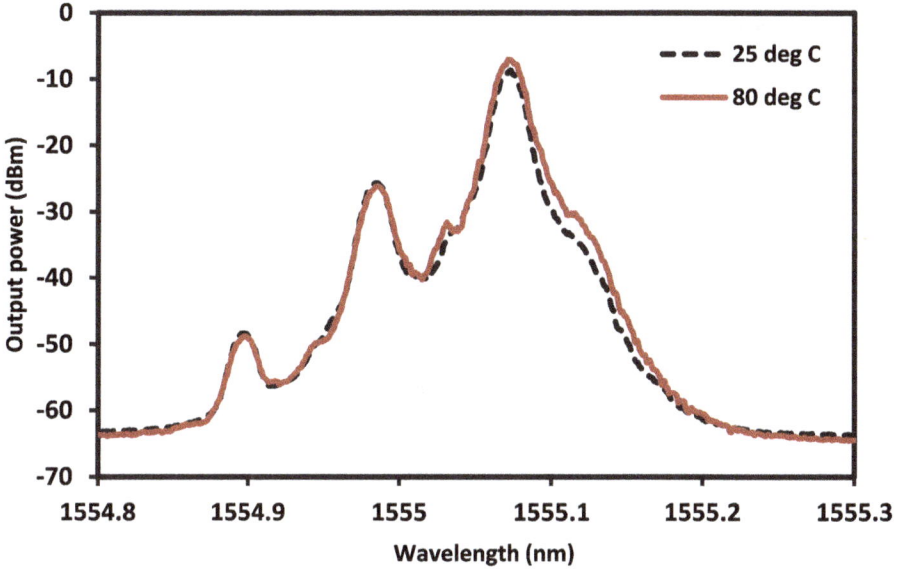

Fig. 14. Stimulated Brillouin scattering monitored by OSA in minimum and maximum used temperature.

$$g_B(v) = g_0 \frac{(\Delta v_B/2)^2}{(v - v_B)^2 + (\Delta v_B/2)^2} \tag{6}$$

where Δv_B is the full-width at half maximum (FWHM). The Brillouin gain spectrum peaks at the Brillouin frequency shift v_B (or the frequency difference between the two beams), and the peak value is given by the Brillouin gain coefficient

$$g_B(v) = g_0 = \frac{2\pi n^7 p_{12}^2}{c \lambda_p^2 \rho_0 V_a \Delta v_B} \tag{7}$$

where p_{12} is the longitudinal elasto-optic coefficient, ρ_0 is the density, λ_p is the pump wavelength and c is the vacuum velocity of light [Heiman et al., 1979; Agrawal, 1989]. The Brillouin linewidth for a spontaneous Brillouin scattering in a standard SMF is reported to be approximately 30 MHz, which can be calculated using Eq. (7). In this work, the Brillouin linewidth is observed to be around 15 MHz as shown in Fig. 13 due to the proposed sensor, which is based on SBS effect. The Brillouin linewidth is also expected to slightly increases with the temperature increase.

The temperature–dependence of the Brillouin gain shift is also investigated for various types of sensing fibers and the result is shown in Fig. 16. The beat frequency is obtained at approximately 10.85, 10.75 and 9.80 GHz with the sensing fibers of SMF, NZ-DSF and PCF, respectively. It is also observed that the central frequency linearly increases with the temperature over a wide temperature range. The slope coefficients of 0.77, 0.56 and 1.45 MHz/oC are obtained for BFLs using SMF, NZ-DSF and PCF, respectively. The error bars

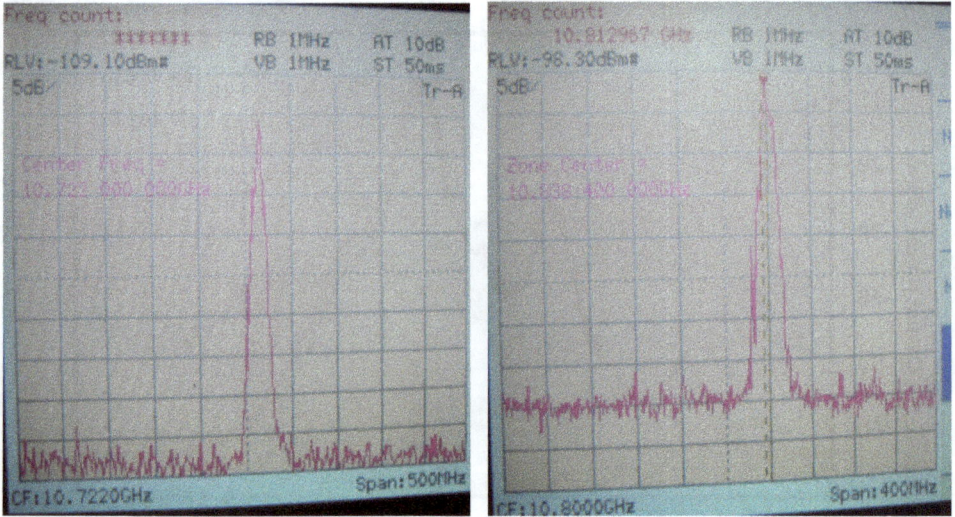

Fig. 15. The typical SBS beat frequency spectra at the minimum and maximum temperatures using the NZ-DSF as the sensing fiber.

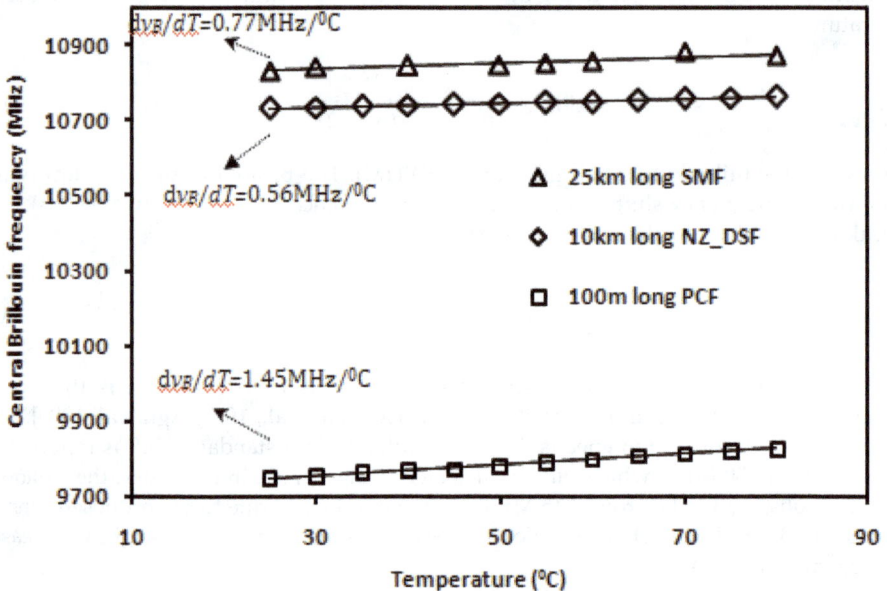

Fig. 16. Brillouin frequency shift as functions of temperature.

are ±3 MHz, corresponding to the BFS fluctuations when the temperature is fixed. All fiber samples are only coated with a thin acrylate micro jacket, so that the coating influence is expected to be kept below 0.1 MHz/ oC. This experiment indicates that the PCF with small core is preferable for this application. As also shown in Fig. 16, the Brillouin shift for PCF is less than that of the other silica fibers. The reason is due to the Germania concentration in the fiber core such that the inhomogeneous distribution within the core of the fiber is responsible for multi peaks in Brillouin gain spectrum. The PCF used is composed of a small-scale solid silica core comprising a Ge-doped center region with multiple air holes arranged in a hexagonal lattice around the core which act as cladding. With the combination of a large refractive index contrast between the silica core and the air-filled microstructure, the PCF can confine the mode tightly in the core that results in a low effective mode area and thereby produce a large nonlinear coefficient. This characteristic avoids the difficulty in distributed fiber sensors that use long SMF and NZ-DSF which contributes to uncertainty in the sensing results. Although the threshold is reduced by the increased length necessary to provide the SBS, it creates other problems like reduced coefficient for sensing temperature.

6. Conclucions

Four different approaches for temperature sensing have been demonstrated; intensity modulated FODS, lifetime measurements, MLR-based and SBS-based sensors. The first sensor uses a fiber-optic displacement sensor based on a multimode POF coupler to measure temperature ranging from 42^0C to 90^0C. The displacement curve has a sensitivity of 0.0005 mV/μm and a linearity of more than 99% within a measurement range between 0 to 1400 μm. By placing the aluminum rod within the linear range, the measured output signal is observed to be a linear function of the aluminum rod temperature with a sensitivity of 0.0044 mV/^{o}C for temperature detection with a linearity of more than 98% and a resolution of 2.4 ^{o}C. The second sensor is based on a lifetime measurement of 90 cm long of EDF, which is diode-pumped by a modulated 980 nm laser. The sensitivity of the sensor is obtained at 0.009 ms/^{o}C with a linearity of more than 94%. This temperature sensor is shown to be effective and useful for its high resolution and precision. The third sensor is based on MLR, which is fabricated from a microfiber, which is derived using a flame heating method. The MLR is constructed by coiling the microfiber to form a loop resonator device. The device is embedded into low refractive index polymer for robustness. The spacing of the transmission comb spectrum of the MLR is observed to be unchanged with the temperature. However, the extinction ratio of the spectrum is observed to be linearly decreasing with the temperature. The slope of the extinction ratio reduction against temperature was about 0.043dB/^{o}C. The dependence of the extinction ratio on temperature is due to the change in the material's refractive index. The temperature-dependence of the Brillouin frequency shift is demonstrated in various types of fiber in the last approach. The beat frequency between the SBS and pump signals is measured using the heterodyne method. The central Brillouin beat frequency increases linearly with increasing temperature with temperature coefficients of 0.77, 0.56 and 1.45MHz/^{o}C. The phenomena are demonstrated for 25 km long SMF, 10 km long NZ-DSF and 100 m PCF, respectively.

7. References

Alahbabi M N, Cho Y T, Newson T P. 150-km-range distributed temperature sensor based on coherent detection of spontaneous Brillouin backscatter and in-line Raman amplification. Journal of the Optical Society of America: B, 2005, 22(6): 1321–1324.

Bao X, Webb D J, Jackson D A. 32-km distributed temperature sensor using Brillouin loss in optical fiber. Optics Letters, 1993, 18(18): 1561–1563.

Choi, H. Y., K. S. Park, et al. (2008). "Miniature fiber-optic high temperature sensor based on a hybrid structured Fabry–Perot interferometer." Optics letters 33(21): 2455-2457.

Culverhouse, D., F. Farahi, et al. (2008). "Potential of stimulated Brillouin scattering as sensing mechanism for distributed temperature sensors." Electronics Letters 25(14): 913-915.

D. Heiman, D. S. Hamilton, and R. W. Hellwarth, "Brillouin scattering measurements on optical glasses," Phys. Rev. B, vol. 19, p. 6583, 1979.

F. Xu, V. Pruneri, V. Finazzi, and G. Brambilla, "High Sensitivity Refractometric Sensor Based on Embedded Optical Microfiber Loop Resonator," in Conference on Lasers and Electro-Optics/Quantum Electronics and Laser Science Conference and Photonic Applications Systems Technologies, OSA Technical Digest (CD) (2008), paper CMJ7.

Golnabi, H. (2000). "Simulation of interferometric sensors for pressure and temperature measurements." Review of scientific Instruments 71: 1608.

G. P. Agrawal, Nonlinear Fiber Optics. Boston, MA: Academic, 1989.

Guo, Q., M. Nishio, et al. (2003). "Temperature dependence of aluminum nitride reflectance spectra in vacuum ultraviolet region." Solid State Communications 126(11): 601-604.

H. Ahmad, W. Y. Chong, K. Thambiratnam, M. Z. Zulklifi, P. Poopalan, M. M. M. Thant and S. W. Harun, "High Sensitivity Fiber Bragg Grating Pressure Sensor Using Thin Metal Diaphragm," IEEE Sensors J., 9 (12), pp. 1654-1659, 2009.

H. C. Nguyen, B. T. Kuhlmey, E. C. Magi, M. J. Steel, P. Domachuk, C. L. Smith, and B. J. Eggleton, "Tapered photonic crystal fibers: Properties, characterization and applications," Appl. Phys. B, vol. 81, pp. 377–387, Jul. 2005.

H. C. Seat, J. H. Sharp, Z. Y. Zhang, K. T. V. Grattan, "Single-crystal ruby fiber temperature sensor," Sensors and Actuators A, Vol. 101, pp. 24-29, 2002.

H. Z. Yang, S. W. Harun and H. Ahmad, "Theoretical and experimental studies on concave mirror-based fiber optic displacement sensor," Sensor Review, Vol. 31, pp. 65-69 (2011).

J. Castrellon-Uribe, "Experimental results of the performance of a laser fiber as a remote sensor of temperature," Optics and Lasers in Engineering, Vol. 43, pp. 633-644, 2005.

J. Villatoro, V. P. Minkovich, and D. Monzón-Hernández, "Temperature independent strain sensor made from tapered holey optical fiber," Opt. Lett., vol. 31, pp. 305–307, Feb. 2006.

K. Amarnath, R. Grover, S. Kanakaraju, P. T. Ho, "Electrically pumped InGaAsP-InP microring optical amplifiers and lasers with surface passivation", IEEE Photon. Technol. Lett., vol. 17, pp. 2280-2282, 2005.

Kee, H. H., G. Lees, et al. (2002). "1.65 m Raman-based distributed temperature sensor." Electronics Letters 35(21): 1869-1871.

K. J. Han, Y. W. Lee, J. Kwon, S. Roh, J. Jung and B. Lee, "Simultaneous measurement of strain and temperature incorporating a long-period fiber grating inscribed on a polarization-maintaining fiber," IEEE Photon. Technol. Lett., Vol. 16, pp. 2114–2116, 2004.

K. S. Lim, S. W. Harun, H. Z. Yang, K. Dimyati and H. Ahmad, "Analytical and experimental studies on asymmetric bundle fiber displacement sensors," J. Modern Optics, 56 (17), pp. 1838-1842, 2009.

L. Li, X. Y. Dong, L. Y. Shao, C. L. Zhao and Y. L. Sun, "Temperature-independent acceleration measurement with a strain-chirped fiber Bragg grating," Optoelectronics and advanced materials – rapid communications, Vol. 4, No. 7, pp. 943-946, 2010.

L. Tong, L. Hu, J. Zhang, J. Qiu, Q. Yang, J. Lou, Y. Shen, J. He, Z. Ye, "Photonic nanowires directly drawn from bulk glasses," Opt. Express, Vol. 14 (1) pp. 82-87, (2006).

M. McSherry, C. Fitzpatrick and E. Lewis, "Review of luminescent based fibre optic temperature sensors," Sensor Review, Vol. 25, pp. 56-62, 2005.

M. Sumetsky, Y. Dulashko, J. M. Fini, A. Hale, and D. J. DiGiovanni, "The Microfiber Loop Resonator: Theory, Experiment, and Application," J. Lightwave Technol. Vol. 24, No. 1, 2006, pp. 242-250.

M. Yasin, S. W. Harun, H. Z. Yang and H. Ahmad, "Title: Fiber optic displacement sensor for measurement of glucose concentration in distilled water," Optoelectronics and advanced materials – rapid communications, Vol. 4, No. 8, pp. 1063-1065, 2010.

M. Yasin, S.W. Harun, W.A. Fawzi, Kusminarto, Karyono, H. Ahmad., "Lateral and axial displacements measurement using fiber optic sensor based on beam-through technique", Microwave And Optical Technology Letters, Vol. 51, pp. 2038-2040 (2009).

Nikles M, Thevenaz L, Robert P. Simple distributed fiber sensor based on Brillouin gain spectrum analysis. Optics Letters, 1996, 21(10): 758–760.

R. Rathod, R. D. Pechstedt, D. A. Jackson, D. J. Webb, "Distributed temperature-change sensor based on Rayleigh backscattering in an optical fiber," Opt. Lett., Vol. 19, pp. 593–595, 1994.

S. Baek, Y. Jeong, J. Nilson, J. K. Sahu, B. Lee, "Temperature-dependent fluorescence characteristics of an ytterbium-sensitized erbium-doped silica fiber for sensor applications," Optical Fiber Technology, Vol. 12, pp. 10-19, 2006.

SUN Xiao-qiang, XU Kun, PEI Yin-qing, FUSong-nian, "Simple distributed optical fiber sensor based on Brillouin amplification of microwave photonic signal" The journal of China Universities of Posts and telecommunications, 2009, 16: 24-28.

S. W. Harun, K. S. Lim, A. A. Jasim, H. Ahmad, "Dual wavelength erbium-doped fiber laser using a tapered fiber, Journal of Modern Optics, Vol. 57, pp. 1362-3044 (2010).

V. Lopez, G. Paez and M. Strojnik, "Sensitivity of a temperature sensor, employing ratio of fluorescence power in a band," Infrared Physics & Technology, Vol. 46, pp. 133–139, 2004.

V. P. Minkovich, J. Villatoro, D. Monzón-Hernández, S. Calixto, A. B. Sotsky, and L. I. Sotskaya, "Holey fiber tapers with resonance transmission for high-resolution refractive index sensing," Opt. Express, vol. 13, pp. 7609–7614, Sep. 2005.

Van Etten, W. and J. Van der Plaats (1991). Fundamentals of optical fiber communications.

Vijay K. Kulkarni, Anandkumar S. Lalasangi, I. I. Pattanashetti, U. S. Raikar, "Fiber Optic Micro-Displacement Sensor Using Coupler," Journal Of Optoelectronics And Advanced Materials Vol. 8, No. 4, August 2006, P. 1610 – 1612

X. Jiang, L. Tong, G. Vienne, X. Guo, A. Tsau, Q. Yang and D. Yang, "Demonstration of optical microfiber knot resonators," Appl. Phys. Lett., Vol. 88 (22), pp. 223501 (2006)

X. Jiang, Q. Yang, G. Vienne, Y. Li, L. Tong, Demonstration of microfiber knot laser Appl. Phys. Lett. 89 (14) (2006) 143513.

Y. Li, L. Tong, Opt. Lett. 33 (4) (2008) 303.

Y. Lou, L. M. Tong, and Z. Z. Ye, "Modeling of silica nanowires for optical sensing," Opt. Express, vol. 13, pp. 2135-2140 (2005)

Zhang, Z., K. Grattan, et al. (2009). "Fiber optic high temperature sensor based on the fluorescence lifetime of alexandrite." Review of scientific Instruments 63(8): 3869-3873.

Permissions

The contributors of this book come from diverse backgrounds, making this book a truly international effort. This book will bring forth new frontiers with its revolutionizing research information and detailed analysis of the nascent developments around the world.

We would like to thank Dr Moh. Yasin, Professor Sulaiman W. Harun and Dr Hamzah Arof, for lending their expertise to make the book truly unique. They have played a crucial role in the development of this book. Without their invaluable contribution this book wouldn't have been possible. They have made vital efforts to compile up to date information on the varied aspects of this subject to make this book a valuable addition to the collection of many professionals and students.

This book was conceptualized with the vision of imparting up-to-date information and advanced data in this field. To ensure the same, a matchless editorial board was set up. Every individual on the board went through rigorous rounds of assessment to prove their worth. After which they invested a large part of their time researching and compiling the most relevant data for our readers. Conferences and sessions were held from time to time between the editorial board and the contributing authors to present the data in the most comprehensible form. The editorial team has worked tirelessly to provide valuable and valid information to help people across the globe.

Every chapter published in this book has been scrutinized by our experts. Their significance has been extensively debated. The topics covered herein carry significant findings which will fuel the growth of the discipline. They may even be implemented as practical applications or may be referred to as a beginning point for another development. Chapters in this book were first published by InTech; hereby published with permission under the Creative Commons Attribution License or equivalent.

The editorial board has been involved in producing this book since its inception. They have spent rigorous hours researching and exploring the diverse topics which have resulted in the successful publishing of this book. They have passed on their knowledge of decades through this book. To expedite this challenging task, the publisher supported the team at every step. A small team of assistant editors was also appointed to further simplify the editing procedure and attain best results for the readers.

Our editorial team has been hand-picked from every corner of the world. Their multi-ethnicity adds dynamic inputs to the discussions which result in innovative outcomes. These outcomes are then further discussed with the researchers and contributors who give their valuable feedback and opinion regarding the same. The feedback is then collaborated with the researches and they are edited in a comprehensive manner to aid the understanding of the subject.

Apart from the editorial board, the designing team has also invested a significant amount of their time in understanding the subject and creating the most relevant covers. They scrutinized every image to scout for the most suitable representation of the subject and create an appropriate cover for the book.

The publishing team has been involved in this book since its early stages. They were actively engaged in every process, be it collecting the data, connecting with the contributors or procuring relevant information. The team has been an ardent support to the editorial, designing and production team. Their endless efforts to recruit the best for this project, has resulted in the accomplishment of this book. They are a veteran in the field of academics and their pool of knowledge is as vast as their experience in printing. Their expertise and guidance has proved useful at every step. Their uncompromising quality standards have made this book an exceptional effort. Their encouragement from time to time has been an inspiration for everyone.

The publisher and the editorial board hope that this book will prove to be a valuable piece of knowledge for researchers, students, practitioners and scholars across the globe.

List of Contributors

Marcelo L. F. Abbade and Eric A. M. Fagotto
Pontifícia Universidade Católica de Campinas, Brazil

Iguatemi E. Fonseca
Universidade Federal da Paraíba, Brazil

Felipe R. Barbosa, André L. A. Costa and Edson Moschim
Universidade Estadual de Campinas, Brazil

Jorge D. Marconi
Universidade Federal do ABC, Brazil

Yury Dudchik
Institute of Applied Physics Problems of Belarus State University, Belarus

Elzbieta Beres-Pawlik, Grzegorz Budzyn and Grzegorz Lis
Institute of Telecommunications, Teleinformatics and Acoustics, Wroclaw University of Technology, Wroclaw, Poland

Carmina del Río Campos and Paloma R. Horche
Escuela Politécnica Superior, Universidad San Pablo CEU, ETSIT, Universidad Politécnica de Madrid, Spain

Albert Gorshtein and Dan Sadot
Ben Gurion University of the Negev, Israel

Paola Frascella and Andrew D. Ellis
Photonics System Group, Tyndall National Institute & Department of Physics, University College Cork, Ireland

Mohammad Syuhaimi Ab-Rahman, Hadi Guna, Mohd Hazwan Harun, Latifah Supian and Kasmiran Jumari
Universiti Kebangsaan Malaysia, Selangor Darul Ehsan, Malaysia

Laszlo Gyongyosi and Sandor Imre
Budapest University of Technology and Economics, Department of Telecommunications, Hungary

Ezra Ip
NEC Labs America, Princeton, NJ, USA

Neng Bai
University of Central Florida, Orlando, FL, USA

Mitsuru Kihara
Technical Assistance and Support Center, NTT East Corporation, Japan

Nazuki Honda
NTT Access Service System Laboratories, NTT Corp., Japan

Konstantinos Manousakis and Emmanouel (Manos) Varvarigos
University of Patras, Department of Computer Engineering and Informatics / Research, Academic Computer Technology Institute, Greece

Aurenice Oliveira
Michigan Technological University, USA

Pablo Sartor Del Giudice and Franco Robledo Amoza
Engineering School - Universidad de la República, Uruguay

J. Torres, R. García, J. Soret, J. Martos, G. Martínez, C. Reig and X. Román
Department of Electronic Engineering, University of Valencia, Valencia, Spain

Javier E. Sierra
GIDATI Research Group, Universidad Pontificia Bolivariana, Medellin, Colombia

H. A. Rahman
Photonic Research Center, University of Malaya, Kuala Lumpur, Malaysia
Department of Electrical Engineering, University of Malaya, Kuala Lumpur, Malaysia
Faculty of Electrical Engineering, Universiti Teknologi MARA (UiTM), Shah Alam, Malaysia

H. Arof
Department of Electrical Engineering, University of Malaya, Kuala Lumpur, Malaysia

H. Ahmad
Photonic Research Center, University of Malaya, Kuala Lumpur, Malaysia

S. W. Harun
Photonic Research Center, University of Malaya, Kuala Lumpur, Malaysia
Department of Electrical Engineering, University of Malaya, Kuala Lumpur, Malaysia

M. Yasin
Photonic Research Center, University of Malaya, Kuala Lumpur, Malaysia
Department of Physics, Faculty of Science and Technology, Airlangga University, Surabaya, Indonesia